Photons and Atoms

T0324137

Photons and Atoms

Introduction to Quantum Electrodynamics

Claude Cohen-Tannoudji
Jacques Dupont-Roc
Gilbert Grynberg

Wiley-VCH Verlag GmbH & Co. KGaA

All books published by Wiley-VCH are carefully produced.
Nevertheless, authors, editors, and publisher do not warrant the information contained in these books, including this book, to be free of errors.
Readers are advised to keep in mind that statements, data, illustrations, procedural details or other items may inadvertently be inaccurate.

Library of Congress Card No.:
Applied for

British Library Cataloging-in-Publication Data:
A catalogue record for this book is available from the British Library

Bibliographic information published by
Die Deutsche Bibliothek
Die Deutsche Bibliothek lists this publication in the Deutsche Nationalbibliografie; detailed bibliographic data is available in the Internet at <http://dnb.ddb.de>.

© 1989 by John Wiley & Sons, Inc.
This is a translation of *Photons et atomes: Introduction à l`éctrodynamique quantique*.
© 1987, InterEditions and Editions du CNRS.
Wiley Professional Paperback Edition Published 1997.
© 2004 WILEY-VCH Verlag GmbH & Co. KGaA, Weinheim

All rights reserved (including those of translation into other languages).
No part of this book may be reproduced in any form – nor transmitted or translated into machine language without written permission from the publishers.
Registered names, trademarks, etc. used in this book, even when not specifically marked as such, are not to be considered unprotected by law.

ISBN-13: 978-0-471-18433-1
ISBN-10: 0-471-18433-0

Contents

Preface XVII

Introduction 1

I
CLASSICAL ELECTRODYNAMICS: THE FUNDAMENTAL EQUATIONS AND THE DYNAMICAL VARIABLES

Introduction 5

A. The Fundamental Equations in Real Space 7
1. The Maxwell–Lorentz Equations 7
2. Some Important Constants of the Motion 8
3. Potentials—Gauge Invariance 8

B. Electrodynamics in Reciprocal Space 11
1. The Fourier Spatial Transformation—Notation 11
2. The Field Equations in Reciprocal Space 12
3. Longitudinal and Transverse Vector Fields 13
4. Longitudinal Electric and Magnetic Fields 15
5. Contribution of the Longitudinal Electric Field to the Total Energy, to the Total Momentum, and to the Total Angular Momentum—*a. The Total Energy. b. The Total Momentum. c. The Total Angular Momentum* 17
6. Equations of Motion for the Transverse Fields 21

C. Normal Variables 23
1. Introduction 23
2. Definition of the Normal Variables 23
3. Evolution of the Normal Variables 24
4. The Expressions for the Physical Observables of the Transverse Field as a Function of the Normal Variables—*a. The Energy H_{trans} of the Transverse Field. b. The Momentum $\mathbf{P}_{\text{trans}}$ and the Angular Momentum $\mathbf{J}_{\text{trans}}$ of the Transverse Field. c. Transverse Electric and Magnetic Fields in Real Space. d. The Transverse Vector Potential $\mathbf{A}_{\perp}(\mathbf{r}, t)$* . . . 26

5. Similarities and Differences between the Normal Variables and the Wave Function of a Spin-1 Particle in Reciprocal Space 30
6. Periodic Boundary Conditions. Simplified Notation 31

D. Conclusion: Discussion of Various Possible Quantization Schemes 33
1. Elementary Approach 33
2. Lagrangian and Hamiltonian Approach 34

COMPLEMENT A_I—THE "TRANSVERSE" DELTA FUNCTION

1. Definition in Reciprocal Space—*a. Cartesian Coordinates. Transverse and Longitudinal Components. b. Projection on the Subspace of Transverse Fields* 36
2. The Expression for the Transverse Delta Function in Real Space—*a. Regularization of $\delta_{ij}^{\perp}(\rho)$. b. Calculation of $g(\rho)$. c. Evaluation of the Derivatives of $g(\rho)$. d. Discussion of the Expression for $\delta_{ij}^{\perp}(\rho)$* 38
3. Application to the Evaluation of the Magnetic Field Created by a Magnetization Distribution. Contact Interaction 42

COMPLEMENT B_I—ANGULAR MOMENTUM OF THE ELECTROMAGNETIC FIELD. MULTIPOLE WAVES

Introduction 45

1. Contribution of the Longitudinal Electric Field to the Total Angular Momentum 45
2. Angular Momentum of the Transverse Field—*a.* $\mathbf{J}_{\text{trans}}$ *in Reciprocal Space. b.* $\mathbf{J}_{\text{trans}}$ *in Terms of Normal Variables. c. Analogy with the Mean Value of the Total Angular Momentum of a Spin-1 Particle* 47
3. Set of Vector Functions of **k** "Adapted" to the Angular Momentum—*a. General Idea. b. Method for Constructing Vector Eigenfunctions for* $\mathbf{J}^2$ *and* J_z*. c. Longitudinal Eigenfunctions. d. Transverse Eigenfunctions* .. 51
4. Application: Multipole Waves in Real Space—*a. Evaluation of Some Fourier Transforms. b. Electric Multipole Waves. c. Magnetic Multipole Waves* 55

COMPLEMENT C_I—EXERCISES

1. H and **P** as Constants of the Motion 61
2. Transformation from the Coulomb Gauge to the Lorentz Gauge 63
3. Cancellation of the Longitudinal Electric Field by the Instantaneous Transverse Field 64

4. Normal Variables and Retarded Potentials 66
5. Field Created by a Charged Particle at Its Own Position. Radiation Reaction 68
6. Field Produced by an Oscillating Electric Dipole 71
7. Cross-section for Scattering of Radiation by a Classical Elastically Bound Electron 74

II
LAGRANGIAN AND HAMILTONIAN APPROACH TO ELECTRODYNAMICS. THE STANDARD LAGRANGIAN AND THE COULOMB GAUGE

Introduction 79

A. Review of the Lagrangian and Hamiltonian Formalism 81
1. Systems Having a Finite Number of Degrees of Freedom—*a. Dynamical Variables, the Lagrangian, and the Action. b. Lagrange's Equations. c. Equivalent Lagrangians. d. Conjugate Momenta and the Hamiltonian. e. Change of Dynamical Variables. f. Use of Complex Generalized Coordinates. g. Coordinates, Momenta, and Hamiltonian in Quantum Mechanics.* 81
2. A System with a Continuous Ensemble of Degrees of Freedom—*a. Dynamical Variables. b. The Lagrangian. c. Lagrange's Equations d. Conjugate Momenta and the Hamiltonian. e. Quantization. f. Lagrangian Formalism with Complex Fields. g. Hamiltonian Formalism and Quantization with Complex Fields* 90

B. The Standard Lagrangian of Classical Electrodynamics 100
1. The Expression for the Standard Lagrangian—*a. The Standard Lagrangian in Real Space. b. The Standard Lagrangian in Reciprocal Space* 100
2. The Derivation of the Classical Electrodynamic Equations from the Standard Lagangian—*a. Lagrange's Equation for Particles. b. The Lagrange Equation Relative to the Scalar Potential. c. The Lagrange Equation Relative to the Vector Potential* 103
3. General Properties of the Standard Lagrangian—*a. Global Symmetries. b. Gauge Invariance. c. Redundancy of the Dynamical Variables* 105

C. Electrodynamics in the Coulomb Gauge 111
1. Elimination of the Redundant Dynamical Variables from the Standard Lagrangian—*a. Elimination of the Scalar Potential. b. The Choice of the Longitudinal Component of the Vector Potential* 111
2. The Lagrangian in the Coulomb Gauge 113

3. Hamiltonian Formalism—*a. Conjugate Particle Momenta. b. Conjugate Momenta for the Field Variables. c. The Hamiltonian in the Coulomb Gauge. d. The Physical Variables* 115
4. Canonical Quantization in the Coulomb Gauge—*a. Fundamental Commutation Relations. b. The Importance of Transversability in the Case of the Electromagnetic Field. c. Creation and Annihilation Operators* 118
5. Conclusion: Some Important Characteristics of Electrodynamics in the Coulomb Gauge—*a. The Dynamical Variables Are Independent. b. The Electric Field Is Split into a Coulomb Field and a Transverse Field. c. The Formalism Is Not Manifestly Covariant. d. The Interaction of the Particles with Relativistic Modes Is Not Correctly Described*.. 121

COMPLEMENT A_{II}—FUNCTIONAL DERIVATIVE. INTRODUCTION AND A FEW APPLICATIONS

1. From a Discrete to a Continuous System. The Limit of Partial Derivatives 126
2. Functional Derivative 128
3. Functional Derivative of the Action and the Lagrange Equations 128
4. Functional Derivative of the Lagrangian for a Continuous System 130
5. Functional Derivative of the Hamiltonian for a Continuous System 132

COMPLEMENT B_{II}—SYMMETRIES OF THE LAGRANGIAN IN THE COULOMB GAUGE AND THE CONSTANTS OF THE MOTION

1. The Variation of the Action between Two Infinitesimally Close Real Motions 134
2. Constants of the Motion in a Simple Case 136
3. Conservation of Energy for the System Charges + Field 137
4. Conservation of the Total Momentum 138
5. Conservation of the Total Angular Momentum 139

COMPLEMENT C_{II}—ELECTRODYNAMICS IN THE PRESENCE OF AN EXTERNAL FIELD

1. Separation of the External Field 141
2. The Lagrangian in the Presence of an External Field—*a. Introduction of a Lagrangian. b. The Lagrangian in the Coulomb Gauge* 142
3. The Hamiltonian in the Presence of an External Field—*a. Conjugate Momenta. b. The Hamiltonian. c. Quantization* 143

COMPLEMENT D_{II}—EXERCISES

1. An Example of a Hamiltonian Different from the Energy 146
2. From a Discrete to a Continuous System: Introduction of the Lagrangian and Hamiltonian Densities 147
3. Lagrange's Equations for the Components of the Electromagnetic Field in Real Space 150
4. Lagrange's Equations for the Standard Lagrangian in the Coulomb Gauge 151
5. Momentum and Angular Momentum of an Arbitrary Field 152
6. A Lagrangian Using Complex Variables and Linear in Velocity 154
7. Lagrangian and Hamiltonian Descriptions of the Schrödinger Matter Field 157
8. Quantization of the Schrödinger Field 161
9. Schrödinger Equation of a Particle in an Electromagnetic Field: Arbitrariness of Phase and Gauge Invariance 167

III
QUANTUM ELECTRODYNAMICS IN THE COULOMB GAUGE

Introduction 169

A. The General Framework 171
1. Fundamental Dynamical Variables. Commutation Relations 171
2. The Operators Associated with the Various Physical Variables of the System 171
3. State Space 175

B. Time Evolution 176
1. The Schrödinger Picture 176
2. The Heisenberg Picture. The Quantized Maxwell–Lorentz Equations—*a. The Heisenberg Equations for Particles. b. The Heisenberg Equations for Fields. c. The Advantages of the Heisenberg Point of View* 176

C. Observables and States of the Quantized Free Field 183
1. Review of Various Observables of the Free Field—*a. Total Energy and Total Momentum of the Field. b. The Fields at a Given Point* **r** *of Space. c. Observables Corresponding to Photoelectric Measurements* . . 183
2. Elementary Excitations of the Quantized Free Field. Photons—*a. Eigenstates of the Total Energy and the Total Momentum.*

b. The Interpretation in Terms of Photons. c. Single-Photon States. Propagation 186
3. Some Properties of the Vacuum—*a. Qualitative Discussion. b. Mean Values and Variances of the Vacuum Field. c. Vacuum Fluctuations* . . 189
4. Quasi-classical States— *a. Introducing the Quasi-classical States. b. Characterization of the Quasi-classical States. c. Some Properties of the Quasi-classical States. d. The Translation Operator for a and a* $^{+}$. 192

D. The Hamiltonian for the Interaction between Particles and Fields 197
1. Particle Hamiltonian, Radiation Field Hamiltonian, Interaction Hamiltonian 197
2. Orders of Magnitude of the Various Interactions Terms for Systems of Bound Particles 198
3. Selection Rules 199
4. Introduction of a Cutoff 200

Complement A_{III}—The Analysis of Interference Phenomena in the Quantum Theory of Radiation

Introduction 204

1. A Simple Model 205
2. Interference Phenomena Observable with Single Photodetection Signals—*a. The General Case. b. Quasi-classical States. c. Factored States. d. Single-Photon States* 206
3. Interference Phenomena Observable with Double Photodetection Signals—*a. Quasi-classical States. b. Single-Photon States. c. Two-Photon States* 209
4. Physical Interpretation in Terms of Interference between Transition Amplitudes 213
5. Conclusion: The Wave–Particle Duality in the Quantum Theory of Radiation 215

Complement B_{III}—Quantum Field Radiated by Classical Sources

1. Assumptions about the Sources 217
2. Evolution of the Fields in the Heisenberg Picture 217
3. The Schrödinger Point of View. The Quantum State of the Field at Time *t* 219

COMPLEMENT C_{III}—COMMUTATION RELATIONS FOR FREE FIELDS AT DIFFERENT TIMES. SUSCEPTIBILITIES AND CORRELATION FUNCTIONS OF THE FIELDS IN THE VACUUM

Introduction 221

1. Preliminary Calculations 222
2. Field Commutators—*a. Reduction of the Expressions in Terms of D. b. Explicit Expressions for the Commutators. c. Properties of the Commutators* 223
3. Symmetric Correlation Functions of the Fields in the Vacuum 227

COMPLEMENT D_{III}—EXERCISES

1. Commutators of **A**, $\mathbf{E}_\perp$, and **B** in the Coulomb Gauge 230
2. Hamiltonian of a System of Two Particles with Opposite Charges Coupled to the Electromagnetic Field 232
3. Commutation Relations for the Total Momentum **P** with H_P, H_R, and H_I 233
4. Bose–Einstein Distribution 234
5. Quasi-Probability Densities and Characteristic Functions 236
6. Quadrature Components of a Single-Mode Field. Graphical Representation of the State of the Field 241
7. Squeezed States of the Radiation Field 246
8. Generation of Squeezed States by Two-Photon Interactions 248
9. Quasi-Probability Density of a Squeezed State 250

IV
OTHER EQUIVALENT FORMULATIONS OF ELECTRODYNAMICS

Introduction 253

A. How to Get Other Equivalent Formulations of Electrodynamics 255
1. Change of Gauge and of Lagrangian 255
2. Changes of Lagrangian and the Associated Unitary Transformation—*a. Changing the Lagrangian. b. The Two Quantum Descriptions. c. The Correspondence between the Two Quantum Descriptions. d. Application to the Electromagnetic Field* 256
3. The General Unitary Transformation. The Equivalence between the Different Formulations of Quantum Electrodynamics 262

B. Simple Examples Dealing with Charges Coupled to an External Field 266
1. The Lagrangian and Hamiltonian of the System 266
2. Simple Gauge Change; Gauge Invariance—*a. The New Description. b. The Unitary Transformation Relating the Two Descriptions—Gauge Invariance* 267
3. The Göppert-Mayer Transformation—*a. The Long-Wavelength Approximation. b. Gauge Change Giving Rise to the Electric Dipole Interaction. c. The Advantages of the New Point of View. d. The Equivalence between the Interaction Hamiltonians* $\mathbf{A} \cdot \mathbf{p}$ *and* $\mathbf{E} \cdot \mathbf{r}$. *e. Generalizations* 269
4. A Transformation Which Does Not Reduce to a Change of Lagrangian: The Henneberger Transformation—*a. Motivation. b. Determination of the Unitary Transformation. Transforms of the Various Operators. c. Physical Interpretation. d. Generalization to a Quantized Field: The Pauli–Fierz–Kramers Transformation* 275

C. The Power–Zienau–Woolley Transformation: The Multipole Form of the Interaction between Charges and Field 280
1. Description of the Sources in Terms of a Polarization and a Magnetization Density—*a. The Polarization Density Associated with a System of Charges. b. The Displacement. c. Polarization Current and Magnetization Current* 280
2. Changing the Lagrangian—*a. The Power–Zienau–Woolley Transformation. b. The New Lagrangian. c. Multipole Expansion of the Interaction between the Charged Particles and the Field* 286
3. The New Conjugate Momenta and the New Hamiltonian—*a. The Expressions for These Quantities. b. The Physical Significance of the New Conjugate Momenta. c. The Structure of the New Hamiltonian* .. 289
4. Quantum Electrodynamics from the New Point of View—*a. Quantization. b. The Expressions for the Various Physical Variables* 293
5. The Equivalence of the Two Points of View. A Few Traps to Avoid .. 296

D. Simplified Form of Equivalence for the Scattering *S*-Matrix 298
1. Introduction of the *S*-Matrix 298
2. The *S*-Matrix from Another Point of View. An Examination of the Equivalence 300
3. Comments on the Use of the Equivalence between the *S*-Matrices ... 302

COMPLEMENT A_{IV}—ELEMENTARY INTRODUCTION TO THE ELECTRIC DIPOLE HAMILTONIAN

Introduction 304

1. The Electric Dipole Hamiltonian for a Localized System of Charges Coupled to an External Field—*a. The Unitary Transformation Suggested*

by the Long-Wavelength Approximation. b. The Transformed Hamiltonian. c. The Velocity Operator in the New Representation 304
2. The Electric Dipole Hamiltonian for a Localized System of Charges Coupled to Quantized Radiation—*a. The Unitary Transformation. b. Transformation of the Physical Variables. c. Polarization Density and Displacement. d. The Hamiltonian in the New Representation* 307
3. Extensions—*a. The Case of Two Separated Systems of Charges. b. The Case of a Quantized Field Coupled to Classical Sources* 312

Complement B_{IV}—One-Photon and Two-Photon Processes: The Equivalence between the Interaction Hamiltonians $\mathbf{A} \cdot \mathbf{p}$ and $\mathbf{E} \cdot \mathbf{r}$

Introduction 316

1. Notations. Principles of Calculations 316
2. Calculation of the Transition Amplitudes in the Two Representations—*a. The Interaction Hamiltonian* $\mathbf{A} \cdot \mathbf{p}$. *b. The Interaction Hamiltonian* $\mathbf{E} \cdot \mathbf{r}$. *c. Direct Verification of the Identity of the Two Amplitudes* 317
3. Generalizations—*a. Extension to Other Processes. b. Nonresonant Processes* 325

Complement C_{IV}—Interaction of Two Localized Systems of Charges from the Power–Zienau–Woolley Point of View

Introduction 328

1. Notation 328
2. The Hamiltonian 329

Complement D_{IV}—The Power–Zienau–Woolley Transformation and the Poincaré Gauge

Introduction 331

1. The Power–Zienau–Woolley Transformation Considered as a Gauge Change 331
2. Properties of the Vector Potential in the New Gauge 332
3. The Potentials in the Poincaré Gauge 333

Complement E_{IV}—Exercises

1. An Example of the Effect Produced by Sudden Variations of the Vector Potential 336
2. Two-Photon Excitation of the Hydrogen Atom. Approximate Results Obtained with the Hamiltonians $\mathbf{A} \cdot \mathbf{p}$ and $\mathbf{E} \cdot \mathbf{r}$ 338
3. The Electric Dipole Hamiltonian for an Ion Coupled to an External Field 342
4. Scattering of a Particle by a Potential in the Presence of Laser Radiation . . 344
5. The Equivalence between the Interaction Hamiltonians $\mathbf{A} \cdot \mathbf{p}$ and $\mathbf{Z} \cdot (\nabla V)$ for the Calculation of Transition Amplitudes 349
6. Linear Response and Susceptibility. Application to the Calculation of the Radiation from a Dipole 352
7. Nonresonant Scattering. Direct Verification of the Equality of the Transition Amplitudes Calculated from the Hamiltonians $\mathbf{A} \cdot \mathbf{p}$ and $\mathbf{E} \cdot \mathbf{r}$ 356

V
INTRODUCTION TO THE COVARIANT FORMULATION OF QUANTUM ELECTRODYNAMICS

Introduction 361

A. Classical Electrodynamics in the Lorentz Gauge 364
1. Lagrangian Formalism—*a. Covariant Notation. Ordinary Notation. b. Selection of a New Lagrangian for the Field. c. Lagrange Equations for the Field. d. The Subsidiary Condition. e. The Lagrangian Density in Reciprocal Space* 364
2. Hamiltonian Formalism—*a. Conjugate Momenta of the Potentials. b. The Hamiltonian of the Field. c. Hamilton–Jacobi Equations for the Free Field* 369
3. Normal Variables of the Classical Field—*a. Definition. b. Expansion of the Potential in Normal Variables. c. Form of the Subsidiary Condition for the Free Classical Field. Gauge Arbitrariness. d. Expression of the Field Hamiltonian* 371

B. Difficulties Raised by the Quantization of the Free Field 380
1. Canonical Quantization —*a. Canonical Commutation Relations. b. Annihilation and Creation Operators. c. Covariant Commutation Relations between the Free Potentials in the Heisenberg Picture* 380
2. Problems of Physical Interpretation Raised by Covariant Quantization —*a. The Form of the Subsidiary Condition in Quantum Theory. b. Problems Raised by the Construction of State Space* 383

C. Covariant Quantization with an Indefinite Metric 387
1. Indefinite Metric in Hilbert Space 387
2. Choice of the New Metric for Covariant Quantization 390
3. Construction of the Physical Kets 393
4. Mean Values of the Physical Variables in a Physical Ket—*a. Mean Values of the Potentials and the Fields. b. Gauge Arbitrariness and Arbitrariness of the Kets Associated with a Physical State. c. Mean Value of the Hamiltonian* 396

D. A Simple Example of Interaction: A Quantized Field Coupled to Two Fixed External Charges 400
1. Hamiltonian for the Problem 400
2. Energy Shift of the Ground State of the Field. Reinterpretation of Coulomb's Law—*a. Perturbative Calculation of the Energy Shift. b. Physical Discussion. Exchange of Scalar Photons between the Two Charges. c. Exact Calculation* 401
3. Some Properties of the New Ground State of the Field—*a. The Subsidiary Condition in the Presence of the Interaction. The Physical Character of the New Ground State. b. The Mean Value of the Scalar Potential in the New Ground State of the Field* 405
4. Conclusion and Generalization 407

COMPLEMENT A_V—AN ELEMENTARY INTRODUCTION TO THE THEORY OF THE ELECTRON–POSITRON FIELD COUPLED TO THE PHOTON FIELD IN THE LORENTZ GAUGE

Introduction 408

1. A Brief Review of the Dirac Equation—*a. Dirac Matrices. b. The Dirac Hamiltonian. Charge and Current Density. c. Connection with the Covariant Notation. d. Energy Spectrum of the Free Particle. e. Negative-Energy States. Hole Theory* 408
2. Quantization of the Dirac Field—*a. Second Quantization. b. The Hamiltonian of the Quantized Field. Energy Levels. c. Temporal and Spatial Translations* 414
3. The Interacting Dirac and Maxwell Fields—*a. The Hamiltonian of the Total System. The Interaction Hamiltonian. b. Heisenberg Equations for the Fields. c. The Form of the Subsidiary Condition in the Presence of Interaction* 418

COMPLEMENT B_V—JUSTIFICATION OF THE NONRELATIVISTIC THEORY IN THE COULOMB GAUGE STARTING FROM RELATIVISTIC QUANTUM ELECTRODYNAMICS

Introduction 424

1. Transition from the Lorentz Gauge to the Coulomb Gauge in Relativistic Quantum Electrodynamics—*a. Transformation on the Scalar Photons Yielding the Coulomb Interaction. b. Effect of the Transformation on the Other Terms of the Hamiltonian in the Lorentz Gauge. c. Subsidiary Condition. Absence of Physical Effects of the Scalar and Longitudinal Photons. d. Conclusion: The Relativistic Quantum Electrodynamics Hamiltonian in the Coulomb Gauge* 425
2. The Nonrelativistic Limit in Coulomb Gauge: Justification of the Pauli Hamiltonian for the Particles—*a. The Dominant Term H_0 of the Hamiltonian in the Nonrelativistic Limit: Rest Mass Energy of the Particles. b. The Effective Hamiltonian inside a Manifold. c. Discussion* 432

COMPLEMENT C_V—EXERCISES

1. Other Covariant Lagrangians of the Electromagnetic Field 441
2. Annihilation and Creation Operators for Scalar Photons: Can One Interchange Their Meanings? 443
3. Some Properties of the Indefinite Metric 445
4. Translation Operator for the Creation and Annihilation Operators of a Scalar Photon 446
5. Lagrangian of the Dirac Field. The Connection between the Phase of the Dirac Field and the Gauge of the Electromagnetic Field 449
6. The Lagrangian and Hamiltonian of the Coupled Dirac and Maxwell Fields 451
7. Dirac Field Operators and Charge Density. A Study of Some Commutation Relations 454

References 457

Index 459

Preface

The spectacular development of new sources of electromagnetic radiation spanning the range of frequencies from rf to the far ultraviolet (lasers, masers, synchrotron sources, etc.) has generated considerable interest in the interaction processes between photons and atoms. New methods have been developed, leading to a more precise understanding of the structure and dynamics of atoms and molecules, to better control of their internal and external degrees of freedom, and also to the realization of novel radiation sources. This explains the growing interest in the low-energy interaction between matter and radiation on the part of an increasing number of researchers drawn from physics, chemistry, and engineering. This work is designed to provide them with the necessary background to understand this area of research, beginning with elementary quantum theory and classical electrodynamics.

Such a program is actually twofold. One has first to set up the theoretical framework for a quantum description of the dynamics of the total system (electromagnetic field and nonrelativistic charged particles), and to discuss the physical content of the theory and its various possible formulations. This is the subject of the present volume, entitled *Photons and Atoms—Introduction to Quantum Electrodynamics*. One has also to describe the interaction processes between radiation and matter (emission, absorption, scattering of photons by atoms, etc.) and to present various theoretical methods which can be used to analyze these processes (perturbative methods, partial resummations of the perturbation series, master equations, optical Bloch equations, the dressed-atom approach, etc). These questions are examined in another volume entitled *Interaction Processes between Photons and Atoms*. The objectives of these two volumes are thus clearly distinct, and according to his interests and to his needs, the reader may use one volume, the other, or both.

An examination of the topics presented here clearly shows that this book is not organized along the same lines as other works treating quantum electrodynamics. In fact, the majority of the latter are addressed to an audience of field theorists for whom such ideas as covariance, relativistic invariance, matter fields, and renormalization, to name a few, are considered as fundamentals. On the other hand, most of the books dealing with quantum optics, and in particular with laser optics, treat the fundamentals of electrodynamics, as well as the problems posed by quantization of radiation, rather succinctly. We have chosen here an approach between these two, since there seems to be a real need for such an intermediate treatment of this subject.

ACKNOWLEDGMENTS

This book is an outcome of our teaching and research, which we have worked at over a period of many years at the College de France, at the University P. et M. Curie, and at the Laboratoire de Physique de l'Ecole Normale Superieure. We would like to express our thanks here to our friends and coworkers who have participated in our research and who have made us the beneficiaries of their ideas.

We want to thank particularly Jean Dalibard, who has been of such great help in the development of the exercises.

Introduction

The electromagnetic field plays a prominent part in physics. Without going back to Maxwell, one can recall for example that it is from the study of light that the Planck constant and the ideas of wave–particle duality arose for the first time in physics. More recently, the electromagnetic field has appeared as the prototype of quantum gauge fields.

It is therefore important to develop a good understanding of the dynamics of the electromagnetic field coupled to charged particles, and in particular of its quantum aspects. To this end, one must explain how the electromagnetic field can be quantized and how the concept of photon arises. One must also specify the observables and the states which describe the various aspects of radiation, and analyze the Hamiltonian which governs the coupled evolution of photons and atoms. It is to the study of these problems that this volume is devoted.

The quantization of the electromagnetic field is the central problem around which the various chapters are organized. Such a quantization requires some caution, owing to the gauge arbitrariness and to the redundancy associated with the vector and scalar potentials. As a result, we will treat these problems at several levels of increasing difficulty.

In Chapter I, we begin with the Maxwell–Lorentz equations which describe the evolution of an ensemble of charged particles coupled to the electromagnetic field and show that a spatial Fourier transformation of the field allows one to see more clearly the actual independent degrees of freedom of the field. We introduce in this way the normal variables which describe the normal vibrational modes of the field in the absence of sources. Quantization then is achieved in an elementary fashion by quantizing the harmonic oscillators associated with each normal mode, the normal variables becoming the creation and annihilation operators for a photon.

The problem is treated again in a more thorough and rigorous fashion in Chapter II, starting with the Lagrangian and the Hamiltonian formulation of electrodynamics. One such approach allows one to define unambiguously the canonically conjugate field variables. This provides also a straightforward method of quantization, the canonical quantization: two operators whose commutator equals $i\hbar$ then represent the two corresponding classical conjugate variables. We show nevertheless that such a theoretical approach is not directly applicable to the most commonly used Lagrangian, the standard Lagrangian. This is due to the fact that the dynamical variables of this Lagrangian, the vector and scalar potentials, are redundant. The most simple way of resolving this problem, and then quantizing the theory, is to choose the Coulomb gauge. Other possibilities exist, each having their advantages and disadvantages; these are examined later in Chapter IV (Poincaré gauge) and Chapter V (Lorentz gauge).

Many of the essential aspects of quantum electrodynamics in the Coulomb gauge are discussed in detail in Chapter III. These include the quantum equations of motion for the coupled system charges + field; the study of the states and observables of the free quantized field, of the properties of the vacuum, and of coherent states; and the analysis of interference and wave–particle duality in the quantum theory of radiation. We also examine in detail the properties of the Hamiltonian which describes the coupling between particles and photons.

This last subject is treated in more detail in Chapter IV, which is devoted to other equivalent formulations of electrodynamics derived from the Coulomb gauge. We show how it is possible to get other descriptions of electrodynamics, better adapted to this or that type of problem, either by changing the gauge or by adding to the standard Lagrangian in the Coulomb gauge the total derivative of a function of the generalized coordinates of the system, or else by directly performing a unitary transformation on the Coulomb-gauge Hamiltonian. Emphasis is placed on the physical significance the various mathematical operators have in the different representations and on the equivalence of the physical predictions derived from these various formulations. It is here that a satisfactory understanding of the fundamentals of quantum electrodynamics is essential if one is to avoid faulty interpretations, concerning for example the interaction Hamiltonians $\mathbf{A} \cdot \mathbf{p}$ or $\mathbf{E} \cdot \mathbf{r}$.

From the point of view adopted in Chapters II and IV, the symmetry between the four components of the potential four-vector is not maintained. The corresponding formulations are thus not adaptable to a covariant quantization of the field. These problems are dealt with in Chapter V, which treats the quantization of the field in the Lorentz gauge. We explain the difficulties which arise whenever the four components of the potential are treated as independent variables. We point out also how it is possible to resolve this problem by selecting, using the Lorentz condition, a subspace of physical states from the space of the radiation states.

We mention finally that, with the exception of the complements of Chapter V, the particles are treated nonrelativistically and are described by Schrödinger wave functions or Pauli spinors. Such an approximation is generally sufficient for the low-energy domain treated here. In addition, the choice of the Coulomb gauge, which explicitly yields the Coulomb interaction between particles which is predominant at low energy, is very convenient for the study of bound states of charged particles, such as atoms and molecules. This advantage holds also for the other formulations derived from the Coulomb gauge and treated in Chapter IV. A quantum relativistic description of particles requires that one consider them as elementary excitations of a relativistic matter field, such as the Dirac field for electrons and positrons. We deal with these problems in two complements in Chapter V. We show in these complements that it is possible to justify the nonrelativistic Hamiltonians used in this volume by considering them as "effective Hamiltonians" acting inside manifolds with a fixed number of particles and derived from the Hamiltonian of relativistic quantum electrodynamics, in which the number of particles, like the number of photons, is indeterminate.

This volume consists of five chapters and nineteen complements. The complements have a variety of objectives. They give more precision to the physical or mathematical concepts introduced in the chapter to which they are joined, or they expand the chapter by giving examples of applications, by introducing other points of view, or by taking up problems not studied in the chapter. The last complement in each chapter contains worked exercises. A short, nonexhaustive bibliography is given, either in the form of general references at the end of the chapter or complement, or in the form of more specialized references at the foot of the page. A detailed list of the books, cited by the author's name alone in the text, appears at the end of the volume.

It is possible to read this volume serially from beginning to end. It is also possible, however, to skip certain chapters and complements in a first study.

If one wishes to get a flavor of field quantization in its simplest form, and to understand the particle and wave aspects of radiation and the dynamics of the system field + particles, one can read Chapter I, then Chapter III and its Complement A_{III}. Reading Complements A_{IV} and B_{IV} can also give one a simple idea of the electric dipole approximation and of the equivalence of the interaction Hamiltonians $\mathbf{A} \cdot \mathbf{p}$ and $\mathbf{E} \cdot \mathbf{r}$ for the study of one- or two-photon processes.

A graduate student or researcher wanting to deepen his understanding of the structure of quantum electrodynamics and of the problems tied to the gauge arbitrariness, should extend his reading to Chapters II, IV, and V and choose those complements which relate best to his needs and his area of interest.

CHAPTER I

Classical Electrodynamics: The Fundamental Equations and the Dynamical Variables

The purpose for this first chapter is to review the basic equations of classical electrodynamics and to introduce a set of dynamical variables allowing one to characterize simply the state of the global system field + particles at a given instant.

The chapter begins (Part A) with a review of the *Maxwell–Lorentz equations* which describe the joint evolution of the electromagnetic field and of a set of charged particles. Some important results concerning the *constants of motion*, the *potentials*, and *gauge invariance* are also reviewed.

With a view to subsequent developments, notably quantization, one then shows (Part B) that classical electrodynamics has a simpler form in reciprocal space, after a Fourier transformation of the field. Such a transformation allows a simple decomposition of the electromagnetic field into its *longitudinal* and *transverse* components. It is then evident that the longitudinal electric field is not a true dynamical variable of the system, since it can be expressed as a function of the positions of the particles.

The following part (Part C) introduces linear combinations of the transverse electric and magnetic fields in reciprocal space which have the important property of evolving independently in the absence of particles and which then describe the *normal vibrational modes* of the free field. These new dynamical variables, called *normal variables*, play a central role in the theory, since they become, after quantization, the *creation and annihilation operators* for photons. All the field observables can be expressed as a function of these normal variables (and the particle variables).

The chapter ends finally (Part D) with a discussion of the various possible strategies for *quantizing* the foregoing theory. One simple, economic method, albeit not very rigorous, consists of quantizing each of the "harmonic oscillators" associated with the various normal modes of vibration of the field. One then gets all the fundamental commutation relations necessary for Chapter III. The problem is approached in a more rigorous manner in Chapter II, beginning with a Lagrangian and Hamiltonian formulation of electrodynamics.

Finally, Complement B_I compiles some results relative to the angular momentum of the electromagnetic field and to the multipole expansion of the field.

A—THE FUNDAMENTAL EQUATIONS IN REAL SPACE

1. The Maxwell–Lorentz Equations

The basic equations are grouped into two sets. First, the *Maxwell equations* relate the electric field $\mathbf{E}(\mathbf{r}, t)$ and the magnetic field $\mathbf{B}(\mathbf{r}, t)$ to the charge density $\rho(\mathbf{r}, t)$ and the current $\mathbf{j}(\mathbf{r}, t)$:

$$\nabla \cdot \mathbf{E}(\mathbf{r}, t) = \frac{1}{\varepsilon_0} \rho(\mathbf{r}, t) \tag{A.1.a}$$

$$\nabla \cdot \mathbf{B}(\mathbf{r}, t) = 0 \tag{A.1.b}$$

$$\nabla \times \mathbf{E}(\mathbf{r}, t) = -\frac{\partial}{\partial t} \mathbf{B}(\mathbf{r}, t) \tag{A.1.c}$$

$$\nabla \times \mathbf{B}(\mathbf{r}, t) = \frac{1}{c^2} \frac{\partial}{\partial t} \mathbf{E}(\mathbf{r}, t) + \frac{1}{\varepsilon_0 c^2} \mathbf{j}(\mathbf{r}, t) . \tag{A.1.d}$$

Next, the *Newton–Lorentz equations* describe the dynamics of each particle α, having mass m_α, charge q_α, position $\mathbf{r}_\alpha(t)$, and velocity $\mathbf{v}_\alpha(t)$, under the influence of electric and magnetic forces exerted by the fields

$$m_\alpha \frac{\mathrm{d}^2}{\mathrm{d}t^2} \mathbf{r}_\alpha(t) = q_\alpha [\mathbf{E}(\mathbf{r}_\alpha(t), t) + \mathbf{v}_\alpha(t) \times \mathbf{B}(\mathbf{r}_\alpha(t), t)] . \tag{A.2}$$

The equations (A.2) are valid only for slow, nonrelativistic particles ($v_\alpha \ll c$).

From (A.1.a) and (A.1.d) one can show that

$$\frac{\partial}{\partial t} \rho(\mathbf{r}, t) + \nabla \cdot \mathbf{j}(\mathbf{r}, t) = 0 . \tag{A.3}$$

Such an equation of continuity expresses the local conservation of the global electric charge,

$$Q = \int \mathrm{d}^3 r \, \rho(\mathbf{r}, t) . \tag{A.4}$$

The expression of ρ and $\mathbf{j}$ as a function of the particle variables is

$$\rho(\mathbf{r}, t) = \sum_\alpha q_\alpha \, \delta[\mathbf{r} - \mathbf{r}_\alpha(t)] \tag{A.5.a}$$

$$\mathbf{j}(\mathbf{r}, t) = \sum_\alpha q_\alpha \, \mathbf{v}_\alpha(t) \, \delta[\mathbf{r} - \mathbf{r}_\alpha(t)] . \tag{A.5.b}$$

One can show that Equations (A.5) satisfy the equation of continuity (A.3).

Equations (A.1) and (A.2) form two sets of coupled equations. The evolution of the field depends on the particles through ρ and $\mathbf{j}$. The motion of the particles depends on the fields $\mathbf{E}$ and $\mathbf{B}$. The equations (A.1) are first-order partial differential equations, while the equations (A.2) are second-order ordinary differential equations. It follows that the state of the global system, field + particles, is determined at some instant t_0 by giving the fields $\mathbf{E}$ and $\mathbf{B}$ at all points $\mathbf{r}$ of space and the position and velocity $\mathbf{r}_\alpha$ and $\mathbf{v}_\alpha$ of each particle α:

$$\{\, \mathbf{E}(\mathbf{r}, t_0), \mathbf{B}(\mathbf{r}, t_0), \mathbf{r}_\alpha(t_0), \mathbf{v}_\alpha(t_0) \,\}\,. \tag{A.6}$$

It is important to note that in the Maxwell equations (A.1), $\mathbf{r}$ is not a dynamical variable (like $\mathbf{r}_\alpha$) but a continuous parameter labeling the field variables.

2. Some Important Constants of the Motion

Starting with Equations (A.1) and (A.2) and the expressions (A.5) for ρ and $\mathbf{j}$, one can show (see Exercise 1) that the following functions of $\mathbf{E}$, $\mathbf{B}$, $\mathbf{r}_\alpha$, and $\mathbf{v}_\alpha$:

$$H = \sum_\alpha \frac{1}{2} m_\alpha \mathbf{v}_\alpha^2(t) + \frac{\varepsilon_0}{2} \int \mathrm{d}^3 r [\mathbf{E}^2(\mathbf{r}, t) + c^2\, \mathbf{B}^2(\mathbf{r}, t)] \tag{A.7}$$

$$\mathbf{P} = \sum_\alpha m_\alpha \mathbf{v}_\alpha(t) + \varepsilon_0 \int \mathrm{d}^3 r\, \mathbf{E}(\mathbf{r}, t) \times \mathbf{B}(\mathbf{r}, t) \tag{A.8}$$

$$\mathbf{J} = \sum_\alpha \mathbf{r}_\alpha(t) \times m_\alpha \mathbf{v}_\alpha(t) + \varepsilon_0 \int \mathrm{d}^3 r\, \mathbf{r} \times [\mathbf{E}(\mathbf{r}, t) \times \mathbf{B}(\mathbf{r}, t)] \tag{A.9}$$

are constants of the motion, that is, independent of t.

H is the *total energy* of the global system field + particles, $\mathbf{P}$ is the *total momentum*, and $\mathbf{J}$ the *total angular momentum*. The fact that these quantities are constants of the motion results from the invariance of the equations of motion with respect to changes in the time origin, the coordinate origin, and the orientation of the coordinate axes. (The connection between the constants of the motion and the invariance properties of the Lagrangian of electrodynamics will be analyzed in Complement $\mathrm{B_{II}}$).

3. Potentials—Gauge Invariance

Equations (A.1.b) and (A.1.c) suggest that the fields $\mathbf{E}$ and $\mathbf{B}$ can always be written in the form

$$\begin{cases} \mathbf{B}(\mathbf{r}, t) = \nabla \times \mathbf{A}(\mathbf{r}, t) & \text{(A.10.a)} \\ \mathbf{E}(\mathbf{r}, t) = -\dfrac{\partial}{\partial t}\mathbf{A}(\mathbf{r}, t) - \nabla U(\mathbf{r}, t) & \text{(A.10.b)} \end{cases}$$

where $\mathbf{A}$ is a vector field, called the *vector potential*, and U a scalar field called the *scalar potential*. A first advantage in introducing $\mathbf{A}$ and U is that the two Maxwell equations (A.1.b) and (A.1.c) are automatically satisfied. Other advantages will appear in the Lagrangian and Hamiltonian formulations of electrodynamics (see Chapter II).

Substituting (A.10) in Maxwell's equations (A.1.a) and (A.1.d), one gets the equations of motion for $\mathbf{A}$ and U

$$\begin{cases} \Delta U(\mathbf{r}, t) = -\dfrac{1}{\varepsilon_0}\rho(\mathbf{r}, t) - \nabla \cdot \dfrac{\partial}{\partial t}\mathbf{A}(\mathbf{r}, t) & \text{(A.11.a)} \\ \left(\dfrac{1}{c^2}\dfrac{\partial^2}{\partial t^2} - \Delta\right)\mathbf{A}(\mathbf{r}, t) = \\ \quad = \dfrac{1}{\varepsilon_0 c^2}\mathbf{j}(\mathbf{r}, t) - \nabla\left[\nabla \cdot \mathbf{A}(\mathbf{r}, t) + \dfrac{1}{c^2}\dfrac{\partial}{\partial t}U(\mathbf{r}, t)\right] & \text{(A.11.b)} \end{cases}$$

which are second-order partial differential equations and no longer first-order as in (A.1). Actually, since $\partial^2 U/\partial t^2$ does not appear in (A.11.a), this equation is not an equation of motion for U, but rather relates U to $\partial \mathbf{A}/\partial t$ at each instant. The state of the field is now fixed by giving $\mathbf{A}(\mathbf{r}, t_0)$ and $\partial \mathbf{A}(\mathbf{r}, t_0)/\partial t$ for all $\mathbf{r}$.

It follows from (A.10) that $\mathbf{E}$ and $\mathbf{B}$ are invariants under the following *gauge transformation*:

$$\mathbf{A}(\mathbf{r}, t) \rightarrow \mathbf{A}'(\mathbf{r}, t) = \mathbf{A}(\mathbf{r}, t) + \nabla F(\mathbf{r}, t) \qquad \text{(A.12.a)}$$

$$U(\mathbf{r}, t) \rightarrow U'(\mathbf{r}, t) = U(\mathbf{r}, t) - \frac{\partial}{\partial t}F(\mathbf{r}, t) \qquad \text{(A.12.b)}$$

where $F(\mathbf{r}, t)$ is an arbitrary function of $\mathbf{r}$ and t. There is then a certain redundancy in these potentials, since the same physical fields $\mathbf{E}$ and $\mathbf{B}$ can be written with many different potentials $\mathbf{A}$ and U. This redundancy can be reduced by the choice of one gauge condition which fixes $\nabla \cdot \mathbf{A}$ (the value of $\nabla \times \mathbf{A}$ is already determined by (A.10.a)).

The two most commonly used gauges are the Lorentz gauge and the Coulomb gauge.

(i) The Lorentz gauge is defined by

$$\nabla \cdot \mathbf{A}(\mathbf{r}, t) + \frac{1}{c^2}\frac{\partial}{\partial t}U(\mathbf{r}, t) = 0 . \qquad \text{(A.13)}$$

One can prove that it is always possible to choose in (A.12) a function F such that (A.13) will be satisfied for $\mathbf{A}'$ and U'. In the Lorentz gauge, the equations (A.11) take a more symmetric form:

$$\begin{cases} \Box U(\mathbf{r}, t) = \dfrac{1}{\varepsilon_0} \rho(\mathbf{r}, t) & \text{(A.14.a)} \\[2ex] \Box \mathbf{A}(\mathbf{r}, t) = \dfrac{1}{\varepsilon_0 c^2} \mathbf{j}(\mathbf{r}, t) & \text{(A.14.b)} \end{cases}$$

where $\Box = \partial^2/c^2\,\partial t^2 - \Delta$ is the d'Alembertian operator. This is due to the fact that the Maxwell's equations on one hand and the Lorentz condition on the other are relativistically invariant, that is, they keep the same form after a Lorentz transformation. Using covariant notation, Equations (A.13) and (A.14) can be written

$$\sum_\mu \partial_\mu A^\mu = 0 \tag{A.15}$$

with $$\partial_\mu = \left\{ \frac{1}{c}\frac{\partial}{\partial t}, \boldsymbol{\nabla} \right\} \qquad A^\mu = \left\{ \frac{U}{c}, \mathbf{A} \right\}$$

and

$$\sum_\nu \partial_\nu\, \partial^\nu A^\mu = \frac{1}{\varepsilon_0 c^2} j^\mu \tag{A.16}$$

with $$j^\mu = \{ c\rho, \mathbf{j} \}$$

where A^μ and j^μ are the four-vectors associated with the potential and the current respectively.

(ii) The Coulomb (or radiation) gauge is defined by

$$\boldsymbol{\nabla} \cdot \mathbf{A}(\mathbf{r}, t) = 0 \tag{A.17}$$

Equations (A.11) then become

$$\begin{cases} \Delta U(\mathbf{r}, t) = -\dfrac{1}{\varepsilon_0} \rho(\mathbf{r}, t) & \text{(A.18.a)} \\[2ex] \Box \mathbf{A}(\mathbf{r}, t) = \dfrac{1}{\varepsilon_0 c^2} \mathbf{j}(\mathbf{r}, t) - \dfrac{1}{c^2} \boldsymbol{\nabla} \dfrac{\partial}{\partial t} U(\mathbf{r}, t)\,. & \text{(A.18.b)} \end{cases}$$

Equation (A.18.a) is Poisson's equation for U. The covariance is lost, but other advantages of the Coulomb gauge will be seen in the subsequent chapters.

B—ELECTRODYNAMICS IN RECIPROCAL SPACE

1. The Fourier Spatial Transformation—Notation

Let $\mathscr{E}(\mathbf{k}, t)$ be the Fourier spatial transform of $\mathbf{E}(\mathbf{r}, t)$. Then $\mathbf{E}$ and $\mathscr{E}$ are related through the following equations:

$$\begin{cases} \mathscr{E}(\mathbf{k}, t) = \dfrac{1}{(2\pi)^{3/2}} \displaystyle\int \mathrm{d}^3 r \, \mathbf{E}(\mathbf{r}, t) \, \mathrm{e}^{-\mathrm{i}\mathbf{k}.\mathbf{r}} & \text{(B.1.a)} \\ \mathbf{E}(\mathbf{r}, t) = \dfrac{1}{(2\pi)^{3/2}} \displaystyle\int \mathrm{d}^3 k \, \mathscr{E}(\mathbf{k}, t) \, \mathrm{e}^{\mathrm{i}\mathbf{k}.\mathbf{r}} . & \text{(B.1.b)} \end{cases}$$

In Table I the notations used for the Fourier transforms of various other physical quantities are shown. Block letters are used for the quantities in real space, and script ones for the same quantities in reciprocal space.

TABLE I

$$\mathbf{E}(\mathbf{r}, t) \leftrightarrow \mathscr{E}(\mathbf{k}, t)$$

$$\mathbf{B}(\mathbf{r}, t) \leftrightarrow \mathscr{B}(\mathbf{k}, t)$$

$$\mathbf{A}(\mathbf{r}, t) \leftrightarrow \mathscr{A}(\mathbf{k}, t)$$

$$U(\mathbf{r}, t) \leftrightarrow \mathscr{U}(\mathbf{k}, t)$$

$$\rho(\mathbf{r}, t) \leftrightarrow \rho(\mathbf{k}, t)$$

$$\mathbf{j}(\mathbf{r}, t) \leftrightarrow \mathscr{j}(\mathbf{k}, t) .$$

Since $\mathbf{E}(\mathbf{r}, t)$ is real, it follows that

$$\mathscr{E}^*(\mathbf{k}, t) = \mathscr{E}(-\mathbf{k}, t) . \tag{B.2}$$

In this treatment one frequently uses the Parseval–Plancherel identity

$$\int \mathrm{d}^3 r \, F^*(\mathbf{r}) \, G(\mathbf{r}) = \int \mathrm{d}^3 k \, \mathscr{F}^*(\mathbf{k}) \, \mathscr{G}(\mathbf{k}) \tag{B.3}$$

where $\mathscr{F}$ and $\mathscr{G}$ are the Fourier transforms of F and G, as well as the fact that the Fourier transform of a product of two functions is proportional to the convolution product of the Fourier transforms of these two functions:

$$\frac{1}{(2\pi)^{3/2}} \int \mathrm{d}^3 r' \, F(\mathbf{r}') \, G(\mathbf{r} - \mathbf{r}') \leftrightarrow \mathscr{F}(\mathbf{k}) \, \mathscr{G}(\mathbf{k}) \tag{B.4}$$

Table II lists some Fourier transforms that are used throughout this book

TABLE II

$$\frac{1}{4\,\pi r} \leftrightarrow \frac{1}{(2\,\pi)^{3/2}}\,\frac{1}{k^2}$$

$$\frac{\mathbf{r}}{4\,\pi r^3} \leftrightarrow \frac{1}{(2\,\pi)^{3/2}}\,\frac{-\,\mathrm{i}\mathbf{k}}{k^2}$$

$$\delta(\mathbf{r}-\mathbf{r}_\alpha) \leftrightarrow \frac{1}{(2\,\pi)^{3/2}}\,\mathrm{e}^{-\,\mathrm{i}\mathbf{k}.\mathbf{r}_\alpha}\,.$$

Finally, to simplify the notation, we write $\dot{\mathbf{r}}_\alpha$ in place of $\mathrm{d}\mathbf{r}_\alpha(t)/\mathrm{d}t$, $\dot{\mathbf{E}}$ in place of $\partial\,\mathbf{E}(\mathbf{r}, t)/\partial t$, $\ddot{\mathscr{E}}$ in place of $\partial^2\mathscr{E}(\mathbf{k}, t)/\partial t^2, \dots,$ whenever there is no chance of confusion.

2. The Field Equations in Reciprocal Space

Since the gradient operator ∇ in real space transforms into multiplication by $\mathrm{i}\mathbf{k}$ in reciprocal space, Maxwell's equations (A.1) in reciprocal space become

$$\left\{\begin{aligned} &\mathrm{i}\mathbf{k}\cdot\mathscr{E} = \frac{1}{\varepsilon_0}\,\rho && \text{(B.5.a)}\\ &\mathrm{i}\mathbf{k}\cdot\mathscr{B} = 0 && \text{(B.5.b)}\\ &\mathrm{i}\mathbf{k}\times\mathscr{E} = -\,\dot{\mathscr{B}} && \text{(B.5.c)}\\ &\mathrm{i}\mathbf{k}\times\mathscr{B} = \frac{1}{c^2}\,\dot{\mathscr{E}} + \frac{1}{\varepsilon_0\,c^2}\,\jmath\,. && \text{(B.5.d)}\end{aligned}\right.$$

It is apparent in (B.5) that $\dot{\mathscr{E}}(\mathbf{k})$ and $\dot{\mathscr{B}}(\mathbf{k})$ depend only on the values of $\mathscr{E}(\mathbf{k})$, $\mathscr{B}(\mathbf{k})$, $\rho(\mathbf{k})$, and $\jmath(\mathbf{k})$ at the *same* point $\mathbf{k}$. Maxwell's equations, which are partial differential equations in real space, become *strictly local* in reciprocal space, which introduces a great simplification.

The equation of continuity (A.3) is now written

$$\mathrm{i}\mathbf{k}\cdot\jmath + \dot{\rho} = 0\,. \tag{B.6}$$

The relationships between the fields and potentials become

$$\left\{\begin{aligned} &\mathscr{B} = \mathrm{i}\mathbf{k}\times\mathscr{A} && \text{(B.7.a)}\\ &\mathscr{E} = -\,\dot{\mathscr{A}} - \mathrm{i}\mathbf{k}\mathscr{U} && \text{(B.7.b)}\end{aligned}\right.$$

the gauge transformation (A.12)

$$\mathscr{A} \rightarrow \mathscr{A}' = \mathscr{A} + \mathrm{i}\mathbf{k}\,\mathscr{F} \tag{B.8.a}$$

$$\mathscr{U} \rightarrow \mathscr{U}' = \mathscr{U} - \dot{\mathscr{F}} \tag{B.8.b}$$

and the equations for the potentials (A.11)

$$\begin{cases} k^2\,\mathscr{U} = \dfrac{1}{\varepsilon_0}\rho + \mathrm{i}\mathbf{k}\cdot\dot{\mathscr{A}} & \text{(B.9.a)} \\ \dfrac{1}{c^2}\ddot{\mathscr{A}} + k^2\,\mathscr{A} = \dfrac{1}{\varepsilon_0 c^2}\,\jmath - \mathrm{i}\mathbf{k}\left(\mathrm{i}\mathbf{k}\cdot\mathscr{A} + \dfrac{1}{c^2}\dot{\mathscr{U}}\right). & \text{(B.9.b)} \end{cases}$$

3. Longitudinal and Transverse Vector Fields

By definition, a longitudinal vector field $\mathbf{V}_{\parallel}(\mathbf{r})$ is a vector field such that

$$\boldsymbol{\nabla} \times \mathbf{V}_{\parallel}(\mathbf{r}) = \mathbf{0}\,. \tag{B.10.a}$$

which, in reciprocal space, becomes

$$\mathrm{i}\mathbf{k} \times \mathscr{V}_{\parallel}(\mathbf{k}) = \mathbf{0}\,. \tag{B.10.b}$$

A transverse vector field $\mathbf{V}_{\perp}(\mathbf{r})$ is characterized by

$$\begin{cases} \boldsymbol{\nabla}\cdot\mathbf{V}_{\perp}(\mathbf{r}) = 0 & \text{(B.11.a)} \\ \mathrm{i}\mathbf{k}\cdot\mathscr{V}_{\perp}(\mathbf{k}) = 0\,. & \text{(B.11.b)} \end{cases}$$

Comparison of (B.10.a) and (B.10.b) or (B.11.a) and (B.11.b) shows that the name longitudinal or transverse has a clear geometrical significance in reciprocal space: for a longitudinal vector field, $\mathscr{V}_{\parallel}(\mathbf{k})$ is parallel to $\mathbf{k}$ for all $\mathbf{k}$; for a transverse vector field, $\mathscr{V}_{\perp}(\mathbf{k})$ is perpendicular to $\mathbf{k}$ for all $\mathbf{k}$.

It is important to note that a vector field is longitudinal [or transverse] if and only if (B.10) [or (B.11)] are satisfied for all $\mathbf{r}$ or all $\mathbf{k}$. For example, in the presence of a point charge at $\mathbf{r}_\alpha$, $\boldsymbol{\nabla}\cdot\mathbf{E}$ is, according to (A.1.a), zero everywhere except at $\mathbf{r}_\alpha$, where the particle is located. In the presence of a charge, $\mathbf{E}$ is therefore not a transverse field. This is even more evident in reciprocal space, since $\mathbf{k}\cdot\mathscr{E}$ is then proportional to $\mathrm{e}^{-i\mathbf{k}\cdot\mathbf{r}_\alpha}$, which is clearly nonvanishing everywhere.

Working in reciprocal space allows also a very simple decomposition of all vector fields into longitudinal and transverse components:

$$\mathscr{V}(\mathbf{k}) = \mathscr{V}_{\parallel}(\mathbf{k}) + \mathscr{V}_{\perp}(\mathbf{k})\,. \tag{B.12}$$

At all points $\mathbf{k}$, $\mathscr{V}_{\parallel}(\mathbf{k})$ is gotten by projection of $\mathscr{V}(\mathbf{k})$ onto the unit vector $\boldsymbol{\kappa}$ in the direction $\mathbf{k}$:

$$\boldsymbol{\kappa} = \mathbf{k}/k \,. \tag{B.13}$$

One thus has

$$\begin{cases} \mathscr{V}_{\parallel}(\mathbf{k}) = \boldsymbol{\kappa}[\boldsymbol{\kappa} \cdot \mathscr{V}(\mathbf{k})] & \text{(B.14.a)} \\ \mathscr{V}_{\perp}(\mathbf{k}) = \mathscr{V}(\mathbf{k}) - \mathscr{V}_{\parallel}(\mathbf{k}) & \text{(B.14.b)} \end{cases}$$

$\mathbf{V}_{\parallel}(\mathbf{r})$ and $\mathbf{V}_{\perp}(\mathbf{r})$ are then gotten by a spatial Fourier transformation of (B.14).

Remarks

(i) In reciprocal space, the relationship which exists between a vector field $\mathscr{V}(\mathbf{k})$ and its longitudinal or transverse components is a local relationship. For example, one can show from (B.14) that

$$\mathscr{V}_{\perp i}(\mathbf{k}) = \sum_j \left(\delta_{ij} - \frac{k_i\, k_j}{k^2} \right) \mathscr{V}_j(\mathbf{k}) \tag{B.15}$$

where $i, j = x, y, z$. Each component of $\mathscr{V}_{\perp}(\mathbf{k})$ at point $\mathbf{k}$ depends only on the components of $\mathscr{V}(\mathbf{k})$ at the same point $\mathbf{k}$. By Fourier transformation, Equation (B.15) then becomes, using (B.4),

$$V_{\perp i}(\mathbf{r}) = \sum_j \int \mathrm{d}^3 r' \; \delta^{\perp}_{ij}(\mathbf{r} - \mathbf{r}') \; V_j(\mathbf{r}') \tag{B.16}$$

where

$$\begin{aligned} \delta^{\perp}_{ij}(\mathbf{r}) &= \frac{1}{(2\,\pi)^3} \int \mathrm{d}^3 k \; \mathrm{e}^{\mathrm{i}\mathbf{k}.\mathbf{r}} \left(\delta_{ij} - \frac{k_i\, k_j}{k^2} \right) \\ &= \delta_{ij}\, \delta(\mathbf{r}) + \frac{\partial^2}{\partial r_i\, \partial r_j} \frac{1}{(2\,\pi)^3} \int \mathrm{d}^3 k \; \mathrm{e}^{\mathrm{i}\mathbf{k}.\mathbf{r}} \frac{1}{k^2} \\ &= \delta_{ij}\, \delta(\mathbf{r}) + \frac{1}{4\,\pi} \frac{\partial^2}{\partial r_i\, \partial r_j} \frac{1}{r} \end{aligned} \tag{B.17.a}$$

$\delta^{\perp}_{ij}(\mathbf{r})$ is called the "transverse δ-function". The presence of the last term in (B.17.a) shows that the relationship between $\mathbf{V}_{\perp}(\mathbf{r})$ and $\mathbf{V}(\mathbf{r})$ is *not local*: $\mathbf{V}_{\perp}(\mathbf{r})$ depends on the values $\mathbf{V}(\mathbf{r}')$ of $\mathbf{V}$ at all other points $\mathbf{r}'$. Note also that the calculation of the last term in (B.17.a) needs special caution at $r = 0$. The second derivative of $1/r$ must be calculated using the theory of distributions and contains a term proportional to $\delta_{ij}\delta(\mathbf{r})$. The calculation, presented in detail in Complement $\mathrm{A_I}$, leads to

$$\delta^{\perp}_{ij}(\mathbf{r}) = \frac{2}{3}\, \delta_{ij}\, \delta(\mathbf{r}) - \frac{1}{4\,\pi r^3} \left(\delta_{ij} - \frac{3\, r_i\, r_j}{r^2} \right) \tag{B.17.b}$$

(ii) The decomposition of a vector field, arising from a four-vector or from an antisymmetric four-tensor, into longitudinal and transverse components is not relativistically invariant. A vector field that appears transverse in a Lorentzian frame is not necessarily transverse in another Lorentzian frame.

(iii) Even though the separation (B.12) introduces nonlocal effects in real space and is no longer relativistically invariant, it is nonetheless interesting in that it simplifies the solution of Maxwell's equations. In effect, as will be seen in the following subsections, two of the four Maxwell equations establish only the longitudinal part of the electric and magnetic fields, whereas the other equations give the rate of variation of the transverse fields. Such an approach then allows one to introduce a convenient set of normal variables for the transverse field.

4. Longitudinal Electric and Magnetic Fields

Return to Maxwell's equations. It is clear now that the first two equations (B.5.a) and (B.5.b) give the longitudinal parts of $\mathscr{E}$ and $\mathscr{B}$. The second equation clearly shows that the magnetic field is purely transverse:

$$\mathscr{B}_{\parallel} = \mathbf{0} = \mathbf{B}_{\parallel} . \tag{B.18}$$

The first equation (B.5.a) relates the longitudinal electric field $\mathscr{E}_{\parallel}(\mathbf{k})$ to the charge distribution $\rho(\mathbf{k})$:

$$\mathscr{E}_{\parallel}(\mathbf{k}) = -\frac{i}{\varepsilon_0}\rho(\mathbf{k})\frac{\mathbf{k}}{k^2} \tag{B.19}$$

and $\mathscr{E}_{\parallel}(\mathbf{k})$ appears then as the product of two functions of $\mathbf{k}$ whose Fourier transforms are

$$\rho(\mathbf{k}) \leftrightarrow \rho(\mathbf{r}) \tag{B.20.a}$$

$$-\frac{i}{\varepsilon_0}\frac{\mathbf{k}}{k^2} \leftrightarrow \frac{(2\pi)^{3/2}}{4\pi\varepsilon_0}\frac{\mathbf{r}}{r^3} . \tag{B.20.b}$$

Using (B.4), one then has

$$\begin{aligned}\mathbf{E}_{\parallel}(\mathbf{r}, t) &= \frac{1}{4\pi\varepsilon_0}\int \mathrm{d}^3r'\, \rho(\mathbf{r}', t)\frac{\mathbf{r} - \mathbf{r}'}{|\mathbf{r} - \mathbf{r}'|^3} \\ &= \frac{1}{4\pi\varepsilon_0}\sum_{\alpha} q_{\alpha}\frac{\mathbf{r} - \mathbf{r}_{\alpha}(t)}{|\mathbf{r} - \mathbf{r}_{\alpha}(t)|^3} .\end{aligned} \tag{B.21}$$

It thus appears that the longitudinal electric field at some time t is the Coulomb field associated with ρ and calculated as if the density of charge ρ were static and assumed to have its value taken at t, i.e., the instantaneous Coulomb field.

It is important to note that this result is independent of the choice of gauge, since it has been derived directly from Maxwell's equations for the field **E** and **B** without reference to the potentials.

The fact that the longitudinal electric field instantly responds to a change in the distribution of charge does not imply the existence of perturbations traveling with a velocity greater than that of light. Actually, only the total electric field has a physical meaning, and one can show that the transverse field $\mathbf{E}_\perp$ also has an instantaneous component which exactly cancels that of $\mathbf{E}_\parallel$, with the result that the total field remains always a purely retarded field. This point will be discussed again later.

Consider now the longitudinal parts of (B.5.c) and (B.5.d). The two terms of (B.5.c) are transverse. The longitudinal part of (B.5.d) is written

$$\dot{\mathcal{E}}_\parallel + \frac{1}{\varepsilon_0} \boldsymbol{j}_\parallel = \mathbf{0} . \tag{B.22}$$

Taking the scalar product of (B.22) with **k**, and using (B.19) and the fact that $\mathbf{k} \cdot \boldsymbol{j}_\parallel = \mathbf{k} \cdot \boldsymbol{j}$, one gets

$$\dot{\rho} + i\mathbf{k} \cdot \boldsymbol{j} = 0 \tag{B.23}$$

which is just the expression of the conservation of charge (B.6) and thus conveys nothing new.

Remarks

(i) From equation (A.10.b) or (B.7.b) connecting the electric field to the potentials, it follows that

$$\mathbf{E}_\perp = - \dot{\mathbf{A}}_\perp \tag{B.24.a}$$

$$\mathbf{E}_\parallel = - \dot{\mathbf{A}}_\parallel - \nabla U . \tag{B.24.b}$$

In the Coulomb gauge, one has $\mathbf{A}_\parallel = \mathbf{0}$, with the result that

$$\mathbf{A}_\parallel = \mathbf{0} \quad \rightarrow \quad \mathbf{E}_\parallel = - \nabla U . \tag{B.25.a}$$

It follows that the longitudinal and transverse parts of **E** are associated, in the Coulomb gauge, with U and **A** respectively. Comparison of (B.25.a) and (B.21) shows that, in the Coulomb gauge, U is nothing more than the Coulomb potential of the charge distribution:

$$\mathbf{A}_\parallel = \mathbf{0} \quad \rightarrow \quad U(\mathbf{r}, t) = \frac{1}{4\pi\varepsilon_0} \int d^3r' \, \frac{\rho(\mathbf{r}', t)}{|\mathbf{r} - \mathbf{r}'|} . \tag{B.25.b}$$

The same result can be gotten directly from Equation (A.18.a). The solution of this Poisson equation, which tends to zero as $|\mathbf{r}| \to \infty$, is nothing more than (B.25.b).

(ii) It is clear from (B.8.a) that a gauge transformation does not change $\mathbf{A}_\perp$. It follows that *the transverse vector potential* $\mathbf{A}_\perp$ *is gauge invariant:*

$$\mathbf{A}'_\perp = \mathbf{A}_\perp . \tag{B.26}$$

(iii) Maxwell's equations are presented here in two sets: (A.1.a) and (A.1.b) give the longitudinal fields, and (A.1.c) and (A.1.d) give the rate of variation of the transverse fields [§B.6]. This grouping is different from the one used in relativity, where Equations (A.1.b) and (A.1.c) on one hand, and (A.1.a) and (A.1.d) on the other, are combined in two covariant equations

$$\partial_\mu F_{\nu\rho} + \partial_\nu F_{\rho\mu} + \partial_\rho F_{\mu\nu} = 0 \qquad \mu \neq \nu \neq \rho \tag{B.27.a}$$

$$\sum_\mu \partial_\mu F^{\mu\nu} = \frac{1}{\varepsilon_0 c^2} j^\nu \tag{B.27.b}$$

where

$$F_{\mu\nu} = \partial_\mu A_\nu - \partial_\nu A_\mu \tag{B.28}$$

is the electromagnetic field tensor, A_μ the potential four-vector, and j_μ the current four-vector.

5. Contribution of the Longitudinal Electric Field to the Total Energy, to the Total Momentum, and to the Total Angular Momentum

One now uses (B.19) for $\mathscr{E}_\parallel(\mathbf{k})$ to evaluate the contribution of the longitudinal electric field to various important physical quantities.

a) The Total Energy

The Parseval–Plancherel identity (B.3) allows one to write

$$\frac{\varepsilon_0}{2} \int \mathrm{d}^3 r \, \mathbf{E} \cdot \mathbf{E} = \frac{\varepsilon_0}{2} \int \mathrm{d}^3 k \, \mathscr{E}^* \cdot \mathscr{E} . \tag{B.29}$$

One then replaces $\mathscr{E}$ by $\mathscr{E}_\parallel + \mathscr{E}_\perp$ and uses $\mathscr{E}_\parallel \cdot \mathscr{E}_\perp = 0$. This yields

$$\frac{\varepsilon_0}{2} \int \mathrm{d}^3 r \, \mathbf{E}^2 = \frac{\varepsilon_0}{2} \int \mathrm{d}^3 k \, | \mathscr{E}_\parallel(\mathbf{k}) |^2 + \frac{\varepsilon_0}{2} \int \mathrm{d}^3 k \, | \mathscr{E}_\perp(\mathbf{k}) |^2 . \tag{B.30}$$

The first term in (B.30) is the contribution H_{long} of the longitudinal electric field to the total energy given in (A.7):

$$H_{\text{long}} = \frac{\varepsilon_0}{2} \int \mathrm{d}^3 k \, | \mathscr{E}_\parallel(\mathbf{k}) |^2 = \frac{\varepsilon_0}{2} \int \mathrm{d}^3 r \, \mathbf{E}_\parallel^2(\mathbf{r}) \tag{B.31.a}$$

while the second, when added to the magnetic energy, gives the contribution H_{trans} of the fields $\mathbf{E}_\perp$ and $\mathbf{B}$:

$$H_{\text{trans}} = \frac{\varepsilon_0}{2}\int \mathrm{d}^3k[|\mathscr{E}_\perp(\mathbf{k})|^2 + c^2|\mathscr{B}(\mathbf{k})|^2]$$

$$= \frac{\varepsilon_0}{2}\int \mathrm{d}^3r[\mathbf{E}_\perp^2(\mathbf{r}) + c^2\,\mathbf{B}^2(\mathbf{r})]. \qquad \text{(B.31.b)}$$

Inserting the expression (B.19) for $\mathscr{E}_\parallel(\mathbf{k})$ in (B.31.a), one gets

$$H_{\text{long}} = \frac{1}{2\,\varepsilon_0}\int \mathrm{d}^3k\;\rho^*(\mathbf{k})\,\frac{\rho(\mathbf{k})}{k^2} \qquad \text{(B.32)}$$

which can finally be written using (B.3) and (B.4) as

$$H_{\text{long}} = \frac{1}{8\,\pi\varepsilon_0}\iint \mathrm{d}^3r\,\mathrm{d}^3r'\,\frac{\rho(\mathbf{r})\,\rho(\mathbf{r}')}{|\mathbf{r}-\mathbf{r}'|} \qquad \text{(B.33)}$$

H_{long} is nothing more than the Coulomb electrostatic energy of the system of charges. Finally, one calculates H_{long} for a system of point charges. For this it is convenient to use the expression

$$\rho(\mathbf{k}) = \sum_\alpha \frac{q_\alpha}{(2\,\pi)^{3/2}}\,\mathrm{e}^{-i\mathbf{k}.\mathbf{r}_\alpha} \qquad \text{(B.34)}$$

for the Fourier transform of the charge distribution given in (A.5.a). Substituting (B.34) in (B.32), one gets

$$H_{\text{long}} = V_{\text{Coul}} = \sum_\alpha \frac{q_\alpha^2}{2\,\varepsilon_0(2\,\pi)^3}\int \frac{\mathrm{d}^3k}{k^2} + \sum_{\alpha\neq\beta}\frac{q_\alpha\,q_\beta}{2\,\varepsilon_0(2\,\pi)^3}\int \mathrm{d}^3k\,\frac{\mathrm{e}^{-i\mathbf{k}.(\mathbf{r}_\alpha-\mathbf{r}_\beta)}}{k^2}. \qquad \text{(B.35)}$$

The first term of (B.35) can be written $\Sigma_\alpha \varepsilon_{\text{Coul}}^\alpha$ where

$$\varepsilon_{\text{Coul}}^\alpha = \frac{q_\alpha^2}{2\,\varepsilon_0(2\,\pi)^3}\int \frac{\mathrm{d}^3k}{k^2} \qquad \text{(B.36)}$$

is the Coulomb self energy of the particle α (in fact, infinite, unless one introduces a cutoff in the integral on k). The second term is nothing more than the Coulomb interaction between pairs of particles (α, β), so that finally

$$H_{\text{long}} = V_{\text{Coul}} = \sum_\alpha \varepsilon_{\text{Coul}}^\alpha + \frac{1}{8\,\pi\varepsilon_0}\sum_{\alpha\neq\beta}\frac{q_\alpha\,q_\beta}{|\mathbf{r}_\alpha - \mathbf{r}_\beta|}. \qquad \text{(B.37)}$$

In conclusion, one has seen in this subsection that the total energy (A.7) of the system can be written

$$H = \sum_\alpha \frac{1}{2} m_\alpha \dot{\mathbf{r}}_\alpha^2 + V_{\text{Coul}} + \text{H}_{\text{trans}} \tag{B.38}$$

and appears as the sum of three energies: the kinetic energy of the particles (first term), their Coulomb energy (second term), and the energy of the transverse field (third term). As in the preceding subsection, these results are independent of the choice of gauge.

b) The Total Momentum

One substitutes $\mathbf{E}_\parallel + \mathbf{E}_\perp$ for **E** in the second term of (A.8). The total momentum of the field appears then as the sum of two contributions, $\mathbf{P}_{\text{long}}$ and $\mathbf{P}_{\text{trans}}$, given by

$$\mathbf{P}_{\text{long}} = \varepsilon_0 \int \mathrm{d}^3 r \, \mathbf{E}_\parallel(\mathbf{r}) \times \mathbf{B}(\mathbf{r}) = \varepsilon_0 \int \mathrm{d}^3 k \, \mathscr{E}_\parallel^*(\mathbf{k}) \times \mathscr{B}(\mathbf{k}) \tag{B.39.a}$$

$$\mathbf{P}_{\text{trans}} = \varepsilon_0 \int \mathrm{d}^3 r \, \mathbf{E}_\perp(\mathbf{r}) \times \mathbf{B}(\mathbf{r}) = \varepsilon_0 \int \mathrm{d}^3 k \, \mathscr{E}_\perp^*(\mathbf{k}) \times \mathscr{B}(\mathbf{k}) . \tag{B.39.b}$$

Using (B.19) for $\mathscr{E}_\parallel$, the relationship (B.7.a) between $\mathscr{B}$ and $\mathscr{A}$, and the identity

$$\mathbf{a} \times (\mathbf{b} \times \mathbf{c}) = (\mathbf{a} \cdot \mathbf{c})\, \mathbf{b} - (\mathbf{a} \cdot \mathbf{b})\, \mathbf{c} \tag{B.40}$$

one can transform (B.39.a) into

$$\begin{aligned} \mathbf{P}_{\text{long}} &= \varepsilon_0 \int \mathrm{d}^3 k \, \frac{i\rho^*}{\varepsilon_0} \frac{\mathbf{k}}{k^2} \times (i\mathbf{k} \times \mathscr{A}) \\ &= \int \mathrm{d}^3 k \, \rho^* [\mathscr{A} - \boldsymbol{\kappa}(\boldsymbol{\kappa} \cdot \mathscr{A})] . \end{aligned} \tag{B.41}$$

The factor in brackets in (B.41) is nothing more than the transverse component of $\mathscr{A}$, with the result that P_{long} takes the simpler form

$$\mathbf{P}_{\text{long}} = \int \mathrm{d}^3 k \, \rho^* \, \mathscr{A}_\perp = \int \mathrm{d}^3 r \, \rho \mathbf{A}_\perp = \sum_\alpha q_\alpha \, \mathbf{A}_\perp(\mathbf{r}_\alpha) \tag{B.42}$$

where (A.5.a) has been used for ρ. As before, this result is independent of the choice of gauge, since $\mathbf{A}_\perp$ is gauge invariant [see (B.26)].

Finally, the total momentum **P** given in (A.8) can be written

$$\mathbf{P} = \sum_\alpha [m_\alpha \dot{\mathbf{r}}_\alpha + q_\alpha \mathbf{A}_\perp(\mathbf{r}_\alpha)] + \mathbf{P}_{\text{trans}} \tag{B.43}$$

and is the sum of the particle mechanical momenta $m_\alpha \dot{\mathbf{r}}_\alpha$, the longitudinal field momentum $\Sigma_\alpha q_\alpha \mathbf{A}_\perp(\mathbf{r}_\alpha)$, and the momentum of the transverse field. Equation (B.43) suggests that one introduce for each particle the quantity

$$\mathbf{p}_\alpha = m_\alpha \dot{\mathbf{r}}_\alpha + q_\alpha \mathbf{A}_\perp(\mathbf{r}_\alpha) \tag{B.44}$$

so that **P** can be written

$$\mathbf{P} = \sum_\alpha \mathbf{p}_\alpha + \mathbf{P}_{\text{trans}}\,. \tag{B.45}$$

In fact, one can show that in the Coulomb gauge, $\mathbf{p}_\alpha$ is the conjugate momentum to $\mathbf{r}_\alpha$ or the generalized momentum of the particle α (see §C.3, Chapter II). One can see then that, in the Coulomb gauge, the difference between the conjugate momentum $\mathbf{p}_\alpha$ and the mechanical momentum $m_\alpha \dot{\mathbf{r}}_\alpha$ of the particle α is nothing more than the momentum associated with the longitudinal field of the particle α.

Remark

Using (B.44), the total energy given in (B.38) can be written

$$H = \sum_\alpha \frac{1}{2\,m_\alpha}[\mathbf{p}_\alpha - q_\alpha \mathbf{A}_\perp(\mathbf{r}_\alpha)]^2 + V_{\text{Coul}} + H_{\text{trans}}\,. \tag{B.46}$$

One can show that H is nothing more than the Hamiltonian of the system in the Coulomb gauge (see §C.3, Chapter II).

c) The Total Angular Momentum

Calculations analogous to the foregoing (see also Complement B_I, §1) show that the total angular momentum **J** given in (A.9) can be written

$$\mathbf{J} = \sum_\alpha \mathbf{r}_\alpha \times \mathbf{p}_\alpha + \mathbf{J}_{\text{trans}} \tag{B.47}$$

where $\mathbf{p}_\alpha$ is defined in (B.44), and where

$$\mathbf{J}_{\text{trans}} = \varepsilon_0 \int \mathrm{d}^3r\, \mathbf{r} \times [\mathbf{E}_\perp(\mathbf{r}) \times \mathbf{B}(\mathbf{r})] \tag{B.48}$$

is the angular momentum of the transverse field.

6. Equations of Motion for the Transverse Fields

One now returns to the second pair of Maxwell's equations (B.5.c) and (B.5.d), and one examines the transverse parts of these two equations, which can be written in the form

$$\begin{cases} \dot{\mathscr{B}} = -\,i\mathbf{k} \times \mathscr{E} = -\,i\mathbf{k} \times \mathscr{E}_\perp & \text{(B.49.a)} \\ \dot{\mathscr{E}}_\perp = ic^2\mathbf{k} \times \mathscr{B} - \dfrac{1}{\varepsilon_0}\mathscr{j}_\perp\,. & \text{(B.49.b)} \end{cases}$$

The second pair of Maxwell's equations then appear as the dynamical equations giving the rate of variation of the transverse fields $\mathscr{B}$ and $\mathscr{E}_\perp$.

It is important to note that the source term appearing in the equation of motion (B.49.b) for $\mathscr{E}_\perp$ is $\mathscr{j}_\perp$, and not $\mathscr{j}$. Since, in real space the relationship between $\mathbf{j}_\perp$ and $\mathbf{j}$ is not local (see Remark i of §B.3 above), the rate of change of $\mathbf{E}_\perp(\mathbf{r}, t)$ at point $\mathbf{r}$ and time t depends on the current $\mathbf{j}(\mathbf{r}', t)$ at all other points $\mathbf{r}'$ at the same time t. It follows that $\mathbf{E}_\perp$ includes, like $\mathbf{E}_\parallel$, instantaneous contributions from the charge distribution. It can be shown (see Exercise 3) that the instantaneous parts of $\mathbf{E}_\perp$ and $\mathbf{E}_\parallel$ compensate each other exactly, so that the total field $\mathbf{E} = \mathbf{E}_\parallel + \mathbf{E}_\perp$ is a purely retarded field.

To conclude this section it is useful to reconsider the definition (A.6) of the "state" of the global system field + particles at time t_0. Since the longitudinal field can, in fact, be expressed totally as a function of $\mathbf{r}_\alpha$ [see (B.21)], the state of the system is completely fixed by giving

$$\{\, \mathscr{E}_\perp(\mathbf{k}, t_0), \mathscr{B}(\mathbf{k}, t_0), \mathbf{r}_\alpha(t_0), \dot{\mathbf{r}}_\alpha(t_0) \,\} \tag{B.50}$$

for all $\mathbf{k}$ and all α. We will see in the next section that it is possible to improve the choice of the dynamical variables characterizing the state of the field.

Remark

In Section B, only the equations (B.5) for the fields have been examined. It is also possible to study the longitudinal and transverse parts of the equations (B.9) for the potentials. Since the last term in (B.9.b) is longitudinal, the transverse component of (B.9.b) can be written

$$\frac{1}{c^2}\ddot{\mathscr{A}}_\perp + k^2\mathscr{A}_\perp = \frac{1}{\varepsilon_0 c^2}\mathscr{j}_\perp \tag{B.51}$$

and this becomes in real space

$$\Box \mathbf{A}_\perp = \frac{1}{\varepsilon_0 c^2}\mathbf{j}_\perp\,. \tag{B.52}$$

This equation is analogous to (A.14.b) except that one now has $\mathbf{A}_\perp$ and $\mathbf{j}_\perp$ in place of $\mathbf{A}$ and $\mathbf{j}$. If one takes the longitudinal part of (B.9.b) and uses (B.9.a) to eliminate $\dot{\mathcal{U}}$, once again one gets the conservation of charge (B.6). As with (B.5.d), the longitudinal part of (B.9.b) gives rise to nothing new. Finally, only (B.9.a) remains, and it can be written

$$k^2 \mathcal{U} = \frac{1}{\varepsilon_0} \rho + i\mathbf{k} \cdot \dot{\mathcal{A}}_\parallel \tag{B.53}$$

(since $\mathbf{k} \cdot \mathcal{A}_\perp = 0$). This equation is not sufficient to fix the motion of $\mathcal{A}_\parallel$ and $\mathcal{U}$. This is not a surprising result, since there is a redundancy in the potentials. To find $\mathcal{A}_\parallel$ and $\mathcal{U}$, it is necessary to have an additional condition, that is, to define the gauge. If one chooses the Coulomb gauge, one makes $\mathcal{A}_\parallel = \mathbf{0}$, and (B.53) then gives $\mathcal{U}$ [see also (A.18.a)]. If one chooses the Lorentz gauge, the supplementary condition (A.13) in reciprocal space is

$$\dot{\mathcal{U}} = -ic^2 \mathbf{k} \cdot \mathcal{A}_\parallel . \tag{B.54}$$

The pair of equations (B.53) and (B.54) then forms a system of two first-order equations giving the evolution of $\mathcal{A}_\parallel$ and $\mathcal{U}$. Other choices of gauge are equally possible.

C—NORMAL VARIABLES

1. Introduction

In ordinary space the rates of change, $\dot{\mathbf{E}}(\mathbf{r})$ and $\dot{\mathbf{B}}(\mathbf{r})$, of the fields **E** and **B** at point **r** depend on the spatial derivatives of **E** and **B** and thus on the values of **E** and **B** in the *neighborhood* of **r**. Maxwell's equations (A.1) are *partial differential* equations.

In going to reciprocal space, one has first of all eliminated $\mathscr{E}_{\parallel}(\mathbf{k})$ which is not really a dynamical variable, since it can be expressed as a function of $\mathbf{r}_\alpha$. One has then seen that the rates of change $\dot{\mathscr{E}}_\perp(\mathbf{k})$ and $\dot{\mathscr{B}}(\mathbf{k})$ depend only on the values of $\mathscr{E}_\perp(\mathbf{k})$ and $\mathscr{B}(\mathbf{k})$ [and on that of $\mathscr{j}_\perp(\mathbf{k})$] at the *same* point **k**. Equations (B.49) give a system of two coupled *differential* equations for each point **k**.

Inspection of this linear system (B.49) suggests that one attempt to introduce two linear combinations of $\mathscr{E}_\perp$ and $\mathscr{B}$ which evolve independently of one another, at least for the free field where $\mathscr{j}_\perp = \mathbf{0}$.

2. Definition of the Normal Variables

To begin, one writes Equations (B.49) in the form

$$\begin{cases} \dot{\mathscr{E}}_\perp = \mathrm{i}c^2\, \mathbf{k} \times \mathscr{B} - \dfrac{1}{\varepsilon_0}\mathscr{j}_\perp & \text{(C.1.a)} \\ \mathbf{k} \times \dot{\mathscr{B}} = \mathrm{i}k^2\, \mathscr{E}_\perp\,. & \text{(C.1.b)} \end{cases}$$

One seeks the eigenfunctions for such a system in the case $\mathscr{j}_\perp = \mathbf{0}$. One then finds from (C.1) that

$$\frac{\partial}{\partial t}(\mathscr{E}_\perp \mp c\boldsymbol{\kappa} \times \mathscr{B}) = \mp\, \mathrm{i}\omega(\mathscr{E}_\perp \mp c\boldsymbol{\kappa} \times \mathscr{B}) \tag{C.2}$$

with

$$\omega = ck \qquad \boldsymbol{\kappa} = \mathbf{k}/k\,. \tag{C.3}$$

One is then led to define, even if $\mathscr{j}_\perp \neq \mathbf{0}$, two new variables $\boldsymbol{\alpha}(\mathbf{k}, t)$ and $\boldsymbol{\beta}(\mathbf{k}, t)$:

$$\begin{cases} \boldsymbol{\alpha}(\mathbf{k}, t) = -\dfrac{\mathrm{i}}{2\,\mathscr{N}(k)}\left[\mathscr{E}_\perp(\mathbf{k}, t) - c\boldsymbol{\kappa} \times \mathscr{B}(\mathbf{k}, t)\right] & \text{(C.4.a)} \\ \boldsymbol{\beta}(\mathbf{k}, t) = -\dfrac{\mathrm{i}}{2\,\mathscr{N}(k)}\left[\mathscr{E}_\perp(\mathbf{k}, t) + c\boldsymbol{\kappa} \times \mathscr{B}(\mathbf{k}, t)\right] & \text{(C.4.b)} \end{cases}$$

where $\mathcal{N}(k)$ is a normalization coefficient which will be chosen later so as to have the simplest and clearest form for the total energy H.

Before proceeding farther, it is important to note that $\boldsymbol{\alpha}$ and $\boldsymbol{\beta}$ are not in fact independent dynamical variables. The real character of $\mathbf{E}_\perp$ and $\mathbf{B}$, which gives rise to equations such as (B.2) for $\boldsymbol{\mathscr{E}}_\perp$ and $\boldsymbol{\mathscr{B}}$, requires that

$$\boldsymbol{\beta}(\mathbf{k}, t) = -\boldsymbol{\alpha}^*(-\mathbf{k}, t). \tag{C.5}$$

Inverting the linear system (C.4) and using (C.5), one then gets

$$\begin{cases} \boldsymbol{\mathscr{E}}_\perp(\mathbf{k}, t) = \mathrm{i}\,\mathcal{N}(k)\left[\boldsymbol{\alpha}(\mathbf{k}, t) - \boldsymbol{\alpha}^*(-\mathbf{k}, t)\right] & \text{(C.6.a)} \\ \boldsymbol{\mathscr{B}}(\mathbf{k}, t) = \dfrac{\mathrm{i}\,\mathcal{N}(k)}{c}\left[\boldsymbol{\kappa}\times\boldsymbol{\alpha}(\mathbf{k}, t) + \boldsymbol{\kappa}\times\boldsymbol{\alpha}^*(-\mathbf{k}, t)\right]. & \text{(C.6.b)} \end{cases}$$

Knowledge of $\boldsymbol{\alpha}(\mathbf{k}, t)$ for all the values of $\mathbf{k}$ is then equivalent to knowing $\boldsymbol{\mathscr{E}}_\perp(\mathbf{k}, t)$ and $\boldsymbol{\mathscr{B}}(\mathbf{k}, t)$. In addition, the $\boldsymbol{\alpha}(\mathbf{k}, t)$ are truly independent variables, since no conditions such as (B.2) exist for $\boldsymbol{\alpha}(\mathbf{k}, t)$. One is able then, for determining the global state of the system, to replace (B.50) with

$$\{\, \boldsymbol{\alpha}(\mathbf{k}, t_0), \mathbf{r}_\alpha(t_0), \dot{\mathbf{r}}_\alpha(t_0) \,\}. \tag{C.7}$$

3. Evolution of the Normal Variables

From Maxwell's equations (C.1) and the definitions (C.4.a) for $\boldsymbol{\alpha}$, one gets

$$\dot{\boldsymbol{\alpha}}(\mathbf{k}, t) + \mathrm{i}\omega\boldsymbol{\alpha}(\mathbf{k}, t) = \frac{\mathrm{i}}{2\,\varepsilon_0\,\mathcal{N}(k)}\,\boldsymbol{\jmath}_\perp(\mathbf{k}, t). \tag{C.8}$$

One notes especially that since $\boldsymbol{\mathscr{E}}_\perp$ and $\boldsymbol{\mathscr{B}}$ are related to $\boldsymbol{\alpha}$ by (C.6), Equation (C.8) is strictly equivalent to Maxwell's equations. It is nevertheless simpler than Maxwell's equations. It resembles the equation of motion of the variable $x + \mathrm{i}(p/m\omega)$ of a fictitious harmonic oscillator with eigenfrequency ω, driven by a source term, due to the particles, proportional to $\boldsymbol{\jmath}_\perp(\mathbf{k}, t)$.

When $\boldsymbol{\jmath}_\perp = \mathbf{0}$ (the case of the free field), the evolutions of the various normal variables $\boldsymbol{\alpha}(\mathbf{k}, t)$ are completely decoupled. The solution of (C.8) is then a pure harmonic oscillation describing a normal vibrational mode of the free field. This is the reason why the $\boldsymbol{\alpha}(\mathbf{k}, t)$ are called "normal variables".

If external sources are introduced, that is to say, sources independent of $\boldsymbol{\alpha}$, the variables $\boldsymbol{\alpha}$ corresponding to different $\mathbf{k}$ continue to evolve independently of one another, each driven by $\boldsymbol{\jmath}_\perp(\mathbf{k}, t)$ (see, for example, Complement $\mathrm{B_{III}}$).

Finally, if the sources are the particles interacting with the field, the motion of $\mathbf{j}_\perp$ depends on $\boldsymbol{\alpha}$, with the result that the evolutions of the various variables $\boldsymbol{\alpha}(\mathbf{k}, t)$ are, in general, coupled through the action of the current $\mathbf{j}_\perp(\mathbf{k}, t)$. It is then necessary to add to (C.8) the equation of motion of $\mathbf{j}_\perp(\mathbf{k}, t)$ [determined from the Newton–Lorentz equation (A.2) and the definition of the current (A.5)] and to solve this coupled set of equations.

To conclude this subsection some new notation is introduced. Since $\boldsymbol{\alpha}$ is (like $\boldsymbol{\mathscr{E}}_\perp$ and $\boldsymbol{\mathscr{B}}$), a transverse vector field, one can, for each value of $\mathbf{k}$, expand $\boldsymbol{\alpha}(\mathbf{k}, t)$ on two unit vectors $\boldsymbol{\varepsilon}$ and $\boldsymbol{\varepsilon}'$, normal to one another and both located in the plane normal to $\boldsymbol{\kappa}$ (Figure 1).

$$\begin{aligned} \boldsymbol{\varepsilon} \cdot \boldsymbol{\varepsilon} = \boldsymbol{\varepsilon}' \cdot \boldsymbol{\varepsilon}' = \boldsymbol{\kappa} \cdot \boldsymbol{\kappa} = 1 \\ \boldsymbol{\varepsilon} \cdot \boldsymbol{\varepsilon}' = \boldsymbol{\varepsilon} \cdot \boldsymbol{\kappa} = \boldsymbol{\varepsilon}' \cdot \boldsymbol{\kappa} = 0 . \end{aligned} \tag{C.9}$$

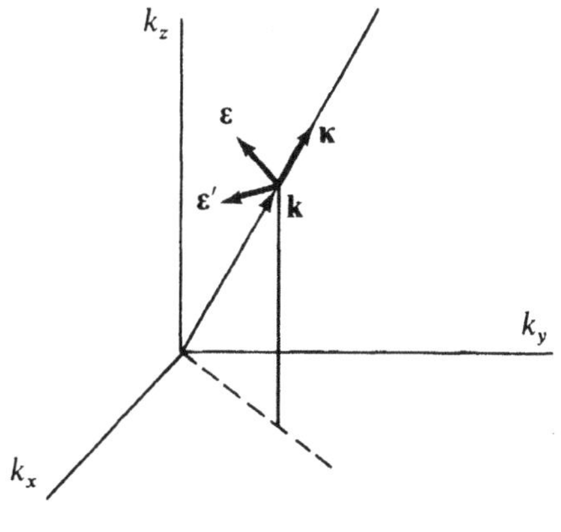

Figure 1. The transverse polarization vectors $\boldsymbol{\varepsilon}$ and $\boldsymbol{\varepsilon}'$.

One thus gets

$$\begin{aligned} \boldsymbol{\alpha}(\mathbf{k}, t) &= \boldsymbol{\varepsilon}\alpha_{\varepsilon}(\mathbf{k}, t) + \boldsymbol{\varepsilon}' \alpha_{\varepsilon'}(\mathbf{k}, t) \\ &= \sum_{\varepsilon} \boldsymbol{\varepsilon}\alpha_{\varepsilon}(\mathbf{k}, t) \end{aligned} \tag{C.10}$$

where

$$\alpha_{\varepsilon}(\mathbf{k}, t) = \boldsymbol{\varepsilon} \cdot \boldsymbol{\alpha}(\mathbf{k}, t) \tag{C.11}$$

is the component of $\boldsymbol{\alpha}$ along $\boldsymbol{\varepsilon}$. The set $\{\alpha_{\varepsilon}(\mathbf{k}, t)\}$ for all $\mathbf{k}$ and $\boldsymbol{\varepsilon}$ forms a complete set of independent variables for the transverse field. The equation of motion for $\alpha_{\varepsilon}(\mathbf{k}, t)$ is

$$\dot{\alpha}_\varepsilon(\mathbf{k}, t) + i\omega\alpha_\varepsilon(\mathbf{k}, t) = \frac{i}{2\,\varepsilon_0\,\mathcal{N}(k)}\,\boldsymbol{\varepsilon}\cdot\boldsymbol{j}(\mathbf{k}, t) \tag{C.12}$$

where one uses $\boldsymbol{\varepsilon}\cdot\boldsymbol{j}_\perp = \boldsymbol{\varepsilon}\cdot\boldsymbol{j}$.

4. The Expressions for the Physical Observables of the Transverse Field as a Function of the Normal Variables

Later on one always uses the normal variables $\alpha_\varepsilon(\mathbf{k}, t)$ (and the corresponding quantum operators) to characterize the state of the transverse field. Thus it is important to have expressions for the various physical observables of the transverse field as a function of the α_ε.

a) The Energy H_{trans} of the Transverse Field

We substitute in (B.31.b) the expressions (C.6) for $\mathscr{E}_\perp$ and $\mathscr{B}$ as a function of $\boldsymbol{\alpha}$ and $\boldsymbol{\alpha}^*_-$ [the more concise notation $\boldsymbol{\alpha}^*_-$ is used for $\boldsymbol{\alpha}^*(-\mathbf{k}, t)$]. In addition, one respects the ordering between $\boldsymbol{\alpha}$ and $\boldsymbol{\alpha}^*$ as it arises in the calculation although $\boldsymbol{\alpha}$ and $\boldsymbol{\alpha}^*$ are numbers which commute. The reason for doing this is that in quantum electrodynamics, $\boldsymbol{\alpha}$ and $\boldsymbol{\alpha}^*$ will be replaced by noncommuting operators. The results obtained in this subsection then remain valid in the quantum case.

From (C.6), one finds

$$\begin{aligned}
\mathscr{E}_\perp^*\cdot\mathscr{E}_\perp &= \mathcal{N}^2(\boldsymbol{\alpha}^* - \boldsymbol{\alpha}_-)\cdot(\boldsymbol{\alpha} - \boldsymbol{\alpha}^*_-)\\
&= \mathcal{N}^2(\boldsymbol{\alpha}^*\cdot\boldsymbol{\alpha} + \boldsymbol{\alpha}_-\cdot\boldsymbol{\alpha}^*_- - \boldsymbol{\alpha}^*\cdot\boldsymbol{\alpha}^*_- - \boldsymbol{\alpha}_-\cdot\boldsymbol{\alpha})\\
c^2\,\mathscr{B}^*\cdot\mathscr{B} &= \mathcal{N}^2(\boldsymbol{\alpha}^* + \boldsymbol{\alpha}_-)\cdot(\boldsymbol{\alpha} + \boldsymbol{\alpha}^*_-)\\
&= \mathcal{N}^2(\boldsymbol{\alpha}^*\cdot\boldsymbol{\alpha} + \boldsymbol{\alpha}_-\cdot\boldsymbol{\alpha}^*_- + \boldsymbol{\alpha}^*\cdot\boldsymbol{\alpha}^*_- + \boldsymbol{\alpha}_-\cdot\boldsymbol{\alpha})
\end{aligned} \tag{C.13}$$

with the result that (B.31.b) becomes

$$H_{\text{trans}} = \varepsilon_0\int d^3k\;\mathcal{N}^2[\boldsymbol{\alpha}^*\cdot\boldsymbol{\alpha} + \boldsymbol{\alpha}_-\cdot\boldsymbol{\alpha}^*_-]\,. \tag{C.14}$$

Changing from $\mathbf{k}$ to $-\mathbf{k}$ in the integral of the second term allows one to replace $\boldsymbol{\alpha}_-\cdot\boldsymbol{\alpha}^*_-$ by $\boldsymbol{\alpha}\cdot\boldsymbol{\alpha}^*$. Let us now take for the normalization coefficient $\mathcal{N}(k)$ the value

$$\mathcal{N}(k) = \sqrt{\frac{\hbar\omega}{2\,\varepsilon_0}} \tag{C.15}$$

chosen so that in the quantum theory the commutation relations between the operators corresponding to α_ε and α_ε^* are simple. Equation (C.14) then takes the more suggestive form

$$H_{\text{trans}} = \int d^3k \sum_\varepsilon \frac{\hbar\omega}{2} [\alpha_\varepsilon^*(\mathbf{k}, t)\, \alpha_\varepsilon(\mathbf{k}, t) + \alpha_\varepsilon(\mathbf{k}, t)\, \alpha_\varepsilon^*(\mathbf{k}, t)]\,. \quad \text{(C.16)}$$

It then appears as the sum of the energies of a set of fictitious harmonic oscillators with an oscillator of frequency $\omega = ck$ being associated with each pair of vectors $\mathbf{k}$, $\boldsymbol{\varepsilon}$ (with $\boldsymbol{\varepsilon}$ normal to $\mathbf{k}$). Such a pair defines a "mode" of the transverse field.

b) The Momentum $\mathbf{P}_{\text{trans}}$ and the Angular Momentum $\mathbf{J}_{\text{trans}}$ of the Transverse Field

A calculation similar to that above allows one to get from (B.39.b)

$$\mathbf{P}_{\text{trans}} = \int d^3k \sum_\varepsilon \frac{\hbar\mathbf{k}}{2} [\alpha_\varepsilon^*(\mathbf{k}, t)\, \alpha_\varepsilon(\mathbf{k}, t) + \alpha_\varepsilon(\mathbf{k}, t)\, \alpha_\varepsilon^*(\mathbf{k}, t)]\,. \quad \text{(C.17)}$$

For the angular momentum $\mathbf{J}_{\text{trans}}$ of the transverse field given in (B.48), the calculations are a little more tedious than for $\mathbf{H}_{\text{trans}}$ and $\mathbf{P}_{\text{trans}}$ (see §2.b of Complement B_I). The following result is obtained:

$$J_a^{\text{trans}} = \frac{\hbar}{2i} \sum_{bcd} \int d^3k [\alpha_d^*\, \varepsilon_{abc}\, k_b\, \partial_c\, \alpha_d + \alpha_b^*\, \varepsilon_{abc}\, \alpha_c - \alpha_d\, \varepsilon_{abc}\, k_b\, \partial_c\, \alpha_d^* - \alpha_b\, \varepsilon_{abc}\, \alpha_c^*] \quad \text{(C.18)}$$

where $a, b, c, d = x, y$, or z, $\partial_c = \partial/\partial k_c$, and ε_{abc} is the antisymmetric tensor.

Remark

The product $\mathbf{E}_\perp \times \mathbf{B}$ appears in the expressions for $\mathbf{P}_{\text{trans}}$ and $\mathbf{J}_{\text{trans}}$. In quantum theory $\mathbf{E}_\perp$ and $\mathbf{B}$ become operators and one can ask if it is not necessary to symmetrize $\mathbf{E}_\perp \times \mathbf{B}$ in the form $(\mathbf{E}_\perp \times \mathbf{B} - \mathbf{B} \times \mathbf{E}_\perp)/2$. In fact, $\mathbf{E}_\perp$ and $\mathbf{B}$ are taken at the same point in space, and we will see in Chapter III (§A.2) that $\mathbf{E}_\perp(\mathbf{r})$ and $\mathbf{B}(\mathbf{r})$ are commuting observables, so that symmetrization is not required.

c) Transverse Electric and Magnetic Fields in Real Space

The expansions of $\mathbf{E}_\perp(\mathbf{r}, t)$ and $\mathbf{B}(\mathbf{r}, t)$ are gotten by taking the Fourier transforms of (C.6.a) and (C.6.b) [in the integral over $\mathbf{k}$ of the last terms of (C.6.a) and (C.6.b) one replaces $\mathbf{k}$ by $-\mathbf{k}$]. This then becomes

$$\mathbf{E}_\perp(\mathbf{r}, t) = i \int d^3k \sum_\varepsilon \mathscr{E}_\omega [\alpha_\varepsilon(\mathbf{k}, t)\, \boldsymbol{\varepsilon}\, e^{i\mathbf{k}.\mathbf{r}} - \alpha_\varepsilon^*(\mathbf{k}, t)\, \boldsymbol{\varepsilon}\, e^{-i\mathbf{k}.\mathbf{r}}] \quad \text{(C.19)}$$

$$\mathbf{B}(\mathbf{r}, t) = \mathrm{i} \int \mathrm{d}^3k \sum_{\varepsilon} \mathscr{B}_\omega [\alpha_\varepsilon(\mathbf{k}, t)\, \boldsymbol{\kappa} \times \boldsymbol{\varepsilon}\, \mathrm{e}^{\mathrm{i}\mathbf{k}.\mathbf{r}} - \alpha_\varepsilon^*(\mathbf{k}, t)\, \boldsymbol{\kappa} \times \boldsymbol{\varepsilon}\, \mathrm{e}^{-\mathrm{i}\mathbf{k}.\mathbf{r}}] \tag{C.20}$$

with

$$\mathscr{E}_\omega = \left[\frac{\hbar\omega}{2\,\varepsilon_0 (2\,\pi)^3}\right]^{1/2} \qquad \mathscr{B}_\omega = \mathscr{E}_\omega / c\,. \tag{C.21}$$

The real character of $\mathbf{E}_\perp$ and $\mathbf{B}$ (*) is obvious in (C.19) and (C.20).

For the free field $\boldsymbol{j}_\perp = \mathbf{0}$, the solution of the equation of motion for α_ε is

$$\alpha_\varepsilon(\mathbf{k}, t) = \alpha_\varepsilon(\mathbf{k})\, \mathrm{e}^{-\mathrm{i}\omega t} \tag{C.22}$$

Substituting (C.22) in (C.19) and (C.20), one gets for $\mathbf{E}_\perp$ and $\mathbf{B}$ an expansion in traveling plane waves $\mathrm{e}^{\mathrm{i}(\mathbf{k}\cdot\mathbf{r}-\omega t)}$, with wave vector $\mathbf{k}$. For example,

$$\mathbf{E}_\perp(\mathbf{r}, t) = \mathrm{i} \int \mathrm{d}^3k \sum_{\varepsilon} \mathscr{E}_\omega\, \alpha_\varepsilon(\mathbf{k})\, \boldsymbol{\varepsilon}\, \mathrm{e}^{\mathrm{i}(\mathbf{k}.\mathbf{r}-\omega t)} + \text{c.c.} \tag{C.23}$$

It is easy to prove that, with the solution (C.22) for a free field, the expressions (C.16), (C.17), and (C.18) for H_{trans}, $\mathbf{P}_{\text{trans}}$, and $\mathbf{J}_{\text{trans}}$ are constants of the motion (independent of t).

It must be emphasized that the definition (C.4) for the normal variables, as well as the expansions (C.16) to (C.20) which follow, are valid in the presence or absence of sources. In contrast, the simple solution (C.22) for $\alpha_\varepsilon(\mathbf{k}, t)$ applies only to a free field. In the presence of sources, the solution of the equation of motion (C.12) is more complicated than (C.22), and the expansions (C.19) and (C.20) for $\mathbf{E}_\perp$ and $\mathbf{B}$ are no longer linear superpositions of traveling plane waves. Likewise, the energy of the transverse field given in (C.16) is no longer a constant of the motion. Energy exchanges take place between the transverse field and the particles, and only the *total* energy given in (B.38) is conserved. The same ideas apply to $\mathbf{P}_{\text{trans}}$ and $\mathbf{J}_{\text{trans}}$.

It is usual to denote by $\mathbf{E}_\perp^{(+)}(\mathbf{r}, t)$ the part of the expansion (C.19) containing only the $\alpha_\varepsilon(\mathbf{k}, t)$ [and not the $\alpha_\varepsilon^*(\mathbf{k}, t)$]:

$$\mathbf{E}_\perp^{(+)}(\mathbf{r}, t) = \mathrm{i} \int \mathrm{d}^3k \sum_{\varepsilon} \mathscr{E}_\omega\, \alpha_\varepsilon(\mathbf{k}, t)\, \boldsymbol{\varepsilon}\, \mathrm{e}^{\mathrm{i}\mathbf{k}.\mathbf{r}}\,. \tag{C.24}$$

(*) The polarization vector $\boldsymbol{\varepsilon}$ is real. It is certainly possible to introduce complex polarization vectors to describe the modes of the transverse field having a circular or elliptical polarization. In that case it is necessary to replace the second $\boldsymbol{\varepsilon}$ in (C.19) and (C.20) by $\boldsymbol{\varepsilon}^*$.

The other part is denoted $\mathbf{E}_{\perp}^{(-)}(\mathbf{r}, t)$:

$$\mathbf{E}_{\perp}^{(-)}(\mathbf{r}, t) = [\mathbf{E}_{\perp}^{(+)}(\mathbf{r}, t)]^* . \qquad \text{(C.25)}$$

For a free field, $\mathbf{E}_{\perp}^{(+)}$ is the "positive-frequency" part of $\mathbf{E}_{\perp}$; $\mathbf{E}_{\perp}^{(-)}$ the "negative-frequency" part.

d) The Transverse Vector Potential $\mathbf{A}_{\perp}(\mathbf{r}, t)$

In what follows it will also be useful to give the expansion of the transverse part of the vector potential **A** in normal variables. Remember that $\mathbf{A}_{\perp}$ is gauge invariant [see Equation (B.26)].

Note first that the transverse fields $\mathbf{E}_{\perp}$ and **B** depend only on $\mathbf{A}_{\perp}$. From (A.10.a) and (B.24.a) it follows that

$$\begin{cases} \mathbf{E}_{\perp}(\mathbf{r}, t) = -\dfrac{\partial}{\partial t}\mathbf{A}_{\perp}(\mathbf{r}, t) & \text{(C.26.a)} \\ \mathbf{B}(\mathbf{r}, t) = \nabla \times \mathbf{A}_{\perp}(\mathbf{r}, t) & \text{(C.26.b)} \end{cases}$$

since $\nabla \times \mathbf{A}_{\parallel} = \mathbf{0}$.

One can show now that $\mathbf{A}_{\perp}(\mathbf{r}, t)$ can be written

$$\mathbf{A}_{\perp}(\mathbf{r}, t) = \int d^3k \sum_{\varepsilon} \mathscr{A}_{\omega}[\alpha_{\varepsilon}(\mathbf{k}, t)\, \boldsymbol{\varepsilon}\, e^{i\mathbf{k}\cdot\mathbf{r}} + \text{c.c.}] \qquad \text{(C.27)}$$

with

$$\mathscr{A}_{\omega} = \mathscr{B}_{\omega}/k = \mathscr{E}_{\omega}/\omega . \qquad \text{(C.28)}$$

Since $\boldsymbol{\varepsilon}$ is normal to **k**, the vector field (C.27) is transverse. If one takes the curl of (C.27), one gets the expression (C.20) for **B**, so that (C.27) verifies (C.26.b). Since $\mathbf{A}_{\perp}$ has a zero divergence, Equation (C.26.b) is sufficient to determine it entirely, and the Maxwell equations insure that (C.26.a) is automatically satisfied.

One can also find the Fourier transform $\mathscr{A}_{\perp}(\mathbf{k}, t)$ of $\mathbf{A}_{\perp}(\mathbf{r}, t)$. After changing **k** to $-\mathbf{k}$ in the second integral of (C.27), we get

$$\mathscr{A}_{\perp}(\mathbf{k}, t) = \sqrt{\frac{\hbar}{2\,\varepsilon_0\,\omega}}\,[\boldsymbol{\alpha}(\mathbf{k}, t) + \boldsymbol{\alpha}^*(-\mathbf{k}, t)] . \qquad \text{(C.29)}$$

Finally, it is possible to combine (C.29) and (C.6.a) to find $\boldsymbol{\alpha}$ as a function of $\mathscr{A}_{\perp}$ and $\mathscr{E}_{\perp}$ [rather than as a function of $\mathscr{E}_{\perp}$ and $\mathscr{B}$ as in (C.4.a)]:

$$\boldsymbol{\alpha}(\mathbf{k}, t) = \sqrt{\frac{\varepsilon_0}{2\,\hbar\omega}}\,[\omega\mathscr{A}_{\perp}(\mathbf{k}, t) - i\mathscr{E}_{\perp}(\mathbf{k}, t)] . \qquad \text{(C.30)}$$

5. Similarities and Differences between the Normal Variables and the Wave Function of a Spin-1 Particle in Reciprocal Space

Consider first the free field. Equation (C.8) can be written

$$i\hbar\dot{\boldsymbol{\alpha}}(\mathbf{k}, t) = \hbar\omega\boldsymbol{\alpha}(\mathbf{k}, t) \tag{C.31}$$

and appears then as a Schrödinger equation relative to a "vector wave function" $\boldsymbol{\alpha}(\mathbf{k}, t)$, the corresponding Hamiltonian being diagonal in the reciprocal space with matrix elements $h\omega\delta(\mathbf{k} - \mathbf{k}')$. Equation (C.16) can also be interpreted as the mean value of such a Hamiltonian in the wave function $\boldsymbol{\alpha}(\mathbf{k}, t)$. Likewise, since in quantum mechanics the momentum operator of a particle is diagonal in reciprocal space with matrix elements $\hbar\mathbf{k}\delta(\mathbf{k} - \mathbf{k}')$, Equation (C.17) can be interpreted as the mean value of the momentum operator in the wave function $\boldsymbol{\alpha}(\mathbf{k}, t)$. Finally, one can show (see §2.c, Complement B_I) that Equation (C.18), giving the angular momentum of the transverse field, coincides with the mean value in the wave function $\boldsymbol{\alpha}(\mathbf{k}, t)$ of $\mathbf{J} = \mathbf{L} + \mathbf{S}$ (where $\mathbf{L}$ and $\mathbf{S}$ are the usual quantum operators for the orbital angular momentum and spin angular momentum). The first term in the bracket of (C.18) corresponds to $\mathbf{L}$, the second to $\mathbf{S}$.

All the preceding results suggest that one interpret $\boldsymbol{\alpha}(\mathbf{k}, t)$ as the wave function in reciprocal space of a particle of spin 1 (*), namely the photon. Such an analogy should not, however, be pushed too far. First of all, one can show that the Fourier transform of $\boldsymbol{\alpha}(\mathbf{k}, t)$ can not be interpreted as the photon wave function in real space and, more generally, that it is impossible to construct a position operator for the photon. (**) Additionally, the equation of motion of $\boldsymbol{\alpha}$ no longer has the form of a Schrödinger equation in the presence of sources: it is not homogeneous. Such a result is not surprising. The Schrödinger equation preserves the norm of the wave function and thus the number of particles. Now it is well known that in the presence of sources, photons can be absorbed or emitted. Thus one cannot introduce a Schrödinger equation for a single photon in the presence of sources. In fact, the electromagnetic field itself must be quantized, and photons then occur as elementary excitations of the quantized field. We will see in the following chapters that the "wave function", or more properly the state vector, of the quantized field is a

(*) The value of 1 for the spin is tied to the vector character of $\boldsymbol{\alpha}$. See Akhiezer and Berestetskii, Chapter I.

(**) See, for example, E. Wigner and T. D. Newton, *Rev. Mod. Phys.*, **21**, 400 (1949); M. H. L. Pryce, *Proc. Roy. Soc.*, **195A**, 62 (1948).

vector in a *Fock space* where the number of photons can vary from zero (vacuum state) to infinity.

The preceding analogy can however be of some use. It suggests, for example, that one examine the transverse eigenfunctions of $\mathbf{J}^2$ and J_z in reciprocal space. This leads to the multipolar expansion of the transverse field (see Complement B_I), more convenient than the plane-wave expansion (given above) in all problems where the angular momentum is important.

6. Periodic Boundary Conditions. Simplified Notation

It is common to consider the fields as being contained in a cube of edge L, and satisfying periodic boundary conditions at the sides of the cube. At the end of the calculation one lets L go to infinity. All the physical predictions (cross-sections, transition probabilities, etc.) should certainly be independent of L.

The advantage of such a procedure is the replacement of the Fourier integrals by Fourier series. In other words, the integrals on $\mathbf{k}$ are replaced by discrete sums over

$$k_{x,y,z} = 2\pi n_{x,y,z}/L \tag{C.32}$$

where $n_{x,y,z}$ are integers (positive, negative, or zero). The variables $\alpha_\varepsilon(\mathbf{k}, t)$ are replaced by the discrete variables $\alpha_{\mathbf{k}\varepsilon}(t)$:

$$\alpha_\varepsilon(\mathbf{k}, t) \rightarrow \alpha_{\mathbf{k}\varepsilon}(t)\,. \tag{C.33.a}$$

One can even use the more concise notation

$$\alpha_{\mathbf{k}_i\varepsilon_i} \rightarrow \alpha_i \tag{C.33.b}$$

where the index i designates the set $(\mathbf{k}_i, \boldsymbol{\varepsilon}_i)$. The correspondence between the two types of sum obeys the following rule:

$$\int \mathrm{d}^3k \sum_{\varepsilon} f(\mathbf{k}, \boldsymbol{\varepsilon}) \leftrightarrow \sum_i \left(\frac{2\pi}{L}\right)^3 f(\mathbf{k}_i, \boldsymbol{\varepsilon}_i)\,. \tag{C.34}$$

In summary, the following are the expansions in α_i and α_i^* of H_{trans}, $\mathbf{P}_{\text{trans}}$, $\mathbf{A}_\perp$, $\mathbf{E}_\perp$, and $\mathbf{B}$:

$$H_{\text{trans}} = \sum_i \frac{\hbar\omega_i}{2}(\alpha_i^*\,\alpha_i + \alpha_i\,\alpha_i^*) \tag{C.35}$$

$$\mathbf{P}_{\text{trans}} = \sum_i \frac{\hbar\mathbf{k}_i}{2}(\alpha_i^*\,\alpha_i + \alpha_i\,\alpha_i^*) \tag{C.36}$$

$$\mathbf{A}_\perp = \sum_i \mathscr{A}_{\omega_i}[\alpha_i\,\boldsymbol{\varepsilon}_i\,\mathrm{e}^{\mathrm{i}\mathbf{k}_i\cdot\mathbf{r}} + \alpha_i^*\,\boldsymbol{\varepsilon}_i\,\mathrm{e}^{-\mathrm{i}\mathbf{k}_i\cdot\mathbf{r}}] \tag{C.37}$$

$$\mathbf{E}_\perp = \mathrm{i}\sum_i \mathscr{E}_{\omega_i}[\alpha_i\, \boldsymbol{\varepsilon}_i\, \mathrm{e}^{\mathrm{i}\mathbf{k}_i.\mathbf{r}} - \alpha_i^*\, \boldsymbol{\varepsilon}_i\, \mathrm{e}^{-\mathrm{i}\mathbf{k}_i.\mathbf{r}}] \tag{C.38}$$

$$\mathbf{B} = \mathrm{i}\sum_i \mathscr{B}_{\omega_i}[\alpha_i\, \boldsymbol{\kappa}_i \times \boldsymbol{\varepsilon}_i\, \mathrm{e}^{\mathrm{i}\mathbf{k}_i.\mathbf{r}} - \alpha_i^*\, \boldsymbol{\kappa}_i \times \boldsymbol{\varepsilon}_i\, \mathrm{e}^{-\mathrm{i}\mathbf{k}_i.\mathbf{r}}] \tag{C.39}$$

with

$$\mathscr{E}_{\omega_i} = \left[\frac{\hbar\omega_i}{2\,\varepsilon_0\, L^3}\right]^{1/2} \qquad \mathscr{B}_{\omega_i} = \frac{\mathscr{E}_{\omega_i}}{c} \qquad \mathscr{A}_{\omega_i} = \frac{\mathscr{E}_{\omega_i}}{\omega_i}. \tag{C.40}$$

In these expressions Σ_i indicates summation on all the modes $\mathbf{k}_i\boldsymbol{\varepsilon}_i$. It is convenient to note also that when going from the Fourier integral to the Fourier series, the factor $(1/2\pi)^{3/2}$ of Equations (B.1) is replaced by $1/L^{3/2}$. This explains why $\mathscr{E}_{\omega_i}$ contains L^3 in place of $(2\pi)^3$ [compare (C.21) and (C.40)]. Finally the evolution of α_i is governed by

$$\dot{\alpha}_i + \mathrm{i}\omega_i\, \alpha_i = \frac{\mathrm{i}}{\sqrt{2\,\varepsilon_0\,\hbar\omega_i}}\, \dot{\jmath}_i \tag{C.41}$$

with

$$\jmath_i = \frac{1}{\sqrt{L^3}}\int \mathrm{d}^3 r\, \mathrm{e}^{-\mathrm{i}\mathbf{k}_i.\mathbf{r}}\, \boldsymbol{\varepsilon}_i \cdot \mathbf{j}(\mathbf{r})\,. \tag{C.42}$$

Remark

The discrete variable α_i has not the same dimensions as the continuous variable $\alpha_\varepsilon(\mathbf{k})$. More precisely,

$$\alpha_i = \left(\frac{2\,\pi}{L}\right)^{3/2} \alpha_{\varepsilon_i}(\mathbf{k}_i)\,. \tag{C.43}$$

D—CONCLUSION: DISCUSSION OF VARIOUS POSSIBLE QUANTIZATION SCHEMES

After this rapid survey of classical electrodynamics, we now face the problem of quantization of the theory. Here we will review various possible strategies for quantization which will clarify the motivation and the organization of the following chapters.

1. Elementary Approach

The formulation given in this chapter lends itself particularly well to an elementary approach. Indeed, we have shown that the global system (electromagnetic field + particles) is formally equivalent to a set of mutually interacting particles and oscillators. The simplest idea which can then be put forth for quantizing such a system is to quantize the particles and the oscillators in the usual way. With the position $\mathbf{r}_\alpha$ and with the momentum $\mathbf{p}_\alpha$ of the particle α we associate operators (*) whose commutator is $i\hbar$, and we replace the normal variables α_i and α_i^* of the oscillator i by the well-known annihilation and creation operators a_i and a_i^+ with commutator equal to 1 taken from the quantum theory of the harmonic oscillator:

$$\alpha_i \to a_i \qquad \alpha_i^* \to a_i^+ . \tag{D.1}$$

All the physical quantities, which can be expressed as functions of $\mathbf{r}_\alpha$, $\mathbf{p}_\alpha$, α_i, and α_i^*, become operators acting in the space of the quantum states of the global system.

Such an approach is, however, heuristic. Since it does not come from a Lagrangian or Hamiltonian formulation, we do not know if $\mathbf{r}_\alpha$ and $\mathbf{p}_\alpha$ on one hand or α_i and α_i^* on the other [more precisely, $q_i = (\alpha_i + \alpha_i^*)/\sqrt{2}$ and $p_i = \mathrm{i}(\alpha_i^* - \alpha_i)/\sqrt{2}$] can be thought of as conjugate dynamical variables with respect to a Hamiltonian which has yet to be written. Certainly, in this chapter the expression for the total energy of the system has been given, but the conditions under which this expression can be considered as the Hamiltonian of the system have not been established.

It is nevertheless possible to avoid this difficulty. One postulates the following expression for the Hamiltonian in the Coulomb gauge:

$$H = \sum_\alpha \frac{1}{2\,m_\alpha}[\mathbf{p}_\alpha - q_\alpha\,\mathbf{A}(\mathbf{r}_\alpha)]^2 + V_{\mathrm{Coul}} + \sum_i \frac{\hbar\omega_i}{2}(a_i^+\,a_i + a_i\,a_i^+) \tag{D.2}$$

(*) To simplify the notation the same symbols are retained to designate the classical variables of the particles $\mathbf{r}_\alpha$ and $\mathbf{p}_\alpha$ and the associated operators.

which is nothing more than Equation (B.46) for the total energy (in the Coulomb gauge $\mathbf{A} = \mathbf{A}_{\perp}$), α_i and α_i^* being replaced in Equation (C.35) for H_{trans} by the operators a_i and a_i^+. One also postulates the following commutation relation for $\mathbf{r}_\alpha$ and $\mathbf{p}_\alpha$ on one hand:

$$\begin{cases} [r_{\alpha i}, r_{\beta j}] = [p_{\alpha i}, p_{\beta j}] = 0 \\ [r_{\alpha i}, p_{\beta j}] = \mathrm{i}\hbar\delta_{\alpha\beta}\,\delta_{ij} \end{cases} \tag{D.3}$$

where $i, j = x, y, z$ (the $\delta_{\alpha\beta}$ indicates that the variables of two different particles commute), and for a_i and a_i^+ on the other:

$$\begin{cases} [a_i, a_j] = [a_i^+, a_j^+] = 0 \\ [a_i, a_j^+] = \delta_{ij}\,. \end{cases} \tag{D.4}$$

The δ_{ij} indicates that the variables of two different modes of the transverse field commute. One can then show (as will be done in §B.2, Chapter III) that the Heisenberg equations derived from the Hamiltonian (D.2) and the commutation relations (D.3) and (D.4) lead to good equations of motion, that is to say, to the Maxwell–Lorentz equations between operators.

The reader ready to accept the foregoing points and wishing to get as quickly and simply as possible to the quantum theory can skip Chapter II, devoted to the Lagrangian and Hamiltonian formulation of electrodynamics, and go directly to Chapter III, which starts from the expressions given in this subsection.

2. Lagrangian and Hamiltonian Approach

This approach consists in showing initially that the basic equations of classical electrodynamics, the Maxwell–Lorentz equations, can be thought of as Lagrange's equations derived variationally from a certain Lagrangian. Canonical quantization of the system is then achieved by associating with each pair formed by a "generalized coordinate" and its "canonically conjugate" momentum two operators with commutator $\mathrm{i}\hbar$.

Although more abstract, such an approach offers a number of advantages. It allows one to identify which field variables are conjugate (for example, in the Coulomb gauge the vector potential and the transverse electric field) and to obtain the Hamiltonian directly without it being necessary to postulate it. This approach also allows a deeper understanding of the problems tied to the choice of gauge. The Coulomb gauge appears then as the most "economical" gauge, allowing one to eliminate most easily the redundant variables in the Lagrangian. Finally, it is well known that two Lagrangians differing only by a total derivative are physically equivalent. It is thus possible to construct many equivalent formulations of quantum electrodynamics and to discuss directly the relations which exist between them.

Chapter II presents classical electrodynamics and its canonical quantization starting from such a point of view. Equations (D.2), (D.3), and (D.4) are therein justified in a rigorous fashion. Changes of Lagrangian and Hamiltonian will be treated in Chapter IV along with the various formulations of electrodynamics to which they give rise.

For certain problems it is important to use a manifestly covariant formulation. This leads one to choose a different gauge from the Coulomb one and complicates the problem of quantization. These questions will be considered in Chapter V.

General references and additional readings

Jackson, Feynman et al (Volume II), Landau and Lifschitz (Volume II), Messiah (Chapter XXI, §III), Akhiezer and Berestetskii (Chapter I), Cohen-Tannoudji (§1).

COMPLEMENT A_I

THE "TRANSVERSE" DELTA FUNCTION

The transverse delta function $\delta_{ij}^{\perp}(\rho)$ allows one to extract from a vector field its transverse component. It is easy to understand why it plays an important role in electrodynamics in the Coulomb gauge, since the transverse fields are favored in this gauge. The purpose of this complement is to establish the expressions and the properties of this function and to illustrate its use in a simple example. The expression for the transverse delta function is particularly simple in reciprocal space. On the other hand, we will see that calculating its Fourier transform to find $\delta_{ij}^{\perp}(\rho)$ presents certain difficulties which justify the detailed treatment given here.

1. Definition in Reciprocal Space

a) Cartesian Coordinates. Transverse and Longitudinal Components

Two different types of basis vectors will be used in reciprocal space to define vector fields: the Cartesian system $\{\mathbf{e}_i\}$ $(i = x, y, z)$, and the system composed of the longitudinal unit vector $\boldsymbol{\kappa} = \mathbf{k}/k$ and the two unit transverse vectors $\boldsymbol{\varepsilon}$ and $\boldsymbol{\varepsilon}'$, introduced in (C.9). The Cartesian components of vector $\boldsymbol{\mathscr{V}}$ are denoted by $\mathscr{V}_i$. One will often have to perform summations on the two transverse polarizations of products of components of $\boldsymbol{\varepsilon}$ and $\boldsymbol{\varepsilon}'$. Their expressions are as follows. Consider first

$$\begin{aligned}
\sum_{\boldsymbol{\varepsilon}\perp\mathbf{k}} \varepsilon_i\,\varepsilon_j &= \varepsilon_i\,\varepsilon_j + \varepsilon_i'\,\varepsilon_j' \\
&= [(\mathbf{e}_i\cdot\boldsymbol{\varepsilon})(\boldsymbol{\varepsilon}\cdot\mathbf{e}_j) + (\mathbf{e}_i\cdot\boldsymbol{\varepsilon}')(\boldsymbol{\varepsilon}'\cdot\mathbf{e}_j) + (\mathbf{e}_i\cdot\boldsymbol{\kappa})(\boldsymbol{\kappa}\cdot\mathbf{e}_j)] - \\
&\qquad - (\mathbf{e}_i\cdot\boldsymbol{\kappa})(\boldsymbol{\kappa}\cdot\mathbf{e}_j) \\
&= \mathbf{e}_i\cdot\mathbf{e}_j - (\mathbf{e}_i\cdot\boldsymbol{\kappa})(\mathbf{e}_j\cdot\boldsymbol{\kappa}) \\
&= \delta_{ij} - \kappa_i\,\kappa_j\,. \qquad (1)
\end{aligned}$$

Another summation is

$$\sum_{\boldsymbol{\varepsilon}\perp\mathbf{k}} \varepsilon_i(\boldsymbol{\kappa}\times\boldsymbol{\varepsilon})_j = \varepsilon_i\,\varepsilon_j' + \varepsilon_i'(-\,\varepsilon_j) \qquad (2)$$

where in writing the right-hand side, we have noted that

$$\boldsymbol{\kappa}\times\boldsymbol{\varepsilon} = \boldsymbol{\varepsilon}' \qquad \boldsymbol{\kappa}\times\boldsymbol{\varepsilon}' = -\,\boldsymbol{\varepsilon}\,. \qquad (3)$$

It appears then that (2) is nothing more than the component of the vector product $\boldsymbol{\varepsilon} \times \boldsymbol{\varepsilon}'$ on $\mathbf{e}_i \times \mathbf{e}_j$. This then becomes

$$\sum_{\boldsymbol{\varepsilon} \perp \mathbf{k}} \varepsilon_i (\boldsymbol{\kappa} \times \boldsymbol{\varepsilon})_j = \sum_l \varepsilon_{ijl} \kappa_l \tag{4}$$

where ε_{ijk} is the antisymmetric tensor. Finally, using (3) and (1), one immediately gets

$$\sum_{\boldsymbol{\varepsilon} \perp \mathbf{k}} (\boldsymbol{\kappa} \times \boldsymbol{\varepsilon})_i (\boldsymbol{\kappa} \times \boldsymbol{\varepsilon})_j = \delta_{ij} - \kappa_i \kappa_j . \tag{5}$$

b) Projection on the Subspace of Transverse Fields

The transverse delta function is closely tied to the projection operator on the subspace of transverse vector fields. In order to see this, consider a vector field $\mathbf{V}(\mathbf{r})$ and its Fourier transform $\mathcal{V}(\mathbf{k})$. In reciprocal space, $\mathcal{V}_\perp(\mathbf{k})$ is easily gotten from $\mathcal{V}(\mathbf{k})$ by projecting $\mathcal{V}(\mathbf{k})$ onto the plane normal to $\mathbf{k}$ at point $\mathbf{k}$:

$$\mathcal{V}_\perp(\mathbf{k}) = \sum_{\boldsymbol{\varepsilon}} \boldsymbol{\varepsilon}(\boldsymbol{\varepsilon} \cdot \mathcal{V}(\mathbf{k})) \tag{6}$$

By projecting on $\mathbf{e}_i$ and using (1), we obtain

$$\mathcal{V}_{\perp i}(\mathbf{k}) = \sum_j (\delta_{ij} - \kappa_i \kappa_j) \, \mathcal{V}_j(\mathbf{k}) . \tag{7}$$

Let us denote by $\Delta^\perp$ the projection operator acting in the space of vector fields and generating the correspondence between $\mathcal{V}$ and $\mathcal{V}_\perp$:

$$| \mathbf{V}_\perp \rangle = \Delta^\perp | \mathbf{V} \rangle . \tag{8}$$

This relation, written between the Cartesian components in reciprocal space, becomes

$$\mathcal{V}_{\perp i}(\mathbf{k}) = \int \mathrm{d}^3 k' \sum_j \Delta^\perp_{ij}(\mathbf{k}, \mathbf{k}') \, \mathcal{V}_j(\mathbf{k}') \tag{9}$$

where $\Delta^\perp_{ij}(\mathbf{k}, \mathbf{k}')$ is the matrix element of the operator $\Delta^\perp$ in the basis $\{ | \mathbf{k}, \mathbf{e}_i \rangle \}$. Comparison with (7) shows that this matrix element is equal to

$$\Delta^\perp_{ij} = \left(\delta_{ij} - \frac{k_i k_j}{k^2} \right) \delta(\mathbf{k} - \mathbf{k}') . \tag{10}$$

In real space, the same relation (8) is written

$$\mathrm{V}_{\perp i}(\mathbf{r}) = \int \mathrm{d}^3 r' \sum_j \Delta^\perp_{ij}(\mathbf{r}, \mathbf{r}') \, \mathrm{V}_j(\mathbf{r}') . \tag{11}$$

The matrix elements of the operator $\Delta^\perp$ in the basis $\{|\mathbf{r}, \mathbf{e}_i\rangle\}$ are given by

$$\begin{aligned}\Delta^\perp_{ij}(\mathbf{r}, \mathbf{r}') &= \int \mathrm{d}^3k \int \mathrm{d}^3k' \frac{\mathrm{e}^{\mathrm{i}\mathbf{k}.\mathbf{r}}}{(2\pi)^{3/2}} \Delta^\perp_{ij}(\mathbf{k}, \mathbf{k}') \frac{\mathrm{e}^{-\mathrm{i}\mathbf{k}'.\mathbf{r}'}}{(2\pi)^{3/2}} \\ &= \int \frac{\mathrm{d}^3k}{(2\pi)^3} \mathrm{e}^{\mathrm{i}\mathbf{k}.(\mathbf{r}-\mathbf{r}')}\left(\delta_{ij} - \frac{k_i k_j}{k^2}\right).\end{aligned} \tag{12}$$

It appears then that the transverse delta function introduced in (B.17),

$$\delta^\perp_{ij}(\boldsymbol{\rho}) = \int \frac{\mathrm{d}^3k}{(2\pi)^3} \mathrm{e}^{\mathrm{i}\mathbf{k}.\boldsymbol{\rho}}\left(\delta_{ij} - \frac{k_i k_j}{k^2}\right) \tag{13}$$

is tied to the matrix element of $\Delta^\perp$ in the basis $\{|\mathbf{r}, \mathbf{e}_i\rangle\}$ by

$$\Delta^\perp_{ij}(\mathbf{r}, \mathbf{r}') = \delta^\perp_{ij}(\mathbf{r} - \mathbf{r}'). \tag{14}$$

Remark

One can likewise introduce the projector Δ^l on the subspace of longitudinal fields which is the complement to $\mathbb{1}$ of $\Delta^\perp$:

$$\Delta^l = \mathbb{1} - \Delta^\perp. \tag{15}$$

2. The Expression for the Transverse Delta Function in Real Space

From the definition (13), it appears that $\delta^\perp_{ij}(\boldsymbol{\rho})$ is the Fourier transform of a function which does not tend to zero when $|\mathbf{k}|$ tends to infinity. The transverse delta function then has a singularity at $\boldsymbol{\rho} = \mathbf{0}$ which one must carefully characterize. To this end, one regularizes this singularity by truncating the spatial frequencies greater than some bound k_M. One later allows k_M to go to infinity. Physically, such a procedure means that one is not interested in variations of the field over infinitesimally short distances, but rather in the mean field over small but finite regions of space.

a) Regularization of $\delta^\perp_{ij}(\boldsymbol{\rho})$

Mathematically, one achieves the regularization by multiplying $\delta_{ij} - (k_i k_j/k^2)$ by $k_M^2/(k^2 + k_M^2)$, which has magnitude 1 for $k \ll k_M$, and decreases as $1/k^2$ at infinity:

$$\delta^\perp_{ij}(\boldsymbol{\rho}) = \int \frac{\mathrm{d}^3k}{(2\pi)^3} \mathrm{e}^{\mathrm{i}\mathbf{k}.\boldsymbol{\rho}}\left(\delta_{ij} - \frac{k_i k_j}{k^2}\right)\frac{k_M^2}{k^2 + k_M^2}. \tag{16}$$

This regularized function can be written, taking account of the properties of the Fourier transformation, as

$$\delta_{ij}^{\perp}(\boldsymbol{\rho}) = \left[\frac{\partial^2}{\partial\rho_i\,\partial\rho_j} - \delta_{ij}\left(\sum_l \frac{\partial^2}{\partial\rho_l^2}\right)\right] g(\boldsymbol{\rho}) \tag{17}$$

where

$$g(\boldsymbol{\rho}) = \int \frac{\mathrm{d}^3k}{(2\,\pi)^3}\,\frac{k_M^2\,\mathrm{e}^{\mathrm{i}\mathbf{k}.\boldsymbol{\rho}}}{k^2(k^2+k_M^2)} \tag{18}$$

b) Calculation of $g(\boldsymbol{\rho})$

One first performs the angular integral on **k**:

$$\begin{aligned} g(\boldsymbol{\rho}) &= \int_0^\infty \frac{k^2\,\mathrm{d}k}{(2\,\pi)^3}\,2\,\pi\int_{-1}^{1}\mathrm{d}u\,\frac{\mathrm{e}^{\mathrm{i}k\rho u}\,k_M^2}{k^2(k^2+k_M^2)} \\ &= \int_0^\infty \frac{\mathrm{d}k}{(2\,\pi)^2}\,\frac{\mathrm{e}^{\mathrm{i}k\rho}-\mathrm{e}^{-\mathrm{i}k\rho}}{\mathrm{i}k\rho}\,\frac{k_M^2}{k^2+k_M^2} \\ &= \frac{k_M^2}{(2\,\pi)^2\,\mathrm{i}\rho}\int_{-\infty}^{+\infty}\mathrm{d}k\,\frac{\mathrm{e}^{\mathrm{i}k\rho}}{k(k^2+k_M^2)}. \end{aligned} \tag{19}$$

This last integral is easily evaluated by the method of residues:

$$\begin{aligned} g(\boldsymbol{\rho}) &= \frac{k_M^2}{(2\,\pi)^2\,\mathrm{i}\rho}\left[2\,\mathrm{i}\pi\,\mathrm{Res}\,(k=\mathrm{i}k_M) + \mathrm{i}\pi\,\mathrm{Res}\,(k=0)\right] \\ &= \frac{k_M^2}{2\,\pi\rho}\left[\frac{\mathrm{e}^{-k_M\rho}}{\mathrm{i}k_M(2\,\mathrm{i}k_M)} + \frac{1}{2\,k_M^2}\right]. \end{aligned}$$

So that finally

$$g(\boldsymbol{\rho}) = \frac{1}{4\,\pi\rho}(1-\mathrm{e}^{-k_M\rho}). \tag{20}$$

Outside the neighborhood of the origin ($\rho \gg 1/k_M$), $g(\boldsymbol{\rho})$ is equal to $1/4\pi\rho$, which is indeed the Fourier transform of $1/k^2$. But as $\rho \to 0$, $g(\boldsymbol{\rho})$ remains finite and tends to $k_M/4\pi$.

c) Evaluation of the Derivatives of $g(\boldsymbol{\rho})$

Equation (17) gives the transverse delta function as a function of the second derivative of the function $g(\boldsymbol{\rho})$. Since this latter is only a function of the modulus of $\boldsymbol{\rho}$, one uses to evaluate its derivatives

$$\frac{\partial}{\partial\rho_i}\,g(\rho) = \frac{\rho_i}{\rho}\,g'(\rho) \tag{21a}$$

$$\frac{\partial}{\partial \rho_j} \rho_i = \delta_{ij} \tag{21b}$$

which yields

$$\begin{aligned}\frac{\partial^2}{\partial \rho_i \, \partial \rho_j} g(\boldsymbol{\rho}) &= \frac{\delta_{ij}}{\rho} g'(\rho) + \frac{\rho_i \, \rho_j}{\rho} \frac{\mathrm{d}}{\mathrm{d}\rho}\left(\frac{g'(\rho)}{\rho}\right) \\ &= \delta_{ij} \frac{g'}{\rho} + \frac{\rho_i \, \rho_j}{\rho^2}\left(g'' - \frac{g'}{\rho}\right)\end{aligned} \tag{21c}$$

Then

$$\sum_l \frac{\partial^2}{\partial \rho_l^2} g(\boldsymbol{\rho}) = g'' + \frac{2}{\rho} g' \tag{21d}$$

Substituting in (17) this gives the expression for $\delta_{ij}^{\perp}(\boldsymbol{\rho})$:

$$\delta_{ij}^{\perp}(\boldsymbol{\rho}) = \frac{\rho_i \, \rho_j}{\rho^2}\left(g'' - \frac{g'}{\rho}\right) - \delta_{ij}\left(g'' + \frac{g'}{\rho}\right) \tag{22}$$

Evaluation of g'' and g'/ρ gives

$$g''(\rho) = \frac{1}{4 \, \pi \rho^3}\left[2 - 2 \, \mathrm{e}^{-k_M \rho}\left(1 + k_M \, \rho + \frac{k_M^2 \, \rho^2}{2}\right)\right] \tag{23a}$$

$$\frac{1}{\rho} g'(\rho) = \frac{-1}{4 \, \pi \rho^3}\left[1 - \mathrm{e}^{-k_M \rho}(1 + k_M \, \rho)\right]. \tag{23b}$$

d) Discussion of the Expression for $\delta_{ij}^{\perp}(\boldsymbol{\rho})$

Equations (21), (22), and (23) lead to the following expression for the regularized transverse delta function (17):

$$\delta_{ij}^{\perp}(\boldsymbol{\rho}) = \gamma_{ij}(\boldsymbol{\rho}) + \frac{1}{4 \, \pi \rho^3}\left[\frac{3 \, \rho_i \, \rho_j}{\rho^2} - \delta_{ij}\right] \eta(\boldsymbol{\rho}) \tag{24}$$

where

$$\gamma_{ij}(\boldsymbol{\rho}) = \frac{k_M^2}{8 \, \pi \rho}\left(\frac{\rho_i \, \rho_j}{\rho^2} + \delta_{ij}\right) \mathrm{e}^{-k_M \rho} \tag{25}$$

$$\eta(\boldsymbol{\rho}) = 1 - \left(1 + k_M \, \rho + \frac{1}{2} k_M^2 \, \rho^2\right) \mathrm{e}^{-k_M \rho} \tag{26}$$

The function $\gamma_{ij}(\boldsymbol{\rho})$ is localized about the origin. At the limit $k_M \to \infty$ it tends to a point-source distribution centered on $\boldsymbol{\rho} = \mathbf{0}$. Such a distribution can a priori be written as a sum of a function $\delta(\boldsymbol{\rho})$ and its derivatives of order $1, 2, \ldots,$ which we are going to evaluate.

Note first that, simply as a result of homogeneity, the integral

$$\int d^3\rho\ \gamma_{ij}(\boldsymbol{\rho}) = I_1 \tag{27}$$

is a number independent of k_M. In contrast, the integral of the same function multiplied by a term of degree m in ρ_i is proportional to $(1/k_M)^m$ and goes to zero when $k_M \to \infty$. Thus, in the integral of the product of $\gamma_{ij}(\boldsymbol{\rho})$ with a function $\psi(\boldsymbol{\rho})$ expandable about $\boldsymbol{\rho} = \mathbf{0}$, only the first term of the series gives a nonzero contribution in the limit $k_M \to \infty$:

$$\int d^3\rho\ \gamma_{ij}(\boldsymbol{\rho})\, \psi(\boldsymbol{\rho}) = I_1\, \psi(\mathbf{0}). \tag{28}$$

It follows that

$$\lim_{k_M \to \infty} \gamma_{ij}(\boldsymbol{\rho}) = I_1\, \delta(\boldsymbol{\rho}) \tag{29}$$

where

$$\begin{aligned} I_1 &= \int d^3\rho\ \gamma_{ij}(\boldsymbol{\rho}) \\ &= \int_0^\infty \frac{k_M^2}{8\pi} \rho\, d\rho \int d\Omega_\rho \left(\delta_{ij} + \frac{\rho_i\, \rho_j}{\rho^2}\right) e^{-k_M \rho}. \end{aligned} \tag{30}$$

The angular average of $\rho_i\rho_j/\rho^2$ is $\delta_{ij}/3$, and the radial integral gives

$$I_1 = \frac{2}{3}\delta_{ij}. \tag{31}$$

At the limit $k_M \to \infty$, $\gamma_{ij}(\boldsymbol{\rho})$ is then simply

$$\gamma_{ij}(\boldsymbol{\rho}) = \frac{2}{3}\delta_{ij}\, \delta(\boldsymbol{\rho}) \tag{32}$$

where it is understood that the function $\delta(\boldsymbol{\rho})$ has an extent of $1/k_M$.

The function $\eta(\rho)$ is a regularizing function which becomes 1 for $\rho \gg 1/k_M$ and which starts as $k_M^3\rho^3/6$ at the origin, with the result that the second term in (24) does not diverge at $\rho = 0$. It behaves like a dipole field regularized at the origin. Such a function has properties in three-dimensional space analogous to those of the principal-part function $\mathcal{P}(1/x)$. Actually, on integrating over a small volume centered at the origin, the second term of (24) gives zero, although it has in this region a value of the order of k_M^3. This property arises from the vanishing of the angular average of $(3\,\rho_i\,\rho_j/\rho^2) - \delta_{ij}$, in the same way that the integral of $\mathcal{P}(1/x)$ is zero as a result of the odd parity of this function.

Finally, the transverse delta function can be written

$$\delta_{ij}^{\perp}(\boldsymbol{\rho}) = \frac{2}{3}\delta_{ij}\,\delta(\boldsymbol{\rho}) + \frac{\eta(\boldsymbol{\rho})}{4\,\pi\rho^3}\left(\frac{3\,\rho_i\,\rho_j}{\rho^2} - \delta_{ij}\right) \tag{33}$$

where $\eta(\boldsymbol{\rho})$ is equal to 1 away from the origin and suppresses the divergence at $\rho = 0$.

Remarks

(i) The factor $\frac{2}{3}$ in (32) can be simply found. Taking the trace on i of (13) gives immediately

$$\sum_i \delta_{ii}^{\perp}(\boldsymbol{\rho}) = 2\,\delta(\boldsymbol{\rho}) \,. \tag{34}$$

In the expression (33) for $\delta_{ij}^{\perp}$, only the first term contributes to the trace and the factor $\frac{2}{3}$ of (32) is necessary to satisfy (34).

(ii) One can ask if the second term of (24) does not give rise, in the limit $k_M \to \infty$, to derivatives of the delta function $\delta(\boldsymbol{\rho})$. In fact, the dimensional argument already developed for $\gamma_{ij}(\boldsymbol{\rho})$ applies: for functions of degree m in ρ_i, the contribution of the neighborhood of the origin to the integral of the product of these functions with the second term of (24) is of the order $1/k_M^m$ and tends to zero when $k_M \to \infty$.

3. Application to the Evaluation of the Magnetic Field Created by a Magnetization Distribution. Contact Interaction

The magnetic field $\mathbf{B}(\mathbf{r})$ created by a magnetization density $\mathbf{M}(\mathbf{r})$ is given by Maxwell's equation

$$\boldsymbol{\nabla} \times \mathbf{B}(\mathbf{r}) = \frac{1}{\varepsilon_0\,c^2}\,\mathbf{j}(\mathbf{r}) \tag{35}$$

where

$$\mathbf{j}(\mathbf{r}) = \boldsymbol{\nabla} \times \mathbf{M}(\mathbf{r}) \tag{36}$$

is the current associated with $\mathbf{M}(\mathbf{r})$. Substituting (36) in (35) and transforming into reciprocal space, this becomes

$$i\mathbf{k} \times \mathscr{B}(\mathbf{k}) = \frac{1}{\varepsilon_0\,c^2}\,i\mathbf{k} \times \mathscr{M}(\mathbf{k}) \,. \tag{37}$$

This equation allows one to find $\mathscr{B}(\mathbf{k})$. Projecting both sides of (37) on $\mathbf{k}$ and using the transverse nature of $\mathscr{B}(\mathbf{k} \cdot \mathscr{B} = 0)$, one gets

$$\mathscr{B}(\mathbf{k}) = \frac{1}{\varepsilon_0\,c^2}\left[\mathscr{M}(\mathbf{k}) - \boldsymbol{\kappa}(\boldsymbol{\kappa} \cdot \mathscr{M}(\mathbf{k}))\right] = \frac{1}{\varepsilon_0\,c^2}\,\mathscr{M}_{\perp}(\mathbf{k}) \tag{38}$$

which shows that $\mathcal{B}$ is (except for a factor $1/\varepsilon_0 c^2$) the transverse component of $\mathcal{M}$. In real space, Equation (38) is written

$$B_i(\mathbf{r}) = \frac{1}{\varepsilon_0 c^2} M_{\perp i}(\mathbf{r}) = \frac{1}{\varepsilon_0 c^2} \sum_j \int d^3r' \, \delta_{ij}^{\perp}(\mathbf{r} - \mathbf{r}') \, M_j(\mathbf{r}') . \tag{39}$$

Using (33) for the transverse delta function appearing in (39) shows that $\mathbf{B}(\mathbf{r})$ is a sum of two contributions. The first one, coming from the first term in (33), is simply proportional to the density $\mathbf{M}(\mathbf{r})$ taken at point $\mathbf{r}$. The second one, coming from the second term of (33), represents physically the dipolar field created at $\mathbf{r}$ by the magnetization density $\mathbf{M}(\mathbf{r}')$ at all other points $\mathbf{r}'$. The presence of the regularization function η in the second term of (33) makes all the expressions finite, and symmetry arguments then allow one to show that the immediate neighborhood of $\mathbf{r}$ does not contribute to the integral on $\mathbf{r}'$ of the product of the second term of (33) with $M_j(\mathbf{r}')$ (see §2.*d* above).

Consider now another magnetization density $\mathbf{M}'(\mathbf{r})$. The interaction energy of $\mathbf{M}'(\mathbf{r})$ with the field $\mathbf{B}(\mathbf{r})$ created by $\mathbf{M}(\mathbf{r})$ is

$$W = - \int d^3r \, \mathbf{M}'(\mathbf{r}) \cdot \mathbf{B}(\mathbf{r}) . \tag{40}$$

Equation (39) allows this to be written in a more symmetric form,

$$W = - \frac{1}{\varepsilon_0 c^2} \sum_i \sum_j \int d^3r \int d^3r' \, M_i'(\mathbf{r}) \, \delta_{ij}^{\perp}(\mathbf{r} - \mathbf{r}') \, M_j(\mathbf{r}') . \tag{41}$$

Using (33) again allows one to separate two contributions in W. The first,

$$W_1 = - \frac{2}{3} \frac{1}{\varepsilon_0 c^2} \int d^3r \, \mathbf{M}'(\mathbf{r}) \cdot \mathbf{M}(\mathbf{r}) \tag{42}$$

which depends on the magnetization densities $\mathbf{M}$ and $\mathbf{M}'$ at the *same* point $\mathbf{r}$, is called for that reason the *contact* interaction. The second,

$$W_2 = - \frac{1}{4\pi\varepsilon_0 c^2} \sum_i \sum_j \int d^3r \int d^3r' \frac{\eta(|\mathbf{r} - \mathbf{r}'|)}{(|\mathbf{r} - \mathbf{r}'|)^3} \times$$
$$\times M_i'(\mathbf{r}) \left[\frac{3(\mathbf{r} - \mathbf{r}')_i (\mathbf{r} - \mathbf{r}')_j}{(\mathbf{r} - \mathbf{r}')^2} - \delta_{ij} \right] M_j(\mathbf{r}') \tag{43}$$

represents the magnetic dipole–dipole interaction between the two densities. As above, the regularization introduced by η and symmetry arguments show that the immediate neighborhood of $|\mathbf{r} - \mathbf{r}'| = 0$ does not contribute to W.

Remark

The foregoing can be applied to the study of the magnetic interaction between the nuclear spin and the electron spin in an atom. One takes as $\mathbf{M}(\mathbf{r})$ the magnetization density of the nucleus. This density can be appreciable only in a volume of the order of r_0^3, where r_0 characterizes the dimensions of the nucleus, which is taken at the coordinate origin. The integral of $\mathbf{M}(\mathbf{r})$ is just the magnetic moment of the nucleus,

$$\boldsymbol{\mu}_N = \int \mathrm{d}^3r\, \mathbf{M}(\mathbf{r}) . \tag{44}$$

One assumes in addition that $\mathbf{M}'(\mathbf{r})$ represents the spin magnetization density of the electron in the state $\psi(\mathbf{r})$, so that

$$\mathbf{M}'(\mathbf{r}) = \boldsymbol{\mu}_e \,|\, \psi(\mathbf{r}) \,|^2 \tag{45}$$

where $\boldsymbol{\mu}_e$ is the spin magnetic moment of the electron. Important simplifications appear in the interaction energy W as a result of the different spatial extensions of $\mathbf{M}(\mathbf{r})$ and $\mathbf{M}'(\mathbf{r})$. Indeed, the spatial extent of $|\psi(\mathbf{r})|^2$ is of the order of the Bohr radius a_0, which is much larger than the nuclear radius. If one ignores the variation of $|\psi(\mathbf{r})|^2$ inside the nucleus, the contact interaction (42) becomes

$$W_1 \simeq -\frac{2}{3}\frac{1}{\varepsilon_0 c^2}\int \mathrm{d}^3r\, \mathbf{M}'(0) \cdot \mathbf{M}(\mathbf{r}) \tag{46}$$

that is, taking into account (44) and (45),

$$W_1 \simeq -\frac{2}{3\,\varepsilon_0 c^2}\boldsymbol{\mu}_N \cdot \boldsymbol{\mu}_e \,|\, \psi(0) \,|^2 . \tag{47}$$

In the same way, one can make a multipole expansion of $\mathbf{M}(\mathbf{r})$ in (43) and only keep the lowest-order term, which yields

$$\mathbf{M}(\mathbf{r}) = \boldsymbol{\mu}_N\, \delta(\mathbf{r}) . \tag{48}$$

Equation (43) then becomes

$$W_2 \simeq -\frac{1}{4\pi\varepsilon_0 c^2}\sum_i\sum_j \int \mathrm{d}^3r'\, \frac{\eta(r')}{r'^3}\, \mu_{Ni}\left(\frac{3\, r_i' r_j'}{r'^2} - \delta_{ij}\right) \mu_{ej} \,|\, \psi(\mathbf{r}') \,|^2 . \tag{49}$$

Finally, regrouping (47) and (49) shows that the magnetic interaction energy between the two spins appears as the mean value in the state $\psi(\mathbf{r})$ of the interaction Hamiltonian

$$\begin{aligned} H &= -\frac{1}{\varepsilon_0 c^2}\sum_i\sum_j \mu_{Ni}\, \delta_{ij}^{\perp}(\mathbf{r})\, \mu_{ej} \\ &= -\frac{1}{4\pi\varepsilon_0 c^2}\left[\frac{8\pi}{3}\boldsymbol{\mu}_e \cdot \boldsymbol{\mu}_N\, \delta(\mathbf{r}) + \frac{3(\boldsymbol{\mu}_e \cdot \mathbf{r})(\boldsymbol{\mu}_N \cdot \mathbf{r})}{r^5} - \frac{\boldsymbol{\mu}_e \cdot \boldsymbol{\mu}_N}{r^3}\right] \end{aligned} \tag{50}$$

where $\mathbf{r}$ is the position of the electron with respect to the nucleus and where it is further understood that the dipole–dipole interaction is regularized at $r = 0$.

COMPLEMENT B_I

ANGULAR MOMENTUM OF THE ELECTROMAGNETIC FIELD. MULTIPOLE WAVES

The first motivation for this complement is to establish some results given without proof in Chapter I and related to the angular momentum of the electromagnetic field. We first show (§1) that the contribution of the longitudinal electric field to the angular momentum of the field can be reexpressed as a function of the particle coordinates $\mathbf{r}_\alpha$ and the transverse vector potential $\mathbf{A}_\perp(\mathbf{r}_\alpha)$ and regrouped with the angular momentum of the particles. We also establish (§2) the expression for the angular momentum of the transverse field as a function of the normal variables $\boldsymbol{\alpha}(\mathbf{k})$. The expression gotten is closely analogous to that giving the mean value for the total angular momentum of a spin-1 particle whose vector wave function is precisely $\boldsymbol{\alpha}(\mathbf{k})$ in reciprocal space.

The foregoing analogy suggests then that one look for functions $\boldsymbol{\alpha}(\mathbf{k})$ suitable for the angular momentum of the transverse field. More precisely, one tries to determine transverse vector functions of $\mathbf{k}$, defined on a sphere of radius k_0, which are also eigenfunctions of $\mathbf{J}^2$ and J_z, where $\mathbf{J}$ is the total angular momentum of a spin-1 particle. Instead of coupling in the usual fashion the orbital angular momentum $\mathbf{L}$ to the spin angular momentum $\mathbf{S}$ of such a particle, we will see in §3 a simpler method for constructing the eigenfunctions of $\mathbf{J}^2$ and J_z which give the longitudinal or transverse eigenfunctions directly.

When the eigenfunctions thus found are substituted for the normal variables in the expansion of the electric and magnetic fields, one then gets, in real space, electromagnetic waves corresponding to photons with well-defined energy, angular momentum, and parity (§4). The second motivation behind this complement is to give a simple derivation of such multipolar waves, which are well suited to all the problems of atomic or nuclear physics where exchanges of angular momentum between matter and radiation play an important role.

1. Contribution of the Longitudinal Electric Field to the Total Angular Momentum

Let

$$\mathbf{J}_{\text{long}} = \varepsilon_0 \int d^3r \, \mathbf{r} \times (\mathbf{E}_{\parallel} \times \mathbf{B}) \tag{1}$$

be the contribution of the longitudinal electric field to the total angular momentum of the system field + particles. Replacing $\mathbf{B}$ by $\nabla \times \mathbf{A}_\perp$ (since $\nabla \times \mathbf{A}_\parallel = \mathbf{0}$) and using the expression for the double vector product to transform $\mathbf{E}_\parallel \times (\nabla \times \mathbf{A}_\perp)$ yields

$$\mathbf{J}_{\text{long}} = \varepsilon_0 \int \mathrm{d}^3 r \left\{ \sum_{a=x,y,z} E_{\parallel a} (\mathbf{r} \times \nabla) A_{\perp a} - \mathbf{r} \times (\mathbf{E}_\parallel \cdot \nabla) \mathbf{A}_\perp \right\}. \tag{2}$$

The last term in (2) can be rewritten by moving $\mathbf{r}$ to the right of ∇:

$$\varepsilon_0 \int \mathrm{d}^3 r \left[-(\mathbf{E}_\parallel \cdot \nabla)(\mathbf{r} \times \mathbf{A}_\perp) + \mathbf{E}_\parallel \times \mathbf{A}_\perp \right]. \tag{3}$$

Integrate the first term of (3) by parts. The integrated term gives a surface integral at infinity which vanishes if the fields go to zero sufficiently quickly. In the remaining term the quantity $\nabla \cdot \mathbf{E}_\parallel$ appears, and this is ρ/ε_0 from Maxwell's equation (A.1.a). Regrouping the expression so obtained for (3) and the first term of (2), and making use of the fact that $\mathbf{E}_\parallel = -\nabla U$, where U is the Coulomb potential, one gets finally

$$\mathbf{J}_{\text{long}} = $$
$$= \int \mathrm{d}^3 r \left\{ \rho (\mathbf{r} \times \mathbf{A}_\perp) - \varepsilon_0 \sum_a (\nabla_a U)(\mathbf{r} \times \nabla) A_{\perp a} - \varepsilon_0 (\nabla U) \times \mathbf{A}_\perp \right\}. \tag{4}$$

One can see now that the last two terms in (4) cancel. Integrating them by parts, one gets

$$\varepsilon_0 \int \mathrm{d}^3 r \left\{ \sum_a U \nabla_a (\mathbf{r} \times \nabla) A_{\perp a} + U (\nabla \times \mathbf{A}_\perp) \right\}. \tag{5}$$

Now

$$\sum_a U \nabla_a (\mathbf{r} \times \nabla) A_{\perp a} = U (\mathbf{r} \times \nabla)(\nabla \cdot \mathbf{A}_\perp) - U (\nabla \times \mathbf{A}_\perp). \tag{6}$$

The first term of (6) vanishes, since $\nabla \cdot \mathbf{A}_\perp = 0$. The second term of (6) cancels with the last term of (5). Only the first term of (4) remains, which, using (A.5.a) for ρ, gives

$$\mathbf{J}_{\text{long}} = \sum_\alpha q_\alpha \mathbf{r}_\alpha \times \mathbf{A}_\perp (\mathbf{r}_\alpha). \tag{7}$$

It is clear then that $\mathbf{J}_{\text{long}}$ can be written as a function of the coordinates $\mathbf{r}_\alpha$ of the particles and of the transverse potential $\mathbf{A}_\perp$. This result is gauge independent, since $\mathbf{A}_\perp$ is gauge invariant [see (B.26)].

One now groups $\mathbf{J}_{\text{long}}$ with the angular momentum of the particles, $\Sigma_\alpha \mathbf{r}_\alpha \times m_\alpha \dot{\mathbf{r}}_\alpha$. This gives

$$\mathbf{J}_{\text{long}} + \sum_\alpha \mathbf{r}_\alpha \times m_\alpha \dot{\mathbf{r}}_\alpha = \sum_\alpha \mathbf{r}_\alpha \times [q_\alpha \mathbf{A}_\perp(\mathbf{r}_\alpha) + m_\alpha \dot{\mathbf{r}}_\alpha]$$

$$= \sum_\alpha \mathbf{r}_\alpha \times \mathbf{p}_\alpha \tag{8}$$

where $\mathbf{p}_\alpha$ is defined by (B.44). In the Coulomb gauge ($\mathbf{A} = \mathbf{A}_\perp$), $\mathbf{p}_\alpha$ is the momentum conjugate with $\mathbf{r}_\alpha$, or equivalently the canonical momentum of particle α. It appears then that in the Coulomb gauge the difference between the quantities $\mathbf{r}_\alpha \times \mathbf{p}_\alpha$ and $\mathbf{r}_\alpha \times m_\alpha \dot{\mathbf{r}}_\alpha$ is just the angular momentum associated with the longitudinal electric field of particle α.

2. Angular Momentum of the Transverse Field

In this section, we transform the expression for the angular momentum of the transverse field,

$$\mathbf{J}_{\text{trans}} = \varepsilon_0 \int \mathrm{d}^3 r\, \mathbf{r} \times (\mathbf{E}_\perp \times \mathbf{B}) \tag{9}$$

a) $\mathbf{J}_{\text{trans}}$ in Reciprocal Space

The calculations at the beginning of §1 above remain valid when one replaces $\mathbf{E}_\parallel$ by $\mathbf{E}_\perp$ throughout. Since $\nabla \cdot \mathbf{E}_\perp = 0$, the integration by parts of the term corresponding to the first term of (3) now gives a zero result, and only the terms corresponding to the first term of (2) and the last term of (3) remain:

$$\mathbf{J}_{\text{trans}} = \varepsilon_0 \int \mathrm{d}^3 r \left\{ \sum_a E_{\perp a} (\mathbf{r} \times \nabla)\, A_{\perp a} + \mathbf{E}_\perp \times \mathbf{A}_\perp \right\}. \tag{10}$$

One now expresses $\mathbf{J}_{\text{trans}}$ as a function of the Fourier spatial transforms $\boldsymbol{\mathscr{E}}_\perp$ and $\boldsymbol{\mathscr{A}}_\perp$ of $\mathbf{E}_\perp$ and $\mathbf{A}_\perp$. For this, one uses the Parseval–Plancherel identity and the following table:

Table I

$$\mathbf{r} \leftrightarrow \mathrm{i}\nabla$$

$$\nabla \leftrightarrow \mathrm{i}\mathbf{k}$$

giving the correspondence between the operators multiplication by $\mathbf{r}$ and gradient with respect to $\mathbf{r}$ in real space on one hand and the operators multiplication by $\mathbf{k}$ and gradient with respect to $\mathbf{k}$ (denoted by $\nabla\!\!\!\!/$ to distinguish it from $\boldsymbol{\nabla}$) in reciprocal space on the other. One then gets

$$\mathbf{J}_{\text{trans}} = \varepsilon_0 \int \mathrm{d}^3k \left\{ \sum_a \mathscr{E}^*_{\perp a}(\mathbf{k} \times \nabla\!\!\!\!/ \,)\mathscr{A}_{\perp a} + \mathscr{E}^*_\perp \times \mathscr{A}_\perp \right\} \tag{11}$$

(using $\mathbf{k} \times \nabla\!\!\!\!/ = - \nabla\!\!\!\!/ \times \mathbf{k}$).

b) $\mathbf{J}_{\text{trans}}$ IN TERMS OF NORMAL VARIABLES

In Part C of this chapter, $\mathscr{E}_\perp$ and $\mathscr{A}_\perp$ have been given in terms of normal variables $\boldsymbol{\alpha}(\mathbf{k})$:

$$\begin{cases} \mathscr{E}_\perp = \mathrm{i}\,\mathscr{N}(\boldsymbol{\alpha} - \boldsymbol{\alpha}^*_-) & (12) \\ \mathscr{A}_\perp = \dfrac{\mathscr{N}}{\omega}(\boldsymbol{\alpha} + \boldsymbol{\alpha}^*_-)\,. & (13) \end{cases}$$

Recall that $\mathscr{N}$ is a normalization coefficient given by (C.15) and that $\boldsymbol{\alpha}_-$ is an abbreviated notation for $\boldsymbol{\alpha}(-k, t)$. Substituting (12) and (13) in (11) and changing $\mathbf{k}$ to $-\mathbf{k}$ in certain terms, one gets

$$\begin{aligned} \mathbf{J}_{\text{trans}} = \varepsilon_0 \int \mathrm{d}^3k \frac{\mathscr{N}^2}{\omega} \Bigg\{ & \sum_a \alpha^*_a (- \mathrm{i}\mathbf{k} \times \nabla\!\!\!\!/ \,)\, \alpha_a - \mathrm{i}\boldsymbol{\alpha}^* \times \boldsymbol{\alpha} - \\ & - \sum_a \alpha_{-a}(- \mathrm{i}\mathbf{k} \times \nabla\!\!\!\!/ \,)\, \alpha_a + \mathrm{i}\boldsymbol{\alpha}_- \times \boldsymbol{\alpha} - \\ & - [\text{same terms where } \boldsymbol{\alpha} \rightleftarrows \boldsymbol{\alpha}^*] \Bigg\}. \end{aligned} \tag{14}$$

The contribution of the second line of (14) to the integral is zero. Firstly, the term $\boldsymbol{\alpha}_- \times \boldsymbol{\alpha}$ is odd in $\mathbf{k}$ and its integral vanishes. Then, changing $\mathbf{k}$ to $-\mathbf{k}$ in the first term of the second line of (14) and integrating it by parts, one sees that the contribution of this term is equal to its opposite and thus vanishes. One has then

$$\mathbf{J}_{\text{trans}} = \frac{\hbar}{2} \int \mathrm{d}^3k \left\{ \left[\sum_\alpha \alpha^*_a(- \mathrm{i}\mathbf{k} \times \nabla\!\!\!\!/ \,)\, \alpha_a - \mathrm{i}\boldsymbol{\alpha}^* \times \boldsymbol{\alpha} \right] - [\boldsymbol{\alpha} \rightleftarrows \boldsymbol{\alpha}^*] \right\} \tag{15}$$

an expression equivalent to Equation (C.18).

In all the calculations we have done, the ordering between $\boldsymbol{\alpha}$ and $\boldsymbol{\alpha}^*$ has always been respected and is as it appears in the equations. If one neglects to take this ordering into account, Equation (15) simplifies and becomes

$$\mathbf{J}_{\text{trans}} = \hbar \int d^3k \left[\sum_a \alpha_a^*(-\mathrm{i}\mathbf{k} \times \nabla)\, \alpha_a - \mathrm{i}\boldsymbol{\alpha}^* \times \boldsymbol{\alpha} \right]. \tag{16}$$

c) Analogy with the Mean Value of the Total Angular Momentum of a Spin-1 Particle

We will now see that it is possible to reinterpret Equation (16) for $\mathbf{J}_{\text{trans}}$. To do this, set aside for the moment the problem of the angular momentum of the electromagnetic field, and consider the quantum mechanics of a spin-1 particle.

There are three possible spin states for such a particle, and the wave function representing the state $|\phi\rangle$ of the particle will be a vector wave function with three components. In reciprocal space these components will be called

$$\langle \mathbf{k}, a \,|\, \phi \rangle = \phi_a(\mathbf{k}). \tag{17}$$

We have taken a spin-state basis $\{|a\rangle\}$ which is not the set of eigenstates $\{|-1\rangle, |0\rangle, |+1\rangle\}$ of S_z, but the set of Cartesian spin states $\{|x\rangle, |y\rangle, |z\rangle\}$ related to it:

$$\begin{cases} |\,x\,\rangle = \dfrac{1}{\sqrt{2}}(|-1\,\rangle - |+1\,\rangle) \\ |\,y\,\rangle = \dfrac{\mathrm{i}}{\sqrt{2}}(|-1\,\rangle + |+1\,\rangle) \\ |\,z\,\rangle = |\,0\,\rangle. \end{cases} \tag{18}$$

One now evaluates the mean value in the state $|\phi\rangle$ of the total angular momentum

$$\mathbf{J} = \mathbf{L} + \mathbf{S} \tag{19}$$

for such a particle, where $\mathbf{L}$ and $\mathbf{S}$ are the orbital and spin angular momenta. Using Table I, $\mathbf{L}$ is given in reciprocal space by the operator

$$\mathbf{L} = \mathrm{i}\,\nabla \times \hbar\mathbf{k} = -\,\mathrm{i}\hbar\mathbf{k} \times \nabla. \tag{20}$$

Since $\mathbf{L}$ does not act on the spin quantum numbers,

$$\langle \phi \,|\, \mathbf{L} \,|\, \phi \rangle = \hbar \int d^3k \sum_a \phi_a^*(\mathbf{k})(-\,\mathrm{i}\mathbf{k} \times \nabla)\, \phi_a(\mathbf{k}). \tag{21}$$

To find the mean value of **S**, let us first give the action of S_x, S_y, and S_z on the Cartesian basis states $|x\rangle$, $|y\rangle$, and $|z\rangle$:

$$S_a \,|\, b\,\rangle = i\hbar \sum_c \varepsilon_{abc} \,|\, c\,\rangle \tag{22}$$

which derives from the known action of S_z and $S_\pm = S_x \pm iS_y$ on the eigenstates of S_z and from equations (18). Since **S** does not act on the quantum numbers **k**, one gets then from (22)

$$\langle\, \phi \,|\, \mathbf{S} \,|\, \phi\,\rangle = -\,i\hbar \int d^3k\, \boldsymbol{\phi}^*(\mathbf{k}) \times \boldsymbol{\phi}(\mathbf{k})\,. \tag{23}$$

Compare (21) and (23) with the two terms appearing on the right-hand side of (16) for $\mathbf{J}_{\text{trans}}$. If one identifies the normal variables $\boldsymbol{\alpha}(\mathbf{k})$ with the vector wave function $\boldsymbol{\phi}(\mathbf{k})$ of a spin-1 particle in reciprocal space, $\mathbf{J}_{\text{trans}}$ appears then as the mean value of the total angular momentum of such a particle, the first and second terms of (16) being associated with the orbital angular momentum and the spin angular momentum respectively.

Remarks

(i) It must be kept in mind that the normal variables $\boldsymbol{\alpha}(\mathbf{k})$ form a *transverse* vector field. If one reinterprets $\boldsymbol{\alpha}(\mathbf{k})$ as the wave function of a photon in reciprocal space, it is equally necessary to constrain the photon wave-function space to belong to the subspace of transverse fields and to consider as physical only the observables leaving such a subspace invariant. Let us show that **L** and **S** are not separately physically observable as $\mathbf{J} = \mathbf{L} + \mathbf{S}$ is. The operator **L** is associated with an "orbital rotation" of the vector field: **L** rotates the point of application **k** of each vector $\boldsymbol{\alpha}(\mathbf{k})$ of the field without rotating $\boldsymbol{\alpha}$ at the same time. In such an operation, the orthogonality between **k** and $\boldsymbol{\alpha}(\mathbf{k})$ is not preserved. Likewise, **S** causes the vector $\boldsymbol{\alpha}$ to rotate without changing its point of application, which also causes the transversality of the field to break down. In contrast, **J** causes the vector and its point of application to rotate at the same time, preserving the angles and thus the transversality. This can also be clarified by the following argument. The spin of a particle represents its total angular momentum in the frame where it is at rest. Such a frame does not exist for the photon, which propagates at the velocity of light, with the result that **S** (and likewise **L**) is not separately observable for a photon.

(ii) Analogous reasoning allows one to understand why **r** cannot be a position operator for the photon: **r** generates translations in reciprocal space, and such operations in general do not retain the orthogonality between $\boldsymbol{\alpha}$ and the vector **k**.

3. Set of Vector Functions of k "Adapted" to the Angular Momentum

a) General Idea

As seen in Part C, giving the complex vector function $\boldsymbol{\alpha}(\mathbf{k})$ completely defines the state of the electromagnetic field. Up to transversality (to which we shall return), this function is isomorphic with the wave function of a spin-1 particle. To each basis in the state space of this particle there corresponds a set of states of the electromagnetic field able to produce through linear combination any state of the field. Thus, the set of functions

$$\boldsymbol{\phi}_{\mathbf{k}_i\,\boldsymbol{\varepsilon}_i}(\mathbf{k}) = \boldsymbol{\varepsilon}_i\,\delta(\mathbf{k} - \mathbf{k}_i) \tag{24}$$

for all $\mathbf{k}_i$ and all $\boldsymbol{\varepsilon}_i$ normal to $\mathbf{k}_i$ form a basis for the space of vector functions, to which corresponds the field expansion in plane waves used in Chapter I. This basis is in fact the basis of eigenstates of the momentum for the associated spin-1 particle. This property explains the simple form of (C.17) for the momentum of the transverse field when using this basis.

In a similar way one can now construct another basis "adapted" to the angular momentum. It is a basis of eigenstates for $\mathbf{J}^2$ and J_z, $\mathbf{J}$ being the total angular momentum (19) of the spin-1 particle which we have associated with the electromagnetic field. It is well known that the orbital-angular-momentum operator $\mathbf{L}$ acts only on the polar angles of vector $\mathbf{k}$, or equivalently on the unit vector $\boldsymbol{\kappa} = \mathbf{k}/k$. The operator for $\mathbf{S}$ does not act on $\mathbf{k}$. Consequently, knowledge of the eigenvalues of $\mathbf{J}^2$ and J_z does not give any information on the radial part of the eigenfunction. If one takes this radial part proportional to $\delta(k - k_0)$, the basis functions are also energy eigenfunctions with eigenvalue $\hbar c k_0$ (since the photon energy depends only on $|\mathbf{k}|$). One thus takes

$$\boldsymbol{\phi}_{k_0 JM}(\mathbf{k}) = \frac{1}{k_0}\,\delta(k - k_0)\,\boldsymbol{\phi}_{JM}(\boldsymbol{\kappa}) \tag{25.a}$$

where the factor $1/k_0$ has been introduced for normalization:

$$\int \mathrm{d}^3k\,\boldsymbol{\phi}^*_{k_0' J'M'}(\mathbf{k}) \cdot \boldsymbol{\phi}_{k_0 JM}(\mathbf{k}) = \delta(k_0 - k_0')\,\delta_{JJ'}\,\delta_{MM'}\,. \tag{25.b}$$

The angular function $\boldsymbol{\phi}_{JM}(\boldsymbol{\kappa})$ is likewise normalized on the sphere of radius 1:

$$\int \mathrm{d}^2\kappa\,\boldsymbol{\phi}^*_{J'M'}(\boldsymbol{\kappa}) \cdot \boldsymbol{\phi}_{JM}(\boldsymbol{\kappa}) = \delta_{JJ'}\,\delta_{MM'} \tag{25.c}$$

and must be fixed with the condition that $\phi_{k_0JM}(\mathbf{k})$ is an eigenfunction of $\mathbf{J}^2$ and $\mathbf{J}_z$ with eigenvalues $J(J+1)\hbar^2$ and $M\hbar$ respectively.

A first method of constructing $\phi_{JM}(\boldsymbol{\kappa})$ is to take the orbital eigenstates for $\mathbf{L}^2$ and L_z (the spherical harmonics) and to couple them to the spin eigenstates for S_z using the usual algebra for combining angular momenta. One thus constructs the vector spherical harmonics. Now these vector functions are in general neither longitudinal nor transverse [$\phi_{JM}(\boldsymbol{\kappa})$ is neither parallel nor perpendicular to $\boldsymbol{\kappa}$]. To get the transverse functions it is necessary to combine the functions with the same quantum numbers J and M but with different eigenvalues $L(L+1)\hbar^2$ of $\mathbf{L}^2$.

Here we are going to use a simpler method, due to Berestetskii, Lifshitz, and Pitayevski, which directly gives the longitudinal and transverse functions.

b) Method for Constructing Vector Eigenfunctions for $\mathbf{J}^2$ and J_z

One can generate vector functions by letting an orbital vector operator $\mathbf{V}$ act on a scalar function $\chi(\boldsymbol{\kappa})$ defined on the sphere of unit radius in reciprocal space. Consider the action of $\mathbf{J}$ on such a function:

$$J_a(\mathbf{V}\chi(\boldsymbol{\kappa})) = L_a(\mathbf{V}\chi(\boldsymbol{\kappa})) + S_a(\mathbf{V}\chi(\boldsymbol{\kappa}))\,. \tag{26}$$

Since $\mathbf{V}$ is an orbital vector operator, it has the following simple commutation relations with $\mathbf{L}$:

$$[L_a, V_b] = i\hbar \sum_c \varepsilon_{abc} V_c \tag{27}$$

which allow one to write

$$L_a(\mathbf{V}\chi(\boldsymbol{\kappa})) = \mathbf{V}(L_a\,\chi(\boldsymbol{\kappa})) - i\hbar\,\mathbf{e}_a \times \mathbf{V}\chi(\boldsymbol{\kappa})\,. \tag{28}$$

To find $S_a(\mathbf{V}\chi(\boldsymbol{\kappa}))$ note first of all that the vector wave function $\mathbf{V}\chi(\boldsymbol{\kappa})$ is associated with the ket $\Sigma_b|b\rangle\,|V_b\chi\rangle$, where $|V_b\chi\rangle$ is the orbital part of the ket and $|b\rangle$ the spin part. Using (22), one gets then

$$S_a \sum_b |\,b\,\rangle\,|\,V_b\,\chi\,\rangle = i\hbar \sum_{c,b} \varepsilon_{abc}\,|\,c\,\rangle\,|\,V_b\,\chi\,\rangle \tag{29}$$

which yields for the associated wave function

$$S_a(\mathbf{V}\chi(\boldsymbol{\kappa})) = i\hbar\,\mathbf{e}_a \times \mathbf{V}\chi(\boldsymbol{\kappa})\,. \tag{30}$$

Add (28) and (30). This gives

$$J_a(\mathbf{V}\chi(\boldsymbol{\kappa})) = \mathbf{V}(L_a\,\chi(\boldsymbol{\kappa}))\,. \tag{31}$$

It follows immediately then that if one selects for $\chi(\boldsymbol{\kappa})$ a spherical harmonic $Y_{JM}(\boldsymbol{\kappa})$ which is an eigenfunction of $\mathbf{L}^2$ and L_z with eigenvalues $J(J+1)\hbar^2$ and $M\hbar$, then $\mathbf{V}Y_{JM}(\boldsymbol{\kappa})$ is a vector eigenfunction of $\mathbf{J}^2$ and J_z corresponding to the same eigenvalues:

$$\begin{cases} \mathbf{J}^2(\mathbf{V}Y_{JM}(\boldsymbol{\kappa})) = J(J+1)\,\hbar^2\,\mathbf{V}Y_{JM}(\boldsymbol{\kappa}) \\ J_z(\mathbf{V}Y_{JM}(\boldsymbol{\kappa})) = M\hbar\;\mathbf{V}Y_{JM}(\boldsymbol{\kappa}). \end{cases} \tag{32}$$

It remains now to select a suitable $\mathbf{V}$ so that $\mathbf{V}Y_{JM}$ will be longitudinal or transverse.

c) Longitudinal Eigenfunctions

As a first choice for $\mathbf{V}$, consider the operator of multiplication by $\boldsymbol{\kappa}$. One gets a vector function, certainly longitudinal, which one calls $\mathbf{N}_{JM}$:

$$\mathbf{N}_{JM}(\boldsymbol{\kappa}) = \boldsymbol{\kappa}\,Y_{JM}(\boldsymbol{\kappa}) \tag{33}$$

and which is normalized on the unit sphere.

Consider the action of the parity operator Π on the function $\mathbf{N}_{JM}(\boldsymbol{\kappa})$. Since $\mathbf{N}_{JM}$ is a polar vector field, the operator Π changes the sign of $\mathbf{N}_{JM}$ at the same time it changes its point of application from $\boldsymbol{\kappa}$ to $-\boldsymbol{\kappa}$:

$$(\Pi\mathbf{N}_{JM})(\boldsymbol{\kappa}) = -\,\mathbf{N}_{JM}(-\,\boldsymbol{\kappa}). \tag{34}$$

Knowing that

$$Y_{JM}(-\,\boldsymbol{\kappa}) = (-\,1)^J\,Y_{JM}(\boldsymbol{\kappa}) \tag{35}$$

one finds

$$(\Pi\mathbf{N}_{JM})(\boldsymbol{\kappa}) = (-\,1)^J\,\mathbf{N}_{JM}(\boldsymbol{\kappa}) \tag{36}$$

$\mathbf{N}_{JM}$ thus has parity $(-1)^J$.

d) Transverse Eigenfunctions

As a second choice for $\mathbf{V}$, take the gradient operator on the unit sphere. Such an operator, denoted ∇_{κ}, acts only on the polar angles of $\boldsymbol{\kappa}$. It is related to the ordinary gradient operator ∇ through

$$\nabla = \boldsymbol{\kappa}\frac{\partial}{\partial k} + \frac{1}{k}\,\nabla_{\kappa}. \tag{37}$$

The first term of (37) gives the radial component of the gradient, and the second the component normal to $\boldsymbol{\kappa}$. The result of this is that $\nabla_{\kappa}Y_{JM}(\boldsymbol{\kappa})$ is in the tangent plane of the unit sphere and is therefore a transverse vector function. We introduce

$$\mathbf{Z}_{JM}(\boldsymbol{\kappa}) = \frac{1}{\sqrt{J(J+1)}} \nabla_{\kappa} Y_{JM}(\boldsymbol{\kappa}) \tag{38}$$

where the factor $1/\sqrt{J(J+1)}$ has been introduced for normalization. The transverse functions $\mathbf{Z}_{JM}$ are orthogonal to the longitudinal ones $\mathbf{N}_{JM}$. They are, additionally, orthogonal among themselves. One has

$$\int \mathrm{d}^2\kappa\, \mathbf{Z}^*_{JM}(\boldsymbol{\kappa}) \cdot \mathbf{Z}_{J'M'}(\boldsymbol{\kappa}) = \frac{1}{\sqrt{J(J+1)\,J'(J'+1)}} \int \mathrm{d}^2\kappa (\nabla_{\kappa} Y_{JM})^* \cdot (\nabla_{\kappa} Y_{J'M'}) . \tag{39}$$

An integration by parts gives $-\Delta_{\kappa} Y_{J'M'}$, which is equal to $J'(J'+1)Y_{J'M'}$, so that, taking into account the orthonormality of the spherical harmonics,

$$\int \mathrm{d}^2\kappa\, \mathbf{Z}^*_{JM}(\boldsymbol{\kappa}) \cdot \mathbf{Z}_{J'M'}(\boldsymbol{\kappa}) = \delta_{JJ'}\, \delta_{MM'} . \tag{40}$$

Note finally that since the operator ∇_{κ} is polar, $\mathbf{Z}_{JM}$ like $\mathbf{N}_{JM}$ has parity $(-1)^J$.

As a third choice for $\mathbf{V}$ consider the operator $\boldsymbol{\kappa} \times \nabla_{\kappa}$ and the functions

$$\mathbf{X}_{JM}(\boldsymbol{\kappa}) = \frac{1}{\sqrt{J(J+1)}} (\boldsymbol{\kappa} \times \nabla_{\kappa})\, Y_{JM}(\boldsymbol{\kappa}) \tag{41}$$

where the factor $1/\sqrt{J(J+1)}$ is again introduced for normalization. The function $\mathbf{X}_{JM}$ is always normal to $\mathbf{N}_{JM}$ and $\mathbf{Z}_{JM}$, and thus orthogonal to both these functions (in the sense of the scalar product of functions of $\boldsymbol{\kappa}$). It is a transverse vector function related to $\mathbf{Z}_{JM}$ by

$$\mathbf{X}_{JM}(\boldsymbol{\kappa}) = \boldsymbol{\kappa} \times \mathbf{Z}_{JM}(\boldsymbol{\kappa}) \tag{42}$$

which can be inverted to give

$$\mathbf{Z}_{JM}(\boldsymbol{\kappa}) = -\,\boldsymbol{\kappa} \times \mathbf{X}_{JM}(\boldsymbol{\kappa}) . \tag{43}$$

Note that there exists a simple connection between $\boldsymbol{\kappa} \times \nabla_{\kappa}$ and the angular momentum operator $\mathbf{L}$. Actually, taking into account (37) and (20), $\boldsymbol{\kappa} \times \nabla_{\kappa}$ is nothing more than $\mathbf{k} \times \nabla$, that is to say, $i\mathbf{L}/\hbar$, so that

$$\mathbf{X}_{JM}(\boldsymbol{\kappa}) = \frac{\mathrm{i}}{\hbar\sqrt{J(J+1)}} \mathbf{L} Y_{JM}(\boldsymbol{\kappa}) . \tag{44}$$

The components of $\mathbf{X}_{JM}$ are easy to find as a function of the spherical harmonics, and likewise, through (43), those of $\mathbf{Z}_{JM}$ as well. The orthonormalization of $\mathbf{X}_{JM}$ immediately results from the orthonormalization relations (40) for $\mathbf{Z}_{JM}$, thanks to Equation (42), which relates the integrals of $\mathbf{X}^*_{JM} \cdot \mathbf{X}_{JM}$ to those of $\mathbf{Z}^*_{JM} \cdot \mathbf{Z}_{JM}$. Finally, since the vector $\boldsymbol{\kappa} \times \boldsymbol{\nabla}_\kappa$ is axial, $\mathbf{X}_{JM}$ has opposite parity to $\mathbf{N}_{JM}$ and $\mathbf{Z}_{JM}$, i.e., $(-1)^{J+1}$.

The three families of functions **N**, **X**, and **Z** at each point on the unit sphere form a rectangular coordinate system on which one can project any vector field; that is, they form a basis for vector functions on the unit sphere. If one keeps only the fields **X** and **Z**, one generates all the transverse fields.

More precisely, the set of functions

$$\boldsymbol{\phi}_{k_0JMX}(\mathbf{k}) = \frac{1}{k_0}\delta(k - k_0)\,\mathbf{X}_{JM}(\boldsymbol{\kappa}) \tag{45.a}$$

$$\boldsymbol{\phi}_{k_0JMZ}(\mathbf{k}) = \frac{1}{k_0}\delta(k - k_0)\,\mathbf{Z}_{JM}(\boldsymbol{\kappa}) \tag{45.b}$$

form a basis for the transverse fields in reciprocal space. Giving the eigenvalues of energy $[\hbar c k_0]$, those of $\mathbf{J}^2$ and J_z $[J(J+1)\hbar^2$ and $M\hbar]$, and the parity $[(-1)^{J+1}$ or $(-1)^J]$ unambiguously specifies the corresponding eigenfunction (45.a) or (45.b). The energy, the total angular momentum ($\mathbf{J}^2$ and J_z), and the parity thus form a complete set of commuting observables for the photon. Every state $\boldsymbol{\alpha}(\mathbf{k})$ of the transverse field can be expanded in only one way on the basis (45):

$$\boldsymbol{\alpha}(\mathbf{k}) = \int_0^\infty dk_0 \sum_{J=1}^{\infty} \sum_{M=-J}^{+J} \times \left\{ \alpha_{k_0JMX}\,\boldsymbol{\phi}_{k_0JMX}(\mathbf{k}) + \alpha_{k_0JMZ}\,\boldsymbol{\phi}_{k_0JMZ}(\mathbf{k}) \right\}. \tag{46}$$

The coefficients α_{k_0JMX} or α_{k_0JMZ} in this expansion (and their complex conjugates) become, after quantization, the destruction (and creation) operators for a photon with energy $\hbar c k_0$, angular momenta $J(J+1)\hbar^2$ and $M\hbar$, and parity $(-1)^{J+1}$ or $(-1)^J$.

Remark

One does not have a function **X** or **Z** with $J = 0$, since Y_{00} is a constant and $\boldsymbol{\nabla}_\kappa Y_{00} = \mathbf{0}$. This is why the sum on J starts with $J = 1$ in (46).

4. Application: Multipole Waves in Real Space

When all the expansion coefficients in (46) are zero except for one, the function $\boldsymbol{\alpha}(\mathbf{k})$ reduces to the eigenfunction $\boldsymbol{\phi}_{k_0JMX}(\mathbf{k})$ or $\boldsymbol{\phi}_{k_0JMZ}(\mathbf{k})$. In this section the structure of the electromagnetic waves gotten by replacing $\boldsymbol{\alpha}(\mathbf{k})$ by $\boldsymbol{\phi}_{k_0JMX}(\mathbf{k})$ or $\boldsymbol{\phi}_{k_0JMZ}(\mathbf{k})$ in (C.19) and (C.20), giving $\mathbf{E}_\perp(\mathbf{r}, t)$ and $\mathbf{B}(\mathbf{r}, t)$ as functions of $\boldsymbol{\alpha}(\mathbf{k})$, will be considered. Such waves, called multipole waves, are the waves associated with photons whose energy, angular momentum, and parity are well defined.

a) Evaluation of Some Fourier Transforms

When one replaces $\Sigma_\varepsilon \alpha_\varepsilon(\mathbf{k})\boldsymbol{\varepsilon} = \boldsymbol{\alpha}(\mathbf{k})$ by the eigenfunction (45a) or (45b) in the equation (C.19) for $\mathbf{E}_\perp$, one sees immediately that one needs the Fourier transforms for these functions. For this it is useful to recall the expression for the expansion of a plane wave in spherical waves (*)

$$e^{i\mathbf{k}.\mathbf{r}} = 4\pi \sum_{l=0}^{\infty} \sum_{m=-l}^{+l} (i)^l j_l(kr)\, Y_{lm}^*(\boldsymbol{\kappa})\, Y_{lm}(\boldsymbol{\rho}) \tag{47}$$

where

$$\boldsymbol{\rho} = \frac{\mathbf{r}}{r} \tag{48}$$

is the unit vector in the direction of $\mathbf{r}$ and where $j_l(kr)$ is the spherical Bessel function of order l. From (47) it follows that

$$\int d^3k\, e^{i\mathbf{k}.\mathbf{r}} \frac{1}{k_0} \delta(k - k_0)\, Y_{lm}(\boldsymbol{\kappa}) = 4\pi(i)^l k_0\, j_l(k_0 r)\, Y_{lm}(\boldsymbol{\rho}) \tag{49}$$

which will be used later.

Let us evaluate first of all the Fourier transform of $\boldsymbol{\phi}_{k_0JMX}(\mathbf{k})$, which is denoted

$$\mathbf{I}_{k_0JMX}(\mathbf{r}) = \text{F.T.}\,\{\,\boldsymbol{\phi}_{k_0JMX}(\mathbf{k})\,\} =$$

$$= \frac{1}{(2\pi)^{3/2}} \int d^3k\, e^{i\mathbf{k}.\mathbf{r}} \frac{1}{k_0} \delta(k - k_0)\,(\boldsymbol{\kappa} \times \nabla_\kappa) \frac{Y_{JM}(\boldsymbol{\kappa})}{\sqrt{J(J+1)}}. \tag{50}$$

Since $\boldsymbol{\kappa} \times \nabla_\kappa$ does not act on k, $\delta(k - k_0)$ can be put to the right of $\boldsymbol{\kappa} \times \nabla_\kappa$, which is written as $\mathbf{k} \times \nabla$ following (37). This yields then, using Table I in §$B_I.2$,

$$\mathbf{I}_{k_0JMX}(\mathbf{r}) = \text{F.T.}\left\{ (\mathbf{k} \times \nabla) \frac{1}{k_0} \delta(k - k_0) \frac{Y_{JM}(\boldsymbol{\kappa})}{\sqrt{J(J+1)}} \right\}$$

$$= \frac{1}{k_0} (\mathbf{r} \times \nabla)\, \text{F.T.} \left\{ \delta(k - k_0) \frac{Y_{JM}(\boldsymbol{\kappa})}{\sqrt{J(J+1)}} \right\}. \tag{51}$$

(*) See, for example, Cohen-Tannoudji, Diu, and Laloe, Complement A_{VIII}.

Using (49), the calculation of the Fourier transform of the expression in braces yields

$$\mathbf{I}_{k_0JMX}(\mathbf{r}) = \frac{k_0}{(2\,\pi)^{3/2}}\,(\mathbf{r}\times\boldsymbol{\nabla})\left[4\,\pi(i)^J\,j_J(k_0\,r)\,\frac{Y_{JM}(\boldsymbol{\rho})}{\sqrt{J(J+1)}}\right]. \qquad (52)$$

Since $\mathbf{r}\times\boldsymbol{\nabla}$ only acts on $\boldsymbol{\rho}$, the expression (41) for $\mathbf{X}_{JM}(\boldsymbol{\rho})$ arises, and finally

$$\mathbf{I}_{k_0JMX}(\mathbf{r}) = \frac{4\,\pi}{(2\,\pi)^{3/2}}\,k_0(\mathrm{i})^J\,j_J(k_0\,r)\,\mathbf{X}_{JM}(\boldsymbol{\rho})\,. \qquad (53)$$

The calculation of

$$\mathbf{I}_{k_0JMZ}(\mathbf{r}) = \mathrm{F.T.}\left\{\frac{1}{k_0}\,\delta(k-k_0)\,\nabla_{\kappa}\,\frac{Y_{JM}(\boldsymbol{\kappa})}{\sqrt{J(J+1)}}\right\} \qquad (54)$$

is a little more complex. Since $Y_{JM}(\boldsymbol{\kappa})$ does not depend on k, the expression in braces can be simplified using (37):

$$\frac{1}{k_0}\,\delta(k-k_0)\,\nabla_{\kappa}\;Y_{JM}(\boldsymbol{\kappa}) = \delta(k-k_0)\,\nabla Y_{JM}(\boldsymbol{\kappa})$$

$$= \nabla\left[\delta(k-k_0)\;Y_{JM}(\boldsymbol{\kappa})\right] - \frac{\mathbf{k}}{k}\,\delta'(k-k_0)\;Y_{JM}(\boldsymbol{\kappa})\,. \qquad (55)$$

This then yields, using Table I in §B_I.2,

$$\mathbf{I}_{k_0JMZ}(\mathbf{r}) = \frac{1}{\sqrt{J(J+1)}}\left\{-\,\mathrm{i}\mathbf{r}\,\mathrm{F.T.}\left[\delta(k-k_0)\;Y_{JM}(\boldsymbol{\kappa})\right] + \right.$$

$$\left. +\;\mathrm{i}\boldsymbol{\nabla}\;\mathrm{F.T.}\left[\frac{1}{k}\,\delta'(k-k_0)\;Y_{JM}(\boldsymbol{\kappa})\right]\right\}. \qquad (56)$$

One next calculates the two Fourier transforms of (56). The first is directly given by (49) to within a factor $k_0/(2\pi)^{3/2}$. To evaluate the second, note that in the integrand of (49) $(1/k_0)\delta(k-k_0)$ can also be written $(1/k)\delta(k-k_0)$. Taking the derivative with respect to k_0 then gives the desired Fourier transform except for the sign:

$$\mathbf{I}_{k_0JMZ}(\mathbf{r}) = \frac{4\,\pi(\mathrm{i})^J}{(2\,\pi)^{3/2}\sqrt{J(J+1)}}\left\{(-\,\mathrm{i}\mathbf{r})\,k_0^2\,j_J(k_0\,r)\;Y_{JM}(\boldsymbol{\rho}) + \right.$$

$$\left. +\;(-\,\mathrm{i}\boldsymbol{\nabla})\,\frac{\partial}{\partial k_0}\left[k_0\,j_J(k_0\,r)\;Y_{JM}(\boldsymbol{\rho})\right]\right\}. \qquad (57)$$

The action of the operator $\boldsymbol{\nabla}$ is seen more explicitly if one separates the action of its radial and angular parts and introduces the dimensionless variable $x = k_0 r$:

$$\boldsymbol{\nabla} = \boldsymbol{\rho}\,\frac{\partial}{\partial r} + \frac{1}{r}\,\boldsymbol{\nabla}_{\rho} = k_0\,\boldsymbol{\rho}\,\frac{\partial}{\partial x} + k_0\,\frac{1}{x}\,\boldsymbol{\nabla}_{\rho} \qquad (58)$$

and

$$\nabla \frac{\partial}{\partial k_0}[k_0 j_J(k_0 r) Y_{JM}(\boldsymbol{\rho})] =$$

$$= \left(k_0 \boldsymbol{\rho} \frac{\partial}{\partial x} + k_0 \frac{1}{x} \nabla_\rho \right) \frac{\partial}{\partial x}(x j_J(x))\, Y_{JM}(\boldsymbol{\rho}) . \tag{59}$$

Referring to (33) and (38), one sees that (57) contains $\mathbf{N}_{JM}(\rho)$ and $\mathbf{Z}_{JM}(\rho)$. The term between the braces in (57) can be written as $-ik_0$ multiplied by

$$\left[x_{jJ}(x) + \frac{\partial^2}{\partial x^2}(x j_J(x)) \right] \mathbf{N}_{JM}(\boldsymbol{\rho}) + \frac{1}{x}\frac{\partial}{\partial x}(x j_J(x))\, \mathbf{Z}_{JM}(\boldsymbol{\rho}) \sqrt{J(J+1)} . \tag{60}$$

Since $j_J(x)$ is the solution of the radial equation

$$j_J''(x) + \frac{2}{x} j_J'(x) + j_J(x) - \frac{J(J+1)}{x^2} j_J(x) = 0 \tag{61}$$

the coefficient of $\mathbf{N}_{JM}(\rho)$ is simply $(1/x) J(J+1) j_J(x)$. After this simplification, Equations (60) and (57) give

$$\mathbf{I}_{k_0 JMZ}(\mathbf{r}) = \frac{4\pi(\mathrm{i})^{J-1}}{(2\pi)^{3/2} r} \left\{ \sqrt{J(J+1)}\, j_J(k_0 r)\, \mathbf{N}_{JM}(\boldsymbol{\rho}) + \right.$$

$$\left. + \frac{\mathrm{d}}{\mathrm{d}(k_0 r)} [k_0 r\, j_J(k_0 r)]\, \mathbf{Z}_{JM}(\boldsymbol{\rho}) \right\} . \tag{62}$$

b) Electric Multipole Waves

These waves correspond to $\boldsymbol{\alpha}(\mathbf{k}) = \boldsymbol{\phi}_{k_0 JMZ}(\mathbf{k})$. From (C.19) and (C.20), $\mathbf{E}_\perp(\mathbf{r}, t)$ and $\mathbf{B}(\mathbf{r}, t)$ are then given by

$$\mathbf{E}_{\perp k_0 JMZ}(\mathbf{r}, t) = \mathrm{i}(2\pi)^{3/2} \mathscr{E}_{\omega_0} \mathbf{I}_{k_0 JMZ}(\mathbf{r})\, \mathrm{e}^{-\mathrm{i}\omega_0 t} + c.c.$$

$$c\mathbf{B}_{k_0 JMZ}(\mathbf{r}, t) = \mathrm{i}(2\pi)^{3/2} \mathscr{E}_{\omega_0} \mathbf{I}_{k_0 JMX}(\mathbf{r})\, \mathrm{e}^{-\mathrm{i}\omega_0 t} + c.c. \tag{63}$$

One assumes the field to be free so that the temporal evolution is purely harmonic, and takes $c\mathbf{B}$ rather than $\mathbf{B}$ so that the dimensions are the same as for $\mathbf{E}_\perp$.

The magnetic field at $\mathbf{r}$ is perpendicular to the vector $\mathbf{r}$, like $\mathbf{I}_{k_0 JMX}(\mathbf{r})$, which is proportional to $\mathbf{X}_{JM}(\rho)$. The expression for the electric field $\mathbf{E}_\perp$ contains both $\mathbf{Z}_{JM}(\rho)$ and $\mathbf{N}_{JM}(\rho)$ [see (62) for $\mathbf{I}_{k_0 JMZ}(\mathbf{r})$]. $\mathbf{E}_\perp(\mathbf{r})$ is then not normal to $\mathbf{r}$. This is indeed necessary if one wants $\mathbf{E}_\perp \times \mathbf{B}$ to have a nonzero moment with respect to the origin. Furthermore, $\mathbf{E}_\perp$ and $\mathbf{B}$ are normal to one another.

Near the origin ($k_0 r \ll 1$)

$$j_J(k_0 r) \simeq \frac{(k_0 r)^J}{(2J+1)!!} \tag{64}$$

so that

$$\mathbf{E}_\perp(\mathbf{r}, t) \simeq 4\pi k_0 \mathscr{E}_{\omega 0} \frac{1}{k_0 r} \frac{(k_0 r)^J}{(2J+1)!!} \times$$
$$\times \left\{ (\mathrm{i})^J [\sqrt{J(J+1)}\, \mathbf{N}_{JM}(\boldsymbol{\rho}) + (J+1)\, \mathbf{Z}_{JM}(\boldsymbol{\rho})]\, \mathrm{e}^{-\mathrm{i}\omega_0 t} + c.c. \right\}$$

$$c\mathbf{B}(\mathbf{r}, t) \simeq 4\pi k_0 \mathscr{E}_{\omega 0} \frac{(k_0 r)^J}{(2J+1)!!} \left\{ (\mathrm{i})^{J+1}\, \mathbf{X}_{JM}(\boldsymbol{\rho})\, \mathrm{e}^{-\mathrm{i}\omega_0 t} + c.c. \right\}. \tag{65}$$

The functions **N**, **X**, and **Z** are on average of the same order of magnitude on the unit sphere. One has then

$$cB \simeq \frac{1}{J} k_0 r E_\perp . \tag{66}$$

Compared to a plane wave with the same electric field, the magnetic field (66) is smaller by a factor of $k_0 r/J$ near the origin. In the neighborhood of the origin such waves are mainly coupled to the electric multipole moments, hence the name electric multipole waves.

At large distance ($k_0 r \gg 1$)

$$j_J(k_0 r) \simeq \frac{1}{k_0 r} \sin\left(k_0 r - J\frac{\pi}{2}\right) \tag{67}$$

and the asymptotic forms of the fields are

$$\mathbf{E}_\perp(\mathbf{r}, t) \simeq 4\pi \mathscr{E}_{\omega 0} \left\{ \left[\frac{1}{r} \cos\left(k_0 r - J\frac{\pi}{2}\right) \mathbf{Z}_{JM}(\boldsymbol{\rho}) + \right.\right.$$
$$\left.\left. + \frac{\sqrt{J(J+1)}}{k_0 r^2} \sin\left(k_0 r - J\frac{\pi}{2}\right) \mathbf{N}_{JM}(\boldsymbol{\rho}) \right] (\mathrm{i})^J\, \mathrm{e}^{-\mathrm{i}\omega_0 t} + c.c. \right\} \tag{68}$$

$$c\mathbf{B}(\mathbf{r}, t) \simeq 4\pi \mathscr{E}_{\omega 0} \left\{ \frac{1}{r} \sin\left(k_0 r - J\frac{\pi}{2}\right) \mathbf{X}_{JM}(\boldsymbol{\rho})\, (\mathrm{i})^{J+1}\, \mathrm{e}^{-\mathrm{i}\omega_0 t} + c.c. \right\}.$$

The radial part of **E** decreases as $1/r^2$, so that at infinity only the wave decreasing as $1/r$ remains, with the structure of a stationary plane wave, transverse in **r**-space.

c) Magnetic Multipole Waves

The waves associated with $\phi_{k_0JMX}(\mathbf{k})$ are called magnetic multipole waves. The corresponding free fields $\mathbf{E}_\perp$ and $\mathbf{B}$ are given by

$$\mathbf{E}_{\perp k_0 JMX}(\mathbf{r}, t) = \mathrm{i}(2\,\pi)^{3/2}\,\mathscr{E}_{\omega_0}\,\mathbf{I}_{k_0 JMX}(\mathbf{r})\,\mathrm{e}^{-\mathrm{i}\omega_0 t} + c.c.$$

$$c\mathbf{B}_{k_0 JMX}(\mathbf{r}, t) = -\,\mathrm{i}(2\,\pi)^{3/2}\,\mathscr{E}_{\omega_0}\,\mathbf{I}_{k_0 JMZ}(\mathbf{r})\,\mathrm{e}^{-\mathrm{i}\omega_0 t} + c.c.\,. \tag{69}$$

In comparing (69) and (63) one discovers that one has simply inverted $\mathbf{E}_\perp$ and $c\mathbf{B}$ and changed the sign of one of the two fields. All of the conclusions of the previous section can be carried over without any difficulty.

General references and further reading

Akhiezer and Berestetskii (Chapter I), Berestetskii, Lifshitz, and Pitayevski (Chapter 1), Jackson (Chapter 16), Blatt and Weisskopf (Appendix B).

COMPLEMENT C_I

EXERCISES

Exercise 1. H and **P** as constants of the motion.
Exercise 2. Transformation from the Coulomb gauge to the Lorentz gauge.
Exercise 3. Cancellation of the longitudinal electric field by the instantaneous transverse field.
Exercise 4. Normal variables and retarded potentials.
Exercise 5. Field created by a charged particle at its own position. Radiation reaction.
Exercise 6. Field produced by an oscillating electric dipole.
Exercise 7. Cross-section for scattering of radiation by a classical elastically bound electron.

1. H AND $\mathbf{P}$ AS CONSTANTS OF THE MOTION

a) Show that the energy of the system particles + electromagnetic field given by

$$H = \sum_\alpha \frac{1}{2} m_\alpha \mathbf{v}_\alpha^2 + \frac{\varepsilon_0}{2} \int \mathrm{d}^3r[\mathbf{E}^2 + c^2 \mathbf{B}^2] \tag{1}$$

is a constant of the motion.

b) Derive the same result using the expansion of the transverse field in normal variables.

c) Show that the total momentum

$$\mathbf{P} = \sum_\alpha m_\alpha \mathbf{v}_\alpha + \varepsilon_0 \int \mathrm{d}^3r\, \mathbf{E} \times \mathbf{B} \tag{2}$$

is also a constant of the motion.

Solution

a) One calculates

$$\frac{\mathrm{d}H}{\mathrm{d}t} = \sum_\alpha m_\alpha \mathbf{v}_\alpha \cdot \frac{\mathrm{d}\mathbf{v}_\alpha}{\mathrm{d}t} + \varepsilon_0 \int \mathrm{d}^3r \left[\mathbf{E} \cdot \frac{\partial \mathbf{E}}{\partial t} + c^2\, \mathbf{B} \cdot \frac{\partial \mathbf{B}}{\partial t}\right] \tag{3}$$

and substitutes for $\mathrm{d}\mathbf{v}_\alpha/\mathrm{d}t$ using the Lorentz equation and for $\partial\mathbf{E}/\partial t$ and $\partial\mathbf{B}/\partial t$ using Maxwell's equations. This gives

$$\frac{\mathrm{d}H}{\mathrm{d}t} = \sum_\alpha \mathbf{v}_\alpha \cdot (q_\alpha \mathbf{E}(\mathbf{r}_\alpha, t)) + \varepsilon_0 \int \mathrm{d}^3r \left[\mathbf{E} \cdot \left(c^2\, \boldsymbol{\nabla} \times \mathbf{B} - \frac{1}{\varepsilon_0}\mathbf{j}\right) - c^2\, \mathbf{B} \cdot (\boldsymbol{\nabla} \times \mathbf{E})\right] \tag{4}$$

since $\mathbf{v}_\alpha \cdot (\mathbf{v}_\alpha \times \mathbf{B}) = 0$.

Substitute (A.5.b) for $\mathbf{j}$ into $\int d^3r\,\mathbf{E}\cdot\mathbf{j}$. One gets

$$\int d^3r\,\mathbf{E}\cdot\mathbf{j} = \sum_\alpha \mathbf{E}(\mathbf{r}_\alpha, t)\cdot(q_\alpha\,\mathbf{v}_\alpha)\,. \tag{5}$$

Now evaluate

$$\int d^3r[\mathbf{B}\cdot(\nabla\times\mathbf{E}) - \mathbf{E}\cdot(\nabla\times\mathbf{B})] = \int d^3r\,\nabla\cdot(\mathbf{E}\times\mathbf{B}) \tag{6}$$

which can be transformed into a surface integral which is zero when the fields are zero at infinity. Using (4), (5), and (6), one gets

$$\frac{dH}{dt} = 0\,. \tag{7}$$

b) In terms of normal variables, the energy is given by

$$H = \sum_\alpha \frac{1}{2} m_\alpha\,\mathbf{v}_\alpha^2 + \int d^3k\,\frac{\rho^*\,\rho}{2\,\varepsilon_0\,k^2} + \sum_\varepsilon \int d^3k\,\frac{\hbar\omega}{2}\,[\alpha_\varepsilon^*\,\alpha_\varepsilon + \alpha_\varepsilon\,\alpha_\varepsilon^*] \tag{8}$$

Calculate dH/dt. Here $d\mathbf{v}_\alpha/dt$ is given by the Lorentz equation. As a result of charge conservation $\dot{\rho}$ is equal to $-i\mathbf{k}\cdot\boldsymbol{j}$. Equation (C.8) gives $\dot{\alpha}_\varepsilon$. One then gets

$$\frac{dH}{dt} = \sum_\alpha q_\alpha\,\mathbf{v}_\alpha\cdot\mathbf{E}(\mathbf{r}_\alpha, t) + \int\frac{d^3k}{2\,\varepsilon_0\,k^2}[i(\mathbf{k}\cdot\boldsymbol{j}^*)\,\rho - i(\mathbf{k}\cdot\boldsymbol{j})\,\rho^*] -$$

$$- \frac{i}{2\,\varepsilon_0}\sum_\varepsilon \int \frac{d^3k}{\mathcal{N}(\mathbf{k})}\,\hbar\omega[(\varepsilon\cdot\boldsymbol{j}^*)\,\alpha_\varepsilon - (\varepsilon\cdot\boldsymbol{j})\,\alpha_\varepsilon^*] \tag{9}$$

Now, using (B.19), we write $\mathscr{E}_\parallel = -i\rho\mathbf{k}/\varepsilon_0 k^2$. The second term of (9) can then be written

$$-\frac{1}{2}\int d^3k(\mathscr{E}_\parallel\cdot\boldsymbol{j}^* + \mathscr{E}_\parallel^*\cdot\boldsymbol{j}) = -\int d^3r\,\mathbf{E}_\parallel\cdot\mathbf{j}\,. \tag{10}$$

To evaluate the last term of (9) note that $\boldsymbol{j}^*(-\mathbf{k}) = \boldsymbol{j}(\mathbf{k})$. Since $\mathcal{N}(-\mathbf{k}) = \mathcal{N}(\mathbf{k})$, one can write the last term as

$$\frac{i}{2\,\varepsilon_0}\sum_\varepsilon \int \frac{d^3k}{\mathcal{N}(\mathbf{k})}\,\hbar\omega[\alpha_\varepsilon^*(\mathbf{k}) - \alpha_\varepsilon(-\mathbf{k})]\,\varepsilon\cdot\boldsymbol{j}(\mathbf{k})\,. \tag{11}$$

Using $\mathcal{N}(\mathbf{k}) = \sqrt{\hbar\omega/2\varepsilon_0}$ and (C.6.a), (11) becomes

$$-\int d^3k\,\mathscr{E}_\perp^*\cdot\boldsymbol{j} = -\int d^3r\,\mathbf{E}_\perp\cdot\mathbf{j}\,. \tag{12}$$

One can see that (10) and (12) cancel the first term of (9).

c) Calculate $d\mathbf{P}/dt$ and use the Maxwell–Lorentz equations to reexpress $d\mathbf{v}_\alpha/dt$, $\partial\mathbf{E}/\partial t$, and $\partial\mathbf{B}/\partial t$:

$$\frac{d\mathbf{P}}{dt} = \sum_\alpha m_\alpha\frac{d\mathbf{v}_\alpha}{dt} + \varepsilon_0\int d^3r\left[\frac{\partial\mathbf{E}}{\partial t}\times\mathbf{B} + \mathbf{E}\times\frac{\partial\mathbf{B}}{\partial t}\right] \tag{13}$$

$$\frac{d\mathbf{P}}{dt} = \sum_\alpha q_\alpha\,\mathbf{E}(\mathbf{r}_\alpha, t) + \sum_\alpha q_\alpha\,\mathbf{v}_\alpha\times\mathbf{B}(\mathbf{r}_\alpha, t) +$$

$$+ \varepsilon_0\int d^3r\left[\left(c^2\,\nabla\times\mathbf{B} - \frac{1}{\varepsilon_0}\mathbf{j}\right)\times\mathbf{B} - \mathbf{E}\times(\nabla\times\mathbf{E})\right] \tag{14}$$

Now Equation (A.5.b) gives

$$-\int d^3r\ \mathbf{j} \times \mathbf{B} = -\sum_\alpha q_\alpha\ \mathbf{v}_\alpha \times \mathbf{B}(\mathbf{r}_\alpha, t) \tag{15}$$

which cancels the second term of (14). In addition, for a vector field **X**

$$\mathbf{X} \times (\nabla \times \mathbf{X}) = \frac{1}{2}\nabla(\mathbf{X}^2) - (\mathbf{X} \cdot \nabla)\,\mathbf{X}$$

$$= \frac{1}{2}\nabla(\mathbf{X}^2) - \sum_j \mathbf{e}_j\,\nabla \cdot (X_j \mathbf{X}) + \mathbf{X}(\nabla \cdot \mathbf{X}) \tag{16}$$

The integral over all space of the first two terms of the right-hand side of (16) can be transformed into a surface integral which is zero if the field decreases sufficiently rapidly at infinity. As for the last term, it is zero for $\mathbf{X} = \mathbf{B}$ and is equal to $\mathbf{E}\rho/\varepsilon_0$ for $\mathbf{X} = \mathbf{E}$. Using (A.5.a),

$$-\,\varepsilon_0 \int d^3r\ \mathbf{E} \times (\nabla \times \mathbf{E}) = -\sum_\alpha q_\alpha\,\mathbf{E}(\mathbf{r}_\alpha, t) \tag{17}$$

which cancels the first term of (14). Finally, Equations (14), (15), and (16) give $d\mathbf{P}/dt = \mathbf{0}$.

2. Transformation from the Coulomb gauge to the Lorentz gauge

Assume the potentials $\mathscr{A}$ and $\mathscr{U}$ are known in the Coulomb gauge. What equation must be satisfied by the gauge function $\mathscr{F}$ which transforms $\mathscr{A}$ and $\mathscr{U}$ into the potentials in the Lorentz gauge?

Solution

Let $\mathscr{A}'$ and $\mathscr{U}'$ be the new potentials. The Lorentz condition (A.13) in reciprocal space is

$$i\mathbf{k} \cdot \mathscr{A}' + \frac{\dot{\mathscr{U}}'}{c^2} = 0 \tag{1}$$

$\mathscr{A}'$ and $\mathscr{U}'$ can be written as functions of $\mathscr{A}$ and $\mathscr{U}$ using (B.8). Combining (1) and (B.8), one gets

$$i\mathbf{k} \cdot (\mathscr{A} + i\mathbf{k}\mathscr{F}) + \frac{1}{c^2}(\dot{\mathscr{U}} - \ddot{\mathscr{F}}) = 0\,. \tag{2}$$

The transversality of $\mathscr{A}$ in the Coulomb gauge gives then

$$\frac{1}{c^2}\ddot{\mathscr{F}} + k^2\,\mathscr{F} = \frac{\dot{\mathscr{U}}}{c^2} \tag{3}$$

so that, using the expression for $\mathscr{U}$ in the Coulomb gauge,

$$\frac{1}{c^2}\ddot{\mathscr{F}} + k^2\,\mathscr{F} = \sum_\alpha \frac{-\,q_\alpha(i\mathbf{k} \cdot \mathbf{v}_\alpha)\,e^{-i\mathbf{k}.\mathbf{r}_\alpha}}{(2\,\pi)^{3/2}\,\varepsilon_0\,c^2\,k^2}\,. \tag{4}$$

It is clear that $\mathscr{F}$ depends on the velocities $\mathbf{v}_\alpha$. It is for this reason that transformation from the Coulomb gauge to the Lorentz gauge is not equivalent to a change in the Lagrangian (see Remark i of §B.3.*b*, Chapter II).

Note finally that Equation (3) can be written

$$\Box F = \frac{\dot{U}}{c^2}\,. \tag{5}$$

in real space.

3. Cancellation of the Longitudinal Electric Field by the Instantaneous Transverse Field

A particle with charge q_α is located at the origin $\mathbf{0}$ of the coordinates. In the interval 0 to T the particle is displaced from $\mathbf{0}$ to $\mathbf{r}_\alpha(T)$ along a path $\mathbf{r}_\alpha(t)$ $(0 \le t \le T)$. Let $\mathbf{r}$ be a point distant from the origin $(r \gg |\mathbf{r}_\alpha(t)|, cT)$. The purpose of this exercise is to prove, starting with Maxwell's equations, that the instantaneous variations of the longitudinal electric field created by the charge q_α at $\mathbf{r}$ are exactly compensated by the instantaneous component of the transverse electric field produced by the displacement of the particle.

a) Calculate, as a function of $\mathbf{r}_\alpha(t)$, the longitudinal electric field $\mathbf{E}_\parallel(\mathbf{r}, t)$ at point $\mathbf{r}$ and time t from charge q_α. Show that $\mathbf{E}_\parallel(\mathbf{r}, t)$ can be written

$$\mathbf{E}_\parallel(\mathbf{r}, t) = \mathbf{E}_\parallel(\mathbf{r}, 0) + \delta\mathbf{E}_\parallel(\mathbf{r}, t)$$

where $\delta\mathbf{E}_\parallel$ is given by a power series in $|\mathbf{r}_\alpha(t)|/r$. Show that the lowest-order term of this expansion can be expressed as a function of $q_\alpha\mathbf{r}_\alpha(t)$ and of the transverse delta function $\delta_{ij}^\perp(\mathbf{r})$.

b) Find the current $\mathbf{j}(\mathbf{r}, t)$ associated with the motion of the particle. Express the transverse current $\mathbf{j}_\perp(\mathbf{r}, t)$ at the point of observation $\mathbf{r}$ as a function of $q_\alpha\dot{\mathbf{r}}_\alpha(t)$ and the transverse delta function $\delta_{ij}^\perp(\mathbf{r} - \mathbf{r}_\alpha(t))$. Show that to the lowest order in $|\mathbf{r}_\alpha(t)|/r$, one can replace $\delta_{ij}^\perp(\mathbf{r} - \mathbf{r}_\alpha(t))$ by $\delta_{ij}^\perp(\mathbf{r})$. Write the Maxwell equation giving $\partial\mathbf{E}_\perp(\mathbf{r}, t)/\partial t$ as a function of $\mathbf{j}_\perp(\mathbf{r}, t)$ and $\mathbf{B}(\mathbf{r}, t)$. Begin by ignoring the contribution of $\mathbf{B}$ to $\partial\mathbf{E}_\perp/\partial t$. Integrate the equation between 0 and t. Show that the transverse electric field $\mathbf{E}_\perp(\mathbf{r}, t)$ produced by $\mathbf{j}_\perp(\mathbf{r}, t)$ compensates exactly (to the lowest order in $|\mathbf{r}_\alpha(t)|/r$) the field $\delta\mathbf{E}_\parallel(\mathbf{r}, t)$ found in part *a*).

c) By eliminating the transverse electric field between the Maxwell equations for the transverse fields, find the equation of motion of the magnetic field $\mathbf{B}$. Show that the source term in this equation can be written in a form which only involves the total current $\mathbf{j}$. Justify the approximation made above of neglecting the contribution of $\mathbf{B}$ to $\partial\mathbf{E}_\perp/\partial t$ over short periods.

Solution

a) From §B.4, $\mathbf{E}_\parallel(\mathbf{r}, t)$ is the Coulomb field created by the charge q_α at time t:

$$\mathbf{E}_\parallel(\mathbf{r}, t) = -\nabla U(\mathbf{r}, t) \tag{1}$$

with

$$U(\mathbf{r}, t) = \frac{q_\alpha}{4\pi\varepsilon_0}\frac{1}{|\mathbf{r} - \mathbf{r}_\alpha(t)|}. \tag{2}$$

Expand U in powers of $|\mathbf{r}_\alpha(t)|/r$:

$$U(\mathbf{r}, t) = \frac{q_\alpha}{4\pi\varepsilon_0} \frac{1}{r - \dfrac{\mathbf{r}\cdot\mathbf{r}_\alpha(t)}{r} + \cdots} =$$

$$= \frac{q_\alpha}{4\pi\varepsilon_0}\left[\frac{1}{r} + \frac{\mathbf{r}\cdot\mathbf{r}_\alpha(t)}{r^3} + \cdots\right]. \tag{3}$$

Substituting (3) in (1), one gets for the ith component of $\mathbf{E}_{\|}(\mathbf{r}, t)$ ($i = x, y, z$)

$$E_{\|i}(\mathbf{r}, t) = E_{\|i}(\mathbf{r}, 0) + \delta E_{\|i}(\mathbf{r}, t) \tag{4}$$

where

$$\delta E_{\|i}(\mathbf{r}, t) = -\frac{q_\alpha}{4\pi\varepsilon_0}\frac{\partial}{\partial r_i}\sum_j \frac{r_j r_{\alpha j}(t)}{r^3} =$$

$$= -\frac{q_\alpha}{4\pi\varepsilon_0}\sum_j\left(\delta_{ij} - \frac{3 r_i r_j}{r^2}\right)\frac{r_{\alpha j}(t)}{r^3}. \tag{5}$$

Comparison of (5) with (B.17.b) gives, taking account of the fact that r is nonzero,

$$\delta E_{\|i}(\mathbf{r}, t) = \frac{1}{\varepsilon_0}\sum_j \delta_{ij}^{\perp}(\mathbf{r})\, q_\alpha\, r_{\alpha j}(t). \tag{6}$$

b) Following (A.5.b),

$$\mathbf{j}(\mathbf{r}, t) = q_\alpha\, \dot{\mathbf{r}}_\alpha(t)\, \delta(\mathbf{r} - \mathbf{r}_\alpha(t)). \tag{7}$$

(B.16) then gives

$$j_{\perp i}(\mathbf{r}, t) = \sum_j \int \mathrm{d}^3 r'\, \delta_{ij}^{\perp}(\mathbf{r} - \mathbf{r}')\, j_j(\mathbf{r}', t)$$

$$= \sum_j \delta_{ij}^{\perp}(\mathbf{r} - \mathbf{r}_\alpha(t))\, q_\alpha\, \dot{r}_{\alpha j}(t) \tag{8}$$

Since $|\mathbf{r} - \mathbf{r}_\alpha(t)|$ is nonzero, $\delta_{ij}^{\perp}(\mathbf{r} - \mathbf{r}_\alpha(t))$ varies as $|\mathbf{r} - \mathbf{r}_\alpha(t)|^{-3}$, which in the lowest order in $|\mathbf{r}_\alpha(t)|/r$ goes as r^{-3}. One can write then to this order

$$j_{\perp i}(\mathbf{r}, t) = \sum_j \delta_{ij}^{\perp}(\mathbf{r})\, q_\alpha\, \dot{r}_{\alpha j}(t). \tag{9}$$

The Maxwell equation (B.49.b) in real space is written

$$\frac{\partial \mathbf{E}_\perp(\mathbf{r}, t)}{\partial t} = c^2\, \nabla \times \mathbf{B}(\mathbf{r}, t) - \frac{1}{\varepsilon_0}\, \mathbf{j}_\perp(\mathbf{r}, t). \tag{10}$$

For $t < 0$, the vectors $\mathbf{E}_\perp$, $\mathbf{B}$, and $\mathbf{j}_\perp$ are zero. If one ignores the contribution of $\mathbf{B}$, the integral of (10) between 0 and t gives

$$E_{\perp i}(\mathbf{r}, t) = -\frac{1}{\varepsilon_0}\int_0^t \mathrm{d}t'\, j_{\perp i}(\mathbf{r}, t') \tag{11}$$

which, taking (9) into account, becomes

$$E_{\perp i}(\mathbf{r}, t) = -\frac{q_\alpha}{\varepsilon_0}\sum_j \delta_{ij}^{\perp}(\mathbf{r})\int_0^t \dot{r}_{\alpha j}(t')\, \mathrm{d}t'$$

$$= -\frac{1}{\varepsilon_0}\sum_j \delta_{ij}^{\perp}(\mathbf{r})\, q_\alpha\, r_{\alpha j}(t). \tag{12}$$

Comparison of (6) and (12) shows that $E_{\perp i}(\mathbf{r}, t)$ at each instant cancels $\delta E_{\|i}(\mathbf{r}, t)$.

c) The elimination of $\mathscr{E}_\perp$ between the equations (B.49) gives

$$\frac{1}{c^2}\ddot{\mathscr{B}} + k^2\,\mathscr{B} = -\frac{1}{\varepsilon_0 c^2}\, i\mathbf{k} \times j_\perp = -\frac{1}{\varepsilon_0 c^2}\, i\mathbf{k} \times j \tag{13}$$

since $\mathbf{k} \times \boldsymbol{j}_{\parallel} = \mathbf{0}$. In real space, (13) becomes

$$\left(\frac{1}{c^2}\frac{\partial^2}{\partial t^2} - \Delta\right)\mathbf{B}(\mathbf{r}, t) = -\frac{1}{\varepsilon_0 c^2}\nabla \times \mathbf{j}(\mathbf{r}, t). \tag{14}$$

Since the source term, which involves $\mathbf{j}$, is localized at $\mathbf{r} = \mathbf{0}$, the solution of (14) is purely retarded and has no instantaneous term. $\mathbf{B}(\mathbf{r}, t)$ is identically zero on the time interval $[0, r/c]$, and it is permissible to ignore the first term of the right-hand side of (10) over this interval.

4. Normal variables and retarded potentials

a) Consider a set of charged particles producing a current $\mathbf{j}(\mathbf{r}, t)$. Integrate the equation of motion of the normal variables of the field, and write the normal variables at time t in the form of an integral of the current between $-\infty$ and t. Derive the value of the vector potential $\mathcal{A}_{\perp}(\mathbf{k}, t)$ in reciprocal space.

b) Calculate, using the results of *a*), the potentials in Coulomb gauge in real space.

c) By again starting with the results of *a*) in reciprocal space, find the total electric field $\mathcal{E}(\mathbf{k}, t)$. Show that the electric field in real space is of the form

$$\mathbf{E}(\mathbf{r}, t) = -\frac{\partial}{\partial t}\left[\frac{1}{4\pi\varepsilon_0 c^2}\int \mathrm{d}^3 r' \frac{\mathbf{j}\left(\mathbf{r}', t - \frac{|\mathbf{r} - \mathbf{r}'|}{c}\right)}{|\mathbf{r} - \mathbf{r}'|}\right] - \nabla\left[\frac{1}{4\pi\varepsilon_0}\int \mathrm{d}^3 r' \frac{\rho\left(\mathbf{r}', t - \frac{|\mathbf{r} - \mathbf{r}'|}{c}\right)}{|\mathbf{r} - \mathbf{r}'|}\right] \tag{1}$$

and that it has no instantaneous Coulomb term.

Solution

a) The solution of Equation (C.8) giving the motion of $\boldsymbol{\alpha}(\mathbf{k}, t)$ is

$$\boldsymbol{\alpha}(\mathbf{k}, t) = \int_{-\infty}^{+\infty} \mathrm{d}t'\, \mathrm{e}^{-\mathrm{i}\omega(t - t')}\,\theta(t - t')\frac{\mathrm{i}}{2\varepsilon_0 \mathcal{N}(k)}\boldsymbol{j}_{\perp}(\mathbf{k}, t) \tag{2}$$

where $\theta(\tau) = 1$ for $\tau > 0$, $\theta(\tau) = 0$ for $\tau < 0$, and where $\mathcal{N}(k) = \sqrt{\hbar\omega/2\varepsilon_0}$. Starting with (C.29), one gets

$$\mathcal{A}_{\perp}(\mathbf{k}, t) = \frac{\mathrm{i}}{2\varepsilon_0\omega}\int_{-\infty}^{+\infty} \mathrm{d}\tau\, \theta(\tau)\,[\mathrm{e}^{-\mathrm{i}\omega\tau}\boldsymbol{j}_{\perp}(\mathbf{k}, t - \tau) - \mathrm{e}^{\mathrm{i}\omega\tau}\boldsymbol{j}_{\perp}^{*}(-\mathbf{k}, t - \tau)] \tag{3}$$

so that, using the fact that $\mathbf{j}(\mathbf{r}, t)$ is real,

$$\mathcal{A}_{\perp}(\mathbf{k}, t) = \frac{1}{\varepsilon_0}\int_{-\infty}^{+\infty} \mathrm{d}\tau\, \theta(\tau)\frac{\sin\omega\tau}{\omega}\boldsymbol{j}_{\perp}(\mathbf{k}, t - \tau). \tag{4}$$

b) The transverse potential is given in real space by

$$\mathbf{A}_\perp(\mathbf{r}, t) = \frac{1}{(2\pi)^{3/2}} \int d^3k \, e^{i\mathbf{k}.\mathbf{r}} \mathcal{A}_\perp(\mathbf{k}, t) \tag{5}$$

—that is, from (4),

$$\mathbf{A}_\perp(\mathbf{r}, t) = \frac{1}{(2\pi)^3 \varepsilon_0} \int d^3k \int_{-\infty}^{+\infty} d\tau \, e^{i\mathbf{k}.\mathbf{r}} \theta(\tau) \frac{\sin \omega\tau}{\omega} \int d^3r' \, \mathbf{j}_\perp(\mathbf{r}', t - \tau) \, e^{-i\mathbf{k}.\mathbf{r}'} . \tag{6}$$

Now, one can show that

$$\frac{1}{(2\pi)^3} \int d^3k \, e^{i\mathbf{k}.(\mathbf{r}-\mathbf{r}')} \frac{\sin \omega\tau}{\omega} \theta(\tau) = \frac{1}{4\pi c^2} \frac{1}{|\mathbf{r} - \mathbf{r}'|} \delta\left(\tau - \frac{|\mathbf{r} - \mathbf{r}'|}{c}\right). \tag{7}$$

(One has just to perform the integration on the polar angles of $\mathbf{k}$ and then the integration on k.) It follows that

$$\mathbf{A}_\perp(\mathbf{r}, t) = \frac{1}{4\pi\varepsilon_0 c^2} \int d^3r' \int_0^{+\infty} d\tau \, \frac{\mathbf{j}_\perp(\mathbf{r}', t - \tau)}{|\mathbf{r} - \mathbf{r}'|} \delta\left(\tau - \frac{|\mathbf{r} - \mathbf{r}'|}{c}\right). \tag{8}$$

In the Coulomb gauge, $\mathbf{A}(\mathbf{r}, t) = \mathbf{A}_\perp(\mathbf{r}, t)$ and the scalar potential is the static Coulomb potential (B.25.b):

$$U(\mathbf{r}, t) = \frac{1}{4\pi\varepsilon_0} \int d^3r' \, \frac{\rho(\mathbf{r}', t)}{|\mathbf{r} - \mathbf{r}'|} \tag{9}$$

U is an instantaneous term. In order that there be a finite velocity of propagation for the E-field, a term in $\mathbf{A}_\perp$ which compensates the contribution of (9) to $\mathbf{E}$ is required. Note that $\mathbf{A}_\perp$ is not purely retarded, since the transverse current $\mathbf{j}_\perp$ which appears in (8) introduces a nonlocal component.

c) The electric field in reciprocal space is given by

$$\mathcal{E} = -i\mathbf{k}\mathcal{U} - \dot{\mathcal{A}}_\perp . \tag{10}$$

Equations (9) and (4) then give

$$\begin{aligned} \mathcal{E}(\mathbf{k}, t) &= -\frac{i\mathbf{k}}{\varepsilon_0 k^2} \rho(\mathbf{k}, t) - \frac{1}{\varepsilon_0} \int_0^{+\infty} d\tau \frac{\sin \omega\tau}{\omega} \dot{j}_\perp(\mathbf{k}, t - \tau) \\ &= -\frac{i\mathbf{k}}{\varepsilon_0 k^2} \rho(\mathbf{k}, t) - \frac{1}{\varepsilon_0} \int_0^{+\infty} d\tau \frac{\sin \omega\tau}{\omega} \dot{j}(\mathbf{k}, t - \tau) + \frac{1}{\varepsilon_0} \int_0^{+\infty} d\tau \frac{\sin \omega\tau}{\omega} \dot{j}_\parallel(\mathbf{k}, t - \tau). \end{aligned} \tag{11}$$

Integrate the last integral by parts:

$$\frac{1}{\varepsilon_0} \int_0^{+\infty} d\tau \frac{\sin \omega\tau}{\omega} \dot{j}_\parallel(\mathbf{k}, t - \tau) = \frac{1}{\varepsilon_0} \int_0^{+\infty} d\tau \cos \omega\tau \, j_\parallel(\mathbf{k}, t - \tau) . \tag{12}$$

The continuity equation (B.6) allows one to transform (12) into

$$\frac{i\mathbf{k}}{\varepsilon_0 k^2} \int_0^{+\infty} d\tau \cos \omega\tau \, \dot{\rho}(\mathbf{k}, t - \tau) = \frac{i\mathbf{k}}{\varepsilon_0 k^2} \rho(\mathbf{k}, t) - \frac{i\mathbf{k}}{\varepsilon_0 k^2} \int_0^{+\infty} d\tau \, \omega \sin \omega\tau \rho(\mathbf{k}, t - \tau) . \tag{13}$$

Substituting (13) in (11), one finds

$$\mathcal{E}(\mathbf{k}, t) = -\frac{1}{\varepsilon_0} \int_0^{+\infty} d\tau \frac{\sin \omega\tau}{\omega} \dot{j}(\mathbf{k}, t - \tau) - \frac{i\mathbf{k}}{\varepsilon_0} c^2 \int_0^{+\infty} d\tau \frac{\sin \omega\tau}{\omega} \rho(\mathbf{k}, t - \tau) \tag{14}$$

which, taking the results of part b) into account, gives Equation (1) in real space as required. The electric field $\mathbf{E}$ is just the field derived from retarded potentials in the Lorentz gauge (*). This proves that the instantaneous Coulomb term [the first term of (11)] has been compensated for by a term arising from $\mathbf{j}_\perp$ [the first term of (13)].

5. Field created by a charged particle at its own position. Radiation reaction

Consider a charged particle with charge q_α, mass m_α, and bound in the neighborhood of the origin by a force $\mathbf{F}$ which derives from an external potential energy $V(\mathbf{r})$: $\mathbf{F} = -\nabla V$. The purpose of this exercise is to calculate, starting with the equation of motion of the normal variables, the field created by this charge at its own position and to get the expression for the "radiation reaction" which results from the interaction of the particle with its own field (**).

a) Find the current associated with the particle in real space and reciprocal space.

b) Write the equation of motion of the normal variables $\alpha_\varepsilon(\mathbf{k}, t)$ describing the state of the transverse field. Let a_0 be the order of the spatial dimensions of the region in which the particle is bound about $\mathbf{0}$. In this exercise only the modes $\mathbf{k}\boldsymbol{\varepsilon}$ of the transverse field for which $k \le k_M$ with $k_M a_0 \ll 1$ will be of interest. Write, to zero order in $k_M a_0$, the equation of motion of $\alpha_\varepsilon(\mathbf{k}, t)$. Integrate this equation between the initial time t_0 and t, and express $\alpha_\varepsilon(\mathbf{k}, t)$ as a function of $\alpha_\varepsilon(\mathbf{k}, t_0)$ and the velocity of the particle, $\dot{\mathbf{r}}_\alpha(t')$, between t_0 and t. What is the physical significance of the two terms one gets?

c) In the mode expansion of the transverse fields $\mathbf{E}_\perp$ and $\mathbf{B}$ one only considers the contribution of the modes $k \le k_M$. Show that in that case one can take $\mathbf{E}_\perp(\mathbf{0}, t)$ and $\mathbf{B}(\mathbf{0}, t)$ as approximations to the transverse fields at point $\mathbf{r}_\alpha$ where the particle is located. Using this approximation (order zero in $k_M a_0$), give, using the results of b), the expression for the transverse fields $\mathbf{E}_{\perp P}(0, t)$ and $\mathbf{B}_P(\mathbf{0}, t)$ produced by the particle at its own position. Find the sum over the polarizations and the integral over the angles of $\mathbf{k}$. Show that $\mathbf{B}_P(\mathbf{0}, t)$ is zero and that $\mathbf{E}_{\perp P}(\mathbf{0}, t)$ can be written

$$\mathbf{E}_{\perp P}(\mathbf{0}, t) = \frac{q_\alpha}{3\pi\varepsilon_0 c^3} \int_0^{t - t_0} \mathrm{d}\tau \, \dot{\mathbf{r}}_\alpha(t - \tau) \, \ddot{\delta}_M(\tau)$$

where

$$\delta_M(\tau) = \frac{1}{2\pi} \int_{-\omega_M}^{+\omega_M} \mathrm{e}^{-i\omega\tau} \, \mathrm{d}\omega$$

is a δ-function of width $1/\omega_M = 1/ck_M$.

(*) See Feynman et al., Vol. II (Chapter 21).

(**) See Feynman et al., Vol. II (Chapter 28), Jackson (Chapter 17).

d) Integrate the expression giving $\mathbf{E}_{\perp P}(\mathbf{0}, t)$ by parts. Assume that $t - t_0$ is very large with respect to $1/\omega_M$ and that $\mathbf{r}_\alpha$ varies slowly on a time interval $1/\omega_M$. Show then that $\mathbf{E}_{\perp P}(\mathbf{0}, t)$ is the sum of two contributions, one proportional to $\ddot{\mathbf{r}}_\alpha(t)$ and the other to $\dddot{\mathbf{r}}_\alpha(t)$.

e) Write the equation of motion for the particle. Show that one of the two terms arising from the interaction of the particle with its own transverse field can be interpreted as describing a modification δm_α of the mass m_α of the particle. Calculate δm_α. What is the physical origin of δm_α?

f) To interpret physically the other term arising from the coupling of the particle with $\mathbf{E}_{\perp P}(\mathbf{0}, t)$, assume that the particle's motion is sinusoidal. Show that this other term then describes a damping of the particle's motion. What is the physical origin of the damping? Justify giving the name radiation reaction to this term.

Solution

a) From (A.5.b)

$$\mathbf{j}(\mathbf{r}, t) = q_\alpha \dot{\mathbf{r}}_\alpha(t)\, \delta(\mathbf{r} - \mathbf{r}_\alpha(t)) \tag{1.a}$$

$$\mathscr{j}(\mathbf{k}, t) = \frac{1}{(2\pi)^{3/2}} q_\alpha \dot{\mathbf{r}}_\alpha(t)\, \mathrm{e}^{-\mathrm{i}\mathbf{k}.\mathbf{r}_\alpha(t)} \tag{1.b}$$

b) The normal variable $\alpha_\varepsilon(\mathbf{k}, t)$ obeys Equation (C.12), which using (C.15) and (1.b) can be written

$$\dot{\alpha}_\varepsilon(\mathbf{k}, t) + \mathrm{i}\omega\alpha_\varepsilon(\mathbf{k}, t) = \frac{\mathrm{i}q_\alpha}{\sqrt{2\,\varepsilon_0\,\hbar\omega(2\pi)^3}}\, \boldsymbol{\varepsilon} \cdot \dot{\mathbf{r}}_\alpha(t)\, \mathrm{e}^{-\mathrm{i}\mathbf{k}.\mathbf{r}_\alpha(t)} \tag{2}$$

To order zero in $k_M a_0$, one can replace $\mathrm{e}^{-\mathrm{i}\mathbf{k}\cdot\mathbf{r}_\alpha(t)}$ by 1 in the last term of (2). Integration of (2) yields

$$\alpha_\varepsilon(\mathbf{k}, t) = \alpha_\varepsilon(\mathbf{k}, t_0)\, \mathrm{e}^{-\mathrm{i}\omega(t-t_0)} + \frac{\mathrm{i}q_\alpha}{\sqrt{2\,\varepsilon_0\,\hbar\omega(2\pi)^3}} \int_{t_0}^{t} \mathrm{d}t'\, \mathrm{e}^{-\mathrm{i}\omega(t-t')}\, \boldsymbol{\varepsilon} \cdot \dot{\mathbf{r}}_\alpha(t'). \tag{3}$$

The first term of (3) represents a field freely evolving between t_0 and t from the initial state $\alpha_\varepsilon(\mathbf{k}, t_0)$. The second term describes the field created by the particle between t_0 and t.

c) The exponentials $\mathrm{e}^{\pm\mathrm{i}\mathbf{k}\cdot\mathbf{r}_\alpha(t)}$ appear in the mode expansion of the fields $\mathbf{E}_\perp$ and $\mathbf{B}$ evaluated at point $\mathbf{r}_\alpha$. If the sums on $\mathbf{k}$ are limited to $k \leq k_M$, it is correct to order zero in $k_M a_0$ to approximate the exponentials by 1, which amounts to identifying the field at $\mathbf{r}_\alpha$ with the field at the origin. To get the fields produced by the particle in its own position, it suffices then to put the last term of (3) and its complex conjugate in the mode expansion of $\mathbf{E}_\perp(\mathbf{0}, t)$ and $\mathbf{B}(\mathbf{0}, t)$. One gets then for the i-component of $\mathbf{E}_{\perp P}$

$$(E_{\perp P}(\mathbf{0}, t))_i = \sum_{j=x,y,z} \int_{k \leq k_M} k^2\, \mathrm{d}k\, \mathrm{d}\Omega\, \times$$

$$\times \sum_{\varepsilon \perp \mathbf{k}} \left(-\frac{q_\alpha}{2\,\varepsilon_0 (2\pi)^3} \right) \varepsilon_i\, \varepsilon_j \int_0^{t-t_0} \mathrm{d}\tau\, \mathrm{e}^{-\mathrm{i}\omega\tau}\, \dot{r}_{\alpha j}(t - \tau) + c.c. \tag{4}$$

Summing over the polarizations gives

$$\sum_{\boldsymbol{\varepsilon}\perp\mathbf{k}} \varepsilon_i\, \varepsilon_j = \delta_{ij} - \frac{k_i\, k_j}{k^2} \tag{5}$$

and the angular integral leads to

$$\int \mathrm{d}\Omega\left(\delta_{ij} - \frac{k_i\, k_j}{k^2}\right) = \frac{8\,\pi}{3}\,\delta_{ij} \tag{6}$$

with the result that

$$\mathbf{E}_{\perp P}(\mathbf{0}, t) = -\frac{q_\alpha}{6\,\varepsilon_0\,\pi^2}\int_0^{t-t_0} \mathrm{d}\tau\;\dot{\mathbf{r}}_\alpha(t-\tau)\int_{-k_M}^{+k_M} \mathrm{d}k\; k^2\, \mathrm{e}^{-\mathrm{i}ck\tau} \tag{7}$$

One evaluates the integral (7) over k. This gives

$$\int_{-k_M}^{+k_M} \mathrm{d}k\; k^2\, \mathrm{e}^{-\mathrm{i}ck\tau} = -\frac{2\,\pi}{c^3}\frac{\mathrm{d}^2}{\mathrm{d}\tau^2}\left[\frac{1}{2\,\pi}\int_{-\omega_M}^{+\omega_M} \mathrm{d}\omega\; \mathrm{e}^{-\mathrm{i}\omega\tau}\right] = -\frac{2\,\pi}{c^3}\,\ddot{\delta}_M(\tau) \tag{8}$$

where

$$\delta_M(\tau) = \frac{1}{2\,\pi}\int^{+\omega_M} \mathrm{d}\omega\; \mathrm{e}^{-\mathrm{i}\omega\tau} \tag{9}$$

is a delta function of width $1/\omega_M$. Finally

$$\mathbf{E}_{\perp P}(\mathbf{0}, t) = \frac{q_\alpha}{3\,\pi\varepsilon_0\, c^3}\int_0^{t-t_0} \mathrm{d}\tau\;\dot{\mathbf{r}}_\alpha(t-\tau)\,\ddot{\delta}_M(\tau)\,. \tag{10}$$

An analogous calculation can be done for $\mathbf{B}_P(\mathbf{0}, t)$. The sum on the polarizations then leads to an odd function of $\boldsymbol{\delta} = \mathbf{k}/k$ whose angular integral is zero. One has then

$$\mathbf{B}_P(\mathbf{0}, t) = \mathbf{0}\,. \tag{11}$$

d) A double integration of (10) by parts gives, taking into account the fact that $\dot{\delta}_M$ is an odd function of τ and that δ_M and $\dot{\delta}_M$ are negligible for $\tau = t - t_0$.

$$\mathbf{E}_{\perp P}(\mathbf{0}, t) = \frac{q_\alpha}{3\,\pi\varepsilon_0\, c^3}\left[-\,\delta_M(0)\,\ddot{\mathbf{r}}_\alpha(t) + \int_0^{t-t_0} \mathrm{d}\tau\;\dddot{\mathbf{r}}_\alpha(t-\tau)\,\delta_M(\tau)\right]. \tag{12}$$

From (9), $\delta_M(0) = \omega_M/\pi$. On the other hand, since $\dddot{\mathbf{r}}_\alpha$ varies slowly on the scale of $1/\omega_M$, one can replace $\dddot{\mathbf{r}}_\alpha(t-\tau)$ by $\dddot{\mathbf{r}}_\alpha(t)$ in the last integral and remove $\dddot{\mathbf{r}}_\alpha(t)$ from the integral, which then is $\frac{1}{2}$ as a result of the even parity of $\delta_M(\tau)$ and the fact that $t - t_0 \gg 1/\omega_M$. One gets finally

$$\mathbf{E}_{\perp P}(\mathbf{0}, t) = -\frac{q_\alpha\,\omega_M}{3\,\pi^2\,\varepsilon_0\, c^3}\,\ddot{\mathbf{r}}_\alpha(t) + \frac{q_\alpha}{6\,\pi\varepsilon_0\, c^3}\,\dddot{\mathbf{r}}_\alpha(t)\,. \tag{13}$$

e) Since $\mathbf{B}_P(\mathbf{0}, t)$ is zero, the equation of motion of the particle is written

$$m_\alpha\,\ddot{\mathbf{r}}_\alpha(t) = -\,\boldsymbol{\nabla} V(\mathbf{r}_\alpha) + q_\alpha\,\mathbf{E}_{\perp P}(\mathbf{0}, t) + q_\alpha\,\mathbf{E}_{\text{free}}(\mathbf{0}, t)\,. \tag{14}$$

The first term describes the effect of the external potential binding the particle near the origin, the second the interaction of the particle with its own transverse field (the interaction of the particle with its own longitudinal field leads to a zero net force through symmetry), and the last, the interaction with the free field [$\mathbf{E}_{\text{free}}$ is associated with the first term of (3) and describes possible incident radiation].

Substitute (13) in (14) and bring the term $\ddot{\mathbf{r}}_\alpha$ to the left-hand side. One gets then

$$(m_\alpha + \delta m_\alpha)\,\ddot{\mathbf{r}}_\alpha(t) = -\,\boldsymbol{\nabla} V(\mathbf{r}_\alpha) + \frac{q_\alpha^2}{6\,\pi\varepsilon_0\, c^3}\,\dddot{\mathbf{r}}_\alpha(t) + q_\alpha\,\mathbf{E}_{\text{free}}(\mathbf{0}, t) \tag{15}$$

where δm_α is given by

$$\delta m_\alpha = \frac{q_\alpha^2 \, k_M}{3 \pi^2 \, \varepsilon_0 \, c^2}. \tag{16}$$

The term in $\ddot{\mathbf{r}}_\alpha$ in (13) then describes a modification of the mass of the particle which can be interpreted as the "electromagnetic inertia" of this particle.

It is possible moreover to relate δm_α to the Coulomb energy $\varepsilon_{\text{Coul}}^\alpha$ of the particle. Indeed, if one introduces the same cutoff at k_M in the integral (B.36) giving $\varepsilon_{\text{Coul}}^\alpha$, one gets

$$\varepsilon_{\text{Coul}}^\alpha = \frac{q_\alpha^2 \, k_M}{4 \pi^2 \, \varepsilon_0} = \frac{3}{4} \delta m_\alpha \, c^2. \tag{17}$$

To within a factor $\frac{4}{3}$, the energy $\delta m_\alpha c^2$ associated with δm_α is the Coulomb energy of the particle.

f) If the particle undergoes an oscillatory motion of frequency ω, one can replace $\dddot{\mathbf{r}}_\alpha$ by $-\omega^2 \dot{\mathbf{r}}_\alpha$. The second term of (15) then describes a frictional force $-(q_\alpha^2 \omega^2 / 6 \pi \varepsilon_0 c^3) \dot{\mathbf{r}}_\alpha$ which dampens the motion of the particle. This force is nothing more than the radiation reaction describing the loss of energy of a charged particle undergoing accelerated motion—loss resulting from the radiation which it emits (see Exercise 7).

6. Field produced by an oscillating electric dipole

A microscopic emitting system is made up of a particle with charge $-q_\alpha$ fixed at the origin and a charge q_α undergoing a motion described by

$$\mathbf{r}_\alpha(t) = \mathbf{a}_0 \cos \omega_0 t. \tag{1}$$

One is interested in the field produced by this system at a great distance from the origin, that is, where $r \gg a_0$ (this corresponds in reciprocal space to wave vectors such that $k a_0 \ll 1$).

a) Find the charge and current densities associated with the system as well as their spatial Fourier transforms. Expand the latter to first order in a_0. Derive expressions for $\mathbf{j}(\mathbf{r}, t)$ and $\rho(\mathbf{r}, t)$ to first order in a_0.

b) By substituting the expressions so found into Equation (1) of exercise 4, show that the field $\mathbf{E}(\mathbf{r}, t)$ radiated by the dipole can be written

$$\mathbf{E}(\mathbf{r}, t) = \frac{q_\alpha}{4 \pi \varepsilon_0} \left[k_0^2 \, \mathbf{a}_0 + \nabla(\mathbf{a}_0 \cdot \nabla) \right] \frac{\cos (k_0 \, r - \omega_0 \, t)}{r}. \tag{2}$$

Show that $\mathbf{E}$ is a sum of three terms in $1/r$, $1/r^2$, and $1/r^3$ respectively (*).

c) Show that $\mathbf{E}(\mathbf{r}, t)$ can also be written as an electric dipole wave such as those defined in Complement B_I with $J = 1$ and $M = 0$, Oz being taken along $\mathbf{a}_0$. The spherical Bessel function of order 1 is given by

$$j_1(x) = \frac{\sin x}{x^2} - \frac{\cos x}{x}. \tag{3}$$

(*) See Feynman et al., Vol. II (Chapter 21).

Solution

a) Using (A.5.a) and (A.5.b), one gets

$$\begin{cases} \boldsymbol{\rho}(\mathbf{r}, t) = q_\alpha\, \delta(\mathbf{r} - \mathbf{r}_\alpha) - q_\alpha\, \delta(\mathbf{r}) & (4.a) \\ \mathbf{j}(\mathbf{r}, t) = q_\alpha\, \dot{\mathbf{r}}_\alpha\, \delta(\mathbf{r} - \mathbf{r}_\alpha)\,. & (4.b) \end{cases}$$

The expansion to first order in a_0 of the spatial Fourier transform of (4) gives

$$\rho(\mathbf{k}, t) = \frac{q_\alpha}{(2\pi)^{3/2}} (\mathrm{e}^{-\mathrm{i}\mathbf{k}.\mathbf{r}_\alpha} - 1) \simeq -\frac{q_\alpha}{(2\pi)^{3/2}} (\mathrm{i}\mathbf{k} \cdot \mathbf{r}_\alpha) \tag{5.a}$$

$$j(\mathbf{k}, t) = \frac{q_\alpha}{(2\pi)^{3/2}}\, \dot{\mathbf{r}}_\alpha\, \mathrm{e}^{-\mathrm{i}\mathbf{k}.\mathbf{r}_\alpha} \simeq \frac{q_\alpha\, \dot{\mathbf{r}}_\alpha}{(2\pi)^{3/2}}\,. \tag{5.b}$$

Returning to real space, one gets

$$\begin{cases} \boldsymbol{\rho}(\mathbf{r}, t) = -\, q_\alpha\, \boldsymbol{\nabla} \cdot \mathbf{r}_\alpha(t)\, \delta(\mathbf{r}) = -\, q_\alpha \cos \omega_0 t\, \boldsymbol{\nabla} \cdot \mathbf{a}_0\, \delta(\mathbf{r}) & (6.a) \\ \mathbf{j}(\mathbf{r}, t) = q_\alpha\, \dot{\mathbf{r}}_\alpha(t)\, \delta(\mathbf{r}) = -\, q_\alpha\, \omega_0 \sin \omega_0 t\, \mathbf{a}_0\, \delta(\mathbf{r}) & (6.b) \end{cases}$$

b) Put (6.a) and (6.b) in Equation (1) of Exercise 4, and consider first the contribution of $\mathbf{j}$. The first integral in r' yields, taking into account (6.b),

$$\frac{1}{4\pi\varepsilon_0 c^2} \int \mathrm{d}^3 r' \frac{\mathbf{j}\left(\mathbf{r}', t - \dfrac{|\mathbf{r} - \mathbf{r}'|}{c}\right)}{|\mathbf{r} - \mathbf{r}'|} = \frac{q_\alpha\, \mathbf{a}_0}{4\pi\varepsilon_0 c^2} \frac{\omega_0 \sin (k_0 r - \omega_0 t)}{r}\,. \tag{7}$$

Following Exercise 4, the contribution of $\mathbf{j}$ to $\mathbf{E}(\mathbf{r}, t)$ is gotten by differentiating (7) with respect to t, which gives

$$\frac{q_\alpha\, \mathbf{a}_0\, k_0^2}{4\pi\varepsilon_0} \frac{\cos (k_0 r - \omega_0 t)}{r}\,. \tag{8}$$

The second integral in $\mathbf{r}'$ in Equation (1) of Exercise 4 can be written using (6.a) as

$$-\frac{q_\alpha}{4\pi\varepsilon_0} \int \mathrm{d}^3 r' f(|\mathbf{r} - \mathbf{r}'|)\, [\boldsymbol{\nabla}_{\mathbf{r}'} \cdot \mathbf{a}_0\, \delta(\mathbf{r}')] \tag{9}$$

where

$$f(u) = \frac{\cos \omega_0 \left(t - \dfrac{u}{c}\right)}{u}\,. \tag{10}$$

Using the identity

$$\lambda \boldsymbol{\nabla} \cdot \mathbf{A} = \boldsymbol{\nabla} \cdot (\lambda \mathbf{A}) - \mathbf{A} \cdot \boldsymbol{\nabla} \lambda \tag{11}$$

allows one to transform (9) into

$$\frac{q_\alpha}{4\pi\varepsilon_0} \int \mathrm{d}^3 r'\, \delta(\mathbf{r}')\, \mathbf{a}_0 \cdot \boldsymbol{\nabla}_{\mathbf{r}'} f(|\mathbf{r} - \mathbf{r}'|) = -\frac{q_\alpha}{4\pi\varepsilon_0} \frac{\mathbf{a}_0 \cdot \mathbf{r}}{r} f'(r) = -\frac{q_\alpha}{4\pi\varepsilon_0}\, \mathbf{a}_0 \cdot \boldsymbol{\nabla} f(r)\,. \tag{12}$$

Following Exercise 4, the contribution of ρ to $\mathbf{E}(\mathbf{r}, t)$ is gotten by applying $-\boldsymbol{\nabla}$ to (12), which gives, using (10),

$$\frac{q_\alpha}{4\pi\varepsilon_0} \boldsymbol{\nabla}(\mathbf{a}_0 \cdot \boldsymbol{\nabla}) \frac{\cos (k_0 r - \omega_0 t)}{r}\,. \tag{13}$$

The sum of (8) and (13) gives Equation (2) as stated.

Calculating the double spatial derivative in (13) gives

$$\mathbf{E}(\mathbf{r}, t) = \frac{q_\alpha}{4\pi\varepsilon_0}\left[3\frac{\mathbf{r}(\mathbf{a}_0\cdot\mathbf{r})}{r^2} - \mathbf{a}_0\right]\frac{\cos(k_0 r - \omega_0 t)}{r^3} + \\ + \frac{q_0}{4\pi\varepsilon_0}\left[3\frac{\mathbf{r}(\mathbf{a}_0\cdot\mathbf{r})}{r^2} - \mathbf{a}_0\right]k_0\frac{\sin(k_0 r - \omega_0 t)}{r^2} + \\ + \frac{q_\alpha}{4\pi\varepsilon_0}\left[\mathbf{a}_0 - \frac{\mathbf{r}(\mathbf{a}_0\cdot\mathbf{r})}{r^2}\right]k_0^2\frac{\cos(k_0 r - \omega_0 t)}{r}. \tag{14}$$

One gets the well-known three terms of dipole radiation, behaving respectively as $1/r$, $1/r^2$, and $1/r^3$.

c) To make the connection with the multipole waves defined in Complement B_I, it is useful to introduce the radial and tangential components of $\mathbf{a}_0$ with respect to $\mathbf{r}$:

$$\mathbf{a}_{0r} = \frac{\mathbf{r}(\mathbf{a}_0\cdot\mathbf{r})}{r^2} \\ \mathbf{a}_{0t} = \mathbf{a}_0 - \mathbf{a}_{0r}. \tag{15}$$

Equation (14) can then be written in the form

$$\mathbf{E}(\mathbf{r}, t) = \frac{q_\alpha}{4\pi\varepsilon_0}\mathbf{a}_{0t}\left[\frac{k_0^2\cos(k_0 r - \omega_0 t)}{r} - \frac{k_0\sin(k_0 r - \omega_0 t)}{r^2} - \frac{\cos(k_0 r - \omega_0 t)}{r^3}\right] + \\ + \frac{q_\alpha}{4\pi\varepsilon_0}\mathbf{a}_{0r}\left[\frac{2k_0\sin(k_0 r - \omega_0 t)}{r^2} + \frac{2\cos(k_0 r - \omega_0 t)}{r^3}\right]. \tag{16}$$

Compare this with the results of Complement B_I. The electric field of an electric dipole wave is, using (63), proportional to $\mathbf{I}_{k_0JMZ}(\mathbf{r})$ with $J = 1$ and $M = 0$, the Oz axis being taken along $\mathbf{a}_0$. We now apply Equation (62) of complement B_I to this particular case. The spherical vector functions $\mathbf{Z}_{10}(\mathbf{r}/r)$ and $\mathbf{N}_{10}(\mathbf{r}/r)$ are derived from the spherical harmonic $Y_{10}(\mathbf{r}/r)$, whose value, to within a constant factor, is

$$Y_{10}\left(\frac{\mathbf{r}}{r}\right) \propto \frac{\mathbf{a}_0\cdot\mathbf{r}}{a_0 r}. \tag{17}$$

Using Equations (33) and (38) of the complement, one finds, with (3),

$$\mathbf{N}_{10}\left(\frac{\mathbf{r}}{r}\right) = \frac{\mathbf{r}(\mathbf{a}_0\cdot\mathbf{r})}{a_0 r^2} = \frac{\mathbf{a}_{0r}}{a_0} \tag{18.a}$$

$$\mathbf{Z}_{10}\left(\frac{\mathbf{r}}{r}\right) = \frac{1}{\sqrt{2}}r\nabla\left(\frac{\mathbf{a}_0\cdot\mathbf{r}}{a_0 r}\right) = \frac{1}{a_0\sqrt{2}}\left[\mathbf{a}_0 - \frac{(\mathbf{a}_0\cdot\mathbf{r})\,\mathbf{r}}{r^2}\right] = \frac{\mathbf{a}_{0t}}{a_0\sqrt{2}}. \tag{18.b}$$

Equation (62) can then be written

$$\mathbf{I}_{k_0 10 Z}(\mathbf{r}) \propto \frac{1}{r}\left\{\frac{\mathbf{a}_{0t}}{a_0\sqrt{2}}\frac{\mathrm{d}}{\mathrm{d}(k_0 r)}\left[-\cos(k_0 r) + \frac{\sin k_0 r}{k_0 r}\right] + \right. \\ \left. + \sqrt{2}\,\frac{\mathbf{a}_{0r}}{a_0}\left[-\frac{\cos k_0 r}{k_0 r} + \frac{\sin k_0 r}{(k_0 r)^2}\right]\right\} \\ \propto \mathbf{a}_{0t}\left[\frac{\cos\left(k_0 r - \frac{\pi}{2}\right)}{r} - \frac{\sin\left(k_0 r - \frac{\pi}{2}\right)}{k_0 r^2} - \frac{\cos\left(k_0 r - \frac{\pi}{2}\right)}{k_0^2 r^3}\right] + \\ + 2\,\mathbf{a}_{0r}\left[\frac{\sin\left(k_0 r - \frac{\pi}{2}\right)}{k_0 r^2} + \frac{\cos\left(k_0 r - \frac{\pi}{2}\right)}{k_0^2 r^3}\right]. \tag{19}$$

Thus to within a multiplicative factor and a phase one gets (16). It is also useful to note that (16) is a traveling wave, while (19) is the amplitude of a spherical standing wave resulting from the superposition of an outgoing and an incoming wave.

7. Cross-section for scattering of radiation by a classical elastically bound electron

A classical electron with charge q and mass m, elastically bound to the origin by a restoring force $-m\omega_0^2\mathbf{r}$, is set into forced motion by an incident monochromatic wave with frequency ω and emits into all space radiation of the same frequency. The purpose of this exercise is to calculate the total scattering cross-section $\sigma(\omega)$ of the electron and to examine its order of magnitude as well as its variation with ω (*).

a) The electron undergoes forced motion along Oz with amplitude a and frequency ω:

$$z = a \cos \omega t \,. \tag{1}$$

One recalls that the total radiated electric field at a distant point M ($OM \gg \lambda = 2\pi c/\omega$) is in the plane ($Oz$, OM), is normal to OM, and has an amplitude (**)

$$E = \frac{qa}{4\pi\varepsilon_0 c^2} \frac{\sin\theta}{r} \omega^2 \cos \omega\left(t - \frac{r}{c}\right) \tag{2}$$

where r is the distance OM and θ is the angle between Oz and OM. The field $\mathbf{B}$ has an amplitude E/c and a direction parallel to $\mathbf{OM} \times \mathbf{E}$.

Find the mean value (over a period $2\pi/\omega$) of the flux of Poynting's vector $\varepsilon_0 c^2 \mathbf{E} \times \mathbf{B}$ through a sphere of very large radius r, and find $\mathrm{d}W/\mathrm{d}t$, the mean energy radiated per unit time.

b) The interaction of the electron with the field it creates at its own position can be described by a force, called the radiation reaction, whose component along Oz has the value (***)

$$R = \frac{q^2}{6\pi\varepsilon_0 c^3} \dddot{z} \,. \tag{3.a}$$

For the forced motion (1), find the mean value (over a period $2\pi/\omega$) of the work done by R on the electron. Compare this result with that from *a*). What is the physical interpretation?

(*) See Jackson (Chapter 17).

(**) These properties can be gotten from the results of Exercise 6.

(***) See Exercise 5 for a demonstration of this result. The interaction of the electron with its own field is also responsible for a change δm in its mass, which is assumed to be included in the mass m used here.

c) One writes the radiation reaction in the form

$$R = \frac{2}{3}\frac{r_0}{\lambda_0}\frac{m\dddot{z}}{\omega_0} \tag{3.b}$$

where r_0 is the classical electron radius given by

$$r_0 = \frac{q^2}{4\pi\varepsilon_0 mc^2} \tag{4}$$

and $\lambda_0 = c/\omega_0$. In the absence of incident radiation, the dynamical equation for the electron is written

$$m\ddot{z} = -m\omega_0^2 z + \frac{q^2}{6\pi\varepsilon_0 c^3}\dddot{z} \tag{5}$$

R (which is proportional to the factor $r_0/\lambda_0 \ll 1$) can be treated as a perturbation. Find the solutions of (5) of the form $e^{i\Omega t}$ and show that, to first order in r_0/λ_0, one has

$$\Omega = \pm\,\omega_0 + i\frac{\gamma_0}{2}. \tag{6}$$

Give the expression for γ_0 as a function of r_0, ω_0, and c. What does the time $\tau_0 = \gamma_0^{-1}$ represent?

d) In the presence of an incident field polarized along Oz whose amplitude at the origin is $E\cos\omega t$, the dynamical equation for the electron is written

$$m\ddot{z} = -m\omega_0^2 z + \frac{q^2}{6\pi\varepsilon_0 c^3}\dddot{z} + qE\cos\omega t. \tag{7}$$

Find the forced oscillatory motion of the electron and, using the results of *a*), the energy radiated per unit time into all space by the electron.

e) Find the energy flux (averaged over one period $2\pi/\omega$) associated with the incident wave, which is assumed to be plane and propagating along Ox. Using the results of *d*), find the total scattering cross-section $\sigma(\omega)$. Express $\sigma(\omega)$ as a function of r_0^2, ω, ω_0, and γ_0.

f) Assume $\omega \ll \omega_0$ (Rayleigh scattering). Show that $\sigma(\omega)$ is then proportional to a power of ω, which should be found.

g) Assume $\omega_0 \ll \omega \ll c/r_0$ (Thomson scattering). Show that $\sigma(\omega)$ is equal to a constant.

h) Assume finally ω near ω_0 (resonant scattering). Show that the variation of $\sigma(\omega)$ with $\omega - \omega_0$ exhibits a resonance. What is the width of the resonance? What is the value of the cross-section $\sigma(\omega_0)$ at resonance?

Solution

a) The flux of the Poynting vector through a sphere of radius r is equal to

$$\phi = \cos^2 \omega\left(t - \frac{r}{c}\right) \int d\Omega \, r^2 \, \varepsilon_0 \left(\frac{qa}{4\pi\varepsilon_0}\right)^2 \frac{\omega^4}{c^3} \frac{\sin^2\theta}{r^2}. \tag{8}$$

After angular integration, one finds for the average value of ϕ over one period

$$\bar{\phi} = \frac{1}{3} \frac{q^2}{4\pi\varepsilon_0} \frac{a^2 \omega^4}{c^3} \tag{9}$$

$\bar{\phi}$ is just the energy radiated per unit time, dW/dt, by the oscillating charge.

b) During time dt, the charge is displaced by $dz = \dot{z}\, dt$, and the work dW' done by R is equal to

$$dW' = R\dot{z}\, dt = \frac{q^2}{6\pi\varepsilon_0 c^3} \dddot{z}\, \dot{z}\, dt. \tag{10}$$

From (10) it follows that

$$\frac{dW'}{dt} = \frac{q^2}{6\pi\varepsilon_0 c^3} \dddot{z}\, \dot{z} = \frac{q^2}{6\pi\varepsilon_0 c^3}\left[\frac{d}{dt}(\dot{z}\,\ddot{z}) - \ddot{z}^2\right] \tag{11}$$

Let us take the average of (11) over one period $2\pi/\omega$. Since $\dot{z}\ddot{z}$ is a periodic function of t with period $2\pi/\omega$, the average of $d(\dot{z}\ddot{z})/dt$ is zero. As for the average of $\ddot{z}^2$, its value, according to (1), is $a^2\omega^4/2$. Finally,

$$\overline{\frac{dW'}{dt}} = -\frac{q^2}{12\pi\varepsilon_0 c^3} a^2 \omega^4 = -\bar{\phi} = -\overline{\frac{dW}{dt}}. \tag{12}$$

The mean work per unit time which R does on the oscillating charge is equal, except for the sign, to the mean energy which this charge radiates into all space per unit time. The radiation reaction thus describes for the electron the dissipative phenomenon associated with the radiative energy loss.

c) Replacing z by $e^{i\Omega t}$ in (5) gives, using (3.b) and $\lambda_0 = c/\omega_0$,

$$\Omega^2 - \omega_0^2 = i\frac{2}{3}\frac{r_0}{\lambda_0}\frac{\Omega^3}{\omega_0}. \tag{13}$$

To zeroth order in r_0/λ_0, $\Omega = \pm\omega_0$. To first order, one can replace Ω^3 by $\pm\omega_0^3$ on the right-hand side of (13), and $\Omega^2 - \omega_0^2$ by $(\Omega - \omega_0)(\Omega + \omega_0) \simeq \pm 2\omega_0(\Omega \mp \omega_0)$, which gives (6) with

$$\gamma_0 = \frac{2}{3}\frac{r_0}{\lambda_0}\omega_0 = \frac{2}{3}\frac{r_0\,\omega_0^2}{c}. \tag{14}$$

If ω_0 is an rf or optical frequency, $r_0 \ll \lambda_0$ and $\gamma_0 \ll \omega_0$. The time $\tau_0 = \gamma_0^{-1}$ is the decay time of the oscillatory energy due to the radiative energy loss. With $\omega_0\tau_0 \gg 1$, the electron undergoes numerous oscillations during decay.

d) Assume

$$z = \Re e(z_0\, e^{i\omega t}). \tag{15}$$

One then gets, substituting (15) in (7),

$$z_0 = \frac{qE}{m} \frac{1}{\omega_0^2 - \omega^2 + i\gamma_0 \dfrac{\omega^3}{\omega_0^2}}. \tag{16}$$

To get the mean energy radiated per unit time in steady state, it is necessary to replace a^2 by $|z_0|^2$ in (9), which then gives

$$\overline{\frac{dW}{dt}} = \frac{1}{3}\frac{q^2}{4\pi\varepsilon_0}\frac{q^2 E^2}{m^2}\frac{\omega^4}{c^3}\frac{1}{[\omega_0^2 - \omega^2]^2 + \gamma_0^2\frac{\omega^6}{\omega_0^4}}. \tag{17}$$

e) The incident flux $\bar{\phi}_i$ is equal to

$$\bar{\phi}_i = \varepsilon_0 c \frac{E^2}{2} \tag{18}$$

which gives for the total cross-section $\sigma(\omega)$

$$\sigma(\omega) = \frac{1}{\bar{\phi}_i}\overline{\frac{dW}{dt}} = \frac{8\pi}{3} r_0^2 \frac{\omega^4}{[\omega_0^2 - \omega^2]^2 + \gamma_0^2\frac{\omega^6}{\omega_0^4}}. \tag{19}$$

f) If $\omega \ll \omega_0$, the denominator of (19) is of the order of ω_0^4, which gives

$$\omega \ll \omega_0 \quad \Rightarrow \quad \sigma(\omega) = \frac{8\pi}{3} r_0^2 \left(\frac{\omega}{\omega_0}\right)^4. \tag{20}$$

The total scattering cross-section varies as the fourth power of the incident frequency.

g) The condition $\omega \ll c/r_0$, which indicates that the wavelength of the incident radiation is very large compared to r_0, shows in view of (14) that $\omega^4 \gg \gamma_0^2\omega^6/\omega_0^4$. If in addition ω is very large with respect to ω_0, the denominator of (19) reduces to ω^4, which shows that

$$\omega_0 \ll \omega \ll \frac{c}{r_0} \quad \Rightarrow \quad \sigma(\omega) = \frac{8\pi}{3} r_0^2 \tag{21}$$

which is just the Thomson scattering cross-section.

h) Near resonance, in (17) one can replace $\gamma_0^2\omega^6/\omega_0^4$ by $\gamma_0^2\omega_0^2$, $(\omega_0^2 - \omega^2)^2$ by $4\omega_0^2(\omega - \omega_0)^2$, and in the numerator ω^4 by ω_0^4, which yields

$$\omega \simeq \omega_0 \quad \Rightarrow \quad \sigma(\omega) = \frac{2\pi}{3} r_0^2 \frac{\omega_0^2}{(\omega - \omega_0)^2 + \frac{\gamma_0^2}{4}}. \tag{22}$$

The resonance variations of $\sigma(\omega)$ are those of a Lorentzian with total width at half maximum

$$\Delta\omega_0 = \gamma_0 . \tag{23}$$

At resonance, the value of $\sigma(\omega_0)$, using (14), is

$$\sigma(\omega_0) = \frac{8\pi}{3}\frac{r_0^2\,\omega_0^2}{\gamma_0^2} = \frac{3\,\lambda_0^2}{2\pi} \tag{24}$$

where

$$\lambda_0 = \frac{2\pi c}{\omega_0} \tag{25}$$

is the resonant wavelength.

For optical radiation, λ_0 is of the order of 5×10^{-7} m, whereas r_0 is of the order of 2.8×10^{-15} m. One has then about 16 orders of magnitude difference between the resonance cross-section (24) and the Thomson cross-section (21).

CHAPTER II

Lagrangian and Hamiltonian Approach to Electrodynamics. The Standard Lagrangian and the Coulomb Gauge

Classical electrodynamics, which has been presented in Chapter I beginning with the Maxwell–Lorentz equations, can also be derived from a very general variational principle, the principle of least action.

In classical mechanics, the principle of least action allows one to select, from all the possible "paths" leading from a *given* initial state to a *given* final state, that path which is indeed followed by the system. (See for example, Figure 1, corresponding to a system with one degree of freedom.) One associates with each path a number, called the action, which is the time integral of an important function, the Lagrangian, and one seeks the path for which the action is an extremum.

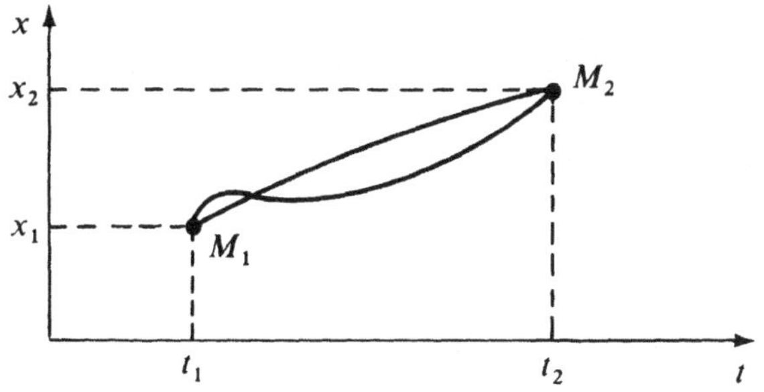

Figure 1. "Paths" connecting the initial state x_1, t_1 with the final state x_2, t_2. The true path, that one followed by the system, is that for which the action is an extremum.

The Lagrangian formulation of classical electrodynamics, using the principle of least action, was first put forward by Schwarzschild in 1903. It required the generalization of the usual Lagrangian formalism to the case where the dynamical system has an infinite number of degrees of freedom. Indeed, to specify the state of the electromagnetic field, it is necessary to give the values of the fields (or the potentials) at all points in space. Beyond its uncontestable compactness and aesthetic character, this formulation has the advantage of introducing all the quantities necessary for *canonical quantization* without ambiguity (Hamiltonian, momentum). In addition, recent progress in quantum field theory, such as the unification of the electromagnetic and weak interactions, is based on such a formulation of the theory.

In Part A of this chapter, we briefly review the essentials of the Lagrangian formalism in the simple case of a system having a discrete set of degrees of freedom. The generalization to systems having a continuum of degrees of freedom is then examined, as well as some results relevant to the use of complex variables in the Lagrangian formalism. It is clear from Chapter I that the equations of electrodynamics have a more transparent structure in reciprocal space. The variables which define the field are then complex.

Part B is devoted to a presentation of the standard Lagrangian formulation of classical electrodynamics with a nonrelativistic treatment of the particles. One shows that the Lagrange equations associated with a certain Lagrangian (the standard Lagrangian) reduce to the Maxwell–Lorentz equations. In fact, the standard Lagrangian uses potentials to describe the electromagnetic field; this results in redundant degrees of freedom, which causes certain difficulties, particularly when one tries to quantize the theory. One method for resolving these difficulties consists in eliminating certain degrees of freedom such as the scalar potential and in using the *Coulomb gauge*.

Part C is precisely devoted to a presentation of electrodynamics in the Coulomb gauge and to a discussion of several important aspects of this theory. One sees how the application of the Lagrangian formalism in this gauge as well as the Hamiltonian formalism allows one to proceed to a canonical quantization of classical electrodynamics and then to justify the commutation relations introduced heuristically in Chapter I.

A—REVIEW OF THE LAGRANGIAN AND HAMILTONIAN FORMALISM

We will review here, without going into detailed calculations, the general Lagrangian and Hamiltonian formalism for a finite number (§1) and then for a continuum of degrees of freedom (§2). The reader who is not familiar with the ideas introduced in §2 is advised to study Exercise 2, dealing with a chain of linear coupled oscillators. The passage to the continuous limit of such a system renders more plausible the results of §2 (the appearance of spatial derivatives in the Lagrangian density, replacement of Kronecker symbols by the Dirac delta function, and so on).

1. Systems Having a Finite Number of Degrees of Freedom

a) Dynamical Variables, the Lagrangian, and the Action

For a system having N degrees of freedom, giving the N generalized coordinates $x_1, \dots, x_N$ and the corresponding velocities $\dot{x}_1, \dots, \dot{x}_N$ at a given time completely determines the subsequent motion. The $2N$ quantities $x_1, \dots, x_N$ and $\dot{x}_1, \dots, \dot{x}_N$ form an ensemble of *dynamical variables*. The accelerations $\ddot{x}_1, \dots, \ddot{x}_N$ can be expressed at any time as a function of these variables. The resulting equations of motion are then second-order differential equations in time. The motion of the system is determined by integrating these equations.

It is equally possible to specify the motion of the system by means of a variational principle. In the Lagrangian approach, one postulates the existence of a function $L(x_j, \dot{x}_j, t)$, called the *Lagrangian*, which depends on the coordinates and the velocities (and perhaps explicitly on time), such that the integral of L between times t_1 and t_2 will be an extremum when $x_j(t)$ corresponds to the *real* path of the system between t_1 and t_2 [the initial and final coordinates $x_j(t_1)$ and $x_j(t_2)$ are assumed to be known]. The integral

$$S = \int_{t_1}^{t_2} L(x_j(t), \dot{x}_j(t), t)\, \mathrm{d}t \tag{A.1}$$

is the *action*, and the corresponding variational principle is called the principle of least action.

In the mechanics of a point particle, the Lagrangian is equal to the difference between the kinetic energy and the potential energy. In particular, for a particle moving in a time independent potential, the Lagrangian does not depend explicitly on time. In the following, we will preserve for isolated systems this time translation invariance and will not show an explicit time dependence for the Lagrangian, which will be written $L(x_j, \dot{x}_j)$.

b) Lagrange's Equations

In the subsection above, two possible approaches to the study of motion have been indicated: one local in time (the equations of motion), and the other global (the principle of least action). We will review briefly how the equations of motion can be derived from the principle of least action.

In expressing the stationarity of the action with respect to variations of path $\delta x_j(t)$ [provided $\delta x_j(t_1) = \delta x_j(t_2) = 0$] about the real path followed by the system, one shows that at each instant the dynamical variables should satisfy on that path relationships which are equivalent to the equations of motion. These are Lagrange's equations:

$$\frac{\mathrm{d}}{\mathrm{d}t}\left(\frac{\partial L}{\partial \dot{x}_j}\right) = \frac{\partial L}{\partial x_j}. \tag{A.2}$$

The explicit derivation of these equations can be found in Complement A_{II} and in a number of books (see the references at the end of the chapter).

c) Equivalent Lagrangians

The Lagrangian of a system is not unique. For example, if one adds to the Lagrangian L the total derivative with respect to time of an arbitrary function f depending on the coordinates x_j and the time, one gets a new function L':

$$L'(x_j, \dot{x}_j, t) = L(x_j, \dot{x}_j, t) + \frac{\mathrm{d}}{\mathrm{d}t} f(x_j, t) \tag{A.3}$$

which has the same properties as the initial Lagrangian L with respect to the principle of least action. The action integral S' relative to L' is written using (A.1) as

$$S' = \int_{t_1}^{t_2} L'\,\mathrm{d}t = S + f(x_j(t_2), t_2) - f(x_j(t_1), t_1). \tag{A.4}$$

Since the initial and final positions are fixed, it follows that S and S' differ only by a constant and thus have the same extremum. L and L' are then equivalent Lagrangians for the study of the dynamics of the system. The transformation (A.3) then allows one to transform a Lagrangian into an equivalent second Lagrangian.

Remark

The function f must not depend on the velocities $\dot{x}_j$, so that the new Lagrangian, like the original, depends only on the dynamical variables x_j and $\dot{x}_j$ and not on the accelerations.

d) Conjugate Momenta and the Hamiltonian

The momentum conjugate to x_j is defined as the partial derivative of the Lagrangian with respect to the velocity $\dot{x}_j$:

$$p_j = \frac{\partial L}{\partial \dot{x}_j} . \tag{A.5}$$

The time derivative of p_j is gotten by using the Lagrange equation (A.2):

$$\dot{p}_j = \frac{\partial L}{\partial x_j} . \tag{A.6}$$

The simplicity of Equation (A.6) suggests the use of the coordinates and momenta as dynamical variables rather than the coordinates and velocities. It is then preferable to substitute for the Lagrangian another function, the *Hamiltonian*, which is considered as a function of x_j and p_j and is defined by

$$H(x_j, p_j) = \sum_j \dot{x}_j p_j - L . \tag{A.7}$$

It suffices then to differentiate (A.7) and to use (A.5) to find that $\mathrm{d}H$ is only a function of $\mathrm{d}x_j$ and $\mathrm{d}p_j$, which leads to the following equations, called *Hamilton's equations*:

$$\begin{cases} \dot{x}_j = \dfrac{\partial H}{\partial p_j} & \text{(A.8.a)} \\[2ex] \dot{p}_j = -\dfrac{\partial H}{\partial x_j} . & \text{(A.8.b)} \end{cases}$$

In order to describe the dynamics of the system, the N Lagrange equations (A.2), which are second-order differential equations, have thus been replaced by a system of $2N$ first-order differential equations (A.8.a) and (A.8.b).

In comparison with the Lagrangian formalism introduced earlier, the Hamiltonian formalism presents several advantages. First of all, if the Lagrangian does not depend explicitly on time, H is a constant of the motion which generally corresponds to the energy (*) and thus has a clear

(*) In certain cases one gets Hamiltonians which are different from the energy (see Exercise 1). This is never the case here.

physical significance. Additionally, the coordinates and momenta play a more symmetric role than the coordinates and velocities [this arises for example in the equations of motion (A.8.a) and (A.8.b)]. It follows that changes of variable are a priori more flexible than in the framework of the Lagrangian formalism, since one can mix the various coordinates and momenta. Finally, the introduction of the momenta and of the Hamiltonian is essential for the quantization of the theory.

e) Change of Dynamical Variables

i) *Change of Generalized Coordinates in the Lagrangian*

It may be useful, for the solution of a problem, to make a change of dynamical variables. In the framework of the Lagrangian formalism, only changes of coordinates which substitute the coordinates $X_1, \dots, X_N$ for $x_1, \dots, x_N$ such that

$$x_j = f_j(X_1 \dots X_N) \tag{A.9}$$

are allowed. The differentiation of (A.9) with respect to time gives the relationship between the old and new velocities. The new Lagrangian is gotten by replacing in $L(x_j, \dot{x}_j)$ the x_j and $\dot{x}_j$ as functions of X_k and $\dot{X}_k$, and the new action is equal to the old one.

The transformation (A.9) does not in general involve the velocities $\dot{X}_k$, since the accelerations $\ddot{X}_k$ would then occur in the Lagrangian.

ii) *The Special Case Where a Velocity Does Not Appear in the Lagrangian*

There is nevertheless a case where a transformation of the type (A.9) including the velocities is possible. This is when one of the velocities does not appear explicitly in the Lagrangian. It is then possible to completely eliminate the corresponding degree of freedom and to substitute for the initial Lagrangian a Lagrangian having fewer dynamical variables.

Assume, for example, that in the Lagrangian L, the velocity $\dot{x}_N$ does not appear explicitly. The Lagrangian L is then a function of N coordinates and $N - 1$ velocities and will be written as $L(x_j, x_N, \dot{x}_j)$. We will show that it is possible to replace this Lagrangian by another Lagrangian $\hat{L}$ which depends only on the $N - 1$ coordinates $x_1, \dots, x_{N-1}$ and the $N - 1$ velocities $\dot{x}_1, \dots, \dot{x}_{N-1}$. The Lagrange equation relative to x_N is

$$\frac{\partial L}{\partial x_N} = 0 \,. \tag{A.10}$$

This equation allows one to express x_N as a function of the $N - 1$ other coordinates x_j and the corresponding velocities $\dot{x}_j$, so that

$$x_N = u(x_j, \dot{x}_j) \,. \tag{A.11}$$

If in L one replaces x_N by its equivalent (A.11), one gets a function of x_j and $\dot{x}_j$ which is the Lagrangian $\hat{L}$:

$$\hat{L}(x_j, \dot{x}_j) = L(x_j, u(x_j, \dot{x}_j), \dot{x}_j) . \tag{A.12}$$

Note first of all that no acceleration appears in $\hat{L}$ in spite of the form of the transformation (A.11). That is due, of course, to the fact that L does not depend on $\dot{x}_N$. To show that $\hat{L}$ has all the properties of a Lagrangian, it is sufficient to note that, if the action relative to L is an extremum for all independent variations of the coordinates x_j and x_N, then that is also the case for the action relative to $\hat{L}$, in which x_N is fixed by (A.11).

Remark

One can verify directly that the Lagrange equations associated with $\hat{L}$ have the expected form. We calculate, for this purpose, the partial derivatives of $\hat{L}$ with respect to x_j and $\dot{x}_j$:

$$\begin{cases} \dfrac{\partial \hat{L}}{\partial x_j} = \dfrac{\partial L}{\partial x_j} + \dfrac{\partial L}{\partial x_N} \dfrac{\partial x_N}{\partial x_j} & \text{(A.13.a)} \\[2ex] \dfrac{\partial \hat{L}}{\partial \dot{x}_j} = \dfrac{\partial L}{\partial \dot{x}_j} + \dfrac{\partial L}{\partial x_N} \dfrac{\partial x_N}{\partial \dot{x}_j} . & \text{(A.13.b)} \end{cases}$$

When Lagrange's equation (A.10) relative to x_N is satisfied, L and $\hat{L}$ have the same partial derivatives with respect to x_j and $\dot{x}_j$. Lagrange's equations relative to x_j and derived from L thus involve those associated with $\hat{L}$. Note also that since the last term of (A.13.b) is zero, the momenta conjugate with x_j in L and $\hat{L}$ are equal.

iii) *Velocity and Momentum Transformation*

It is interesting to establish the transformations for the velocities and momenta when one changes variables in the Lagrangian. The equations (A.9), allowing one to go from the old coordinates x_j to the new coordinates X_k, give by differentiation

$$\dot{x}_j = \sum_k a_{jk} \dot{X}_k \tag{A.14}$$

with

$$a_{jk} = \frac{\partial f_j}{\partial X_k} . \tag{A.15}$$

We denote by p_j and P_k the momenta conjugate with x_j and X_k respectively. One gets

$$P_k = \frac{\partial L}{\partial \dot{X}_k} = \sum_j \frac{\partial L}{\partial \dot{x}_j} \frac{\partial \dot{x}_j}{\partial \dot{X}_k} \tag{A.16}$$

that is, on using (A.14),

$$P_k = \sum_k p_j a_{jk} . \tag{A.17}$$

The equations (A.14) and (A.17) can be rewritten in matrix form. Denoting by A the $N \times N$ matrix whose elements are a_{jk}, and by $(\dot{x})$, $(\dot{X})$, (p), and (P) the column vectors whose elements are $\dot{x}_j$, $\dot{X}_k$, p_j, and P_k respectively, one finds

$$(\dot{x}) = A(\dot{X}) \tag{A.18.a}$$

$$(P) = A^t(p) . \tag{A.18.b}$$

The transposed matrix A^t is assumed to be invertible, so that the relationship between the momenta can be rewritten by going, as with the velocities, from the new to old variables:

$$(p) = (A^t)^{-1} (P) . \tag{A.19}$$

It is clear then that the momenta transform like the velocities only if the matrix A is orthogonal.

iv) *Changes of Variables in the Hamiltonian Formalism*

The changes of variables in the framework of the Hamiltonian formalism are inherently broader than the transformations of the form (A.9), which depend only on the coordinates. Certain transformations of the form

$$X_i = g_i(x_1 \dots x_N, p_1 \dots p_N) \tag{A.20.a}$$

$$P_j = h_j(x_1 \dots x_N, p_1 \dots p_N) \tag{A.20.b}$$

are possible a priori. The equations of motion for the new variables X and P have a form analogous to (A.8) only if certain restrictions are imposed. One can show that the corresponding conditions are

$$\{ X_i, X_j \} = 0 \tag{A.21.a}$$

$$\{ P_i, P_j \} = 0 \tag{A.21.b}$$

$$\{ X_i, P_j \} = \delta_{ij} \tag{A.21.c}$$

where the Poisson bracket $\{a, b\}$ of the two functions a and b is defined by

$$\{ a, b \} = \sum_k \left(\frac{\partial a}{\partial x_k} \frac{\partial b}{\partial p_k} - \frac{\partial a}{\partial p_k} \frac{\partial b}{\partial x_k} \right) . \tag{A.22}$$

Quantities X and P satisfying (A.21) are called *canonically conjugate*.

f) Use of Complex Generalized Coordinates

i) *Introduction*

Up to this point we have considered only the case where the coordinates and velocities are real quantities. It can be useful to introduce complex quantities for the resolution of certain problems. Consider, for example, a Lagrangian depending on two coordinates x_1 and x_2 and their velocities, and introduce the complex variable

$$X = \frac{1}{\sqrt{2}}(x_1 + \mathrm{i}x_2) . \tag{A.23}$$

The Lagrangian is now a function of X and X^* as well as their derivatives with respect to time.

We will show below that the Lagrangian and Hamiltonian formalisms can be generalized to complex coordinates X and X^* [related to x_1 and x_2 by (A.23)], and that all the results obtained above remain valid (with certain amendments to the definitions, notably that for momentum) provided that X and X^* are considered formally as independent variables.

Remark

One could consider from the beginning a Lagrangian depending on complex dynamical variables. One has however to keep in mind that the action must be real, since the principle of least action involves finding a minimum of the action. The Lagrangian is thus a real quantity and depends on both X and X^*.

ii) *Lagrange's Equations*

The passage from variables x_1 and x_2 to X and X^* is linear and can be inverted. One deduces the following relations:

$$\frac{\partial L}{\partial X} = \frac{1}{\sqrt{2}}\left[\frac{\partial L}{\partial x_1} - \mathrm{i}\frac{\partial L}{\partial x_2}\right] \tag{A.24.a}$$

$$\frac{\partial L}{\partial X^*} = \frac{1}{\sqrt{2}}\left[\frac{\partial L}{\partial x_1} + \mathrm{i}\frac{\partial L}{\partial x_2}\right] \tag{A.24.b}$$

and the similar expressions for the derivatives relative to the velocities. The Lagrange equations for the complex variables are gotten then by combining the Lagrange equations relative to x_1 and x_2. One then gets two Lagrange equations, one relative to X and the other to X^*, which have the normal form (A.2).

iii) *Conjugate Momenta*

Rather than define the momentum conjugate to X by $\partial L/\partial \dot{X}$, we prefer to take

$$P = \left(\frac{\partial L}{\partial \dot{X}}\right)^* \tag{A.25}$$

To justify such a choice, note that the laws of transformation of the partial derivatives (A.24) give

$$P = \frac{1}{\sqrt{2}}(p_1 + ip_2)\,. \tag{A.26}$$

Comparison of (A.23) and (A.26) shows then that the real and imaginary parts of the momentum P correspond respectively to the momenta conjugate to the real and imaginary parts of the generalized coordinate X, which justifies the choice of (A.25). Equations similar to (A.24) for velocities show in addition that

$$\left(\frac{\partial L}{\partial \dot{X}}\right)^* = \frac{\partial L}{\partial \dot{X}^*}\,. \tag{A.27}$$

In fact, this property is general and is a consequence of the real nature of the Lagrangian. An equivalent definition of P is then

$$P = \frac{\partial L}{\partial \dot{X}^*}\,. \tag{A.28}$$

iv) *The Hamiltonian*

The Hamiltonian introduced in (A.7) depends on the quantity $\Sigma_j \dot{x}_j p_j$. By applying (A.23) and (A.26), one easily sees that

$$\dot{x}_1 p_1 + \dot{x}_2 p_2 = \dot{X}P^* + \dot{X}^* P\,. \tag{A.29}$$

It follows that the Hamiltonian H, expressed as a function of complex variables, is

$$H = \dot{X}P^* + \dot{X}^* P - L \tag{A.30}$$

H is clearly real.

v) *Change of Complex Variables*

It is possible as above to imagine a change in complex variables transforming an ensemble of complex coordinates $X_1, \ldots, X_N$, $X_1^*, \ldots, X_N^*$ to another ensemble $Z_1, \ldots, Z_N, Z_1^*, \ldots, Z_N^*$. If one requires that the momenta (defined in A.25) transform like velocities, there are certain constraints on the matrix of partial derivatives of the old coordinates with respect to the new ones. A development analogous to that in §A.1.*e*.iii shows that this matrix must be unitary.

g) Coordinates, Momenta, and Hamiltonian in Quantum Mechanics

The various physical quantities of a system become operators in quantum mechanics. These operators act in an abstract space called the state space, which has the properties of a Hilbert space. The canonical commutation relations between Cartesian components of the position and momentum operators **x** and **p** are equal to

$$\begin{cases} [x_n, x_m] = 0 & \text{(A.31.a)} \\ [p_n, p_m] = 0 & \text{(A.31.b)} \\ [x_n, p_m] = i\hbar\, \delta_{nm}\,. & \text{(A.31.c)} \end{cases}$$

In quantum mechanics, the state of a system is described by a vector $|\psi\rangle$ of Hilbert space. One can at this stage adopt essentially two points of view, either assuming that the state vector is fixed and the operators evolve with time (Heisenberg) or assuming that the operators are fixed and the state vector evolves with time (Schrödinger).

In the first point of view the evolution of a physical quantity G is described by the Heisenberg equation

$$i\hbar \frac{\mathrm{d}}{\mathrm{d}t} G = [G, H] \tag{A.32}$$

where H is the quantum operator associated with the Hamiltonian. In the case where the operator G corresponds to a coordinate or a momentum, the equations derived from (A.32) are the quantum equivalents of Hamilton's equations (A.8).

In the second point of view, the operators are fixed and the evolution of the state vector is determined by Schrödinger's equation

$$i\hbar \frac{\mathrm{d}}{\mathrm{d}t} |\,\psi\,\rangle = H\,|\,\psi\,\rangle\,. \tag{A.33}$$

Mathematically, the correspondence between these two points of view is via a unitary transformation on the state vector.

Remark

In the case where a velocity does not appear in the initial Lagrangian, the conjugate momentum associated with the corresponding coordinate is identically zero. This poses a serious problem for quantization, since it is then impossible to postulate the canonical quantization relation (A.31). One can resolve this problem by eliminating from the Lagrangian the coordinate associated with this velocity (see §A.1.*e*.ii). The conjugate momenta are then calculated from the new Lagrangian (with a reduced number of dynamical variables), and the canonical quantization relations (A.31) are then imposed.

The complex variables have been introduced in §A.1.*f*. We now examine the canonical commutation relations in this case. Equations (A.23) and (A.26) show that

$$[X, P] = \frac{1}{2}[x_1 + ix_2, p_1 + ip_2] \tag{A.34.a}$$

$$[X, P^+] = \frac{1}{2}[x_1 + ix_2, p_1 - ip_2]\,. \tag{A.34.b}$$

It follows then from (A.31) that

$$[X, P] = 0 \tag{A.35.a}$$
$$[X^+, P^+] = 0 \tag{A.35.b}$$
$$[X, P^+] = i\hbar \tag{A.35.c}$$
$$[X^+, P] = i\hbar \tag{A.35.d}$$

(the other commutators, between two X or between two P, are zero). The definition (A.25), which we have taken for the conjugate momentum, leads to a nonzero commutation relation between the operator X and the adjoint of the operator associated with the conjugate momentum.

2. A System with a Continuous Ensemble of Degrees of Freedom

a) Dynamical Variables

The state of the system is now determined by a set—no longer discrete, but *continuous*—of dynamical variables. This extension is necessary in so far as one wishes to study the electromagnetic field, which is defined by its value at all points of space. We thus consider generalized coordinates which depend on a continuous index (denoted by **r**, a point in three-dimensional space, in anticipation of the application to the electromagnetic field) and a discrete index j (which varies from 1 to N).

As with the discrete case, the coordinates $A_j(\mathbf{r})$ and the velocities $\dot{A}_j(\mathbf{r})$ defined by

$$\dot{A}_j(\mathbf{r}, t) = \frac{\partial}{\partial t} A_j(\mathbf{r}, t) \tag{A.36}$$

form an ensemble of dynamical variables for the system. It is important to stress the fact that in the Lagrangian formalism developed below, **r** is not a dynamical variable but an index (of the same nature as j).

b) The Lagrangian

The *Lagrangian* L, which is a function of the dynamical variables $A_j(\mathbf{r})$ and $\dot{A}_j(\mathbf{r})$ (where j and **r** take on all possible values), can have a very large variety of structures. We assume here that one can write

$$L = \int \mathrm{d}^3 r \, \mathscr{L} \tag{A.37}$$

where the function $\mathscr{L}$ is called the *Lagrangian density*. $\mathscr{L}$ is a real function of the coordinates $A_j(\mathbf{r})$, the velocities $\dot{A}_j(\mathbf{r})$, and also the spatial derivatives (denoted $\partial_i A_j$ with $\partial_i = \partial_x, \partial_y, \partial_z$) whose presence simply shows that the motion of the coordinate $A_j(\mathbf{r}, t)$ is coupled to the motion at a neighboring point in the same way as, in a problem with discrete variables q_j, the motion of q_j depends on q_{j-1} and q_{j+1} (see Exercise 2). It should be clear that these spatial derivatives are not new independent dynamical variables, but rather linear combinations of generalized coordinates. One can include a priori in the Lagrangian density the spatial derivatives of all orders (remember, though, that only the first-order time derivative is allowed). Now taking into account the later application to the electromagnetic field, we will only study Lagrangian densities of the form $\mathscr{L}(A_j, \dot{A}_j, \partial_i A_j)$.

Remarks

(i) One can imagine an explicit dependence of $\mathscr{L}$ on the point **r** and the time t. We will not show that, to prevent overburdening the notation.

(ii) The Lagrangian density that is used in electrodynamics contains spatial derivatives. Such a structure can easily be understood. Maxwell's equations describe the motion of fields coupled from point to point in space, and the absence of spatial derivatives in the Lagrangian density would lead to a theory where the field evolves independently at each point in space. The fact that the Maxwell equations involve the spatial derivatives of the field requires taking a Lagrangian density that likewise depends on the spatial derivatives. This suggests studying the Lagrangian density in reciprocal space rather than in real space, since it has been seen in Chapter I that the Maxwell equations are strictly local in reciprocal space. We will return to this point in Part B of this chapter.

c) Lagrange's Equations

In going from the discrete case to the continuous one, most of the equations written in §A.1 remain formally valid. However, certain operations (the derivative of the Lagrangian with respect to continuous variables, for example) are not mathematically obvious, and it will be useful to explain certain points in more detail.

Note first of all that the action S is, as in the discrete case, the time integral of the Lagrangian. By reason of the form postulated for the Lagrangian density, we can then write

$$S = \int_{t_1}^{t_2} \mathrm{d}t \int \mathrm{d}^3r \, \mathscr{L}(A_j, \dot{A}_j, \partial_i A_j) . \tag{A.38}$$

The principle of least action is of course unchanged; S is an extremum when $A_j(\mathbf{r}, t)$ corresponds to the actual motion of the field between times t_1 and t_2.

To establish the equations of motion, one uses the principle of least action. The same steps can be followed as in the discrete case; one studies the modification of S when the field is varied by the quantity $\delta A_j(\mathbf{r}, t)$ with respect to the path for which S is extremal [$\delta A_j(\mathbf{r}, t)$ being zero at the temporal limits t_1 and t_2 of the integral and likewise when $|\mathbf{r}|$ tends toward infinity]. By stating that S is extremal one then gets (see Complement $\mathrm{A_{II}}$) Lagrange's equations, which can be written in the form

$$\frac{\mathrm{d}}{\mathrm{d}t} \frac{\partial \mathscr{L}}{\partial \dot{A}_j} = \frac{\partial \mathscr{L}}{\partial A_j} - \sum_{i=x,y,z} \partial_i \frac{\partial \mathscr{L}}{\partial(\partial_i A_j)} . \tag{A.39}$$

Remark

Equation (A.39) uses the Lagrangian density $\mathscr{L}$, and not the Lagrangian L as in the discrete case (A.2). However, it is possible to write (A.39) in a form identical to (A.2). To do this, the notion of a "functional derivative" must be introduced (the extension of the idea of the partial derivative in the continuous case), and this is discussed in Complement $\mathrm{A_{II}}$. The introduction of the Lagrangian density is mathematically convenient in the sense that, since $\mathscr{L}$ depends only on a finite set of variables, the use of the partial derivative is perfectly clear.

The Lagrangian of a continuous system, like the Lagrangian of a discrete system, is not unique. One can add to the Lagrangian density the time derivative of a function and the divergence of an arbitrary field (but one which tends to 0 sufficiently quickly at infinity), possibly depending on the generalized coordinates $A_j(\mathbf{r})$:

$$\mathscr{L}' = \mathscr{L} + \frac{\mathrm{d}}{\mathrm{d}t} f_0(A_j(\mathbf{r}), \mathbf{r}, t) + \sum_{i=x,y,z} \partial_i f_i(A_j(\mathbf{r}), \mathbf{r}, t) . \tag{A.40}$$

To calculate the new Lagrangian L', it is necessary to integrate $\mathscr{L}'$ over space. The integral of $\boldsymbol{\nabla} \cdot \mathbf{f}$ then transforms into a surface integral at infinity, which vanishes by hypothesis, and L' differs from L only by the time derivative of some function. L' is then equivalent to L.

d) Conjugate Momenta and the Hamiltonian

The conjugate momentum is defined in the continuous case by generalizing the equations gotten in the discrete case. For a Lagrangian of the type (A.37), the conjugate momentum associated with the variable $A_j(\mathbf{r})$ has a simple form as a function of the Lagrangian density:

$$\Pi_j(\mathbf{r}) = \frac{\partial \mathscr{L}}{\partial \dot{A}_j(\mathbf{r})}. \tag{A.41}$$

Remark

It is easy to understand why (A.41) is the generalization to the continuous case of (A.5) for the discrete case. To see this we transform the integral (A.37) defining the Lagrangian to a sum over small spatial elements of volume a^3 centered on the points $\mathbf{r}_m(x_m, y_m, z_m)$:

$$L_a = \sum_m a^3 \mathscr{L}\left[A_j(\mathbf{r}_m), \dot{A}_j(\mathbf{r}_m), \frac{A_j(x_{m+1}, y_m, z_m) - A_j(\mathbf{r}_m)}{a}\right]. \tag{A.42}$$

The final argument in the bracket symbolizes the various quantities which, in the limit $a \to 0$, tend to the partial derivatives $\partial_i A_j$. The conjugate momentum associated with the variable $A_j(\mathbf{r}_m)$ can be found, for the Lagrangian L_a, as in the discrete case and is

$$\Pi_j^a(\mathbf{r}_m) = \frac{\partial L_a}{\partial \dot{A}_j(\mathbf{r}_m)} = a^3 \frac{\partial \mathscr{L}}{\partial \dot{A}_j(\mathbf{r}_m)}. \tag{A.43}$$

It appears then that the conjugate momentum $\Pi_j(\mathbf{r})$ defined in (A.41) is equal to the limiting value of $a^{-3}\Pi_j^a(\mathbf{r}_m)$ when a goes to zero. Now the limit of $a^{-3}\,\partial L_a/\partial \dot{A}_j(\mathbf{r}_m)$ when a goes to zero is nothing but the functional derivative of the Lagrangian $\eth L/\eth \dot{A}_j(\mathbf{r})$ (see Complement A_{II} for more details). Equation (A.41) can then be written in the equivalent form

$$\Pi_j(\mathbf{r}) = \frac{\eth L}{\eth \dot{A}_j(\mathbf{r})}. \tag{A.44}$$

A step analogous to that used in the discrete case allows one to go from the coordinate–velocity pairs of dynamical variables $\left(A_j(\mathbf{r}), \dot{A}_j(\mathbf{r})\right)$ to another pair made up of the coordinate $A_j(\mathbf{r})$ and its conjugate momentum $\Pi_j(\mathbf{r})$ and then to introduce the Hamiltonian H and the Hamiltonian density $\mathscr{H}$:

$$H = \int \mathrm{d}^3r \sum_j \Pi_j(\mathbf{r})\,\dot{A}_j(\mathbf{r}) - L = \int \mathrm{d}^3r\, \mathscr{H} \tag{A.45.a}$$

$$\mathscr{H} = \sum_j \Pi_j(\mathbf{r})\, \dot{A}_j(\mathbf{r}) - \mathscr{L}\,. \tag{A.45.b}$$

Remark

It is similarly possible to introduce the momentum and the momentum density of the field as well as the angular momentum and the angular momentum density (see Exercise 5).

As in the discrete case, the introduction of the Hamiltonian formalism introduces new dynamical equations (Hamilton's equations). These can be written simply using the Hamiltonian density $\mathscr{H}$ as

$$\begin{cases} \dot{A}_j = \dfrac{\partial \mathscr{H}}{\partial \Pi_j} & \text{(A.46.a)} \\[2ex] \dot{\Pi}_j = -\dfrac{\partial \mathscr{H}}{\partial A_j} + \sum_i \partial_i \dfrac{\partial \mathscr{H}}{\partial(\partial_i A_j)}\,. & \text{(A.46.b)} \end{cases}$$

These equations can be written directly as a function of the Hamiltonian H with the aid of the functional derivative (see Complement A_{II}). They are then identical to Hamilton's equations in the discrete case.

e) Quantization

As in the discrete case, the fundamental commutation relation is imposed between the operators associated with a coordinate and its conjugate momentum. In the case where the field is expressed as a function of its Cartesian coordinates and where the three coordinates are independent dynamical variables, the canonical commutation relations can be written

$$\begin{cases} [A_n(\mathbf{r}), A_m(\mathbf{r}')] = 0 & \text{(A.47.a)} \\ [\Pi_n(\mathbf{r}), \Pi_m(\mathbf{r}')] = 0 & \text{(A.47.b)} \\ [A_n(\mathbf{r}), \Pi_m(\mathbf{r}')] = i\hbar\, \delta_{nm}\, \delta(\mathbf{r} - \mathbf{r}')\,. & \text{(A.47.c)} \end{cases}$$

Remarks

(i) In the continuous case, the Dirac distribution $\delta(\mathbf{r} - \mathbf{r}')$ has replaced the Kronecker symbol of the discrete case. This can be understood if one recalls the situation considered in the Remark of the preceding subsection (§A.2.*d*), where the space was divided into cells of dimension a^3. In that case, the rules of the discrete case give

$$[A_n(\mathbf{r}_k), \Pi_m^a(\mathbf{r}_l)] = i\hbar\, \delta_{nm}\, \delta_{kl}\,. \tag{A.48}$$

Since $\Pi_m(\mathbf{r}_l)$ is the limit of $a^{-3}\Pi_m^a(\mathbf{r}_l)$ when a goes to zero, it follows that

$$[A_n(\mathbf{r}_k), \Pi_m(\mathbf{r}_l)] = i\hbar\, \delta_{nm} \lim_{a \to 0} \frac{\delta_{kl}}{a^3} \tag{A.49}$$

δ_{kl} is 1 if $\mathbf{r}_k$ and $\mathbf{r}_l$ belong to the same cell of volume a^3, and 0 otherwise. It appears then that, given as a function of $\mathbf{r}_k$ and $\mathbf{r}_l$, the limiting value of the second term of (A.49) is just the Dirac distribution $\delta(\mathbf{r}_k - \mathbf{r}_l)$.

(ii) The commutation relations (A.47) are valid only if the three components of the field A_x, A_y and A_z are independent dynamical variables. We will see that in electrodynamics there are situations where this is not the case. It is important then to identify the truly independent dynamical variables in order to write the commutation relations correctly.

The state of a system is described by a vector in state space, and—as in the discrete case—one can treat the dynamics from the Heisenberg point of view or from the Schrödinger point of view.

f) Lagrangian Formalism with Complex Fields

The generalization of the foregoing results to the case of a complex field is particularly important, since in electrodynamics it is often most interesting to study the equations in reciprocal space.

Consider now a Lagrangian L and a Lagrangian density $\mathscr{L}$ dependent on the complex fields $\mathscr{A}_j$ and their velocities $\dot{\mathscr{A}}_j$. Since L must be real, $\mathscr{L}$ must then depend on $\mathscr{A}_j^*$ and $\dot{\mathscr{A}}_j^*$, with the result that

$$L = \int d^3k\, \mathscr{L}(\mathscr{A}_j, \dot{\mathscr{A}}_j, \partial_i\mathscr{A}_j, \mathscr{A}_j^*, \dot{\mathscr{A}}_j^*, \partial_i\mathscr{A}_j^*) \tag{A.50}$$

—the integration variable being now denoted as $\mathbf{k}$ in anticipation of later applications, and ∂_i denoting $\partial/\partial k_i$.

Remark

The electrodynamic Lagrangian is simpler than (A.50). Since Maxwell's equations are strictly local in reciprocal space, so is the Lagrangian density. One then does not get the derivatives $\partial/\partial k_i$ in the electrodynamic Lagrangian (on the other hand, one has an explicit dependence on k which arises from the Fourier transform of the spatial derivatives). However, we will retain the form of the Lagrangian (A.50) for the general considerations in this subsection, the results providing on one hand application to other physical situations (see Exercise 7, where the Schrödinger equation is derived from a variational principle) and allowing on the other hand a clearer comparison with the real-field case.

Consider the following linear combinations:

$$\begin{cases} \mathscr{A}_j^R(\mathbf{k}) = \dfrac{\mathscr{A}_j(\mathbf{k}) + \mathscr{A}_j^*(\mathbf{k})}{\sqrt{2}} & \text{(A.51.a)} \\ \mathscr{A}_j^I(\mathbf{k}) = \dfrac{\mathscr{A}_j(\mathbf{k}) - \mathscr{A}_j^*(\mathbf{k})}{\mathrm{i}\sqrt{2}} & \text{(A.51.b)} \end{cases}$$

which allow the replacement in (A.50) of the complex variables with the real variables $\mathscr{A}_j^R(\mathbf{k})$ and $\mathscr{A}_j^I(\mathbf{k})$. It is then possible to develop the Lagrangian and Hamiltonian formalism previously set forth using these real variables then to restate the equations one gets as a function of $\mathscr{A}_j$ and $\mathscr{A}_j^*$. Such a procedure has already been carried out in the discrete case (§A.1.*f*), and we have seen that one gets the same result by considering at once the complex dynamical variables and their complex conjugates as independent dynamical variables. One establishes in this way two Lagrange equations relative to $\mathscr{A}_j$ and $\mathscr{A}_j^*$:

$$\frac{\mathrm{d}}{\mathrm{d}t}\frac{\partial \mathscr{L}}{\partial \dot{\mathscr{A}}_j} = \frac{\partial \mathscr{L}}{\partial \mathscr{A}_j} - \sum_i \partial_i \frac{\partial \mathscr{L}}{\partial(\partial_i \mathscr{A}_j)} \tag{A.52.a}$$

$$\frac{\mathrm{d}}{\mathrm{d}t}\frac{\partial \mathscr{L}}{\partial \dot{\mathscr{A}}_j^*} = \frac{\partial \mathscr{L}}{\partial \mathscr{A}_j^*} - \sum_i \partial_i \frac{\partial \mathscr{L}}{\partial(\partial_i \mathscr{A}_j^*)}. \tag{A.52.b}$$

As for the momentum conjugate with the variable $\mathscr{A}_j(\mathbf{k})$, it is defined in a fashion similar to the discrete case:

$$\pi_j(\mathbf{k}) = \left(\frac{\partial \mathscr{L}}{\partial \dot{\mathscr{A}}_j(\mathbf{k})}\right)^* \tag{A.53}$$

This choice assures one, as in the discrete case, that the real and imaginary parts of $\pi_j(\mathbf{k})$ are clearly the momenta conjugate with $\mathscr{A}_j^R$ and $\mathscr{A}_j^I$. It is for this reason (see the Remark below) that we have chosen the definition (A.53) in preference to the usual convention [where one does not have the complex conjugate in the right hand member of (A.53)].

The fact that $\mathscr{L}$ is real shows that (A.53) can finally be written

$$\pi_j(\mathbf{k}) = \frac{\partial \mathscr{L}}{\partial \dot{\mathscr{A}}_j^*(\mathbf{k})} \tag{A.54}$$

which shows that the momentum conjugate with $\mathscr{A}_j^*$ is π_j^*.

Remark

The definition (A.53) of the conjugate momentum has another advantage when the fields $\mathscr{A}_j(\mathbf{k})$ are the Fourier transforms of the fields $A_j(\mathbf{r})$. The $\pi_j(\mathbf{k})$ are then the Fourier transforms of the momenta $\Pi_j(\mathbf{r})$ conjugate with the variables $A_j(\mathbf{r})$. To understand this result, first consider going from the coordinates $A_j(\mathbf{r})$ to the coordinates $\mathscr{A}_j(\mathbf{k})$ as a change of variables to which the results of §A.1.*f* can be applied. It has been stated there, for discrete variables, that the momenta transform like velocities if the transformation matrix is unitary. This property can be generalized to the continuous case and is clearly fulfilled by the Fourier transformation. In addition, the transformation being linear, the coordinates transform in the same way as the velocities and momenta. It is also possible to give a direct proof of this result by using the definition of the conjugate momentum in terms of the functional derivative (A.44). One can write

$$\frac{\partial L}{\partial \dot{\mathscr{A}}_j(\mathbf{k})} = \int \mathrm{d}^3 r \, \frac{\partial L}{\partial \dot{A}_j(\mathbf{r})} \, \frac{\partial \dot{A}_j(\mathbf{r})}{\partial \dot{\mathscr{A}}_j(\mathbf{k})} . \tag{A.55}$$

Now, by differentiating the relation connecting the fields in real space with that in reciprocal space [Chapter I, Equation (B.1)], one gets

$$\frac{\partial \dot{A}_j(\mathbf{r})}{\partial \dot{\mathscr{A}}_j(\mathbf{k})} = \frac{\mathrm{e}^{i\mathbf{k}.\mathbf{r}}}{(2\,\pi)^{3/2}} . \tag{A.56}$$

Since Equation (A.53) can also be written

$$\pi_j(\mathbf{k}) = \left(\frac{\partial L}{\partial \dot{\mathscr{A}}_j(\mathbf{k})} \right)^* \tag{A.57}$$

it follows that

$$\pi_j(\mathbf{k}) = \frac{1}{(2\,\pi)^{3/2}} \int \mathrm{d}^3 r \, \Pi_j(\mathbf{r}) \, \mathrm{e}^{-i\mathbf{k}.\mathbf{r}} \tag{A.58}$$

which demonstrates that the conjugate momenta are transformed like the variables.

g) Hamiltonian Formalism and Quantization with Complex Fields

To find the relationships involving the Hamiltonian density and Hamilton's equations or the canonical commutation relations, it is necessary to start with the expressions found in §§A.2.*d* and A.2.*e* for the real fields $\mathscr{A}_j^R$ and $\mathscr{A}_j^I$, and to combine them to get the corresponding expressions for the complex field. This has been done in the discrete case (§§A.1.*f* and A.1.*g*) and here the results will be given without the intermediate steps.

For the Hamiltonian density one finds

$$\mathscr{H} = \sum_j (\pi_j \dot{\mathscr{A}}_j^* + \pi_j^* \, \dot{\mathscr{A}}_j) - \mathscr{L} \tag{A.59}$$

which generalizes the expression (A.30) relative to the discrete case.

Hamilton's equations are written

$$\dot{\mathscr{A}}_j^* = \frac{\partial \mathscr{H}}{\partial \pi_j} \tag{A.60.a}$$

$$\dot{\pi}_j^* = -\frac{\partial \mathscr{H}}{\partial \mathscr{A}_j} + \sum_i \partial_i \frac{\partial \mathscr{H}}{\partial(\partial_i \mathscr{A}_j)}. \tag{A.60.b}$$

The canonical commutation relations for the quantized fields are finally

$$[\mathscr{A}_n(\mathbf{k}),\ \pi_m(\mathbf{k}')] = 0 \tag{A.61.a}$$

$$[\mathscr{A}_n(\mathbf{k}),\ \pi_m^+(\mathbf{k}')] = \mathrm{i}\hbar\, \delta_{mn}\, \delta(\mathbf{k} - \mathbf{k}'). \tag{A.61.b}$$

The other commutators between $\mathscr{A}_n(\mathbf{k})$ and $\mathscr{A}_m(\mathbf{k}')$ or between $\pi_n(\mathbf{k})$ and $\pi_m(\mathbf{k}')$ are zero. As in the discrete case, the field operator and the adjoint of the operator associated with the momentum do not commute.

Remarks

(i) In the foregoing, quantization has been accomplished by associating with the dynamical variables and their conjugate momenta operators which satisfy the commutation relations (A.61). In fact, the fundamental requirement with respect to quantum theory involves the quantum equations governing the evolution of the variables $\mathscr{A}_n$ and π_n. These equations, written

$$\begin{cases} \dot{\mathscr{A}}_n = \dfrac{1}{\mathrm{i}\hbar}[\mathscr{A}_n, H] \\ \dot{\pi}_n = \dfrac{1}{\mathrm{i}\hbar}[\pi_n, H] \end{cases} \tag{A.62}$$

mean that the Hamiltonian is the generator of time translations. They must have a form analogous to that of the classical equations

$$\begin{cases} \dot{\mathscr{A}}_n = \partial H/\partial \pi_n^* \\ \dot{\pi}_n = -\, \partial H/\partial \mathscr{A}_n^*. \end{cases} \tag{A.63}$$

Such a condition is simply satisfied if one postulates the commutation relations (A.61) between $\mathscr{A}_n$ and π_n, since these relations imply

$$\begin{cases} [\mathscr{A}_n, H] = \mathrm{i}\hbar\, \partial H/\partial \pi_n^+ \\ [\pi_n, H] = -\, \mathrm{i}\hbar\, \partial H/\partial \mathscr{A}_n^+. \end{cases} \tag{A.64}$$

For certain quadratic Hamiltonians, it is equally possible to satisfy the same requirement by replacing the *commutators* $[A, B] = AB - BA$ with the *anti-commutators* $[A, B]_+ = AB + BA$ in (A.61):

$$\begin{cases} [\mathscr{A}_n(\mathbf{k}),\ \pi_m(\mathbf{k}')]_+ = 0 \\ [\mathscr{A}_n(\mathbf{k}),\ \pi_m^+(\mathbf{k}')]_+ = \mathrm{i}\hbar\, \delta_{mn}\, \delta(\mathbf{k} - \mathbf{k}') \end{cases} \tag{A.65}$$

the other anticommutators being zero. For example, we show in Exercise 8 that the quantization of the Schrödinger equation, considered as the equation of motion of a classical field $\psi(\mathbf{r})$, can be effected in a coherent fashion either with commutators or with anticommutators. In both cases, the Heisenberg equation for the quantum field $\Psi(\mathbf{r})$ associated with the classical field $\psi(\mathbf{r})$,

$$\dot{\Psi}(\mathbf{r}) = \frac{1}{i\hbar}[\Psi(\mathbf{r}), H] \tag{A.66}$$

has the form of a Schrödinger equation

$$i\hbar\dot{\Psi}(\mathbf{r}) = \left[-\frac{\hbar^2}{2m}\Delta + V(\mathbf{r})\right]\Psi(\mathbf{r}). \tag{A.67}$$

Note that the rules concerning the measurement of physical quantities are unchanged. For example, two physical quantities relative to the quantized field can be measured simultaneously only if the corresponding operators commute, whether the theory is quantized with the commutation relations (A.61) or with the anticommutation relations (A.65). However, it is necessary to mention here that the fields themselves are not necessarily physical variables. Thus, in the example of the quantization of the Schrödinger equation with anticommutators, one finds that it is not possible to give physical meaning to the operator $\Psi(\mathbf{r})$ (which has real and imaginary parts) as one can to an electric or magnetic field. Only the quadratic Hermitian functions of Ψ represent physical quantities, to which one then applies the measurement postulates. For example, $q\Psi^+(\mathbf{r})\Psi(\mathbf{r})$ is the operator associated with the charge density at point $\mathbf{r}$. The fact that $\Psi(\mathbf{r})$ is not a physical variable renders less troublesome certain of its properties—for example, the fact that $\Psi(\mathbf{r})$ anticommutes with itself.

(ii) Depending on whether the quantization of the field rests on commutators or anticommutators, the particles associated with the elementary excitations of the quantized field are bosons or fermions (see for example Exercise 8). When the field is relativistic, a link exists between the "spin" of the field and the statistics of the particles associated with it. Very general considerations (relativistic invariance, causality, positive energy) allow one to show that the quantization of a relativistic field of integer spin can only occur in a satisfactory way (that is, without violating the principles above) if it depends on commutators. In contrast, if the spin is half-integer, it is necessary to use anticommutators (*). Thus, the electromagnetic field, which is a vector field and has spin 1, must be quantized with commutators, with the result that the particles associated with it are bosons. In contrast, the Dirac field has spin $\frac{1}{2}$, and the particles associated with it (electrons and positrons) are fermions. Complement A_V of Chapter V gives an idea of the connection which exists in this case between the requirement for positive energy and the quantization by anticommutators.

(*) This result is known as the "spin–statistics" theorem; it is due to W. Pauli, *Phys. Rev.*, **58**, 716 (1940).

B—THE STANDARD LAGRANGIAN OF CLASSICAL ELECTRODYNAMICS

In this part we begin (§B.1) by giving the expression for the Lagrangian most generally used in classical electrodynamics, and which we call "the standard Lagrangian". We will show then (§B.2) that the Maxwell–Lorentz equations arise naturally as the Lagrange equations for one such Lagrangian. Finally (§B.3) we analyze general properties of the standard Lagrangian, namely, symmetry properties, gauge invariance, and redundancy of the dynamical field variables.

1. The Expression for the Standard Lagrangian

a) The Standard Lagrangian in Real Space

The Lagrangian for the system made up of particles interacting with the electromagnetic field is given as a function of the dynamical variables relative to each of the subsystems. The dynamical variables of the particles form a discrete set involving the components of the position $\mathbf{r}_\alpha$ and of the velocity $\dot{\mathbf{r}}_\alpha$ for the particles denoted by the index α. For the electromagnetic field, it is the potentials and not the fields which appear as "good" generalized coordinates in the Lagrangian formalism. This is not surprising, since the equations of motion for the potentials are second order in time, as the Lagrange equations, while the Maxwell equations for the field are first order. At each point $\mathbf{r}$, four generalized coordinates are required, these being the three components $A_j(\mathbf{r})$ of the vector potential $\mathbf{A}(\mathbf{r})$ and the scalar potential $U(\mathbf{r})$ and the four corresponding velocities $\dot{A}_j(\mathbf{r})$ and $\dot{U}(\mathbf{r})$, so that the field dynamical variables are

$$\{\mathbf{A}(\mathbf{r}), U(\mathbf{r}); \dot{\mathbf{A}}(\mathbf{r}), \dot{U}(\mathbf{r})\} \qquad \text{for all } \mathbf{r}. \tag{B.1}$$

The dynamics of the system particles + electromagnetic field can be derived from the standard Lagrangian

$$L = \sum_\alpha \frac{1}{2} m_\alpha \dot{\mathbf{r}}_\alpha^2 + \frac{\varepsilon_0}{2} \int \mathrm{d}^3 r [\mathbf{E}^2(\mathbf{r}) - c^2 \mathbf{B}^2(\mathbf{r})] + \sum_\alpha [q_\alpha \dot{\mathbf{r}}_\alpha \cdot \mathbf{A}(\mathbf{r}_\alpha) - q_\alpha U(\mathbf{r}_\alpha)] \tag{B.2}$$

the fields $\mathbf{E}$ and $\mathbf{B}$ being given as a function of the potentials $\mathbf{A}$ and U:

$$\mathbf{E}(\mathbf{r}) = -\nabla U(\mathbf{r}) - \dot{\mathbf{A}}(\mathbf{r}) \tag{B.3.a}$$

$$\mathbf{B}(\mathbf{r}) = \nabla \times \mathbf{A}(\mathbf{r}) . \tag{B.3.b}$$

We will show below that this Lagrangian gives back the Maxwell–Lorentz equations, which will justify a posteriori the choice (B.2). This Lagrangian has three terms: the Lagrangian for the particles, L_P [first term of (B.2)]; the Lagrangian of the electromagnetic field, L_R [second term of (B.2)], and the interaction Lagrangian L_I [last term of (B.2)]:

$$L = L_P + L_R + L_I \tag{B.4.a}$$

$$L_P = \sum_\alpha \frac{1}{2} m_\alpha \dot{\mathbf{r}}_\alpha^2 \tag{B.4.b}$$

$$L_R = \frac{\varepsilon_0}{2} \int \mathrm{d}^3r [\mathbf{E}^2(\mathbf{r}) - c^2 \mathbf{B}^2(\mathbf{r})] \tag{B.4.c}$$

$$L_I = \sum_\alpha [q_\alpha \dot{\mathbf{r}}_\alpha \cdot \mathbf{A}(\mathbf{r}_\alpha) - q_\alpha U(\mathbf{r}_\alpha)] \,. \tag{B.4.d}$$

By using the charge density $\rho(\mathbf{r})$ and the current $\mathbf{j}(\mathbf{r})$ introduced in Chapter I [see Equations (A.5.a) and (A.5.b)], one can then rewrite L_I in the form

$$L_I = \int \mathrm{d}^3r [\mathbf{j}(\mathbf{r}) \cdot \mathbf{A}(\mathbf{r}) - \rho(\mathbf{r})\, U(\mathbf{r})] \tag{B.4.e}$$

Finally, regrouping (B.4.c) and (B.4.e) leads to the introduction of the Lagrangian density $\mathscr{L}$:

$$\mathscr{L}(\mathbf{r}) = \frac{\varepsilon_0}{2} [\mathbf{E}^2(\mathbf{r}) - c^2 \mathbf{B}^2(\mathbf{r})] + \mathbf{j}(\mathbf{r}) \cdot \mathbf{A}(\mathbf{r}) - \rho(\mathbf{r})\, U(\mathbf{r}) \tag{B.5.a}$$

and the following form for the standard Lagrangian:

$$L = \sum_\alpha \frac{1}{2} m_\alpha \dot{\mathbf{r}}_\alpha^2 + \int \mathrm{d}^3r\, \mathscr{L}(\mathbf{r}) \tag{B.5.b}$$

Note that the interaction term (B.4.e) is local; the current density (or the charge density) at point $\mathbf{r}$ is multiplied by the vector (or scalar) potential at the same point. In the field Lagrangian (B.4.c), spatial derivatives of the potentials arise through $\mathbf{E}$ and $\mathbf{B}$, which expresses a coupling between the field variables from point to point. This coupling is the origin of the propagation of the free field.

b) The Standard Lagrangian in Reciprocal Space

We have seen in Chapter I that the Maxwell equations are much simpler in reciprocal space. In the same way, it is interesting to express the standard Lagrangian as a function of the potentials in reciprocal space.

The Parseval–Plancherel equality allows one to rewrite (B.5) in the form

$$L = \sum_\alpha \frac{1}{2} m_\alpha \dot{\mathbf{r}}_\alpha^2 + \frac{\varepsilon_0}{2} \int d^3k [| \mathscr{E}(\mathbf{k}) |^2 - c^2 | \mathscr{B}(\mathbf{k}) |^2] +$$

$$+ \int d^3k [\jmath^*(\mathbf{k}) \cdot \mathscr{A}(\mathbf{k}) - \rho^*(\mathbf{k}) \mathscr{U}(\mathbf{k})] \qquad \text{(B.6)}$$

Equation (B.6) suggests choosing as dynamical variables the components of the potentials in reciprocal space as well as their velocities. However, it is necessary to take several precautions: going from real space to reciprocal space corresponds to a change of variables which transforms real quantities into complex quantities. The new variables then have twice as many degrees of freedom as the old variables. But there are constraint relationships tied to the fact that $\mathbf{A}(\mathbf{r})$ and $U(\mathbf{r})$ are real:

$$\mathscr{A}(\mathbf{k}) = \mathscr{A}^*(-\mathbf{k}) \qquad \text{(B.7.a)}$$

$$\mathscr{U}(\mathbf{k}) = \mathscr{U}^*(-\mathbf{k}) . \qquad \text{(B.7.b)}$$

If the potentials are known in a "reciprocal half space", they are known everywhere. One is then led to take as independent variables the potentials and their complex conjugates in only half of reciprocal space. The equalities

$$\mathscr{E}(-\mathbf{k}) \cdot \mathscr{E}^*(-\mathbf{k}) = \mathscr{E}^*(\mathbf{k}) \cdot \mathscr{E}(\mathbf{k}) \qquad \text{(B.8.a)}$$

$$\jmath^*(-\mathbf{k}) \cdot \mathscr{A}(-\mathbf{k}) = \jmath(\mathbf{k}) \cdot \mathscr{A}^*(\mathbf{k}) \qquad \text{(B.8.b)}$$

which follow from (B.7), allow the rewriting of the Lagrangian (B.6) as a function of the field variables in a half space. Denoting by $\int d^3k$ the integral extended over a half volume of the reciprocal space and by $\overline{\mathscr{L}}$ the Lagrangian density in the reciprocal space, one gets

$$L = \sum_\alpha \frac{1}{2} m_\alpha \dot{\mathbf{r}}_\alpha^2 + \int d^3k \, \overline{\mathscr{L}} \qquad \text{(B.9.a)}$$

$$\overline{\mathscr{L}} = \varepsilon_0 [| \mathscr{E}(\mathbf{k}) |^2 - c^2 | \mathscr{B}(\mathbf{k}) |^2] +$$

$$+ [\jmath^*(\mathbf{k}) \cdot \mathscr{A}(\mathbf{k}) + \jmath(\mathbf{k}) \cdot \mathscr{A}^*(\mathbf{k}) - \rho^*(\mathbf{k}) \mathscr{U}(\mathbf{k}) - \rho(\mathbf{k}) \mathscr{U}^*(\mathbf{k})] \qquad \text{(B.9.b)}$$

or, again expressing $\mathscr{E}$ and $\mathscr{B}$ as functions of $\mathscr{A}$ and $\mathscr{U}$,

$$\mathscr{E} = -\dot{\mathscr{A}} - i\mathbf{k}\mathscr{U} \qquad \text{(B.10.a)}$$

$$\mathscr{B} = i\mathbf{k} \times \mathscr{A} \qquad \text{(B.10.b)}$$

$$\overline{\mathscr{L}} = \varepsilon_0[|\dot{\mathscr{A}}(\mathbf{k}) + \mathrm{i}\mathbf{k}\mathscr{U}(\mathbf{k})|^2 - c^2|\mathbf{k} \times \mathscr{A}(\mathbf{k})|^2] +$$
$$+ [j^*(\mathbf{k}) \cdot \mathscr{A}(\mathbf{k}) + j(\mathbf{k}) \cdot \mathscr{A}^*(\mathbf{k}) - \rho^*(\mathbf{k})\mathscr{U}(\mathbf{k}) - \rho(\mathbf{k})\mathscr{U}^*(\mathbf{k})]. \quad (B.11)$$

This new form, equivalent to the standard Lagrangian, presents certain advantages. First of all, the Lagrangian density is strictly local in **k**. The derivatives of $\mathscr{A}$ and $\mathscr{U}$ with respect to **k** do not appear (there is no coupling between neighboring points as in real space). Additionally, in (B.9), the contribution of the various modes of the field appear explicitly. As we shall see below, it is then very easy to separate the contributions of the nonrelativistic modes, or those of the long-wavelength modes, for which an electric dipole approximation is possible.

2. The Derivation of the Classical Electrodynamic Equations from the Standard Lagrangian

a) Lagrange's Equation for Particles

Since the particle variables are discrete, we apply Lagrange's equation (A.2) to the standard Lagrangian (B.2). One calculates first $\partial L/\partial(\mathbf{r}_\alpha)_i$ and $\partial L/\partial(\dot{\mathbf{r}}_\alpha)_i$:

$$\frac{\partial L}{\partial(\mathbf{r}_\alpha)_i} = -q_\alpha \frac{\partial}{\partial(\mathbf{r}_\alpha)_i} U(\mathbf{r}_\alpha, t) + q_\alpha \frac{\partial}{\partial(\mathbf{r}_\alpha)_i}[\dot{\mathbf{r}}_\alpha \cdot \mathbf{A}(\mathbf{r}_\alpha, t)] \quad (B.12)$$

which, using the vector identity

$$\nabla(\mathbf{A} \cdot \mathbf{B}) = (\mathbf{B} \cdot \nabla)\mathbf{A} + (\mathbf{A} \cdot \nabla)\mathbf{B} + \mathbf{B} \times (\nabla \times \mathbf{A}) + \mathbf{A} \times (\nabla \times \mathbf{B}) \quad (B.13)$$

becomes

$$\frac{\partial L}{\partial(\mathbf{r}_\alpha)_i} = -q_\alpha \frac{\partial}{\partial(\mathbf{r}_\alpha)_i} U(\mathbf{r}_\alpha, t) + q_\alpha(\dot{\mathbf{r}}_\alpha \cdot \nabla) A_i(\mathbf{r}_\alpha, t) +$$
$$+ q_\alpha[\dot{\mathbf{r}}_\alpha \times (\nabla \times \mathbf{A}(\mathbf{r}_\alpha, t))]_i. \quad (B.14)$$

In addition

$$\frac{\partial L}{\partial(\dot{\mathbf{r}}_\alpha)_i} = m_\alpha(\dot{\mathbf{r}}_\alpha)_i + q_\alpha A_i(\mathbf{r}_\alpha, t). \quad (B.15)$$

The Lagrange equation describing the motion of particle α is gotten by differentiating (B.15) with respect to time:

$$\frac{\mathrm{d}}{\mathrm{d}t}\frac{\partial L}{\partial(\dot{\mathbf{r}}_\alpha)_i} = m_\alpha(\ddot{\mathbf{r}}_\alpha)_i + q_\alpha \frac{\partial}{\partial t} A_i(\mathbf{r}_\alpha, t) + q_\alpha(\dot{\mathbf{r}}_\alpha \cdot \nabla) A_i(\mathbf{r}_\alpha, t) \quad (B.16)$$

and by setting that expression equal to (B.14). One gets finally

$$m_\alpha \ddot{\mathbf{r}}_\alpha = q_\alpha \left[- \frac{\partial \mathbf{A}(\mathbf{r}_\alpha, t)}{\partial t} - \nabla U(\mathbf{r}_\alpha, t) \right] + q_\alpha \dot{\mathbf{r}}_\alpha \times (\nabla \times \mathbf{A}(\mathbf{r}_\alpha, t)) \quad \text{(B.17.a)}$$

$$m_\alpha \ddot{\mathbf{r}}_\alpha = q_\alpha \mathbf{E}(\mathbf{r}_\alpha) + q_\alpha \dot{\mathbf{r}}_\alpha \times \mathbf{B}(\mathbf{r}_\alpha) \quad \text{(B.17.b)}$$

which is the Lorentz equation.

b) The Lagrange Equation Relative to the Scalar Potential

For the equations relative to the field, one can use the Lagrangian density in real space or in reciprocal space. Here we take the second option, since it gives the quickest result. Starting with (B.11), one gets

$$\frac{\partial \bar{\mathscr{L}}}{\partial \mathscr{U}^*} = - \varepsilon_0 \, i\mathbf{k} \cdot [\dot{\mathscr{A}} + i\mathbf{k}\mathscr{U}] - \rho \, . \quad \text{(B.18)}$$

In addition, since $\dot{\mathscr{U}}^*$ does not appear in $\bar{\mathscr{L}}$,

$$\frac{\partial \bar{\mathscr{L}}}{\partial \dot{\mathscr{U}}^*} = 0 \, . \quad \text{(B.19)}$$

The Lagrange equation (A.52.b) is then written as

$$- i\mathbf{k} \cdot [\dot{\mathscr{A}} + i\mathbf{k}\mathscr{U}] = \frac{\rho}{\varepsilon_0} \quad \text{(B.20)}$$

which is finally

$$i\mathbf{k} \cdot \mathscr{E} = \frac{\rho}{\varepsilon_0} \quad \text{(B.21)}$$

and is nothing but one of Maxwell's equations written in reciprocal space [see (B.5.a) of Chapter I].

c) The Lagrange Equation Relative to the Vector Potential

Starting with (B.11) for $\bar{\mathscr{L}}$ and using the identity

$$(\mathbf{k} \times \mathscr{A}) \cdot (\mathbf{k} \times \mathscr{A}^*) = [(\mathbf{k} \times \mathscr{A}) \times \mathbf{k}] \cdot \mathscr{A}^* \quad \text{(B.22)}$$

one can derive

$$\frac{\partial \bar{\mathscr{L}}}{\partial \mathscr{A}_i^*} = \varepsilon_0 \, c^2 [\mathbf{k} \times (\mathbf{k} \times \mathscr{A})]_i + \mathscr{J}_i \, . \quad \text{(B.23.a)}$$

In addition

$$\frac{\partial \overline{\mathscr{L}}}{\partial \dot{\mathscr{A}}_i^*} = \varepsilon_0[\dot{\mathscr{A}}_i + ik_i \mathscr{U}]. \tag{B.23.b}$$

The Lagrange equation relative to $\mathscr{A}_i^*$ is then

$$\varepsilon_0[\ddot{\mathscr{A}}_i + ik_i \dot{\mathscr{U}}] = -\varepsilon_0 c^2[\mathbf{ik} \times (\mathbf{ik} \times \mathscr{A})]_i + j_i \tag{B.24}$$

which is finally, using (B.10),

$$\mathbf{ik} \times \mathscr{B} = \frac{1}{c^2}\dot{\mathscr{E}} + \frac{1}{\varepsilon_0 c^2} j. \tag{B.25}$$

One has here another of Maxwell's equations in reciprocal space [see (B.5.d) of Chapter I].

In conclusion, the application of the principle of least action to the standard Lagrangian has given us on the one hand the Lorentz equation for a particle in an electromagnetic field, and on the other the second pair of Maxwell equations which relates the fields to their sources. [The first pair of Maxwell equations results directly from Equations (B.10) relating the fields $\mathscr{E}$ and $\mathscr{B}$ to the potentials $\mathscr{A}$ and $\mathscr{U}$.]

3. General Properties of the Standard Lagrangian

a) Global Symmetries

The form of the Lagrangian is invariant under certain geometric transformations: translation and rotation with respect to the system of axes to which the particles and the field are referred. The Lagrangian is also invariant under a change of the time origin. From these invariance properties it is possible to derive expressions for a certain number of conserved quantities, namely, the momentum, the angular momentum, and the total energy of the system field + particles. (This is done in Complement B_{II}, on the form which the standard Lagrangian takes in the Coulomb gauge.)

The standard Lagrangian (B.2) does not transform simply under a Lorentz transformation. Indeed, it is clear that the standard Lagrangian does not treat the particles in a relativistic way, the Lagrangian of the particles, equal to $\Sigma_\alpha m_\alpha \dot{\mathbf{r}}_\alpha^2/2$, being purely Galilean. We are now going to show that the Lagrangian (B.2) can be gotten in the classical low-velocity limit ($v/c \ll 1$), starting from a relativistic Lagrangian, that is, one with a relativistically invariant action.

We note initially that the Lagrangian density of the electromagnetic field is a relativistic scalar field. Indeed, it is a function of the electromagnetic tensor field $F^{\mu\nu}$ [see Chapter I, Equation (B.28)] of the following form:

$$\mathscr{L}_R = -\varepsilon_0 \frac{c^2}{4} \sum_{\mu\nu} F_{\mu\nu} F^{\mu\nu} \tag{B.26}$$

which is manifestly invariant under a Lorentz transformation. The contribution to the action of the Lagrangian density of the free field is written

$$S_R = \int \mathrm{d}t \int \mathrm{d}^3r \, \mathscr{L}_R . \tag{B.27}$$

Now, $\mathscr{L}_R$ on one hand and the volume element $\mathrm{d}t\,\mathrm{d}^3r$ on the other are relativistic invariants. It is clear then that the action S_R is a relativistic invariant.

We will show now that the interaction Lagrangian between the particles and the field contributes equally to the action in a covariant fashion. For this it is sufficient to note that the infinitesimal variation of the action relative to the interaction of particle α with the field arises as the scalar product of the four-vector $\mathrm{d}x_\alpha^\mu$ with the four-potential A_μ:

$$\mathrm{d}S_I = L_I \, \mathrm{d}t = \sum_\alpha q_\alpha [\mathrm{d}\mathbf{r}_\alpha \cdot \mathbf{A}(\mathbf{r}_\alpha, t) - \mathrm{d}t \, U(\mathbf{r}_\alpha, t)] \tag{B.28}$$

$$\mathrm{d}S_I = -\sum_{\alpha\mu} q_\alpha \, \mathrm{d}x_\alpha^\mu \, A_\mu . \tag{B.29}$$

Finally, it suffices to transform the Lagrangian of the particles, L_P, to get a relativistic Lagrangian. To this end, we replace L_P by

$$L_P^{\mathrm{rel}} = -\sum_\alpha m_\alpha c^2 \sqrt{1 - \frac{\dot{\mathbf{r}}_\alpha^2}{c^2}} . \tag{B.30}$$

The infinitesimal variation of the action corresponding to (B.30) is then written

$$\mathrm{d}S = L_P^{\mathrm{rel}} \, \mathrm{d}t = -\sum_\alpha m_\alpha c^2 \sqrt{1 - \frac{\dot{\mathbf{r}}_\alpha^2}{c^2}} \, \mathrm{d}t = -\sum_\alpha m_\alpha c^2 \, \mathrm{d}\tau_\alpha \tag{B.31}$$

where

$$\mathrm{d}\tau_\alpha = \mathrm{d}t \sqrt{1 - \frac{\dot{\mathbf{r}}_\alpha^2}{c^2}} \tag{B.32}$$

is the proper time of the particle α. Since $\mathrm{d}\tau_\alpha$ and $\mathrm{d}S$ are relativistic invariants, the Lagrangian (B.30) is also a relativistic Lagrangian. Additionally, expansion of (B.30) in powers of $\dot{\mathbf{r}}_\alpha^2/c^2$ gives, to within a constant term $-\Sigma_\alpha m_\alpha c^2$, the Lagrangian L_P given in (B.4.b).

In conclusion, it appears possible to introduce a fully relativistic Lagrangian

$$L = -\sum_\alpha m_\alpha c^2 \sqrt{1 - \frac{\dot{\mathbf{r}}_\alpha^2}{c^2}} + \frac{\varepsilon_0}{2}\int d^3r[\mathbf{E}^2(\mathbf{r}) - c^2\,\mathbf{B}^2(\mathbf{r})] + \\ + \sum_\alpha [q_\alpha \dot{\mathbf{r}}_\alpha \cdot \mathbf{A}(\mathbf{r}_\alpha) - q_\alpha U(\mathbf{r}_\alpha)] \quad \text{(B.33)}$$

which can serve as the basis for classical electrodynamics. However, if one proceeds in this fashion, difficulties arise in quantization of the theory, primarily as a result of the impossibility of constructing a relativistic quantum theory for a fixed number of particles.

Remarks

(i) It turns out that the correct procedure for constructing a relativistic quantum theory involves starting from a classical theory where the particles are described, like radiation, as a relativistic field (Klein–Gordon field, Dirac field, etc.) coupled to the Maxwell field. Then when such a theory is quantized, the particles, indeterminate in number, appear as elementary excitations of the quantum matter field and interact with the photons, which are the elementary excitations of the quantized Maxwell field (see Complement A_V).

(ii) It is possible to justify the use of the standard nonrelativistic Lagrangian (B.2), and, as a result, of the Hamiltonian in the Coulomb gauge which we will derive below, by starting from relativistic quantum electrodynamics and examining the low-energy limit of this theory. One finds to the lowest order in v/c the dynamics described by (B.2). One also gets the interaction terms tied to the spins of the particles (see Complement B_V).

b) Gauge Invariance

The Maxwell–Lorentz theory of electrodynamics is manifestly invariant under a change of gauge, since only the electric and magnetic fields appear in the basic equations. Gauge invariance is less evident for the Lagrangian theory, which uses the potentials as variables to describe the field. It is thus appropriate to examine the consequences of a gauge change in the Lagrangian formalism.

Following Equations (A.12) from Chapter I, a gauge change is defined by

$$\mathbf{A}'(\mathbf{r}, t) = \mathbf{A}(\mathbf{r}, t) + \nabla F(\mathbf{r}, t) \quad \text{(B.34.a)}$$

$$U'(\mathbf{r}, t) = U(\mathbf{r}, t) - \frac{\partial}{\partial t} F(\mathbf{r}, t) \quad \text{(B.34.b)}$$

F can be an explicit function of $\mathbf{r}$ and t, but can also depend on the field variables, which are themselves functions of $\mathbf{r}$ and t.

In the transformation (B.34), the Lagrangian of the particles is evidently not modified; nor is the Lagrangian of the field, which depends only on the electric and magnetic fields. Only the interaction Lagrangian is changed. The gauge change amounts to adding to the Lagrangian density $\mathscr{L}$ of the field given by (B.5.a) the quantity

$$\mathscr{L}_1 = \mathbf{j} \cdot \nabla F + \rho \frac{\partial F}{\partial t} \tag{B.35}$$

which one can write in the form

$$\mathscr{L}_1 = \nabla(\mathbf{j}F) + \frac{\partial}{\partial t}(\rho F) - \left(\nabla \cdot \mathbf{j} + \frac{\partial \rho}{\partial t}\right) F \tag{B.36}$$

The first two terms add to the Lagrangian density a divergence and a time derivative. According to (A.40) this transforms the Lagrangian into an equivalent Lagrangian (see however the Remark below). As for the last term of (B.36), it is zero as a result of charge conservation. It then appears clear that charge conservation is a necessary condition for gauge invariance.

Remarks

(i) There is not total equivalence between the changes in the Lagrangian and the gauge transformations. For example, in (B.34), F can depend on $\mathbf{A}$, U, $\dot{\mathbf{A}}$, and $\dot{U}$, which are themselves functions of $\mathbf{r}$ and t. All transformations which leave Maxwell's equations and the fields $\mathbf{E}$ and $\mathbf{B}$ invariant are gauge transformations. On the other hand, it is only when F does not depend on the velocities $\dot{\mathbf{A}}$ and $\dot{U}$ that it also corresponds to a change in the Lagrangian, since otherwise the accelerations $\ddot{\mathbf{A}}$ and $\ddot{U}$ would appear in the Lagrangian. Conversely, the changes in the Lagrangian density defined by (A.40) do not necessarily correspond to a gauge transformation. By comparing (A.40) and (B.36), one sees that for that it is necessary that a function F exist such that

$$\mathbf{f} = \mathbf{j}F \tag{B.37.a}$$

$$f_0 = \rho F\,. \tag{B.37.b}$$

Now it is not possible in general to satisfy both these conditions.

(ii) In the gauge field theories, gauge invariance plays a much more fundamental role. Starting with the fields representing particles, one requires that the theory be invariant under a *local* change of phase of the fields. To realize this invariance, it appears necessary to introduce a vector field (the electromagnetic field) coupled to the field of the particles in such a way that the phase changes in the matter field entail gauge transformations of the vector field (see Exercise 9). One introduces in this way a fundamental relationship between the change of phase of the matter field and the gauge change of the electromagnetic field.

c) Redundancy of the Dynamical Variables

In the description of electrodynamics through the standard Lagrangian, the field is described at each point **r** by the potentials **A** and U and the corresponding velocities (B.1). Thus, the dynamical variables are eight in number at each point in space. Now the approach to electrodynamics in Chapter I, resting on the Maxwell–Lorentz equations, introduces six degrees of freedom for each point [the three components of the electric and magnetic fields **E**(**r**) and **B**(**r**)]. Besides this, writing the Maxwell equations in reciprocal space allows one to show that the longitudinal components $\mathscr{E}_{\parallel}(\mathbf{k})$ and $\mathscr{B}_{\parallel}(\mathbf{k})$ are fixed by algebraic equations [Equations (B.5.a) and (B.5.b) from Chapter I]; the evolution of the four other dynamical variables (the transverse electric and magnetic fields) is described by differential equations which are first order in time [Equations (B.49.a) and (B.49.b) of Chapter I]. It is thus evident that in describing the electromagnetic field by the potentials $\mathscr{A}$ and $\mathscr{U}$ one has introduced an overabundance of degrees of freedom. Thus constraint relations must exist between the dynamical field variables.

We now examine how these constraints appear. An analysis of the Lagrangian density $\bar{\mathscr{L}}$ in (B.11) shows that $\dot{\mathscr{U}}$ does not appear in this Lagrangian density. This implies, on one hand, that the conjugate momentum associated with the variable $\mathscr{U}$ is identically zero, and on the other hand, that the Lagrange equation (B.20) associated with $\mathscr{U}$ relates $\mathscr{U}$ to the other dynamical variables by an algebraic equation. This type of problem has already been considered in §A.1.e. When the velocity associated with a generalized coordinate does not appear in the Lagrangian, this coordinate can be eliminated by expressing it as a function of the other dynamical variables, giving a reduced Lagrangian. Here such a step allows the elimination of the scalar potential $\mathscr{U}$, and one gets a Lagrangian where only the three components of the vector potential $\mathscr{A}$ and their time derivatives appear. One can further reduce the number of degrees of freedom of the electromagnetic field through the choice of gauge. It follows from equations (B.8.a) and (B.26) of Chapter I that a choice of gauge amounts to fixing the longitudinal component of the vector potential $\mathscr{A}_{\parallel}$, which is otherwise arbitrary. This then leads to a satisfactory physical situation where the field has at each point four independent physical variables which correspond to the two transverse orthogonal components of the vector potential $\mathscr{A}_\varepsilon(\mathbf{k})$ and $\mathscr{A}_{\varepsilon'}(\mathbf{k})$ and to their time derivatives $\dot{\mathscr{A}}_\varepsilon(\mathbf{k})$ and $\dot{\mathscr{A}}_{\varepsilon'}(\mathbf{k})$.

Remarks

(i) $\mathscr{A}_\varepsilon(\mathbf{k})$ is taken in a reciprocal half space and satisfies

$$\mathscr{A}_\varepsilon(\mathbf{k}) = \boldsymbol{\varepsilon} \cdot \mathscr{A}(\mathbf{k}) \tag{B.38}$$

where $\boldsymbol{\varepsilon}$ is one of the two (real) transverse vectors. In the other half space we define

$$\mathscr{A}_\varepsilon(-\mathbf{k}) = \boldsymbol{\varepsilon} \cdot \mathscr{A}(-\mathbf{k}) \tag{B.39}$$

where $\boldsymbol{\varepsilon}$ is the same vector for $\mathbf{k}$ and $-\mathbf{k}$.

(ii) Since the fields in reciprocal space are complex, one could imagine that the component $\mathscr{A}_\varepsilon(\mathbf{k})$, for example, corresponds to two real degrees of freedom (the real and imaginary parts). In fact, since

$$\mathscr{A}_\varepsilon(-\mathbf{k}) = \boldsymbol{\varepsilon} \cdot \mathscr{A}(-\mathbf{k}) = \boldsymbol{\varepsilon} \cdot \mathscr{A}^*(\mathbf{k}) = \mathscr{A}_\varepsilon^*(\mathbf{k}) \tag{B.40}$$

one has for the set of points $\mathbf{k}$ and $-\mathbf{k}$ two real degrees of freedom, i.e. one at each point.

The step described above results in a reduced Lagrangian where the field is described only by four dynamical variables, and the symmetry between the four components of the four-potential in the standard Lagrangian is now destroyed. It is of course tempting to try to quantize the theory without going through the reduced Lagrangian, keeping the four components of the four-potential as independent variables. However, such a procedure is impossible if one starts from the standard Lagrangian, since the conjugate momentum $\pi_{\mathscr{U}}$ associated with $\mathscr{U}$ is identically zero according to (B.19). It is thus impossible to impose upon the operators associated with $\mathscr{U}$ and $\pi_{\mathscr{U}}$ the canonical commutation relations (A.61.b). The conservation of symmetry between the four components of the four-potential is then possible only through the use of another Lagrangian (see Chapter V).

The natural step to quantize the theory starting from the standard Lagrangian consists then in eliminating $\mathscr{U}$ to get a reduced Lagrangian, and then choosing a gauge by fixing $\mathscr{A}_{\parallel}$. The simplest possible choice corresponds to the Coulomb gauge (see Chapter I, §A.3). One is then naturally led to study electrodynamics in the Coulomb gauge. Other choices of gauge corresponding to other values of $\mathscr{A}_{\parallel}$ can of course be considered (examples of this are given in Chapter IV).

C—ELECTRODYNAMICS IN THE COULOMB GAUGE

In this final part, we will show how one eliminates the redundant dynamical variables in the standard Lagrangian (§C.1). This will lead to the Lagrangian in the Coulomb gauge, the properties of which will be examined in §C.2. We will then pass to the Hamiltonian formalism (§C.3) and to the canonical quantization of the theory (§C.4). Finally we will discuss the important characteristics of this theory (§C.5).

1. Elimination of the Redundant Dynamical Variables from the Standard Lagrangian

a) Elimination of the Scalar Potential

Following the route sketched in §B.3.c, we will use Lagrange's equation relative to $\mathscr{U}$ to express the scalar potential as a function of the other dynamical variables and thus get a Lagrangian depending on a smaller number of degrees of freedom.

Lagrange's equation (B.20) allows one to write

$$\mathscr{U} = \frac{1}{k^2}\left[ik\dot{\mathscr{A}}_{\parallel} + \frac{\rho}{\varepsilon_0}\right] \tag{C.1}$$

where

$$\dot{\mathscr{A}}_{\parallel} = \boldsymbol{\kappa}\cdot\dot{\mathscr{A}}_{\parallel}\,. \tag{C.2}$$

By replacing $\mathscr{U}$ with (C.1) in the standard Lagrangian (B.11), one gets a Lagrangian depending on a reduced number of dynamical variables (the components of the vector potential and the associated velocities) which we still call L

$$L = \sum_{\alpha}\frac{1}{2}m_{\alpha}\dot{\mathbf{r}}_{\alpha}^2 + \varepsilon_0\int \mathrm{d}^3k\left[\dot{\mathscr{A}}_{\perp}^*\cdot\dot{\mathscr{A}}_{\perp} + \frac{\rho^*\rho}{\varepsilon_0^2k^2} - c^2(\mathbf{k}\times\mathscr{A}_{\perp}^*)\cdot(\mathbf{k}\times\mathscr{A}_{\perp})\right]$$
$$+\int \mathrm{d}^3k\left[\mathscr{j}^*\cdot\mathscr{A} + \mathscr{j}\cdot\mathscr{A}^* - 2\frac{\rho^*\rho}{\varepsilon_0k^2} - \frac{\mathrm{i}}{k}(\rho^*\dot{\mathscr{A}}_{\parallel} - \rho\dot{\mathscr{A}}_{\parallel}^*)\right]. \tag{C.3}$$

In the same way, one eliminates $\mathscr{U}$ from the expressions for all the physical variables which depend on it. Thus, the electric field in reciprocal space (B.10.a) is now written as

$$\mathscr{E}(\mathbf{k}) = -\dot{\mathscr{A}}_{\perp}(\mathbf{k}) - \frac{\mathrm{i}}{\varepsilon_0}\rho\frac{\mathbf{k}}{k^2}\,. \tag{C.4}$$

It depends on the field variables ($\dot{\mathscr{A}}_{\perp}$) and the positions of the particles (which appear in ρ).

Returning to the Lagrangian (C.3) and grouping the terms, one gets

$$L = \sum_\alpha \frac{1}{2} m_\alpha \dot{\mathbf{r}}_\alpha^2 - \int \mathrm{d}^3k \frac{\rho^* \rho}{\varepsilon_0 k^2} + \varepsilon_0 \int \mathrm{d}^3k [\dot{\mathcal{A}}_\perp^* \cdot \dot{\mathcal{A}}_\perp - c^2 k^2 \mathcal{A}_\perp^* \cdot \mathcal{A}_\perp] + \\ + \int \mathrm{d}^3k [\mathcal{j}^* \cdot \mathcal{A}_\perp + \mathcal{j} \cdot \mathcal{A}_\perp^*] + \int \mathrm{d}^3k \Big[\mathcal{j}_\parallel^* \mathcal{A}_\parallel + \mathcal{j}_\parallel \mathcal{A}_\parallel^* - \\ - \frac{\mathrm{i}}{k}(\rho^* \dot{\mathcal{A}}_\parallel - \rho \dot{\mathcal{A}}_\parallel^*)\Big]. \quad (\mathrm{C.5})$$

b) THE CHOICE OF THE LONGITUDINAL COMPONENT OF THE VECTOR POTENTIAL

The longitudinal component $\mathcal{A}_\parallel$ of $\mathcal{A}$ appears only in the density $\bar{\mathcal{L}}_\parallel$ arising in the last term of (C.5):

$$\overline{\mathcal{L}}_\parallel = \mathcal{j}_\parallel^* \mathcal{A}_\parallel + \mathcal{j}_\parallel \mathcal{A}_\parallel^* - \frac{\mathrm{i}}{k}(\rho^* \dot{\mathcal{A}}_\parallel - \rho \dot{\mathcal{A}}_\parallel^*). \quad (\mathrm{C.6})$$

The Lagrange equation for $\mathcal{A}_\parallel$ derived from (C.6) is written

$$\dot{\rho} = -\,\mathrm{i}k\mathcal{j}_\parallel \quad (\mathrm{C.7})$$

and is just the well-known equation for charge conservation (in reciprocal space). Clearly, this is not an equation of motion for $\mathcal{A}_\parallel$, so that $\mathcal{A}_\parallel$ can take any value.

This last point is even clearer if the equation of charge conservation (C.7) is used to express $\mathcal{j}_\parallel$ as a function of $\dot{\rho}$. One then finds that (C.6) can be written

$$\overline{\mathcal{L}}_\parallel = \frac{\mathrm{i}}{k}[\dot{\rho}\mathcal{A}_\parallel^* + \rho\dot{\mathcal{A}}_\parallel^* - \rho^* \dot{\mathcal{A}}_\parallel - \dot{\rho}^* \mathcal{A}_\parallel] \\ = \frac{\mathrm{i}}{k}\frac{\mathrm{d}}{\mathrm{d}t}[\rho\mathcal{A}_\parallel^* - \rho^* \mathcal{A}_\parallel] \quad (\mathrm{C.8})$$

which gives for the Lagrangian (C.5)

$$L = \sum_\alpha \frac{1}{2} m_\alpha \dot{\mathbf{r}}_\alpha^2 - \int \mathrm{d}^3k \frac{\rho^* \rho}{\varepsilon_0 k^2} + \varepsilon_0 \int \mathrm{d}^3k [\dot{\mathcal{A}}_\perp^* \cdot \dot{\mathcal{A}}_\perp - c^2 k^2 \mathcal{A}_\perp^* \cdot \mathcal{A}_\perp] + \\ + \int \mathrm{d}^3k [\mathcal{j}^* \cdot \mathcal{A}_\perp + \mathcal{j} \cdot \mathcal{A}_\perp^*] + \frac{\mathrm{d}}{\mathrm{d}t}\int \mathrm{d}^3k \frac{\mathrm{i}}{k}[\rho\mathcal{A}_\parallel^* - \rho^* \mathcal{A}_\parallel]. \quad (\mathrm{C.9})$$

Since two Lagrangians which differ only in a total time derivative of a function of the coordinates are equivalent, it appears that the evolution of the system does not depend on the value of $\mathscr{A}_{\parallel}$, which appears exclusively in such a total derivative. $\mathscr{A}_{\parallel}$ is not a *true dynamical variable*, since its value can be arbitrarily chosen without changing the dynamics of the system.

Remark

The possibility of choosing $\mathscr{A}_{\parallel}$ arbitrarily is evidently related to gauge invariance. On changing the gauge, $\mathscr{A}_{\perp}$ does not change and $\mathscr{A}_{\parallel}$ becomes [see (B.8), Chapter I]

$$\mathscr{A}'_{\parallel} = \mathscr{A}_{\parallel} + ik\,\mathscr{F}. \tag{C.10}$$

On changing the gauge, only the last term of the Lagrangian (C.9) is changed, and this is a total derivative with respect to time.

One can imagine various possible choices for the longitudinal component of the vector potential. The simplest choice obviously is

$$\mathscr{A}_{\parallel} = 0 \tag{C.11}$$

which requires that $\nabla \cdot \mathbf{A}$ be zero in the entire real space and thus selects the Coulomb gauge.

Starting from this point, unless otherwise stated we will work in the Coulomb gauge, where the vector potential is purely transverse:

$$\mathscr{A} = \mathscr{A}_{\perp}\,. \tag{C.12}$$

To simplify the notation, we will henceforth omit the index $\perp$.

2. The Lagrangian in the Coulomb Gauge

The Lagrangian in reciprocal space in the Coulomb gauge derives from (C.9):

$$L = \sum_{\alpha} \frac{1}{2} m_\alpha \dot{\mathbf{r}}_\alpha^2 - \int d^3k \frac{|\rho|^2}{\varepsilon_0 k^2} + \int d^3k \overline{\mathscr{L}}_C \tag{C.13.a}$$

$$\overline{\mathscr{L}}_C = \varepsilon_0[\dot{\mathscr{A}}^* \cdot \dot{\mathscr{A}} - c^2 k^2 \mathscr{A}^* \cdot \mathscr{A}] + \jmath^* \cdot \mathscr{A} + \jmath \cdot \mathscr{A}^*. \tag{C.13.b}$$

The dynamical variables of the particles, $\mathbf{r}_\alpha$ and $\dot{\mathbf{r}}_\alpha$, appear not only in the term $\Sigma_\alpha m_\alpha \dot{\mathbf{r}}_\alpha^2/2$ but also in the charge density ρ and in the current $\jmath$. The second term of (C.13.a) can be transformed into an integral over all space, thanks to the reality condition $\rho(\mathbf{k}) = \rho^*(-\mathbf{k})$. One then finds precisely the integral (B.32) from Chapter I, which is nothing but the Coulomb energy of a system of charges:

$$\int d^3k \frac{|\rho|^2}{\varepsilon_0 k^2} = V_{\text{Coul}} = \sum_\alpha \varepsilon^\alpha_{\text{Coul}} + \sum_{\alpha>\beta} \frac{q_\alpha q_\beta}{4\pi\varepsilon_0 |\mathbf{r}_\alpha - \mathbf{r}_\beta|} \tag{C.14}$$

$\varepsilon^\alpha_{\text{Coul}}$ being the energy of the Coulomb field of particle α, defined by (B.36) of Chapter I.

The Lagrangian of the field and the interaction Lagrangian can also be expressed as functions of the fields in real space. After transforming the integrals on a reciprocal half space into integrals on a full space and applying the Parseval–Plancherel equality, one gets

$$L = \sum_\alpha \frac{1}{2} m_\alpha \dot{\mathbf{r}}_\alpha^2 - V_{\text{Coul}} + \int d^3r\, \mathscr{L}_C \tag{C.15.a}$$

$$\mathscr{L}_C = \frac{\varepsilon_0}{2} [\dot{\mathbf{A}}^2 - c^2(\nabla \times \mathbf{A})^2] + \mathbf{j} \cdot \mathbf{A}. \tag{C.15.b}$$

One should remember now that the vector potential $\mathbf{A}$ has only two degrees of freedom at each point of space $\mathbf{r}$. The constraint relations, which have a simple expression in reciprocal space (see C.11), are more cumbersome in real space, since they have a differential form:

$$\nabla \cdot \mathbf{A} = 0. \tag{C.16}$$

As a result we will most often use the Lagrangian density in reciprocal space. For example, in Exercise 4 Lagrange's equations are derived directly from the Lagrangian (C.13).

Remark

None of the transformations which allow one to go from the standard Lagrangian (B.2) to Lagrangians (C.13) or (C.15) in the Coulomb gauge ever involve the Lagrangian of the particles $\Sigma_\alpha m_\alpha \dot{\mathbf{r}}_\alpha^2/2$. The same procedure could be applied to the relativistic Lagrangian (B.33). It follows that the Lagrangian

$$L = -\sum_\alpha m_\alpha c^2 \sqrt{1 - \frac{\dot{\mathbf{r}}_\alpha^2}{c^2}} - V_{\text{Coul}} + \int d^3k\, \bar{\mathscr{L}}_C \tag{C.17}$$

where $\bar{\mathscr{L}}_C$ has the same form as in (C.13.b), is relativistic for the particles as well as for the field. On the other hand, such a Lagrangian is not manifestly covariant; that property is tied to the choice of gauge (C.16), which is not invariant under Lorentz transformation.

3. Hamiltonian Formalism

a) Conjugate Particle Momenta

To find the conjugate momentum associated with the variable $\mathbf{r}_\alpha$, one uses (C.15) for the Lagrangian, where the current $\mathbf{j}$ is replaced by its expression in terms of $\mathbf{r}_\alpha$ and $\dot{\mathbf{r}}_\alpha$:

$$\mathbf{j} = \sum_\alpha q_\alpha \, \dot{\mathbf{r}}_\alpha \, \delta(\mathbf{r} - \mathbf{r}_\alpha(t)) . \tag{C.18}$$

The only terms in the Lagrangian depending on the particle velocities are

$$\sum_\alpha \frac{1}{2} m_\alpha \, \dot{\mathbf{r}}_\alpha^2 + \sum_\alpha q_\alpha \, \dot{\mathbf{r}}_\alpha \cdot \mathbf{A}(\mathbf{r}_\alpha) .$$

It follows that the momentum associated with the discrete variable $(\mathbf{r}_\alpha)_i$ is

$$(\mathbf{p}_\alpha)_i = \frac{\partial L}{\partial (\dot{\mathbf{r}}_\alpha)_i} = m_\alpha (\dot{\mathbf{r}}_\alpha)_i + q_\alpha \, A_i(\mathbf{r}_\alpha) . \tag{C.19}$$

The conjugate momentum associated with $\mathbf{r}_\alpha$ is different from the mechanical momentum $m_\alpha \dot{\mathbf{r}}_\alpha$. We shall return to this point.

Remark

One can also calculate the conjugate momentum starting with the relativistic Lagrangian (C.17). One gets

$$\mathbf{p}_\alpha = \frac{m_\alpha \, \dot{\mathbf{r}}_\alpha}{\sqrt{1 - \dfrac{\dot{\mathbf{r}}_\alpha^2}{c^2}}} + q_\alpha \, \mathbf{A}(\mathbf{r}_\alpha) \tag{C.20}$$

which agrees well with (C.19) in the limit $v/c \ll 1$.

b) Conjugate Momenta for the Field Variables

To calculate the conjugate momenta associated with the field variables, one uses the Lagrangian (C.13). At each point $\mathbf{k}$, the field has two independent generalized coordinates $\mathscr{A}_\varepsilon$ and $\mathscr{A}_{\varepsilon'}$. The definition of the conjugate momentum in reciprocal space (A.53) leads to

$$\pi_\varepsilon(\mathbf{k}) = \varepsilon_0 \, \dot{\mathscr{A}}_\varepsilon(\mathbf{k}) . \tag{C.21}$$

The complex vector $\boldsymbol{\pi}$ whose two components are π_ε and $\pi_{\varepsilon'}$:

$$\boldsymbol{\pi} = \pi_\varepsilon \, \boldsymbol{\varepsilon} + \pi_{\varepsilon'} \, \boldsymbol{\varepsilon}' \tag{C.22}$$

is the momentum conjugate with $\mathscr{A}$ and can then be written using (C.21) as

$$\boldsymbol{\pi}(\mathbf{k}) = \varepsilon_0\,\dot{\mathscr{A}}(\mathbf{k})\,. \tag{C.23}$$

Remarks

(i) Rigorously, we have calculated $\boldsymbol{\pi}(\mathbf{k})$ only in a half space, namely the one corresponding to the integration domain of (C.13.a). One can define $\boldsymbol{\pi}(\mathbf{k})$ for all $\mathbf{k}$ by extending the equality (C.23) to all space. The fact that $\dot{\mathbf{A}}$ is real requires then that

$$\boldsymbol{\pi}(-\mathbf{k}) = \boldsymbol{\pi}^*(\mathbf{k}) \tag{C.24}$$

or equivalently, taking the derivative of (B.40) with respect to time and using (C.23),

$$\pi_\varepsilon(-\mathbf{k}) = \pi_\varepsilon^*(\mathbf{k})\,. \tag{C.25}$$

(ii) The conjugate momentum in real space is gotten by Fourier transformation of (C.23) defined over all space [see (A.58)]. One then gets

$$\boldsymbol{\Pi}(\mathbf{r}) = \varepsilon_0\,\dot{\mathbf{A}}(\mathbf{r})\,. \tag{C.26}$$

Note that this result could also be gotten directly by differentiating the Lagrangian density (C.15.b) with respect to $\dot{\mathbf{A}}$. However, such a step is not rigorously correct, since the three components of $\mathbf{A}$ are related by the constraint relation (C.16) and cannot be varied independently.

c) The Hamiltonian in the Coulomb Gauge

The procedure of §§A.2.*d* and A.2.*g* applied to the Lagrangian (C.13) leads to the following Hamiltonian:

$$H = \sum_\alpha \mathbf{p}_\alpha \cdot \dot{\mathbf{r}}_\alpha + \int \mathrm{d}^3k\,[\boldsymbol{\pi}\cdot\dot{\mathscr{A}}^* + \boldsymbol{\pi}^*\cdot\dot{\mathscr{A}}] - L\,. \tag{C.27}$$

Using (C.19) and (C.23) to eliminate the velocities, one gets

$$H = \sum_\alpha \frac{1}{2\,m_\alpha}[\mathbf{p}_\alpha - q_\alpha\,\mathbf{A}(\mathbf{r}_\alpha)]^2 + V_{\mathrm{Coul}} + {} + \varepsilon_0 \int \mathrm{d}^3k \left[\frac{\boldsymbol{\pi}^*\cdot\boldsymbol{\pi}}{\varepsilon_0^2} + c^2 k^2\,\mathscr{A}^*\cdot\mathscr{A}\right]. \tag{C.28}$$

It is also possible to express the Hamiltonian as a function of the fields in real space:

$$H = \sum_\alpha \frac{1}{2\,m_\alpha}[\mathbf{p}_\alpha - q_\alpha\,\mathbf{A}(\mathbf{r}_\alpha)]^2 + V_{\mathrm{Coul}} + \frac{\varepsilon_0}{2}\int \mathrm{d}^3r\left[\left(\frac{\boldsymbol{\Pi}}{\varepsilon_0}\right)^2 + c^2(\boldsymbol{\nabla}\times\mathbf{A})^2\right]. \tag{C.29}$$

Using (C.19) and (C.26), the Hamiltonian in the Coulomb gauge appears as the sum of the kinetic energy of the particles, their Coulomb energy, and the energy of the transverse fields.

Remark

Using (C.27) and the relativistic expression (C.20) for the conjugate momentum $\mathbf{p}_\alpha$, one gets a Hamiltonian

$$H = \sum_\alpha m_\alpha c^2 \left[1 + \frac{[\mathbf{p}_\alpha - q_\alpha \mathbf{A}(\mathbf{r}_\alpha)]^2}{m_\alpha^2 c^2} \right]^{1/2} + \\ + V_{\text{Coul}} + \frac{\varepsilon_0}{2} \int \mathrm{d}^3 r \left[\left(\frac{\mathbf{\Pi}}{\varepsilon_0} \right)^2 + c^2 (\nabla \times \mathbf{A})^2 \right] \quad \text{(C.30)}$$

whose nonrelativistic limit is (C.29).

d) The Physical Variables

In this subsection, various physical variables related to the particles and to the field are expressed as functions of the variables and their conjugate momenta.

The velocities of the particles derive easily from (C.19):

$$\dot{\mathbf{r}}_\alpha = \frac{1}{m_\alpha} [\mathbf{p}_\alpha - q_\alpha \mathbf{A}(\mathbf{r}_\alpha)] . \quad \text{(C.31)}$$

Expressing $\dot{\mathscr{A}}$ as a function of the conjugate momentum $\boldsymbol{\pi}$ in (C.4), we find

$$\mathscr{E}(\mathbf{k}) = \mathscr{E}_{\parallel}(\mathbf{k}) + \mathscr{E}_{\perp}(\mathbf{k}) \quad \text{(C.32.a)}$$

$$\mathscr{E}_{\parallel}(\mathbf{k}) = -\frac{\mathrm{i}}{\varepsilon_0} \rho \frac{\mathbf{k}}{k^2} \quad \text{(C.32.b)}$$

$$\mathscr{E}_{\perp}(\mathbf{k}) = -\frac{1}{\varepsilon_0} \boldsymbol{\pi}(\mathbf{k}) . \quad \text{(C.32.c)}$$

In real space this same expression becomes, following Fourier transformation,

$$\mathbf{E}(\mathbf{r}) = \mathbf{E}_{\parallel}(\mathbf{r}) + \mathbf{E}_{\perp}(\mathbf{r}) \quad \text{(C.33.a)}$$

$$\mathbf{E}_{\parallel}(\mathbf{r}) = \frac{1}{4\pi\varepsilon_0} \sum_\alpha q_\alpha \frac{\mathbf{r} - \mathbf{r}_\alpha}{|\mathbf{r} - \mathbf{r}_\alpha|^3} \quad \text{(C.33.b)}$$

$$\mathbf{E}_{\perp}(\mathbf{r}) = -\frac{1}{\varepsilon_0} \mathbf{\Pi}(\mathbf{r}) . \quad \text{(C.33.c)}$$

For the magnetic field, the equations do not depend on the conjugate momenta and it is sufficient to give (B.10.b) and (B.3.b):

$$\mathscr{B} = \mathrm{i}\mathbf{k} \times \mathscr{A} \tag{C.34.a}$$

$$\mathbf{B} = \nabla \times \mathbf{A}\,. \tag{C.34.b}$$

Finally, one can express the momentum **P** and the angular momentum **J** of the global system particles + radiation [defined in Chapter I by (A.8) and (A.9)] as functions of the variables and their conjugate momenta. One finds

$$\mathbf{P} = \sum_\alpha \mathbf{p}_\alpha - \mathrm{i}\int \mathrm{d}^3k[\pi^* \times (\mathbf{k} \times \mathscr{A}) - \pi \times (\mathbf{k} \times \mathscr{A}^*)] \tag{C.35.a}$$

$$\mathbf{P} = \sum_\alpha \mathbf{p}_\alpha - \int \mathrm{d}^3r[\Pi \times (\nabla \times \mathbf{A})] \tag{C.35.b}$$

$$\mathbf{J} = \sum_\alpha \mathbf{r}_\alpha \times \mathbf{p}_\alpha - \int \mathrm{d}^3r\, \mathbf{r} \times [\Pi \times (\nabla \times \mathbf{A})]\,. \tag{C.36}$$

Note also that the expressions (C.28), (C.29), (C.35), and (C.36) gotten above for **H**, **P**, and **J** can be derived directly from the symmetry properties of the Lagrangian in the Coulomb gauge (see Complement B_{II}).

4. Canonical Quantization in the Coulomb Gauge

a) Fundamental Commutation Relations

The general principles of canonical quantization have been examined in §§A.1.*g* and A.2.*g* for the cases of a discrete number and a continuum of degrees of freedom respectively. It suffices then to apply these results to the particular case of the variables and conjugate momenta introduced in the subsection above.

For the particles, the fundamental quantization relations use the operators associated with $(\mathbf{r}_\alpha)_i$ and $(\mathbf{p}_\beta)_j$:

$$[(\mathbf{r}_\alpha)_i, (\mathbf{p}_\beta)_j] = \mathrm{i}\hbar\, \delta_{ij}\, \delta_{\alpha\beta}\,. \tag{C.37}$$

For the electromagnetic field, we have shown that at every point **k** in a reciprocal half space there are two independent complex dynamical variables $\mathscr{A}_\varepsilon(\mathbf{k})$ and $\mathscr{A}_{\varepsilon'}(\mathbf{k})$, associated with which are two conjugate momenta $\pi_\varepsilon(\mathbf{k})$ and $\pi_{\varepsilon'}(\mathbf{k})$. The commutation relations between the operators associated with these variables derive from (A.61) and are written

$$\begin{cases} [\mathscr{A}_\varepsilon(\mathbf{k}),\ \pi_{\varepsilon'}(\mathbf{k}')] = 0 & \text{(C.38.a)} \\ [\mathscr{A}_\varepsilon(\mathbf{k}),\ \pi^+_{\varepsilon'}(\mathbf{k}')] = i\hbar\,\delta_{\varepsilon\varepsilon'}\,\delta(\mathbf{k}-\mathbf{k}') & \text{(C.38.b)} \end{cases}$$

the other commutators being zero.

Remark

In (C.38), **k** and **k**′ belong to the same half space. It is possible to generalize (C.38) to all reciprocal space using the relationships between operators given by

$$\mathscr{A}_\varepsilon(-\mathbf{k}) = \mathscr{A}_\varepsilon^{+}(\mathbf{k}) \qquad \text{(C.39.a)}$$

$$\pi_\varepsilon(-\mathbf{k}) = \pi_\varepsilon^{+}(\mathbf{k}) \qquad \text{(C.39.b)}$$

which follow from (B.40) and (C.25) between classical quantities. Equations (C.38) then become

$$[\mathscr{A}_\varepsilon(\mathbf{k}),\ \pi_{\varepsilon'}(\mathbf{k}')] = i\hbar\,\delta_{\varepsilon\varepsilon'}\,\delta(\mathbf{k}+\mathbf{k}') \qquad \text{(C.40.a)}$$

$$[\mathscr{A}_\varepsilon(\mathbf{k}),\ \pi^+_{\varepsilon'}(\mathbf{k}')] = i\hbar\,\delta_{\varepsilon\varepsilon'}\,\delta(\mathbf{k}-\mathbf{k}') \qquad \text{(C.40.b)}$$

all other commutators being zero.

To prove (C.40.a) it suffices to note that if **k** and **k**′ are in the same reciprocal half space, **k** + **k**′ can never be zero, so that (C.40.a) reduces to (C.38.a). On the other hand, if **k** and **k**′ are not in the same half space, it suffices, taking account of (C.39.b), to replace $\pi_{\varepsilon'}(\mathbf{k}')$ by $\pi^+_{\varepsilon'}(-\mathbf{k}')$ and to use (C.38.b) (which then applies since **k** and −**k**′ are in the same half space) to prove (C.40.a). A similar procedure serves to prove (C.40.b).

b) The Importance of Transversality in the Case of the Electromagnetic Field

The commutation relations (C.38) and (C.40) arise as a natural consequence of the steps leading to quantization. It is necessary to stress the fact that one of the most important stages in this procedure has been to isolate the truly independent dynamical field variables. In particular, the fact that the vector potential is transverse (zero divergence in real space) implies that the three components $A_i(\mathbf{r})$ ($i = x, y, z$) are not independent. This constraint also appears when the commutation relation between $A_i(\mathbf{r})$ and $\Pi_j(\mathbf{r}')$ is considered. We will see indeed that

$$[A_i(\mathbf{r}), \Pi_j(\mathbf{r}')] \neq i\hbar\,\delta_{ij}\,\delta(\mathbf{r}-\mathbf{r}'). \qquad \text{(C.41)}$$

So one must not crudely apply the commutation relations (A.47) when the independent dynamical field variables have not been clearly identified.

To determine the commutator between $A_i(\mathbf{r})$ and $\Pi_j(\mathbf{r})$, one begins with the commutator between the corresponding quantities in reciprocal space. The Cartesian component $\mathscr{A}_i(\mathbf{k})$ as a function of the transverse components is

$$\mathscr{A}_i(\mathbf{k}) = \varepsilon_i \, \mathscr{A}_\varepsilon(\mathbf{k}) + \varepsilon'_i \, \mathscr{A}_{\varepsilon'}(\mathbf{k}) \tag{C.42.a}$$

with

$$\varepsilon_i = \mathbf{e}_i \cdot \boldsymbol{\varepsilon} \, . \tag{C.42.b}$$

The Cartesian component $\pi_j(\mathbf{k}')$ is defined by equations analogous to (C.42). Using (C.40.b) one gets

$$[\mathscr{A}_i(\mathbf{k}), \; \pi_j^{\,+}(\mathbf{k}')] = i\hbar(\varepsilon_i \, \varepsilon_j + \varepsilon'_i \, \varepsilon'_j) \, \delta(\mathbf{k} - \mathbf{k}') \, . \tag{C.43}$$

Since $\{\boldsymbol{\varepsilon}, \boldsymbol{\varepsilon}', \boldsymbol{\kappa}\}$ forms a basis for the space, (C.43) can be replaced by

$$[\mathscr{A}_i(\mathbf{k}), \; \pi_j^{+}(\mathbf{k}')] = i\hbar\left[\delta_{ij} - \frac{k_i \, k_j}{k^2}\right] \delta(\mathbf{k} - \mathbf{k}') \, . \tag{C.44}$$

We have seen above that (C.40.b) is true for all values of $\mathbf{k}$ and of $\mathbf{k}'$. The same result holds for (C.44). Multiplying both sides of (C.44) by $e^{-i\mathbf{k}\cdot\mathbf{r}} e^{i\mathbf{k}'\cdot\mathbf{r}'}/(2\pi)^3$ and integrating twice over all reciprocal space, one gets by Fourier transformation, on the left, $[A_i(\mathbf{r}), \Pi_j^{+}(\mathbf{r}')]$ and on the right the function $\delta_{ij}^{\perp}(\mathbf{r} - \mathbf{r}')$ introduced in Chapter I [see (B.17) in Chapter I and Complement A_{I}]. Finally, using the fact that $\Pi_j(\mathbf{r})$ is Hermitian one finds

$$[A_i(\mathbf{r}), \Pi_j(\mathbf{r}')] = i\hbar \, \delta_{ij}^{\perp}(\mathbf{r} - \mathbf{r}') \, . \tag{C.45}$$

c) Creation and Annihilation Operators

In Chapter I, the classical normal variables $\boldsymbol{\alpha}(\mathbf{k})$ were introduced as linear combinations of $\mathscr{A}(\mathbf{k})$ and of $\mathscr{E}_\perp(\mathbf{k})$ [see (C.30) of Chapter I]. Using (C.32.c), one can express $\mathscr{E}_\perp(\mathbf{k})$ as a function of the conjugate momentum $\boldsymbol{\pi}(\mathbf{k})$ and then get the following expression for $\alpha_\varepsilon(\mathbf{k}) = \boldsymbol{\varepsilon} \cdot \boldsymbol{\alpha}(\mathbf{k})$:

$$\alpha_\varepsilon(\mathbf{k}) = \sqrt{\frac{\varepsilon_0}{2\,\hbar\omega}}\left[\omega \mathscr{A}_\varepsilon(\mathbf{k}) + \frac{i}{\varepsilon_0}\,\pi_\varepsilon(\mathbf{k})\right]. \tag{C.46}$$

As has already been seen in Chapter I, the interest in the normal variables is that they evolve independently in the absence of sources. Additionally, one does not have a relationship between $\boldsymbol{\alpha}(-\mathbf{k})$ and $\boldsymbol{\alpha}^*(\mathbf{k})$ analogous to the constraint relations (B.7.a) and (C.24) (which arise because $\mathbf{A}$ and $\boldsymbol{\Pi}$

are real), with the result that the normal variables at each point $\mathbf{k}$ are independent of the normal variables at point $-\mathbf{k}$.

After quantization $\mathscr{A}_\varepsilon(\mathbf{k})$ and $\pi_\varepsilon(\mathbf{k})$ become operators. The same linear combination of these operators, analogous to (C.46), allows one to define the operators $a_\varepsilon(\mathbf{k})$, which are the quantum analogues of the normal variables $\alpha_\varepsilon(\mathbf{k})$ (the operators a_ε^+ being associated with α_ε^*):

$$a_\varepsilon(\mathbf{k}) = \sqrt{\frac{\varepsilon_0}{2\hbar\omega}}\left[\omega\mathscr{A}_\varepsilon(\mathbf{k}) + \frac{i}{\varepsilon_0}\,\pi_\varepsilon(\mathbf{k})\right] \tag{C.47.a}$$

$$a_\varepsilon^+(\mathbf{k}) = \sqrt{\frac{\varepsilon_0}{2\hbar\omega}}\left[\omega\mathscr{A}_\varepsilon^+(\mathbf{k}) - \frac{i}{\varepsilon_0}\,\pi_\varepsilon^+(\mathbf{k})\right]. \tag{C.47.b}$$

To find the commutation relations for the operators $a_\varepsilon(\mathbf{k})$ and $a_\varepsilon^+(\mathbf{k})$, it is necessary to use their definition (C.47) and the commutation relations (C.40) (which remain valid whatever $\mathbf{k}$ and $\mathbf{k}'$ may be). Such a procedure gives

$$[a_\varepsilon(\mathbf{k}), a_{\varepsilon'}(\mathbf{k}')] = 0 \tag{C.48.a}$$

$$[a_\varepsilon^+(\mathbf{k}), a_{\varepsilon'}^+(\mathbf{k}')] = 0 \tag{C.48.b}$$

$$[a_\varepsilon(\mathbf{k}), a_{\varepsilon'}^+(\mathbf{k}')] = \delta_{\varepsilon\varepsilon'}\,\delta(\mathbf{k} - \mathbf{k}')\,. \tag{C.48.c}$$

These relations are identical to those (D.4) postulated in Chapter I. They show that the operators $a_\varepsilon(\mathbf{k})$ and $a_\varepsilon^+(\mathbf{k})$ are the annihilation and creation operators for the harmonic oscillator associated with the mode $\mathbf{k}\boldsymbol{\varepsilon}$. We will see in Chapter III that $a_\varepsilon(\mathbf{k})$ and $a_\varepsilon^+(\mathbf{k})$ destroy and create a photon $\mathbf{k}\boldsymbol{\varepsilon}$.

Finally, it is possible to give all the observables of the field as a function of the operators a and a^+ and to prove that one then gets the same expressions as those given in Chapter I (§C.4).

5. Conclusion: Some Important Characteristics of Electrodynamics in the Coulomb Gauge

To conclude this section devoted to electrodynamics in the Coulomb gauge we now review a few important characteristics of the theory which has been elaborated above.

a) The Dynamical Variables Are Independent

A great advantage of the theory developed above is that one has eliminated the redundant variables of the field to get a simple situation where the number of dynamical variables is the minimum necessary to describe the field dynamics. As a result, the field can be quantized with a great economy in the formalism.

b) The Electric Field Is Split into a Coulomb Field and a Transverse Field

Equations (C.4) and (C.33) clearly show that the electric field appears as the sum of two terms; the first, depending on the particle coordinates, is the Coulomb field, and the other is the transverse field. Similarly, in the expressions for other physical quantities, such as the Hamiltonian, the terms arising from the Coulomb field are separated from those from the transverse field. This separation seems at first glance to raise a problem, since the Coulomb interaction is instantaneous (it depends only on the positions of particles α and β at t). It is clear, however, that the real interaction between particles α and β takes place with a retardation associated with the propagation of the field with velocity c. In fact, the retarded character of the electromagnetic interactions results from an exact compensation between two instantaneous parts coming from the Coulomb field and the transverse field respectively. This point has already been discussed in Chapter I (see §B.6 and Exercise 3). Globally the retarded character of the interactions is retained, but it is not obvious.

On the other hand, this separation has an immense advantage for atomic and molecular physics in easily isolating the Coulomb interaction. For particles moving with low velocities ($v/c \ll 1$), this is an excellent approximation to the interaction between particles and allows, in particular, a simple treatment of bound states. The terms corresponding to the transverse field then appear as corrections, in general of the order v^2/c^2 with respect to the Coulomb interaction, and describing the effects of the retardation or magnetism. This separation is thus truly advantageous for physics at low energies.

c) The Formalism Is Not Manifestly Covariant

In a manifestly covariant formalism the three components of the vector potential and the scalar potential form a four-vector. By eliminating U from the Lagrangian and by taking $\mathbf{A}_{\parallel} = \mathbf{0}$ we have deliberately abandoned such a point of view. However, it is necessary to be aware that abandonment of the manifest covariance does not mean loss of relativistic invariance. The predictions for the electromagnetic field are in agreement with the theory of relativity even within the framework of the Coulomb gauge.

There are however situations where it is necessary to retain covariance in the expressions, for example, to unambiguously eliminate the infinities in renormalization. Under those conditions U must be retained. However, since the momentum conjugate with U is zero in the standard Lagrangian, it is necessary to modify the standard Lagrangian to preserve the symmetry between U and $\mathbf{A}$ (see Chapter V). Quantization of radiation is then carried out in a space where the field has more degrees of freedom than are physically necessary. This leads to the imposition of constraint conditions on the possible states in this overly large space. Furthermore, if one wants to treat the particles properly, it is necessary to think of them as excitations of a relativistic matter field (Dirac field, Klein–Gordon field, etc.).

Finally, for physical problems at low energies such as are discussed in this book, such an approach would introduce several formal complications without improving our understanding of the physical processes.

d) The Interaction of the Particles with Relativistic Modes Is Not Correctly Described

In the Coulomb-gauge Lagrangian (C.13), the particles can interact with arbitrarily high-frequency modes. Analogously, the Coulomb interaction is assumed exact at arbitrarily small distances. Now it is well known that in relativistic quantum theory the interaction with modes with frequency ν such that $h\nu \gtrsim mc^2$ involves effects (creation of electron–positron pairs, vacuum polarization, relativistic recoil) which are not included in the theory developed here. These same effects (vacuum polarization) introduce a modification of the Coulomb interaction for small distances of the order h/mc.

To recapitulate, resonant interaction with the optical or rf modes of the field ($h\nu \ll mc^2$) is correctly described by the theory described in this chapter. For processes involving virtual emission or absorption of photons by an isolated atom, only the contribution of low-frequency modes is correctly evaluated. In contrast, the Lagrangian (C.13) cannot satisfactorily treat the effects associated with high-frequency modes.

Remark

Rather than retain the interaction terms with the high-frequency modes in (C.13), which would lead to erroneous results, it is possible to decouple the particles from these modes by making a "cutoff" in order to annul these interactions. We thus give up all couplings with high-frequency modes. To account correctly for these it would be necessary to develop a more elaborate theory, relativistic quantum electrodynamics, which is beyond the scope of this book.

To cut off the interaction with the high-frequency modes, we multiply the current $\mathbf{j}$ and the charge density ρ of the standard Lagrangian (B.11) by a function $g(k)$ whose behavior is shown in Figure 2. Thus through the introduction of the function $g(k)$, the integrals of the interaction Lagrangian over k are limited to values less than k_c, which is chosen of the order of $mc/\hbar$. The particles then cannot interact with relativistic modes. It is also possible to visualize the effect of this cutoff in real space; the Fourier transform of $\rho(k)g(k)$, for example, arises as a function whose contours are diluted on a

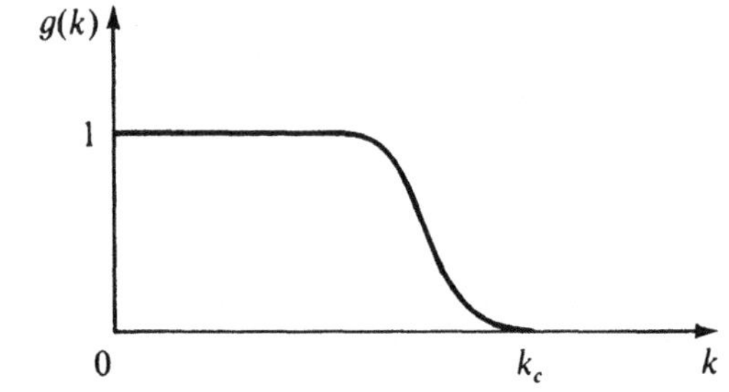

Figure 2. The cutoff function $g(k)$ introduced into the standard Lagrangian to eliminate the interaction with high-frequency modes.

distance of the order of $1/k_c$. A point particle then appears with this cutoff with a charge distributed over dimensions of the order of $1/k_c$. This is not physically real, but is only an expedient for disregarding the effects taking place at distances smaller than $1/k_c$.

We now look at how the derivation leading from the standard Lagrangian to the Lagrangian in the Coulomb gauge is modified by the introduction of the cutoff. To eliminate $\mathscr{U}$ it is necessary to write Lagrange's equation relative to this variable. Starting with the Lagrangian following from (B.11),

$$L = \sum_\alpha \frac{1}{2} m_\alpha \dot{\mathbf{r}}_\alpha^2 + \int \mathrm{d}^3k \, \overline{\mathscr{L}} \tag{C.49.a}$$

$$\overline{\mathscr{L}} = \varepsilon_0[|\dot{\mathscr{A}} + \mathrm{i}\mathbf{k}\mathscr{U}|^2 - c^2|\mathbf{k} \times \mathscr{A}|^2] + \\ + (j^* \cdot \mathscr{A} + \quad \cdot \mathscr{A}^* - \rho^* \mathscr{U} - \rho \mathscr{U}^*) \, g(k) . \tag{C.49.b}$$

one deduces

$$- \mathrm{i}\mathbf{k} \cdot [\dot{\mathscr{A}} + \mathrm{i}\mathbf{k}\mathscr{U}] = \frac{1}{\varepsilon_0} \rho g(k) \tag{C.50}$$

which differs from (B.20) by a factor $g(k)$. Eliminating $\mathscr{U}$ and going to the Coulomb gauge, one gets finally

$$L = \sum_\alpha \frac{1}{2} m_\alpha \dot{\mathbf{r}}_\alpha^2 - \int \mathrm{d}^3k \frac{|\rho|^2}{\varepsilon_0 k^2} [g(k)]^2 + \varepsilon_0 \int \mathrm{d}^3k [\dot{\mathscr{A}}^* \cdot \dot{\mathscr{A}} - c^2 k^2 \mathscr{A}^* \cdot \mathscr{A}] \\ + \int \mathrm{d}^3k [\mathscr{j}^* \cdot \mathscr{A} + \mathscr{j} \cdot \mathscr{A}^*] \, g(k) . \tag{C.51}$$

Thus we find that in the term describing the interaction with the transverse field [last term of (C.51)] the cutoff function $g(k)$ decouples the particles from the high-frequency transverse modes. The function $g(k)$ arises also in the Coulomb interaction terms [second term of (C.51)], which can be written

$$\tilde{V}_{\mathrm{Coul}} = \sum_\alpha \tilde{\varepsilon}^\alpha_{\mathrm{Coul}} + \sum_{\alpha > \beta} \tilde{V}_{\alpha\beta} \tag{C.52.a}$$

$$\tilde{\varepsilon}^{\alpha}_{\text{Coul}} = \frac{q_\alpha^2}{4\,\pi^2\,\varepsilon_0}\int_0^\infty [g(k)]^2\,\mathrm{d}k \sim \frac{q_\alpha^2}{4\,\pi^2\,\varepsilon_0}k_c \qquad \text{(C.52.b)}$$

$$\tilde{V}_{\alpha\beta} = \frac{q_\alpha\,q_\alpha}{4\,\pi\varepsilon_0\,|\,\mathbf{r}_\alpha - \mathbf{r}_\beta\,|}[1 - O(k_c\,|\,\mathbf{r}_\alpha - \mathbf{r}_\beta\,|)] \qquad \text{(C.52.c)}$$

where $O(x)$ is a function which goes to zero when x goes to infinity. The results in (C.52) conform to the picture which we give for the effect of the cutoff in real space. For distances large with respect to $1/k_c \simeq \hbar/mc$, the interaction between charged particles is the Coulomb interaction with excellent precision. This clearly justifies the use of the Coulomb interaction in the treatment of a system like the hydrogen atom, where the Bohr radius is large with respect to the Compton wavelength. The contribution of the low-frequency modes to the Coulomb self-energy is given by (C.52.b). Its order of magnitude is that of the Coulomb energy of a charged sphere with charge q_α and radius $\hbar/mc$.

General References and Useful Reading

— For the Lagrangian description of systems with a finite number of degrees of freedom see Landau and Lifschitz (Vol. I), Goldstein.

— For the Lagrangian description of continuous systems and canonical field quantization see Goldstein (Chapter XI), Schiff (Chapter XIII), Messiah (Chapter XXI, §I), Roman (Chapter I).

— For the Lagrangian description of the electromagnetic field see Sommerfeld (§32), Landau and Lifschitz (Vol. II, Chapter IV), Schiff (Chapter XIV), Power (Chapter 6), Kroll, Healy (Chapter 3).

— More advanced texts treating the relativistic fields and using covariant Lagrangians are cited in Chapter V.

COMPLEMENT A_{II}

FUNCTIONAL DERIVATIVE. INTRODUCTION AND A FEW APPLICATIONS

The functional derivative generalizes the partial derivative for functions depending on a continuous infinity of variables (§§A_{II}.1 and A_{II}.2). This tool is most useful when a physical law can be derived from a variational principle. For example, the principle of least action is expressed mathematically by the vanishing of the functional derivatives of the action. The explicit evaluation of these functional derivatives gives back the Lagrange equations (§A_{II}.3). Another source of interest in the functional derivative is that it allows one to write the equations of motion of a continuous system (such as a field) in a form analogous to that of a discrete system in the Lagrangian formalism (§A_{II}.4) as well as in the Hamiltonian one (§A_{II}.5).

1. From a Discrete to a Continuous System. The Limit of Partial Derivatives

Consider a chain of N point particles with mass m separated by a distance a from one another along the x-axis and moving in the same direction y. Let v_n be the velocity of the nth particle. Its kinetic energy is

$$T_n = \frac{1}{2} m v_n^2 \tag{1}$$

and the total kinetic energy is equal to

$$E_c = \sum_{n=1}^{N} T_n \tag{2}$$

E_c appears then as a function of N real variables v_n (the velocities of the particles). We now calculate the differential of E_c with respect to the v_n. It can be given in various forms:

$$\begin{aligned} \mathrm{d}E_c &= \sum_{n=1}^{N} \frac{\partial E_c}{\partial v_n} \mathrm{d}v_n = \sum_{n=1}^{N} \frac{\mathrm{d}T_n}{\mathrm{d}v_n} \mathrm{d}v_n \\ &= \sum_{n=1}^{N} m v_n \, \mathrm{d}v_n \,. \end{aligned} \tag{3}$$

Let us examine what occurs when the number of particles increases but their mass and separation decreases in such a fashion that the mass per unit length $\mu = m/a$ and the length $l = Na$ remains constant. Schematically, one goes from a chain to a string moving in the y-direction. In going to the continuous limit, Equations (2) and (1) become respectively

$$E_c = \int_0^l \mathrm{d}x\, T(x) \tag{4}$$

$$T(x) = \frac{1}{2}\mu v^2(x)\,. \tag{5}$$

The discrete index n has been replaced by a continuous index x characterizing the position of the length element under consideration. The kinetic energy T_n is replaced by the kinetic energy density $T(x)$, and the total kinetic energy becomes a function of a continuous set of variables $\{v(x)\}$.

We will now examine what happens to Equation (3), giving the differential of E_c, in the continuous limit. For the independent variations $\mathrm{d}v_n$ of the N velocities (Figure 1a) we now substitute a function $\delta v(x)$ describing the small variations of v which can take on arbitrary values at the different points x (Figure 1b).

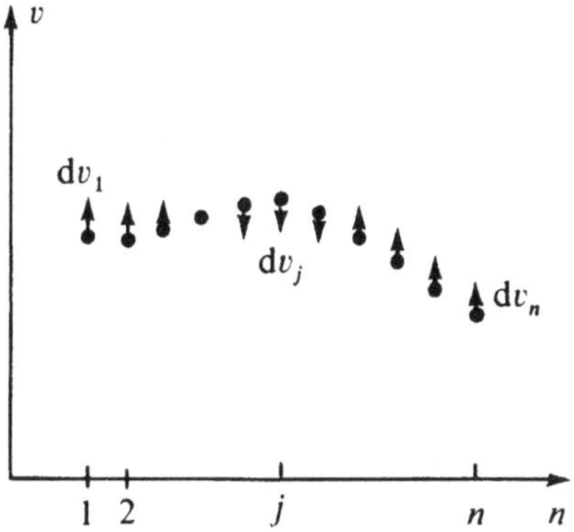

Figure 1a

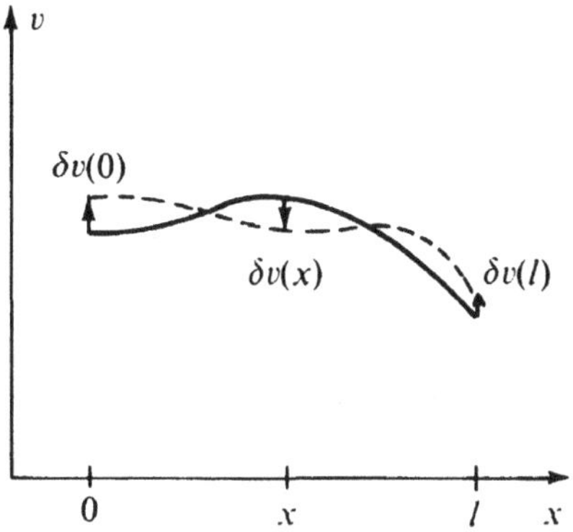

Figure 1b

The generalization of (3) is then

$$\delta E_c = \int_0^l \mathrm{d}x \frac{\partial E_c}{\partial v(x)}\,\delta v(x) = \int_0^l \mathrm{d}x \frac{\mathrm{d}T(x)}{\mathrm{d}v(x)}\,\delta v(x)$$
$$= \int_0^l \mathrm{d}x\, \mu v(x)\, \delta v(x)\,. \tag{6}$$

The notation $\partial E_c / \partial v(x)$ represents the functional derivative of E_c with respect to $v(x)$. This functional derivative describes how E_c varies when v varies by the amount $\delta v(x)$ on the interval $[x, x + \mathrm{d}x]$. More precisely,

$\eth E_c / \eth v(x)$ is the coefficient of proportionality between the increase δE_c and the cause $\delta v(x)\mathrm{d}x$ of this increase. Such a functional derivative, which generalizes the entire set of partial derivatives $\partial E_c/\partial v_n$ when n becomes infinitly large, is thus a function of x which is equal to $\mu v(x)$ in this particular example. Note, finally, that in (6), $\mathrm{d}T(x)/\mathrm{d}v(x)$ is a derivative in the usual sense, since from (5) the kinetic energy density $T(x)$ appears as a function of the real variable $v(x)$.

2. Functional Derivative

The ideas above can be generalized in an elementary way. We will consider an application which associates a real number $\phi(u)$ with every function $u(x)$ of the real variable x. Such an application ϕ is called a *functional.* It is called differentiable if, for all infinitesimal variations $\delta u(x)$ of $u(x)$, the corresponding variation $\delta\phi$ of ϕ can be expressed as a linear functional of $\delta u(x)$ in the form

$$\begin{aligned} \delta\phi &= \phi(u + \delta u) - \phi(u) \\ &= \int \mathrm{d}x\, D(x)\, \delta u(x) + O(\delta u)^2 . \end{aligned} \tag{7}$$

If one assumes that the variations $\delta u(x)$ at the different points can be chosen independently, the functional derivative of ϕ with respect to $u(x)$ is

$$\frac{\eth\phi}{\eth u(x)} = D(x) . \tag{8}$$

Intuitively, $\eth\phi / \eth u(x)$ is the coefficient of proportionality between the increase $\delta\phi$ and the variation $\delta u(x)\mathrm{d}x$ which gives rise to it.

Remarks

(i) The definition of the functional derivative through Equation (8) assumes that the space of the functions $u(x)$ is sufficiently large that one can vary $u(x)$ independently at the different points. A precise analogue exists in the definition of partial derivatives of a function of several variables $\phi(x_1, x_2, \ldots, x_N)$, where the variations $\mathrm{d}x_1, \mathrm{d}x_2, \ldots, \mathrm{d}x_N$ must be independent.

(ii) The ideas above can be easily generalized to functions u of several real variables x, y, z and to the case where u is a vector in $\mathbb{R}_n$ or $\mathbb{C}_n$. In the latter case, $D(x)$ is a vector of the same dimension.

3. Functional Derivative of the Action and the Lagrange Equations

The functional derivative is particularly useful when a physical law is described in terms of a variational principle—in particular, in the framework of the Lagrangian formulation of mechanics, where the true motion appears as an extremum of the action

$$S = \int_{t_1}^{t_2} dt\, L(x_j(t), \dot{x}_j(t), t) . \tag{9}$$

(We adopt here the same notation as in §A.1). The functional S here depends on N independent functions $x_j(t)$. In effect, giving the $x_j(t)$, with t varying from t_1 to t_2, automatically determines the values of the velocities $\dot{x}_j(t)$ on the same domain of variation of t. The principle of least action states that the increase δS of S, which can be written following §A_{II}.2 above as

$$\delta S = \sum_{j=1}^{N} \int_{t_1}^{t_2} \mathrm{d}t \frac{\eth S}{\eth x_j(t)} \delta x_j(t) \tag{10}$$

must be zero no matter how the variations $\delta x_j(t)$ about the true path are chosen. This is simply stated by the equation

$$\frac{\eth S}{\eth x_j(t)} = 0 . \tag{11}$$

We will show that this equation is just Lagrange's equation relative to x_j. To this end one restates δS as a function of L. Using (9) one finds

$$\delta S = \int_{t_1}^{t_2} \mathrm{d}t \sum_{j=1}^{N} \left[\frac{\partial L}{\partial x_j} \delta x_j + \frac{\partial L}{\partial \dot{x}_j} \delta \dot{x}_j \right] . \tag{12}$$

The variations δx_j and $\delta \dot{x}_j = \mathrm{d}(\delta x_j)/\mathrm{d}t$ are not independent in (12). To reexpress the $\delta \dot{x}_j$ as functions of δx_j, one integrates the second term in the brackets by parts. This yields

$$\delta S = \int_{t_1}^{t_2} \mathrm{d}t \sum_{j=1}^{N} \left[\frac{\partial L}{\partial x_j} - \frac{\mathrm{d}}{\mathrm{d}t} \frac{\partial L}{\partial \dot{x}_j} \right] \delta x_j(t) + \left[\frac{\partial L}{\partial \dot{x}_j} \delta x_j(t) \right]_{t_1}^{t_2} . \tag{13}$$

In the framework of the least-action principle, the endpoints of the trajectory are assumed fixed. It follows that one must be restricted to variations such that $\delta x_j(t_1) = \delta x_j(t_2) = 0$. The second term of (13) then vanishes. Comparison of (10) and (11) then yields

$$\frac{\eth S}{\eth x_j(t)} = \left[\frac{\partial L}{\partial x_j} - \frac{\mathrm{d}}{\mathrm{d}t} \frac{\partial L}{\partial \dot{x}_j} \right] = 0 . \tag{14}$$

The condition (11) on the functional derivative of the action thus gives back Lagrange's equations for mechanics.

4. Functional Derivative of the Lagrangian for a Continuous System

In the case of a dynamical system depending on a continuous set of degrees of freedom, the Lagrangian itself appears as a functional. More precisely, using the notation of §A.2, L appears as a functional of $A_j(\mathbf{r})$ and $\dot{A}_j(\mathbf{r})$ which can be written

$$L = \int \mathrm{d}^3 r\, \mathscr{L}(A_j, \dot{A}_j, \partial_i A_j) . \tag{15}$$

Note that even if the Lagrangian density depends on spatial derivatives $\partial_i A_j$, these must not be considered as independent variables of A_j in the functional L. Actually, giving $A_j(\mathbf{r})$ automatically specifies $\partial_i A_j(\mathbf{r})$. In contrast, at a given instant, $A_j(\mathbf{r}, t)$ and $\dot{A}_j(\mathbf{r}, t)$ are independent variables. We will evaluate the differential of L,

$$\delta L = \int \mathrm{d}^3 r \sum_{i,j} \left[\frac{\partial \mathscr{L}}{\partial A_j} \delta A_j(\mathbf{r}) + \frac{\partial \mathscr{L}}{\partial \dot{A}_j} \delta \dot{A}_j(\mathbf{r}) + \frac{\partial \mathscr{L}}{\partial(\partial_i A_j)} \delta(\partial_i A_j(\mathbf{r})) \right] . \tag{16}$$

To reexpress $\delta(\partial_i A_j(\mathbf{r}))$ as a function of δA_j, integrate the last term in the brackets by parts and use the fact that the $A_j(\mathbf{r})$ correspond, in practice, to fields which vanish at infinity. We get

$$\delta L = \int \mathrm{d}^3 r \sum_{i,j} \left\{ \left[\frac{\partial \mathscr{L}}{\partial A_j} - \partial_i \left(\frac{\partial \mathscr{L}}{\partial(\partial_i A_j)} \right) \right] \delta A_j(\mathbf{r}) + \frac{\partial \mathscr{L}}{\partial \dot{A}_j} \delta \dot{A}_j(\mathbf{r}) \right\} . \tag{17}$$

By using the functional derivatives of L one can also write the differential of L in the form

$$\delta L = \int \mathrm{d}^3 r \sum_j \left[\frac{\delta L}{\delta A_j(\mathbf{r})} \delta A_j(\mathbf{r}) + \frac{\delta L}{\delta \dot{A}_j(\mathbf{r})} \delta \dot{A}_j(\mathbf{r}) \right] . \tag{18}$$

Equating (17) and (18) shows that

$$\frac{\delta L}{\delta A_j(\mathbf{r})} = \frac{\partial \mathscr{L}}{\partial A_j} - \sum_i \partial_i \left(\frac{\partial \mathscr{L}}{\partial(\partial_i A_j)} \right) \tag{19}$$

$$\frac{\delta L}{\delta \dot{A}_j(\mathbf{r})} = \frac{\partial \mathscr{L}}{\partial \dot{A}_j} . \tag{20}$$

We will now derive the Lagrange equations for a continuous system, starting from the principle of least action. The action

$$S = \int_{t_1}^{t_2} \mathrm{d}t\, L(t) = \int_{t_1}^{t_2} \mathrm{d}t \int \mathrm{d}^3 r\, \mathscr{L}(A_j(\mathbf{r}, t), \dot{A}_j(\mathbf{r}, t), \partial_i A_j(\mathbf{r}, t)) \tag{21}$$

is a functional of $A_j(\mathbf{r}, t)$. In effect, giving $A_j(\mathbf{r}, t)$ between t_1 and t_2 automatically fixes the values of $\dot{A}_j(\mathbf{r}, t)$ on the same time interval. Even though the Lagrangian L defined at each instant by (15) is a functional of A_j and $\dot{A}_j$, the action S defined in (21) is a functional of A_j only. Its differential δS is equal to

$$\delta S = \int_{t_1}^{t_2} \mathrm{d}t \int \mathrm{d}^3 r \, \frac{\eth S}{\eth A_j(\mathbf{r}, t)} \, \delta A_j(\mathbf{r}, t) \tag{22}$$

and the principle of least action is expressed by

$$\frac{\eth S}{\eth A_j(\mathbf{r}, t)} = 0 \, . \tag{23}$$

To find these derivatives, substitute in δS the functional derivatives of the Lagrangian

$$\delta S = \int_{t_1}^{t_2} \mathrm{d}t \, \delta L = \int_{t_1}^{t_2} \mathrm{d}t \int \mathrm{d}^3 r \sum_j \left[\frac{\eth L}{\eth A_j(\mathbf{r})} \, \delta A_j(\mathbf{r}) + \frac{\eth L}{\eth \dot{A}_j(\mathbf{r})} \, \delta \dot{A}_j(\mathbf{r}) \right] \tag{24}$$

and restate $\delta \dot{A}_j = \mathrm{d}(\delta A_j)/\mathrm{d}t$ as a function of δA_j by integrating the last term by parts with respect to time. Using the fact that the principle of least action is used with zero variations of δA_j at times t_1 and t_2, one finds

$$\delta S = \int_{t_1}^{t_2} \mathrm{d}t \int \mathrm{d}^3 r \sum_j \left[\frac{\eth L}{\eth A_j(\mathbf{r})} - \frac{\mathrm{d}}{\mathrm{d}t} \frac{\eth L}{\eth \dot{A}_j(\mathbf{r})} \right] \delta A_j(\mathbf{r}) \, . \tag{25}$$

The differential of the action being zero along a true trajectory, one gets the Lagrange equations

$$\frac{\eth S}{\eth A_j(\mathbf{r}, t)} = \frac{\eth L}{\eth A_j(\mathbf{r})} - \frac{\mathrm{d}}{\mathrm{d}t} \frac{\eth L}{\eth \dot{A}_j(\mathbf{r})} = 0 \, . \tag{26}$$

Using functional derivatives thus leads to equations formally identical to those gotten in the discrete case (14). To get the Lagrange equations using the Lagrangian density (A.39) it is necessary to replace $\eth L / \eth A_j(\mathbf{r})$ and $\eth L / \eth \dot{A}_j(\mathbf{r})$ by their expressions (19) and (20). It is however advantageous within the framework of formal calculations to keep the functional derivative of the Lagrangian, since the equations (and the proofs) are then very similar to those developed in the discrete case. The following paragraph shows how this method is used with the Hamiltonian.

5. Functional Derivative of the Hamiltonian for a Continuous System

In the case where the Lagrangian depends on a discrete set of variables, the momentum associated with the variable x_n is the partial derivative of the Lagrangian with respect to $\dot{x}_n$ [cf. (A.5)]. The natural extension of this to the case of a continuum of variables is

$$\Pi_j(\mathbf{r}) = \frac{\eth L}{\eth \dot{A}_j(\mathbf{r})}. \tag{27}$$

Note that in the example of §A_{II}.1 such a definition leads to a momentum equal to $\mu v(x)$, which is just the limiting value of the momentum per unit length, $a^{-1}mv_n$, of the discrete system when a tends to zero. Note also that this definition is identical to (A.41) as a result of (20).

The Hamiltonian is introduced by an expression identical to that of the discrete case,

$$H = \int \mathrm{d}^3r \sum_j \Pi_j(\mathbf{r})\, \dot{A}_j(\mathbf{r}) - L. \tag{28}$$

This expression can be transformed [see (A.45.a) and (A.45.b)] in order to introduce a Hamiltonian density $\mathscr{H}$. It is easy now to directly derive Hamilton's dynamical equations with the form (28) of the Hamiltonian. To this end, one finds the differential of H, with H considered as a function of $A_j(\mathbf{r})$ and $\Pi_j(\mathbf{r})$. On one hand we have

$$\delta H = \int \mathrm{d}^3r \sum_j \left[\frac{\eth H}{\eth A_j(\mathbf{r})}\,\delta A_j(\mathbf{r}) + \frac{\eth H}{\eth \Pi_j(\mathbf{r})}\,\delta \Pi_j(\mathbf{r})\right] \tag{29}$$

and on the other hand, using (28),

$$\delta H = \int \mathrm{d}^3r \sum_j \Big[\Pi_j(\mathbf{r})\,\delta\dot{A}_j(\mathbf{r}) + \dot{A}_j(\mathbf{r})\,\delta\Pi_j(\mathbf{r}) - \\ - \frac{\eth L}{\eth A_j(\mathbf{r})}\,\delta A_j(\mathbf{r}) - \frac{\eth L}{\eth \dot{A}_j(\mathbf{r})}\,\delta\dot{A}_j(\mathbf{r})\Big]. \tag{30}$$

The definition (27) for $\Pi_j(\mathbf{r})$ implies that the linear terms in $\delta\dot{A}_j$ of (30) vanish. In addition, by virtue of (26) and (27), one can replace $\eth L/\eth A_j(\mathbf{r})$ by $\dot{\Pi}_j(\mathbf{r})$. Comparing the forms (29) and (30) of the differential δH, one finds then

$$\dot{\Pi}_j(\mathbf{r}) = -\frac{\eth H}{\eth A_j(\mathbf{r})} \tag{31.a}$$

$$\dot{A}_j(\mathbf{r}) = \frac{\eth H}{\eth \Pi_j(\mathbf{r})}. \tag{31.b}$$

These equations can be rewritten using the Hamiltonian density $\mathcal{H}$ [just as Lagrange's equation (26) can be rewritten with the aid of the Lagrangian density thanks to (19) and (20)]. By proceeding in that fashion, one gets Equations (A.46.a) and (A.46.b). Here again, the use of functional derivatives of H allows one to write the Hamilton–Jacobi equations of a continuous system in a form analogous to that of the discrete case.

COMPLEMENT B_{II}

SYMMETRIES OF THE LAGRANGIAN IN THE COULOMB GAUGE AND THE CONSTANTS OF THE MOTION

In this complement we will show that it is possible to derive from the Lagrangian in the Coulomb gauge the expression for the conserved quantities relative to the system particles + field. To find the expression for these constants of the motion, we will calculate the action along the actual path and relate the infinitesimal variations of the action to the Hamiltonian and to the momentum. We will then use the invariance of the Lagrangian (and thus that of the action) under certain transformations (time translation, spatial translation, spatial rotation) to find the constants of the motion (energy, total momentum, total angular momentum).

1. The Variation of the Action between Two Infinitesimally Close Real Motions

The principle of least action requires that the integral of the Lagrangian between two times t_1 and t_2 be an extremum when $x(t)$ corresponds to the real evolution of the generalized coordinate x between t_1 and t_2, the values at the endpoints $x(t_1)$ and $x(t_2)$ being initially fixed. From such a point of view, one evaluates the action S on many *virtual* paths in the (x, t) plane and seeks that which minimizes S. A second possibility is to consider all the possible *real* paths starting from $x(t_1) = x_1$ and to study the variation $S' - S$ of the action when one varies the other extremity from (x_2, t_2) to (x'_2, t'_2) (Figure 1). It is this procedure which we will follow here.

We will calculate $S' - S$:

$$S' - S = \int_{t_1}^{t'_2} L(x', \dot{x}') \, dt - \int_{t_1}^{t_2} L(x, \dot{x}) \, dt \tag{1}$$

$$S' - S = \int_{t_1}^{t_2} [L(x', \dot{x}') - L(x, \dot{x})] \, dt + \int_{t_2}^{t'_2} L(x', \dot{x}') \, dt . \tag{2}$$

Consider two infinitesimally close paths, and take

$$dS = S' - S \tag{3.a}$$

$$dt_2 = t'_2 - t_2 \tag{3.b}$$

$$dx_2 = x'_2 - x_2 = x'(t_2 + dt_2) - x(t_2) \tag{3.c}$$

$$\delta x(t) = x'(t) - x(t) . \tag{3.d}$$

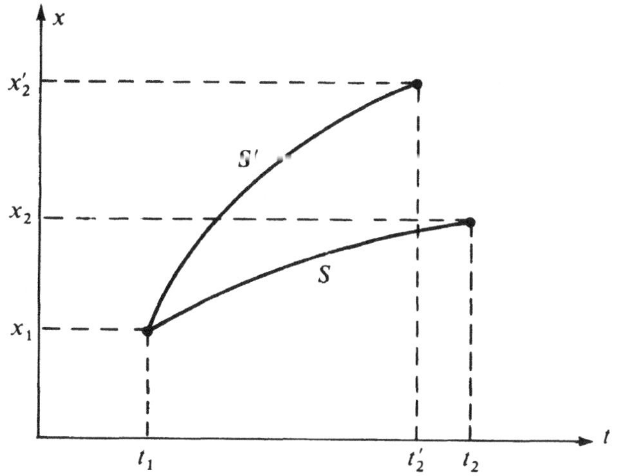

Figure 1. Two real motions of the system, both starting from (x_1, t_1) and ending respectively at (x_2, t_2) and (x_2', t_2'). The difference $S' - S$ is the variation of the action from one path to the other.

Up to second-order terms, the last integral of (2) is

$$L(x(t_2), \dot{x}(t_2))\,\mathrm{d}t_2 = L(2)\,\mathrm{d}t_2\,. \tag{4}$$

The first integral of (2) can be transformed using Equation (13) of Complement A_{II}, with the result that

$$\mathrm{d}S = \int_{t_1}^{t_2} \mathrm{d}t \left[\frac{\partial L}{\partial x} - \frac{\mathrm{d}}{\mathrm{d}t}\frac{\partial L}{\partial \dot{x}}\right]\delta x(t) + \left[\frac{\partial L}{\partial \dot{x}}\,\delta x\right]_{t_1}^{t_2} + L(2)\,\mathrm{d}t_2\,. \tag{5}$$

Since the path $x(t)$ corresponds to an actual motion, the Lagrange equations are satisfied and the first term of (5) is identically zero, so that $\mathrm{d}S$ can be simply given in the form

$$\mathrm{d}S = p(t_2)\,\delta x(t_2) + L(2)\,\mathrm{d}t_2\,. \tag{6}$$

We have used the definition (A.5) of the momentum and the fact that $\delta x(t_1) = 0$. Finally, we will relate $\delta x(t_2)$ to $\mathrm{d}x_2$. Taking (3.d) and (3.c) into account, we have

$$\begin{aligned} \delta x(t_2) &= x'(t_2) - x(t_2) \\ &\simeq x'(t_2 + \mathrm{d}t_2) - \dot{x}'(t_2)\,\mathrm{d}t_2 - x(t_2) \\ &\simeq \mathrm{d}x_2 - \dot{x}(t_2)\,\mathrm{d}t_2\,. \end{aligned} \tag{7}$$

Finally, combining (6) and (7), we get

$$\begin{aligned} \mathrm{d}S &= p(t_2)\,\mathrm{d}x_2 - [p(t_2)\,\dot{x}(t_2) - L(2)]\,\mathrm{d}t_2 \\ &= p(t_2)\,\mathrm{d}x_2 - H(2)\,\mathrm{d}t_2 \end{aligned} \tag{8}$$

where H is the Hamiltonian.

More generally, if one varies not only x_2 and t_2 but also x_1 and t_1, and if there are many generalized coordinates x_j, then dS is given by

$$dS = \sum_{j=1}^{N} p_j(t_2)\, dx_{2j} - \sum_{j=1}^{N} p_j(t_1)\, dx_{1j} - H(2)\, dt_2 + H(1)\, dt_1 . \quad (9)$$

We will now show how in a simple case the preceding equations can be applied to find the conservation laws.

2. Constants of the Motion in a Simple Case

We will consider here an ensemble of interacting particles having as their Lagrangian

$$L = \sum_{\alpha} \frac{1}{2} m_\alpha \dot{\mathbf{r}}_\alpha^2 - \sum_{\alpha<\beta} V(|\mathbf{r}_\alpha - \mathbf{r}_\beta|) . \quad (10)$$

Such a Lagrangian is clearly invariant under the transformation $t \to t + \varepsilon$ (translation by ε in the time dimension). The action will be unchanged if the times t_1 and t_2 are displaced by the same amount ε, the initial and final coordinates of the path being fixed. For an infinitesimal translation $dt_1 = dt_2 = \varepsilon$ we then get on one hand

$$dS = 0 \quad (11)$$

and on the other from (9)

$$dS = [H(1) - H(2)]\, \varepsilon . \quad (12)$$

Thus $H(1) = H(2)$. H is conserved for a time-independent Lagrangian. The Hamiltonian in this circumstance corresponds to the total energy, which is thereby a constant of the motion.

The Lagrangian (10) is also invariant under all spatial translations ($\mathbf{r}_\alpha \to \mathbf{r}_\alpha + \boldsymbol{\eta}$). Such a transformation changes neither the velocities of the particles (and therefore their kinetic energy) nor their relative positions (and thus their potential energy). For an infinitesimal spatial translation $d\mathbf{r}_{1\alpha} = d\mathbf{r}_{2\alpha} = \boldsymbol{\eta}$ (for all α), $dt_1 = dt_2 = 0$, the infinitesimal variation of S is given [using (9)] by

$$dS = \left[\sum_{\alpha} \mathbf{p}_\alpha(t_2) - \sum_{\alpha} \mathbf{p}_\alpha(t_1) \right] \cdot \boldsymbol{\eta} . \quad (13)$$

Now since dS is zero as a result of the invariance of L with respect to translation, it follows that the total momentum $\Sigma_\alpha \mathbf{p}_\alpha$ is a constant of the motion.

Finally, the Lagrangian (10) is invariant under all rotations. If one considers an infinitesimal rotation $d\varphi$ about an axis $\mathbf{n}$ ($\mathbf{r}_\alpha \to \mathbf{r}_\alpha + d\varphi\, \mathbf{n} \times \mathbf{r}_\alpha$ for all α), an argument analogous to that above leads to

$$dS = \left[\sum_\alpha \mathbf{p}_\alpha(t_2) \cdot (\mathbf{n} \times \mathbf{r}_\alpha(t_2)) - \sum_\alpha \mathbf{p}_\alpha(t_1) \cdot (\mathbf{n} \times \mathbf{r}_\alpha(t_1))\right] d\varphi = 0 . \quad (14)$$

The axis $\mathbf{n}$ being arbitrary, one concludes that the angular momentum $\sum_\alpha \mathbf{r}_\alpha \times \mathbf{p}_\alpha$ is a constant of the motion.

Thus we have been able to associate a constant of the motion with each continuous symmetry transformation group of the Lagrangian L.

3. Conservation of Energy for the System Charges + Field

The procedure which we will follow in this section and those following is conceptually identical to those introduced above in the discrete case. Equation (9) when extended to the case of electrodynamics in the Coulomb gauge becomes

$$\begin{aligned} dS = {} & \sum_\alpha \mathbf{p}_\alpha(t_2) . d\mathbf{r}_{2\alpha} + \int d^3k [\pi^*(\mathbf{k}, t_2) . d\mathscr{A}_2(\mathbf{k}) + \pi(\mathbf{k}, t_2) . d\mathscr{A}_2^*(\mathbf{k})] - \\ & - \sum_\alpha \mathbf{p}_\alpha(t_1) . d\mathbf{r}_{1\alpha} - \int d^3k [\pi^*(\mathbf{k}, t_1) . d\mathscr{A}_1(\mathbf{k}) + \pi(\mathbf{k}, t_1) . d\mathscr{A}_1^*(\mathbf{k})] \\ & - H(2)\, dt_2 + H(1)\, dt_1 \end{aligned} \quad (15)$$

where H is the Hamiltonian of the system particles + field, $\mathbf{r}_\alpha$ and $\mathbf{p}_\alpha$ the coordinates and conjugate momenta of the particles α, and $\pi(\mathbf{k})$ the conjugate momenta associated with the dynamical variables $\mathscr{A}(\mathbf{k})$ of the transverse field. dS is the variation of the action between two "real motions" whose initial and final positions differ respectively by $d\mathbf{r}_{1\alpha}$, $d\mathscr{A}_1(\mathbf{k})$, dt_1 and by $d\mathbf{r}_{2\alpha}$, $d\mathscr{A}_2(\mathbf{k})$, dt_2.

Now the Lagrangian in the Coulomb gauge was introduced in Part C and is given by

$$L = \sum_\alpha \frac{1}{2} m_\alpha \dot{\mathbf{r}}_\alpha^2 - \int d^3k \frac{|\rho|^2}{\varepsilon_0 k^2} + \int d^3k\, \overline{\mathscr{L}_c} \quad (16.a)$$

$$\overline{\mathscr{L}_c} = \varepsilon_0[\dot{\mathscr{A}}^* \cdot \dot{\mathscr{A}} - c^2 k^2 \mathscr{A}^* \cdot \mathscr{A}] + j^* \cdot \mathscr{A} + j \cdot \mathscr{A}^* \quad (16.b)$$

It is obviously invariant under time translation. Reasoning analogous to that above allows one to show that $dS = 0$ if $dt_1 = dt_2$ and if all of the other variations $d\mathbf{r}_{1\alpha}$, $d\mathbf{r}_{2\alpha}$, $d\mathscr{A}_1(\mathbf{k})$, and $d\mathscr{A}_2(\mathbf{k})$ are zero, which gives [using (15)] $H(1) = H(2)$. Thus the invariance of the Lagrangian (16) under time translation implies that the Hamiltonian of the system particles + field is a constant of the motion. One will recall that such a Hamiltonian is written [see (C.28)]

$$H = \sum_\alpha \frac{1}{2\,m_\alpha}[\mathbf{p}_\alpha - q_\alpha\,\mathbf{A}(\mathbf{r}_\alpha)]^2 + \\ + V_{\text{Coul}} + \varepsilon_0 \int \mathrm{d}^3k \left[\frac{\boldsymbol{\pi}^* \cdot \boldsymbol{\pi}}{\varepsilon_0^2} + c^2\,k^2\,\mathscr{A}^* \cdot \mathscr{A}\right] \tag{17}$$

and represents the sum of the kinetic energies of the particles, their Coulomb energies, and the energy of the transverse field.

4. Conservation of the Total Momentum

The electrodynamic Lagrangian is also invariant under spatial translations. We will verify this using its expression (16.b) in reciprocal space. Assume that one translates the particles and the field simultaneously by the same amount $\boldsymbol{\eta}$. The new coordinates of the particles are

$$\mathbf{r}'_\alpha = \mathbf{r}_\alpha + \boldsymbol{\eta} \tag{18}$$

while the translated field $\mathbf{A}'(\mathbf{r}, t)$ is equal to

$$\mathbf{A}'(\mathbf{r}, t) = \mathbf{A}(\mathbf{r} - \boldsymbol{\eta}, t) \tag{19}$$

so that in reciprocal space

$$\mathscr{A}'(\mathbf{k}, t) = \mathscr{A}(\mathbf{k}, t)\,\mathrm{e}^{-\mathrm{i}\mathbf{k}.\boldsymbol{\eta}}\,. \tag{20}$$

(Note that $\mathscr{A}'$ is transverse just like $\mathscr{A}$.) The Lagrangian of the particles [similar to (10)] is unchanged under the transformation (18). The Lagrangian of the field, which involves quantities of the type $\mathscr{A}'^* \cdot \mathscr{A}'$, is also unchanged on account of (20). The interaction Lagrangian is not modified for the same reason, the terms $\mathscr{j}'^* \cdot \mathscr{A}'$ being unaltered because the law of transformation for $\mathscr{j}'$ has the same form as (20). The invariance of L under translation implies then that S does not vary under a translation.

We will now find $\mathrm{d}S$ using (15). To this end, it is necessary first to relate $\mathrm{d}\mathscr{A}$ to $\boldsymbol{\eta}$ using (20) for infinitesimal $\boldsymbol{\eta}$:

$$\mathrm{d}\mathscr{A} = -\,\mathrm{i}(\mathbf{k} \cdot \boldsymbol{\eta})\,\mathscr{A} \tag{21}$$

By substituting this value in (15), replacing $\boldsymbol{\pi}$ by $-\varepsilon_0\mathscr{E}_\perp$, and extending the integral over $\mathbf{k}$ to all reciprocal space, one gets the following conserved quantity:

$$\mathbf{P} = \sum_\alpha \mathbf{p}_\alpha + \varepsilon_0 \int \mathrm{d}^3k(\mathscr{E}_\perp^* \cdot \mathscr{A})\,\mathrm{i}\mathbf{k}\,. \tag{22}$$

Thanks to the transversality of $\mathscr{E}_\perp$, the last term of (22) can be rewritten as a double vector product, and we get finally

$$\mathbf{P} = \sum_\alpha \mathbf{p}_\alpha + \varepsilon_0 \int d^3k \, \mathscr{E}_\perp^* \times (i\mathbf{k} \times \mathscr{A}) . \tag{23}$$

This expression, giving the total momentum, coincides with that given in Chapter I [(B.45) and (B.39.b)]. The translational invariance of the Lagrangian then implies the conservation of the total momentum of the system particles + field.

5. Conservation of the Total Angular Momentum

Consider a rotation $\mathscr{R}$ of the ensemble particles + field through an angle $d\varphi$ about the axis $\mathbf{n}$. The new coordinates of the particles are

$$\mathbf{r}'_\alpha = \mathbf{r}_\alpha + d\varphi \, \mathbf{n} \times \mathbf{r}_\alpha . \tag{24}$$

The field $\mathbf{A}'$ after rotation is equal to

$$\mathbf{A}'(\mathbf{r}, t) = \mathscr{R} \mathbf{A}(\mathscr{R}^{-1} \mathbf{r}, t) \tag{25}$$

so that in reciprocal space

$$\mathscr{A}'(\mathbf{k}, t) = \mathscr{R} \mathscr{A}(\mathscr{R}^{-1} \mathbf{k}, t) . \tag{26}$$

[To go from (25) to (26) it is sufficient to note that the scalar product $\mathbf{k} \cdot \mathbf{r}$ appearing in the Fourier transformation is unchanged under a simultaneous rotation of $\mathbf{k}$ and $\mathbf{r}$. Note also that (26) insures the transversality of $\mathscr{A}'$.] Finally, for an infinitesimal rotation, (26) becomes

$$\mathscr{A}'(\mathbf{k}, t) = \mathscr{A}(\mathbf{k} - d\varphi \, \mathbf{n} \times \mathbf{k}, t) + d\varphi \, \mathbf{n} \times \mathscr{A}(\mathbf{k} - d\varphi \, \mathbf{n} \times \mathbf{k}, t) \tag{27}$$

so that to first order in $d\varphi$,

$$d\mathscr{A}(\mathbf{k}, t) = d\varphi \, \{ -[(\mathbf{n} \times \mathbf{k}) \cdot \nabla] \mathscr{A} + \mathbf{n} \times \mathscr{A} \} \tag{28}$$

where ∇ is the gradient operator in reciprocal space.

Under the rotation, the Lagrangian of the particles is invariant, since the relative positions of the particles are unaltered (Coulomb interaction). The Lagrangian of the transverse field involves integrals over reciprocal space of the type $\mathscr{A}'^*(\mathbf{k}, t) \cdot \mathscr{A}'(\mathbf{k}, t)$, which with (26) are also equal to $\mathscr{A}^*(\mathscr{R}^{-1}\mathbf{k}, t) \cdot \mathscr{A}(\mathscr{R}^{-1}\mathbf{k}, t)$. Taking $\mathbf{k}' = \mathscr{R}^{-1}\mathbf{k}$ as the variable of integration shows the invariance of this term. Finally, the interaction Lagrangian between the particles and field contains the scalar product

$j'^*(\mathbf{k}, t) \cdot \mathcal{A}'(\mathbf{k}, t)$. The current $\mathbf{j}$ transforms under rotation like $\mathbf{A}$, and the invariance of this term is seen in an identical fashion. Finally, the total Lagrangian (16) is invariant under a rotation by angle $d\varphi$ around $\mathbf{n}$. It follows that the differential of the action corresponding to this coordinate transformation is zero.

We will now evaluate dS using (15). To this end, we will replace $d\mathbf{r}_\alpha$ and $d\mathcal{A}$ by their values given as functions of $d\varphi$. Additionally, expressing $\boldsymbol{\pi}$ as a function of $\mathcal{E}_\perp$ and extending the integral to all reciprocal space, one sees that the quantity

$$\sum_\alpha \mathbf{p}_\alpha \cdot (\mathbf{n} \times \mathbf{r}_\alpha) + \int d^3k \, \varepsilon_0 [\mathcal{E}_\perp^* \cdot \{(\mathbf{n} \times \mathbf{k}) \cdot \nabla\} \mathcal{A} - \mathcal{E}_\perp^* \cdot (\mathbf{n} \times \mathcal{A})] \tag{29}$$

is conserved. Equation (29) can then be written

$$\mathbf{n} \cdot \left[\sum_\alpha \mathbf{r}_\alpha \times \mathbf{p}_\alpha + \varepsilon_0 \int d^3k \left\{ \sum_{a=x,y,z} \mathcal{E}_{\perp a}^* (\mathbf{k} \times \nabla) \mathcal{A}_a + \mathcal{E}_\perp^* \times \mathcal{A} \right\} \right]. \tag{30}$$

This quantity being conserved for any direction of the vector $\mathbf{n}$, one finds a new constant of the motion,

$$\mathbf{J} = \sum_\alpha \mathbf{r}_\alpha \times \mathbf{p}_\alpha + \varepsilon_0 \int d^3k \left\{ \sum_a \mathcal{E}_{\perp a}^* (\mathbf{k} \times \nabla) \mathcal{A}_a + \mathcal{E}_\perp^* \times \mathcal{A} \right\}. \tag{31}$$

One recognizes in (31) the total angular momentum, which is the angular momentum of the particles and of the longitudinal field [first term as in (8), Complement B_I] plus the angular momentum of the transverse field [see (11) of Complement B_I].

General references and additional reading

The results derived in this complement constitute an example of the application of Noether's theorem, which relates the symmetries of the Lagrangian to the constants of the motion. The reader interested in this theorem can refer to more advanced works on field theory such as Bogoliubov and Shirkov (Chapter I) or Itzykson and Zuber (Chapter I).

COMPLEMENT C_{II}

ELECTRODYNAMICS IN THE PRESENCE OF AN EXTERNAL FIELD

In Chapter II the particles and the electromagnetic field have been treated as forming an isolated system. Now it is often necessary to describe the dynamics of the system particles + electromagnetic field in the presence of an externally applied field. We will first show (§C_{II}.1) how to distinguish the external field in this formalism. We will then give (§C_{II}.2) the Lagrangian and then (§C_{II}.3) the Hamiltonian of the system made up of the particles and the electromagnetic field in the presence of an externally applied field.

1. Separation of the External Field

In the electrodynamic theory developed earlier, the electromagnetic field is a dynamical variable whose value can not be fixed a priori. If one wishes to study the evolution of the system particles + electromagnetic field in an externally applied field, one proceeds in the theory as in practice and lays out around the system external sources which are determined experimentally to give the desired field. The particles then evolve under the simultaneous action of the external field and the rest of the electromagnetic field—in particular, the field created by their own dynamics. On the other hand, we assume that the sources of the external field are not part of the dynamical system. The charge and current densities of these sources, $\rho_e(\mathbf{r}, t)$ and $\mathbf{j}_e(\mathbf{r}, t)$, will be considered as given functions of time, fixed independently of the field of the particles. We thus assume either that the reaction of the field back on the external sources is negligible or that it is compensated for by the experimental arrangement.

The external field can be described by the potentials $\mathbf{A}_e(\mathbf{r}, t)$ and $U_e(\mathbf{r}, t)$, which are solutions of the equations [analogous to (A.11) of Chapter I]

$$\Delta U_e + \nabla \cdot \dot{\mathbf{A}}_e = -\frac{\rho_e}{\varepsilon_0} \tag{1.a}$$

$$\ddot{\mathbf{A}}_e - c^2\,\Delta\mathbf{A}_e + c^2\,\nabla(\nabla \cdot \mathbf{A}_e) + \nabla\dot{U}_e = \frac{\mathbf{j}_e}{\varepsilon_0}. \tag{1.b}$$

(Note that it is not necessary to choose the Coulomb gauge to describe the field of the sources.)

Let $\mathbf{A}_t$ and U_t be the potentials corresponding to the total electromagnetic field. It is useful to introduce new dynamical field variables $\mathbf{A}$ and U such that

$$\mathbf{A}_t(\mathbf{r}, t) = \mathbf{A}_e(\mathbf{r}, t) + \mathbf{A}(\mathbf{r}, t) \tag{2.a}$$

$$U_t(\mathbf{r}, t) = U_e(\mathbf{r}, t) + U(\mathbf{r}, t)\,. \tag{2.b}$$

The structure of Equations (1) and (2) clearly shows that $\mathbf{A}$ and U are the potentials corresponding to the sum of the free field and the field created by the particles. As a result of the linearity of Maxwell's equations, $\mathbf{A}$ and U will satisfy equations analogous to (1) but where the charge and current densities correspond to the particles of the system only. As for the dynamics of the particles, it is fixed by the Lorentz equations, where the field is the total field corresponding to the potentials $\mathbf{A}_t$ and U_t.

We will introduce in the following section a Lagrangian whose Lagrange equations correspond precisely to these specifications.

2. The Lagrangian in the Presence of an External Field

a) Introduction of a Lagrangian

We will consider the following Lagrangian, where the dynamical variables of the field are $\{\mathbf{A}(\mathbf{r}), U(\mathbf{r}), \dot{\mathbf{A}}(\mathbf{r}), \dot{U}(\mathbf{r})\}$:

$$L = \sum_\alpha \frac{1}{2} m_\alpha \dot{\mathbf{r}}_\alpha^2 + \frac{\varepsilon_0}{2} \int \mathrm{d}^3r [(- \nabla U - \dot{\mathbf{A}})^2 - c^2(\nabla \times \mathbf{A})^2] + $$
$$+ \int \mathrm{d}^3r [\mathbf{j}_P \cdot (\mathbf{A}_e + \mathbf{A}) - \rho_P(U_e + U)]\,. \tag{3}$$

In this Lagrangian $\rho_P(\mathbf{r}, t)$ and $\mathbf{j}_P(\mathbf{r}, t)$ are the charge and current densities of the particles of the system only. Calculations analogous to those of §B.2 give the following Lagrange equations for the particles and the field respectively:

$$m_\alpha \ddot{\mathbf{r}}_\alpha = q_\alpha[\mathbf{E}_e(\mathbf{r}_\alpha, t) + \mathbf{E}(\mathbf{r}_\alpha, t)] + q_\alpha \dot{\mathbf{r}}_\alpha \times [\mathbf{B}_e(\mathbf{r}_\alpha, t) + \mathbf{B}(\mathbf{r}_\alpha, t)] \tag{4.a}$$

$$\Delta U + \nabla \cdot \dot{\mathbf{A}} = - \frac{\rho_P}{\varepsilon_0} \tag{4.b}$$

$$\ddot{\mathbf{A}} - c^2 \Delta \mathbf{A} + c^2 \nabla(\nabla \cdot \mathbf{A}) + \nabla \dot{U} = \frac{\mathbf{j}_P}{\varepsilon_0}\,. \tag{4.c}$$

The dynamics of the particles is thus determined by the total electromagnetic field ($\mathbf{E}_t = \mathbf{E}_e + \mathbf{E}$, $\mathbf{B}_t = \mathbf{B}_e + \mathbf{B}$), while the dynamical variables of the field **A** and U are only related to the charge and current densities of the particles.

b) The Lagrangian in the Coulomb Gauge

The elimination of U and the choice of the Coulomb gauge for the dynamical field variables is made in the same way as in Part C. We then get a Lagrangian similar to that given by (C.13),

$$L = \sum_\alpha \frac{1}{2} m_\alpha \dot{\mathbf{r}}_\alpha^2 - V_{\text{Coul}} + \int \mathrm{d}^3k\, \overline{\mathscr{L}}_C \tag{5.a}$$

$$\overline{\mathscr{L}}_C = \varepsilon_0[\dot{\mathscr{A}} \cdot \dot{\mathscr{A}}^* - c^2 k^2 \mathscr{A} \cdot \mathscr{A}^*] + j_P^* \cdot (\mathscr{A} + \mathscr{A}_e) + \\ + j_P \cdot (\mathscr{A}^* + \mathscr{A}_e^*) - \rho_P \mathscr{U}_e^* - \rho_P^* \mathscr{U}_e . \tag{5.b}$$

In this Lagrangian the potential $\mathscr{A}$ is transverse (from the Coulomb gauge choice):

$$\mathscr{A}_{\parallel} = 0 . \tag{6}$$

On the other hand, we have not necessarily chosen the same gauge for the external field. That is why $\mathscr{U}_e$ and $\mathscr{A}_{e\parallel}$ can appear in (5.b). In the same way, Equations (2), giving the potentials of the total electromagnetic field, now take in the reciprocal space the following form:

$$\mathscr{A}_{t\perp} = \mathscr{A}_{e\perp} + \mathscr{A}_{\perp} \tag{7.a}$$

$$\mathscr{A}_{t\parallel} = \mathscr{A}_{e\parallel} \tag{7.b}$$

$$\mathscr{U}_t = \mathscr{U}_e + \frac{\rho_P}{\varepsilon_0 k^2} . \tag{7.c}$$

3. The Hamiltonian in the Presence of an External Field

a) Conjugate Momenta

To find the conjugate momenta, one follows the procedure of §C.3. We then get immediately the generalization of (C.19) for the particles,

$$\mathbf{p}_\alpha = m_\alpha \dot{\mathbf{r}}_\alpha + q_\alpha[\mathbf{A}(\mathbf{r}_\alpha) + \mathbf{A}_e(\mathbf{r}_\alpha, t)] \tag{8}$$

and the expression (C.21) for the field,

$$\pi_\varepsilon(\mathbf{k}) = \varepsilon_0 \dot{\mathscr{A}}_\varepsilon(\mathbf{k}) \tag{9}$$

where $\pi_\varepsilon = \partial\overline{\mathscr{L}}_C/\partial\dot{\mathscr{A}}_\varepsilon^*$ is the momentum associated with the dynamical variable $\mathscr{A}_\varepsilon$.

b) THE HAMILTONIAN

The Hamiltonian of the system in the Coulomb gauge is found as in §C.3. Using (C.27), we find

$$H = \sum_{\alpha} \frac{1}{2\,m_{\alpha}} \{ \mathbf{p}_{\alpha} - q_{\alpha}[\mathbf{A}(\mathbf{r}_{\alpha}) + \mathbf{A}_e(\mathbf{r}_{\alpha}, t)] \}^2 + \\ + V_{\text{Coul}} + \int \mathrm{d}^3k [\rho_P^* \, \mathcal{U}_e + \rho_P \, \mathcal{U}_e^*] \\ + \varepsilon_0 \int \mathrm{d}^3k \left[\frac{\boldsymbol{\pi} \cdot \boldsymbol{\pi}^*}{\varepsilon_0^2} + c^2 k^2 \boldsymbol{\mathscr{A}} \cdot \boldsymbol{\mathscr{A}}^* \right]. \tag{10}$$

It is possible to decompose H into three terms

$$H = H_P + H_R + H_I \tag{11}$$

which are described below. These terms correspond respectively to the Hamiltonian H_P of the particles evolving in the Coulomb field and in the external field, to the Hamiltonian H_R of the free field, and to the interaction Hamiltonian H_I describing the coupling between the radiation field and the particles evolving in the external field:

$$H_P = \sum_{\alpha} \frac{(\mathbf{p}_{\alpha}^e)^2}{2\,m_{\alpha}} + V_{\text{Coul}} + \int \mathrm{d}^3r \; \rho_P(\mathbf{r}) \; U_e(\mathbf{r}, t) \tag{12}$$

where

$$\mathbf{p}_{\alpha}^e = \mathbf{p}_{\alpha} - q_{\alpha} \, \mathbf{A}_e(\mathbf{r}_{\alpha}, t) \, . \tag{13}$$

We emphasize that the momentum conjugate with $\mathbf{r}_{\alpha}$ is $\mathbf{p}_{\alpha}$ and not $\mathbf{p}_{\alpha}^e$, which is a useful physical variable to which it is related.

H_R and H_I are equal to

$$H_R = \varepsilon_0 \int \mathrm{d}^3k \left[\frac{\boldsymbol{\pi} \cdot \boldsymbol{\pi}^*}{\varepsilon_0^2} + c^2 k^2 \boldsymbol{\mathscr{A}} \cdot \boldsymbol{\mathscr{A}}^* \right] \tag{14}$$

$$H_I = - \frac{q_{\alpha}}{m_{\alpha}} \sum_{\alpha} \mathbf{p}_{\alpha}^e \cdot \mathbf{A}(\mathbf{r}_{\alpha}) + \frac{q_{\alpha}^2}{2\,m_{\alpha}} \mathbf{A}^2(\mathbf{r}_{\alpha}) \tag{15}$$

where $\mathbf{p}_{\alpha}^e$ has been introduced in (13).

c) QUANTIZATION

The quantization of the theory is carried out like that of §C.4 and results in the following commutation relations for the particles and the field:

$$[(\mathbf{r}_\alpha)_i, (\mathbf{p}_\beta)_j] = i\hbar\, \delta_{ij}\, \delta_{\alpha\beta} \tag{16.a}$$

$$[\mathscr{A}_\varepsilon(\mathbf{k}), \pi_{\varepsilon'}(\mathbf{k}')] = i\hbar\, \delta_{\varepsilon\varepsilon'}\, \delta(\mathbf{k} + \mathbf{k}') \tag{16.b}$$

$$[\mathscr{A}_\varepsilon(\mathbf{k}), \pi^+_{\varepsilon'}(\mathbf{k}')] = i\hbar\, \delta_{\varepsilon\varepsilon'}\, \delta(\mathbf{k} - \mathbf{k}') . \tag{16.c}$$

It is important to note that the commutation relations for the particles are given in terms of $\mathbf{p}_\alpha$ and not $\mathbf{p}^e_\alpha$. Note further that the Cartesian components of $\mathbf{p}^e_\alpha$ do not commute among themselves:

$$[(\mathbf{p}^e_\alpha)_i, (\mathbf{p}^e_\beta)_j] = i\hbar\, \delta_{\alpha\beta}\, q_\alpha \sum_l \varepsilon_{ijl}[\mathbf{B}_e(\mathbf{r}_\alpha, t)]_l \, . \tag{17}$$

Finally, we want to emphasize that in the absence of particles, the total field does not reduce to that which we call the external field. Actually, we will show in Chapter III that the quantum field associated with the vector potential $\mathscr{A}_\perp$ appearing in (7.a), whose dynamics is described by (14), is zero only in its mean value. The fluctuations of $\mathscr{A}_\perp$ constitute what one calls the vacuum fluctuations. In the absence of particles, the total field is thus the superposition of the external field and the vacuum fluctuations.

COMPLEMENT D_{II}

EXERCISES

Exercise 1. An example of a Hamiltonian different from the energy.
Exercise 2. From a discrete to a continuous system: Introduction of the Lagrangian and Hamiltonian densities.
Exercise 3. Lagrange's equations for the components of the electromagnetic field in real space.
Exercise 4. Lagrange's equations for the standard Lagrangian in the Coulomb gauge.
Exercise 5. Momentum and angular momentum of an arbitrary field.
Exercise 6. A Lagrangian using complex variables and linear in velocity.
Exercise 7. Lagrangian and Hamiltonian descriptions of the Schrödinger matter field.
Exercise 8. Quantization of the Schrödinger field.
Exercise 9. Schrödinger equation of a particle in an electromagnetic field: Arbitrariness of phase and gauge invariance.

1. An example of a Hamiltonian different from the energy

Consider the Lagrangian

$$L = m\dot{x}\dot{y} - m\omega_0^2\, xy\,. \tag{1}$$

a) Write the Lagrange equations associated with L. With what physical system is L associated? What is a priori the energy E for such a system?

b) Find the Hamiltonian H associated with L. Compare E and H.

Solution

a) From (1) one gets

$$\frac{\partial L}{\partial \dot{x}} = m\dot{y} \quad (2.\text{a}) \qquad \frac{\partial L}{\partial x} = -\, m\omega_0^2\, y \quad (2.\text{b})$$

$$\frac{\partial L}{\partial \dot{y}} = m\dot{x} \quad (2.\text{c}) \qquad \frac{\partial L}{\partial y} = -\, m\omega_0^2\, x\,. \quad (2.\text{d})$$

The Lagrange equations

$$m\ddot{y} = -\, m\omega_0^2\, y \quad (3.\text{a}) \qquad m\ddot{x} = -\, m\omega_0^2\, x \quad (3.\text{b})$$

are those of two harmonic oscillators with the same frequency ω_0. The total energy E for these two oscillators is a priori

$$E = \frac{1}{2} m\dot{x}^2 + \frac{1}{2} m\dot{y}^2 + \frac{1}{2} m\omega_0^2 x^2 + \frac{1}{2} m\omega_0^2 y^2 \tag{4}$$

b) The conjugate momenta associated with x and y are gotten from (2.a) and (2.c) and lead to the following Hamiltonian:

$$H = m\dot{x}\dot{y} + m\omega_0^2 xy \tag{5}$$

which is different from the energy (4).

2. From a discrete to a continuous system: Introduction of the Lagrangian and Hamiltonian densities

The purpose of this exercise is to introduce intuitively the concept of the Lagrangian and Hamiltonian densities of a continuous system by studying how a discrete mechanical system can be transformed into a continuous one.

(i) *The discrete system.* Consider an infinite set of point particles with mass m aligned along the x-axis with equilibrium spacing a. The displacement along the x-axis of the nth particle (whose equilibrium position is na) is called q_n. The state of the system at t is fixed by giving the dynamical variables $q_n(t)$ and $\dot{q}_n(t)$. The potential energy of the system of n particles depends on their separations and is equal to

$$V = \frac{1}{2} m\omega_1^2 \sum_n (q_{n+1} - q_n)^2 . \tag{1}$$

a) Write the Lagrangian of this system. Derive the equations of motion (Lagrange's equations).

b) Look for a solution of the form

$$q_n(t) = \delta \, e^{i(kna - \omega t)} . \tag{2}$$

What relationship is there between ω and k? Denote by v the phase velocity $v = \omega/k$. Give the value v_0 of v in the long-wavelength limit ($k \to 0$).

c) Calculate the conjugate momentum p_n for the coordinate q_n, and find the Hamiltonian. Give the canonical commutation relation between q_n and p_n.

(ii) *The continuous system gotten by passing to the limit.* Let the distance a between two adjacent particles and the mass m of each particle go to zero in such a way that the mass per unit length, $\mu = m/a$, is kept constant. Similarly, let ω_1 vary in such a way that when a goes to 0, v_0

remains constant. One gets then in this limit a continuous string with mass per unit length μ and where the velocity of sound is v_0.

The discrete dynamical variable $q_n(t)$ which represents the displacement of the point na becomes a continuous variable $q(x, t)$ giving the displacement of a point on the string whose equilibrium position is x. Similarly, $\dot{q}_n(t)$ becomes $\partial q(x, t)/\partial(t)$ in the continuous limit. One then moves from a discrete index n to a continuous index x.

a) Show that at the continuous limit, the Lagrangian can be put in the form $L = \int \mathrm{d}x\, \mathscr{L}$, where $\mathscr{L}$ is a function to be determined.

b) What is the continuous limit of the equations of motion gotten from the discrete case? Is this result identical to that gotten using the Lagrangian density and Lagrange's equations for a continuous system?

c) The momentum $\Pi(x)$ conjugate with the continuous variable $q(x)$ is defined by $\Pi(x) = \partial \mathscr{L}/\partial \dot{q}(x)$. Show that $\Pi(x)$ corresponds to the limit of p_n/a when a goes to zero. Write the Hamiltonian of the continuous system in the form $H = \int \mathrm{d}x\, \mathscr{H}$, and give the expression for the Hamiltonian density $\mathscr{H}$.

d) Show how the commutator $[q_n, p_{n'}] = \mathrm{i}\hbar\delta_{nn'}$ from the discrete case becomes $[q(x), \Pi(x')] = \mathrm{i}\hbar\delta(x - x')$ in the continuous case.

Solution

(i)

a) The Lagrangian of this system of point particles is

$$L = \sum_n \frac{1}{2} m\dot{q}_n^2 - \sum_n \frac{1}{2} m\omega_1^2 (q_{n+1} - q_n)^2 . \tag{3}$$

The relations

$$\frac{\partial L}{\partial \dot{q}_n} = m\dot{q}_n \tag{4.a}$$

$$\frac{\partial L}{\partial q_n} = m\omega_1^2(q_{n+1} - q_n) - m\omega_1^2(q_n - q_{n-1}) \tag{4.b}$$

lead to the equations of motion

$$\ddot{q}_n = \omega_1^2(q_{n+1} - q_n) - \omega_1^2(q_n - q_{n-1}) . \tag{5}$$

b) Equation (2) for q_n is the solution of (5) if

$$-\omega^2 = \omega_1^2[(\mathrm{e}^{\mathrm{i}ka} - 1) - (1 - \mathrm{e}^{-\mathrm{i}ka})] = -4\,\omega_1^2 \sin^2 \frac{ka}{2} . \tag{6}$$

We calculate the phase velocity and its limit when $k \to 0$:

$$v = \frac{2\,\omega_1}{k} \sin \frac{ka}{2} \tag{7}$$

$$v_0 = \omega_1\, a . \tag{8}$$

c) Using the results of §A.1 of Chapter II, one gets immediately

$$p_n = m\dot{q}_n \tag{9.a}$$

$$H = \sum_n p_n \dot{q}_n - L$$

$$= \sum_n \frac{1}{2} m\dot{q}_n^2 + \sum_n \frac{1}{2} m\omega_1^2(q_{n+1} - q_n)^2 \tag{9.b}$$

$$[q_n, p_{n'}] = i\hbar\, \delta_{nn'} . \tag{9.c}$$

(ii)

a) The Lagrangian (3) can be written

$$L = \mu \sum_n a\left[\frac{\dot{q}_n^2}{2} - \frac{v_0^2}{2}\frac{(q_{n+1} - q_n)^2}{a^2}\right] \tag{10}$$

since $\mu = m/a$ and $v_0 = \omega_1 a$. When a goes to zero, $(q_{n+1} - q_n)/a$ tends to $\partial q(x)/\partial x$ and Equation (10) becomes

$$L = \int \mathrm{d}x \frac{\mu}{2}\left[(\dot{q}(x))^2 - v_0^2\left(\frac{\partial q(x)}{\partial x}\right)^2\right]. \tag{11.a}$$

The Lagrangian density $\mathscr{L}$ is then given by

$$\mathscr{L} = \frac{\mu}{2}\left[(\dot{q}(x))^2 - v_0^2\left(\frac{\partial q(x)}{\partial x}\right)^2\right] \tag{11.b}$$

b) The equation of motion (5) can be rewritten in the form

$$\ddot{q}_n = \omega_1^2\, a^2 \frac{[(q_{n+1} - q_n)/a] - [(q_n - q_{n-1})/a]}{a}. \tag{12}$$

When $a \to 0$,

$$(q_{n+1} - q_n)/a \to \frac{\partial q(x)}{\partial x} \tag{13.a}$$

$$(q_n - q_{n-1})/a \to \frac{\partial q(x - a)}{\partial x} \tag{13.b}$$

and thus Equation (12) tends to

$$\ddot{q}(x) = v_0^2 \frac{\partial^2 q(x)}{\partial x^2} \tag{14}$$

which can also be gotten by using (11.b) and Lagrange's equation (A.39).

c) Using (11), one gets

$$\Pi(x) = \mu\dot{q}(x) \tag{15}$$

which corresponds to the limit of p_n/a when a tends to 0. The Hamiltonian H of the discrete system

$$H = \sum_n p_n \dot{q}_n - L = a\sum_n \frac{p_n}{a}\dot{q}_n - L \tag{16.a}$$

has as its limit when $a \to 0$

$$H = \int \mathrm{d}x\, \mathscr{H} \tag{16.b}$$

$$\mathscr{H} = \Pi(x)\, \dot{q}(x) - \mathscr{L} \tag{16.c}$$

so that, using (11.b),

$$\mathscr{H} = \frac{\Pi^2(x)}{2\,\mu} + \mu\,\frac{v_0^2}{2}\left(\frac{\partial q(x)}{\partial x}\right)^2 \tag{17}$$

d) The commutation relation (9.c) divided by a becomes

$$[q_n, p_{n'}/a] = i\hbar\,\delta_{nn'}/a\,. \tag{18}$$

The function on the right is zero for $n \neq n'$ and has, in the limit $a \to 0$, an infinite value for $n = n'$. Furthermore, one has

$$\sum_{n'} a\,\frac{\delta_{n,n'}}{a} = 1 \tag{19}$$

which corresponds in the continuous limit to $\int \mathrm{d}x'\,\delta(x - x') = 1$. One concludes that the continuous limit of (18) is

$$[q(x), \Pi(x')] = i\hbar\,\delta(x - x') \tag{20}$$

which is effectively the quantum canonical commutation relation (A.47.c) for a continuous system.

3. Lagrange's equations for the components of the electromagnetic field in real space

Starting from the expression for the standard Lagrangian

$$L = \sum_\alpha \frac{1}{2}\,m\dot{\mathbf{r}}_\alpha^2 + \frac{\varepsilon_0}{2}\int \mathrm{d}^3r\,[\mathbf{E}^2(\mathbf{r}) - c^2\,\mathbf{B}^2(\mathbf{r})] + \int \mathrm{d}^3r\,[\mathbf{j}(\mathbf{r}).\mathbf{A}(\mathbf{r}) - \rho(\mathbf{r})\,U(\mathbf{r})]. \tag{1}$$

find the Lagrange equations for the field and show that they agree with Maxwell's equations. Use the field components in real space (not in reciprocal space, since that has been done in §B.2 of this chapter). The dynamical variables of the fields are the potentials $U(\mathbf{r})$ and $\mathbf{A}(\mathbf{r})$, and their time derivatives are $\dot{U}(\mathbf{r})$ and $\dot{\mathbf{A}}(\mathbf{r})$.

Solution

Replace $\mathbf{E}$ and $\mathbf{B}$ in (1) by

$$\mathbf{E}(\mathbf{r}) = -\,\dot{\mathbf{A}}(\mathbf{r}) - \nabla U(\mathbf{r}) \quad (2.\text{a}) \qquad \mathbf{B}(\mathbf{r}) = \nabla \times \mathbf{A}(\mathbf{r})\,. \quad (2.\text{b})$$

The Lagrange equation relative to U is found by evaluating $\partial\mathscr{L}/\partial\dot{U}$, $\partial\mathscr{L}/\partial U$, $\partial\mathscr{L}/\partial(\partial_i U)$, where $\mathscr{L}$ is the Lagrangian density

$$\frac{\partial\mathscr{L}}{\partial\dot{U}} = 0 \quad (3.\text{a}) \qquad \frac{\partial\mathscr{L}}{\partial U} = -\,\rho \quad (3.\text{b}) \qquad \frac{\partial\mathscr{L}}{\partial(\partial_i U)} = \frac{\partial\mathscr{L}}{\partial E_i}\,\frac{\partial E_i}{\partial(\partial_i U)} = -\,\varepsilon_0\,E_i \quad (3.\text{c})$$

from which one gets Lagrange's equation

$$-\,\rho + \varepsilon_0 \sum_i \partial_i E_i = 0 \tag{4}$$

that is,

$$\nabla \cdot \mathbf{E} = \frac{\rho}{\varepsilon_0}\,. \tag{5}$$

The Lagrange equation relative to A_k requires knowing $\partial\mathscr{L}/\partial\dot{A}_k$, $\partial\mathscr{L}/\partial A_k$, $\partial\mathscr{L}/\partial(\partial_j A_k)$:

$$\frac{\partial\mathscr{L}}{\partial\dot{A}_k} = \frac{\partial\mathscr{L}}{\partial E_k}\frac{\partial E_k}{\partial\dot{A}_k} = -\varepsilon_0 E_k \tag{6.a}$$

$$\frac{\partial\mathscr{L}}{\partial A_k} = j_k \tag{6.b}$$

$$\frac{\partial\mathscr{L}}{\partial(\partial_j A_k)} = \sum_i \frac{\partial\mathscr{L}}{\partial B_i}\frac{\partial B_i}{\partial(\partial_j A_k)} = -\varepsilon_0 c^2 \sum_i \varepsilon_{ijk} B_i \tag{6.c}$$

Hence Lagrange's equation is

$$-\varepsilon_0 \dot{E}_k = j_k + \varepsilon_0 c^2 \sum_{i,j} \varepsilon_{jki}\,\partial_j B_i \tag{7}$$

which is the projection on $\mathbf{e}_k$ of the equation

$$\boldsymbol{\nabla}\times\mathbf{B} = \frac{1}{c^2}\dot{\mathbf{E}} + \frac{1}{\varepsilon_0 c^2}\mathbf{j}. \tag{8}$$

4. Lagrange's equations for the standard Lagrangian in the Coulomb gauge

Consider the standard Lagrangian in Coulomb gauge,

$$L = \sum_\alpha \frac{1}{2} m_\alpha \dot{\mathbf{r}}_\alpha^2 - \sum_\alpha \varepsilon_{\text{Coul}}^\alpha - \sum_{\alpha>\beta} \frac{q_\alpha q_\beta}{4\pi\varepsilon_0 |\mathbf{r}_\alpha - \mathbf{r}_\beta|} + L_C \tag{1}$$

with

$$L_C = \int \mathrm{d}^3k\, \overline{\mathscr{L}_C} = \int \mathrm{d}^3r\, \mathscr{L}_C \tag{2.a}$$

$$\overline{\mathscr{L}_C} = \varepsilon_0[\dot{\mathscr{A}}^* \cdot \dot{\mathscr{A}} - c^2 k^2 \mathscr{A}^* \cdot \mathscr{A}] + j^* \cdot \mathscr{A} + j \cdot \mathscr{A}^* \tag{2.b}$$

$$\mathscr{L}_C = \frac{\varepsilon_0}{2}[\dot{\mathbf{A}}^2 - c^2(\boldsymbol{\nabla}\times\mathbf{A})^2] + \mathbf{j}\cdot\mathbf{A}. \tag{2.c}$$

Write Lagrange's equations and find the equations of motion for the dynamical variables $(\mathbf{r}_\alpha)_i$ of the particles and $\mathscr{A}_\varepsilon(\mathbf{k})$ of the electromagnetic field. Show, in particular, that the source term of the equation of evolution of $\mathbf{A}(\mathbf{r})$ is the transverse component of the current, $\mathbf{j}_\perp(\mathbf{r})$.

Solution

Consider first the particle variables, and calculate $\partial L/\partial(\mathbf{r}_\alpha)_i$ and $\partial L/\partial(\dot{\mathbf{r}}_\alpha)_i$. Using (1), (2.a), and (2.c), one gets

$$\frac{\partial L}{\partial(\mathbf{r}_\alpha)_i} = \sum_{\beta\neq\alpha} \frac{q_\alpha q_\beta}{4\pi\varepsilon_0}\frac{(\mathbf{r}_\alpha - \mathbf{r}_\beta)_i}{|\mathbf{r}_\alpha - \mathbf{r}_\beta|^3} + q_\alpha \frac{\partial}{\partial(\mathbf{r}_\alpha)_i}[\dot{\mathbf{r}}_\alpha \cdot \mathbf{A}(\mathbf{r}_\alpha, t)] \tag{3.a}$$

$$\frac{\partial L}{\partial(\dot{\mathbf{r}}_\alpha)_i} = m_\alpha(\dot{\mathbf{r}}_\alpha)_i + q_\alpha \mathbf{A}_i(\mathbf{r}_\alpha, t) \tag{3.b}$$

The last terms of (3) arise from replacement of $\mathbf{j}$ by its explicit form. One gets the Lagrange equation associated with $\mathbf{r}_\alpha$ using a procedure analogous to that of §B.2.*a*. Rearranging the terms (as in §B.2.*a*), one finds the following equation of motion:

$$m_\alpha \ddot{\mathbf{r}}_\alpha = q_\alpha \left[\sum_{\beta \neq \alpha} \frac{q_\beta (\mathbf{r}_\alpha - \mathbf{r}_\beta)}{4 \pi \varepsilon_0 \, | \mathbf{r}_\alpha - \mathbf{r}_\beta |^3} - \frac{\partial \mathbf{A}(\mathbf{r}_\alpha, t)}{\partial t} \right] + q_\alpha \dot{\mathbf{r}}_\alpha \times \mathbf{B}(\mathbf{r}_\alpha, t) . \tag{4}$$

Consider now the field variables $\mathscr{A}_\varepsilon(\mathbf{k})$ and $\mathscr{A}_{\varepsilon'}(\mathbf{k})$. Calculate $\partial \bar{\mathscr{L}}_C / \partial \mathscr{A}_\varepsilon^*$ and $\partial \bar{\mathscr{L}}_C / \partial \dot{\mathscr{A}}_\varepsilon^*$ using (2.b):

$$\frac{\partial \overline{\mathscr{L}_C}}{\partial \mathscr{A}_\varepsilon^*} = - \varepsilon_0 \, c^2 \, k^2 \, \mathscr{A}_\varepsilon + j_\varepsilon \tag{5.a}$$

$$\frac{\partial \overline{\mathscr{L}_C}}{\partial \dot{\mathscr{A}}_\varepsilon^*} = \varepsilon_0 \, \dot{\mathscr{A}}_\varepsilon . \tag{5.b}$$

The Lagrange equation (A.52.b) then gives the equation describing the dynamics of $\mathscr{A}_\varepsilon$:

$$\ddot{\mathscr{A}}_\varepsilon + c^2 \, k^2 \, \mathscr{A}_\varepsilon = \frac{j_\varepsilon}{\varepsilon_0} . \tag{6}$$

Combining this equation and that associated with $\mathscr{A}_{\varepsilon'}$, one gets, in real space,

$$\Box \mathbf{A} = \frac{1}{\varepsilon_0 \, c^2} \, \mathbf{j}_\perp . \tag{7}$$

The source term in the equation of evolution of $\mathbf{A}$ is $\mathbf{j}_\perp$, which depends on $\mathbf{j}$ in a nonlocal fashion [see (B.16) of Chapter I].

5. Momentum and angular momentum of an arbitrary field

Let a field be defined by its components $A_j(\mathbf{r})$ $(j = 1, 2, \ldots, n)$ assumed independent. Its dynamics is described by a Lagrangian density $\mathscr{L}[\ldots A_j(\mathbf{r}) \ldots ; \ldots \dot{A}_j(\mathbf{r}) \ldots ; \ldots \partial_i A_j(\mathbf{r})]$ $(\partial_i = \partial/\partial x, \partial/\partial y, \partial/\partial z)$.

a) Consider a real motion going from the field state defined by $A_j^{(1)}(\mathbf{r})$ at time t_1 to that defined by $A_j^{(2)}(\mathbf{r})$ at t_2. Let the corresponding action integral be

$$S = \int_{t_1}^{t_2} \mathrm{d}t \int \mathrm{d}^3 r \, \mathscr{L}(\ldots A_j \ldots ; \ldots \dot{A}_j \ldots ; \ldots \partial_i A_j \ldots) \tag{1}$$

with
$$A_j(\mathbf{r}) \, |_{t = t_1} = A_j^{(1)}(\mathbf{r}) \quad \text{and} \quad A_j(\mathbf{r}) \, |_{t = t_2} = A_j^{(2)}(\mathbf{r}) . \tag{2}$$

Another real, infinitesimally close motion goes in the same time interval from a state $A_j'^{(1)}(\mathbf{r})$ to $A_j'^{(2)}(\mathbf{r})$. S' is the corresponding action integral.

Take

$$\mathrm{d}S = S' - S \tag{3}$$

$$\mathrm{d}A_j^{(1)}(\mathbf{r}) = A_j'^{(1)}(\mathbf{r}) - A_j^{(1)}(\mathbf{r}); \qquad \text{similarly for } \mathrm{d}A_j^{(2)}(\mathbf{r}). \tag{4}$$

Using the method of Complement B_{II}, find $\mathrm{d}S$ as a function of $\mathrm{d}A_j^{(1)}(\mathbf{r})$ and $\mathrm{d}A_j^{(2)}(\mathbf{r})$.

b) Find the variation dA_j of A_j corresponding to an infinitesimal translation of the field by an amount $\boldsymbol{\eta}$. Show that, if $\mathscr{L}$ is invariant with respect to spatial translation,

$$\mathbf{P} = \int d^3r \sum_j \frac{\partial \mathscr{L}}{\partial \dot{A}_j} (\nabla A_j) \tag{5}$$

is a constant of the motion. By definition, the physical quantity $\mathbf{P}$ is called the field momentum (even if $\mathscr{L}$ is no longer invariant with respect to spatial translation). Here also look to Complement B_{II} for direction.

c) In an analogous fashion show that the angular momentum of the field is

$$\mathbf{J} = -\int d^3r \sum_j \frac{\partial \mathscr{L}}{\partial \dot{A}_j} [(\mathbf{r} \times \nabla) A_j]. \tag{6}$$

Solution

a) Let

$$dA_j(\mathbf{r}, t) = A_j'(\mathbf{r}, t) - A_j(\mathbf{r}, t)$$

$$dS = \sum_j \int_{t_1}^{t_2} dt \int d^3r \left\{ \frac{\partial \mathscr{L}}{\partial A_j} dA_j(\mathbf{r}, t) + \frac{\partial \mathscr{L}}{\partial \dot{A}_j} d\dot{A}_j(\mathbf{r}, t) + \sum_i \frac{\partial \mathscr{L}}{\partial(\partial_i A_j)} \partial_i \, dA_j(\mathbf{r}, t) \right\}. \tag{7}$$

Integrate by parts the second term with respect to t and the last term with respect to x_i. Since the field vanishes at infinity, this becomes

$$dS = \sum_j \int_{t_1}^{t_2} dt \int d^3r \left\{ \frac{\partial \mathscr{L}}{\partial A_j} - \frac{d}{dt}\left(\frac{\partial \mathscr{L}}{\partial \dot{A}_j}\right) - \sum_i \partial_i\left(\frac{\partial \mathscr{L}}{\partial(\partial_i A_j)}\right) \right\} dA_j(\mathbf{r}, t) + \\ + \sum_j \int d^3r \frac{\partial \mathscr{L}}{\partial \dot{A}_j} dA_j(\mathbf{r}, t) \Big|_{t_1}^{t_2}. \tag{8}$$

The integrand of the remaining integral is the product of $dA_j(\mathbf{r}, t)$ by a term identically zero for a real motion obeying the Lagrange equations. Only the last term of (8) remains:

$$dS = \sum_j \int d^3r \frac{\partial \mathscr{L}}{\partial \dot{A}_j} dA_j(\mathbf{r}, t) \Big|_{t_1}^{t_2}$$

$$= \sum_j \int d^3r \left\{ \frac{\partial \mathscr{L}}{\partial \dot{A}_j}\Big|_{t=t_2} dA_j^{(2)} - \frac{\partial \mathscr{L}}{\partial \dot{A}_j}\Big|_{t=t_1} dA_j^{(1)} \right\}. \tag{9}$$

b) Under an infinitesimal translation $\boldsymbol{\eta}$, the field A_j becomes

$$A_j'(\mathbf{r}) = A_j(\mathbf{r} - \boldsymbol{\eta}) \tag{10}$$

so that

$$dA_j = A_j'(\mathbf{r}) - A_j(\mathbf{r}) = -\boldsymbol{\eta} \cdot \nabla A_j(\mathbf{r}). \tag{11}$$

In particular

$$dA_j^{(2)} = -\boldsymbol{\eta} \cdot \nabla A_j^{(2)}(\mathbf{r}) \qquad dA_j^{(1)} = -\boldsymbol{\eta} \cdot \nabla A_j^{(1)}(\mathbf{r}). \tag{12}$$

If the form of $\mathscr{L}$ is invariant under translation, $\mathrm{d}S$ is identically zero for such a transformation. Equation (9) then implies that

$$- \boldsymbol{\eta} \cdot \int \mathrm{d}^3 r \left\{ \left. \frac{\partial \mathscr{L}}{\partial \dot{A}_j} \right|_{t_2} \nabla A_j^{(2)}(\mathbf{r}) - \left. \frac{\partial \mathscr{L}}{\partial \dot{A}_j} \right|_{t_1} \nabla A_j^{(1)}(\mathbf{r}) \right\} = 0 . \tag{13}$$

The quantity defined by (5) has thus the same value at t_1 and t_2. It is a constant of the motion.

c) The variation of the field in a rotation $\mathrm{d}\varphi$ about the axis $\mathbf{u}$ is

$$\begin{aligned} \mathrm{d}A_j &= - [(\mathbf{u}\, \mathrm{d}\varphi \times \mathbf{r}) \cdot \nabla]\, A_j(\mathbf{r}) \\ &= - \mathbf{u}\, \mathrm{d}\varphi \cdot (\mathbf{r} \times \nabla)\, A_j(\mathbf{r}) . \end{aligned} \tag{14}$$

The same type of proof as in *b*) shows that if $\mathscr{L}$ is invariant under rotation, the angular momentum of the field given in (6) is a constant of the motion.

We note that the angular momentum of the field is expressed by an angular momentum density $\mathscr{J}$ just as the momentum is the sum of a momentum density $\mathscr{P}$:

$$\mathscr{P}(\mathbf{r}) = \sum_j - \frac{\partial \mathscr{L}}{\partial \dot{A}_j} (\nabla A_j) \tag{15.a}$$

$$\mathscr{J}(\mathbf{r}) = \sum_j - \frac{\partial \mathscr{L}}{\partial \dot{A}_j} [(\mathbf{r} \times \nabla)\, A_j] \tag{15.b}$$

$\mathscr{P}(\mathbf{r})$ and $\mathscr{J}(\mathbf{r})$ are given as functions of the field, its gradient, and its conjugate momentum at the same point $\mathbf{r}$.

6. A Lagrangian using complex variables and linear in velocity

This exercise examines a Lagrangian L dependent on one complex variable z. Since the structure of L is analogous to that of the Lagrangian density used to describe the dynamics of the Dirac and Schrödinger matter fields, the results established here will be useful for the following three exercises as well as Exercises 5 and 6 of Chapter V.

Consider the following Lagrangian L which depends on the complex variable z:

$$L = \frac{i\hbar}{2} (z^* \dot{z} - \dot{z}^* z) - f(z, z^*) \tag{1}$$

where f is a real function of z and z^*.

a) Write the Lagrange equations relative to L and show that by giving $z(t_0)$ the evolution of the system is completely determined.

b) Find $\partial L / \partial \dot{z}$ and $\partial L / \partial \dot{z}^*$ as well as

$$\tilde{H} = \dot{z} \frac{\partial L}{\partial \dot{z}} + \dot{z}^* \frac{\partial L}{\partial \dot{z}^*} - L . \tag{2}$$

c) Using the results above, show that there are redundant dynamical variables in the Lagrangian (1).

d) Set $z = x + \mathrm{i}y$. Write the Lagrangian as a function of x, y, $\dot{x}$, and $\dot{y}$. [Take $f(z, z^*) = g(x, y)$.]

e) Show that by adding to L the total derivative with respect to time of a function of x and y, one can eliminate $\dot{y}$ from the Lagrangian. Call the new Lagrangian L'.

f) Show, without going into detailed calculations, that it is possible to find a Lagrangian $\hat{L}'$ depending only on x and $\dot{x}$. (One will eliminate y from L' using the Lagrange equation relative to y.) Show that

$$\frac{\partial \hat{L}'}{\partial \dot{x}} = \frac{\partial L'}{\partial \dot{x}}. \tag{3}$$

What is the momentum conjugate with x in $\hat{L}'$? Find the Hamiltonian H' associated with $\hat{L}'$.

g) Proceed to the quantization of the theory based on $\hat{L}'$. Find the commutator $[x, y]$.

h) Express the preceding results as a function of the initial complex variables z and z^*. First show that H' coincides with $\tilde{H}$. Next find the value of the commutator $[z, z^+]$ resulting from quantization. Show that the result obtained differs from that which would have been gotten by quantizing the theory hastily, taking $\partial L/\partial \dot{z}^*$ as the momentum conjugate with z.

Solution

a) Find $\partial L/\partial z^*$ and $\partial L/\partial \dot{z}^*$:

$$\frac{\partial L}{\partial z^*} = \frac{\mathrm{i}\hbar}{2}\dot{z} - \frac{\partial f}{\partial z^*} \tag{4.a}$$

$$\frac{\partial L}{\partial \dot{z}^*} = -\frac{\mathrm{i}\hbar}{2}z. \tag{4.b}$$

From Lagrange's equation one gets the following equation of motion:

$$\mathrm{i}\hbar\dot{z} = \frac{\partial f}{\partial z^*} \tag{5}$$

which is first order in time. Giving $z(t_0)$ then fixes the subsequent behavior of the system.

b) $\partial L/\partial \dot{z}^*$ is given in (4.b) and coincides to within a multiplicative factor with z. Likewise, $\partial L/\partial \dot{z}$ is proportional to z^*. Using these expressions, we get

$$\dot{z}\frac{\partial L}{\partial \dot{z}} + \dot{z}^*\frac{\partial L}{\partial \dot{z}^*} = \frac{\mathrm{i}\hbar}{2}(\dot{z}z^* - \dot{z}^* z) \tag{6}$$

$$\tilde{H} = f(z, z^*). \tag{7}$$

c) In the Lagrangian formalism the evolution of a system is determined by giving the coordinates and velocities at the initial time. The fact that giving $z(t_0)$ suffices to determine the evolution of the system proves that one cannot consider z, z^*, $\dot{z}$, and $\dot{z}^*$ as independent variables. Likewise, the conjugate momenta of z and z^* coincide (to within a multiplicative factor) with these variables and cannot be considered as independent variables, which prevents basing a Hamiltonian theory on $\tilde{H}$.

d) Replacing z by $x + iy$, one finds

$$z^* \dot{z} - \dot{z}^* z = -2\,\mathrm{i}(y\dot{x} - x\dot{y}) \tag{8}$$

$$L = \hbar(y\dot{x} - x\dot{y}) - g(x, y)\,. \tag{9}$$

e) Consider $L' = L + \mathrm{d}u(x, y)/\mathrm{d}t$ with

$$u(x, y) = \hbar xy \tag{10}$$

One gets immediately

$$L' = 2\,\hbar y\dot{x} - g(x, y)\,. \tag{11}$$

f) The Lagrange equation relative to y is written

$$\frac{\partial L'}{\partial y} = 0 \tag{12}$$

that is to say

$$2\,\hbar\dot{x} - \frac{\partial}{\partial y} g(x, y) = 0\,. \tag{13}$$

Solution of this equation fixes y as a function of x and $\dot{x}$. If this is now substituted in L', one gets a new Lagrangian $\hat{L}'$ which is a function of x and $\dot{x}$ only. Calculate $\partial\hat{L}'/\partial\dot{x}$:

$$\frac{\partial\hat{L}'}{\partial\dot{x}} = \frac{\partial L'}{\partial\dot{x}} + \frac{\partial L'}{\partial y}\frac{\partial y}{\partial\dot{x}}\,. \tag{14}$$

The equality (3) derives directly from (14) and (12). This equality gives with (11)

$$p_x = \frac{\partial\hat{L}'}{\partial\dot{x}} = \frac{\partial L'}{\partial\dot{x}} = 2\,\hbar y \tag{15}$$

where y is the function of x and $\dot{x}$ fixed by (13). The Hamiltonian H' associated with $\hat{L}'$ is equal to

$$\begin{aligned} H' &= \dot{x}p_x - \hat{L}' \\ &= 2\,\hbar\dot{x}y - (2\,\hbar\dot{x}y - g(x, y)) \\ &= g(x, y) = g(x, p_x/2\,\hbar)\,. \end{aligned} \tag{16}$$

g) Knowing that $[x, p_x] = i\hbar$, we get using (15)

$$[x, y] = \frac{\mathrm{i}}{2}\,. \tag{17}$$

h) Comparison of (16) and (7) shows that $\tilde{H}$ and H' are equal. We find the commutator $[z, z^+]$ using (17):

$$[z, z^+] = [x + \mathrm{i}y, x - \mathrm{i}y] = \mathrm{i}[y, x] - \mathrm{i}[x, y] \tag{18}$$

so that

$$[z, z^+] = 1\,. \tag{19}$$

If we had taken $\partial L/\partial\dot{z}^*$ as the momentum conjugate with z, we would have gotten by applying (A.35.c) a commutator $[z, z^+]$ equal to 2 rather than 1. This shows that one must be careful in quantizing a theory where there is an overabundance of dynamical variables. In summary, for a Lagrangian having the structure given in (1), it is possible to find the Hamiltonian using (2) providing that one remembers that z^+ and z satisfy (19).

7. Lagrangian and Hamiltonian descriptions of the Schrödinger matter field

Consider the Lagrangian $L = \int d^3r\, \mathscr{L}$ with

$$\mathscr{L} = \frac{i\hbar}{2}(\psi^* \dot{\psi} - \dot{\psi}^* \psi) - \frac{\hbar^2}{2m}\nabla\psi^* \cdot \nabla\psi - V(r)\,\psi^*\psi\,. \tag{1}$$

In this Lagrangian $\psi(\mathbf{r})$ is considered as a complex classical field called the Schrödinger matter field. The purpose of the exercise is first to derive the Schrödinger equation from the Lagrangian (1), and then to show that the physical quantities associated with the field (energy, momentum) arise as the mean values in the state $\psi(\mathbf{r})$ of the usual quantum-mechanical operators.

a) Show that the Lagrange equations associated with (1) coincide with the Schrödinger equation.

b) Since the Schrödinger equation is a first-order equation in time, giving $\psi(\mathbf{r}, t_0)$ is sufficient to fix the future evolution of the system. It follows that the Lagrangian L taken as a function of $\psi(\mathbf{r})$ and $\psi^*(\mathbf{r})$ and their time derivatives contains an excess of dynamical variables. In order to find the conjugate momenta and to pass to the Hamiltonian formalism, it is necessary then to eliminate from (1) the redundant dynamic variables. The procedure here is close to that introduced in Exercise 6 for the discrete case, and the reader should have previously studied that exercise.

i) Let ψ_r and ψ_i be the real and imaginary parts of ψ:

$$\psi(\mathbf{r}) = \psi_r(\mathbf{r}) + i\psi_i(\mathbf{r})\,. \tag{2}$$

Give $\mathscr{L}$ as a function of $\psi_r(\mathbf{r})$, $\psi_i(\mathbf{r})$, and their temporal derivatives.

ii) Show that by adding to L the time derivative of a function (to be found) of $\psi_r(\mathbf{r})$ and $\psi_i(\mathbf{r})$, one gets a new Lagrangian L' equivalent to L and no longer containing $\dot{\psi}_i(\mathbf{r})$.

iii) Without detailing the calculations, show that it is possible to use the Lagrange equation relative to $\psi_i(\mathbf{r})$ to eliminate $\psi_i(\mathbf{r})$ from the Lagrangian L'. The new Lagrangian which is obtained is a function only of $\psi_r(\mathbf{r})$ and $\dot{\psi}_r(\mathbf{r})$ and is denoted $\hat{L}'$.

iv) Show that, for a real motion

$$\frac{\partial L'}{\partial \dot{\psi}_r(\mathbf{r})} = \frac{\partial \hat{L}'}{\partial \dot{\psi}_r(\mathbf{r})} \tag{3}$$

and derive the conjugate momentum $\Pi_r(\mathbf{r})$ of $\psi_r(\mathbf{r})$. Give $\psi(\mathbf{r})$ as a function of $\psi_r(\mathbf{r})$ and $\Pi_r(\mathbf{r})$.

c) Show that the Hamiltonian H associated with $\hat{L}'$ can be written

$$H = \int d^3r\, \psi^*(\mathbf{r}) \left[-\frac{\hbar^2}{2m} \Delta + V(r) \right] \psi(\mathbf{r}) . \tag{4}$$

d) One recalls (see Exercise 5) that the total momentum of a field $\psi_r(\mathbf{r})$ is equal to

$$\mathbf{P} = - \int d^3r\, \Pi_r(\mathbf{r})\, \nabla\psi_r(\mathbf{r}) . \tag{5}$$

Show that the total momentum of the Schrödinger matter field can be written

$$\mathbf{P} = \int d^3r\, \psi^*(\mathbf{r}) \frac{\hbar}{i} \nabla\psi(\mathbf{r}) . \tag{6}$$

One will note that the expressions (4) and (6) coincide with the mean values of the Hamiltonian and momentum operators in quantum mechanics.

e) To proceed to the canonical quantization of the theory one replaces the classical fields $\psi(\mathbf{r})$ and $\psi^*(\mathbf{r})$ by operators $\Psi(\mathbf{r})$ and $\Psi^+(\mathbf{r})$. Calculate, using the canonical commutation relations between the operators associated with $\psi_r(\mathbf{r})$ and $\Pi_r(\mathbf{r})$, the commutators $[\Psi(\mathbf{r}), \Psi^+(\mathbf{r}')]$ and $[\Psi(\mathbf{r}), \Psi(\mathbf{r}')]$.

Solution

a) We calculate the partial derivatives of the Lagrangian density appearing in the Lagrange equation (A.52.b)

$$\frac{\partial \mathcal{L}}{\partial \dot{\psi}^*} = -\frac{i\hbar}{2}\psi \qquad (7.\mathrm{a}) \qquad\qquad \frac{\partial \mathcal{L}}{\partial \psi^*} = \frac{i\hbar}{2}\dot{\psi} - V(r)\,\psi \qquad (7.\mathrm{b})$$

$$\frac{\partial \mathcal{L}}{\partial(\partial_j\psi)^*} = -\frac{\hbar^2}{2m}\partial_j\psi . \tag{7.c}$$

Substituting these expressions in (A.52.b), we get

$$i\hbar\dot{\psi} - V(r)\,\psi + \frac{\hbar^2}{2m}\Delta\psi = 0 \tag{8}$$

which is just the Schrödinger equation of a particle of mass m in a potential $V(r)$.

b) To solve the Lagrange equations, it is necessary to know the coordinates and their velocities at the initial time. The fact that giving only $\psi(\mathbf{r}, t_0)$ suffices to fix the subsequent development of the system shows that one has an excess of dynamical variables.

i) Taking account of (2), the Lagrangian density (1) becomes

$$\mathcal{L} = \hbar(\psi_i \dot{\psi}_r - \psi_r \dot{\psi}_i) - \frac{\hbar^2}{2m}[(\nabla\psi_r)^2 + (\nabla\psi_i)^2] - V(r)(\psi_r^2 + \psi_i^2) . \tag{9}$$

ii) Consider the following Lagrangian L':

$$L' = L + \frac{d}{dt}\int d^3r\, \hbar(\psi_r(\mathbf{r})\, \psi_i(\mathbf{r})) . \tag{10}$$

It is clear that L' and L are equivalent Lagrangians, since their difference is the derivative with respect to time of a function of the coordinates. Starting from (9) and (10), one gets for the Lagrangian density $\mathscr{L}'$ associated with L'

$$\mathscr{L}' = 2\,\hbar\psi_i\,\dot{\psi}_r - \frac{\hbar^2}{2\,m}[(\nabla\psi_r)^2 + (\nabla\psi_i)^2] - V(r)\,(\psi_r^2 + \psi_i^2) \tag{11}$$

which does not depend on $\dot{\psi}_i$.

iii) The Lagrange equation relative to ψ_i,

$$\frac{\delta L'}{\delta\psi_i(\mathbf{r})} = 0 \tag{12}$$

$$\hbar\dot{\psi}_r - V(r)\,\psi_i + \frac{\hbar^2}{2\,m}\,\Delta\psi_i = 0 \tag{13}$$

allows one to express $\psi_i(\mathbf{r})$ at each instant as a function of $\dot{\psi}_r(\mathbf{r})$. Substituting this expression for $\psi_i(\mathbf{r})$ in L', one gets a Lagrangian $\hat{L}'$ which depends only on $\psi_r(\mathbf{r})$ and $\dot{\psi}_r(\mathbf{r})$.

iv) We calculate $\partial\hat{L}'/\,\partial\dot{\psi}_r(\mathbf{r})$ as a function of the functional derivatives of L':

$$\frac{\delta\hat{L}'}{\delta\dot{\psi}_r(\mathbf{r})} = \frac{\delta L'}{\delta\dot{\psi}_r(\mathbf{r})} + \frac{\delta L'}{\delta\psi_i(\mathbf{r})}\,\frac{\delta\psi_i(\mathbf{r})}{\delta\dot{\psi}_r(\mathbf{r})} \tag{14}$$

which, taking into account the Lagrange equation (12), leads to Equation (3). The conjugate momentum $\Pi_r(\mathbf{r})$ of $\psi_r(\mathbf{r})$ is then equal to

$$\Pi_r(\mathbf{r}) = \frac{\delta\hat{L}'}{\delta\dot{\psi}_r(\mathbf{r})} = \frac{\delta L'}{\delta\dot{\psi}_r(\mathbf{r})} \tag{15}$$

—that is, following (11),

$$\Pi_r(\mathbf{r}) = 2\,\hbar\psi_i(\mathbf{r}) . \tag{16}$$

Equations (2) and (16) then yield

$$\psi(\mathbf{r}) = \psi_r(\mathbf{r}) + \frac{i}{2\,\hbar}\,\Pi_r(\mathbf{r}) . \tag{17}$$

c) The Hamiltonian density $\mathscr{H}$ is given by (A.45.b), which, taking into account (11) and (16), gives

$$\begin{aligned}\mathscr{H} &= \Pi_r(\mathbf{r})\,\dot{\psi}_r(\mathbf{r}) - \mathscr{L}'[\psi_r, \dot{\psi}_r, \psi_i(\dot{\psi}_r)] \\ &= \frac{\hbar^2}{2\,m}[(\nabla\psi_r)^2 + (\nabla\psi_i)^2] + V(r)\,[\psi_r^2 + \psi_i^2] .\end{aligned} \tag{18}$$

The expression for H can be transformed using integration by parts:

$$\int d^3r(\nabla\psi_r)^2 = -\int d^3r\,\psi_r\,\Delta\psi_r \tag{19}$$

since the fields ψ are equal to zero at infinity. One then gets

$$H = \int d^3r\left[-\frac{\hbar^2}{2\,m}(\psi_r\,\Delta\psi_r + \psi_i\,\Delta\psi_i) + V(r)\,(\psi_r^2 + \psi_i^2)\right] . \tag{20}$$

Noting that

$$\int d^3r(\psi_r\,\Delta\psi_i - \psi_i\,\Delta\psi_r) = 0 \tag{21}$$

(which can be demonstrated with two integrations by parts) one can reexpress (20) as a function of ψ and ψ^*, and then one gets (4) as stated.

d) Replace $\Pi_r(\mathbf{r})$ in (5) by its value (16). One then gets

$$\mathbf{P} = -\int d^3r\, 2\,\hbar\psi_i(\mathbf{r})\,\nabla\psi_r(\mathbf{r})\,. \tag{22}$$

By expressing $\psi_r(\mathbf{r})$ and $\psi_i(\mathbf{r})$ as functions of $\psi(\mathbf{r})$ and $\psi^*(\mathbf{r})$, one then finds

$$\mathbf{P} = \frac{\hbar}{2\mathrm{i}}\left[\int d^3r\,\psi^*\,\nabla\psi - \int d^3r\,\psi\,\nabla\psi^* - \int d^3r\,\psi\,\nabla\psi + \int d^3r\,\psi^*\,\nabla\psi^*\right]. \tag{23}$$

One integration by parts allows one to show that the second integral is the negative of the first. As for the last two integrals, they lead to a zero result, since the term to integrate is of the form $\nabla\psi^2/2$. Finally one finds

$$\mathbf{P} = \int d^3r\,\psi^*(\mathbf{r})\,\frac{\hbar}{\mathrm{i}}\,\nabla\psi(\mathbf{r}) \tag{24}$$

which agrees with (6).

e) One starts with the canonical commutation relation

$$[\Psi_r(\mathbf{r}), \Pi_r(\mathbf{r}')] = \mathrm{i}\hbar\,\delta(\mathbf{r} - \mathbf{r}') \tag{25}$$

In the commutator $[\Psi(\mathbf{r}), \Psi^+(r')]$, one expresses the operators $\Psi(\mathbf{r})$ and $\Psi^+(r)$ as functions of $\Psi_r(\mathbf{r})$ and $\Pi_r(\mathbf{r})$ using (17). This yields

$$\begin{aligned}[\Psi(\mathbf{r}), \Psi^+(\mathbf{r}')] = {} & [\Psi_r(\mathbf{r}), \Psi_r(\mathbf{r}')] + \frac{1}{4\,\hbar^2}[\Pi_r(\mathbf{r}), \Pi_r(\mathbf{r}')] - \\ & - \frac{\mathrm{i}}{2\,\hbar}[\Psi_r(\mathbf{r}), \Pi_r(\mathbf{r}')] + \frac{\mathrm{i}}{2\,\hbar}[\Pi_r(\mathbf{r}), \Psi_r(\mathbf{r}')]\,. \end{aligned} \tag{26}$$

The first two commutators are zero. The last two give, using (25),

$$[\Psi(\mathbf{r}), \Psi^+(\mathbf{r}')] = \delta(\mathbf{r} - \mathbf{r}')\,. \tag{27}$$

The commutator $[\Psi(\mathbf{r}), \Psi(\mathbf{r}')]$ can be found in an analogous fashion, the result being

$$[\Psi(\mathbf{r}), \Psi(\mathbf{r}')] = 0\,. \tag{28}$$

The properties of the quantized Schrödinger field are studied in Exercise 8. In the first part of this exercise, we start from the commutator (27) and show that the elementary excitations of the quantized field thus gotten correspond to bosons. In the second part we substitute for the canonical commutation relation (25) an anticommutation relation, which leads to replacing (27) and (28) by

$$[\Psi(\mathbf{r}), \Psi^+(\mathbf{r}')]_+ = \delta(\mathbf{r} - \mathbf{r}') \tag{29.a}$$

$$[\Psi(\mathbf{r}), \Psi(r')]_+ = 0\,. \tag{29.b}$$

We then show that the particles obtained after quantization of the theory are fermions.

8. Quantization of the Schrödinger field

Consider now the Schrödinger matter field whose classical properties were examined in Exercise 7. After quantization of the theory, the classical expression for the energy and momentum become operators given by

$$H = \int d^3r\, \Psi^+(\mathbf{r}) \left(-\frac{\hbar^2}{2\,m} \Delta + V(r) \right) \Psi(\mathbf{r}) \tag{1}$$

$$\mathbf{P} = \int d^3r\, \Psi^+(\mathbf{r}) \frac{\hbar}{i} \nabla \Psi(\mathbf{r}) \tag{2}$$

where $\Psi(\mathbf{r})$ is the field operator at point $\mathbf{r}$. The main purpose of this exercise is to show that the quantization of the field can be accomplished starting either from the commutation relations between the operators and the conjugate momenta [part (i)] or from the anticommutation relations [part (ii)]. A second purpose is to show that the particles associated with the field are bosons in the first case and fermions in the second. (*)

(i) We postulate here that the fields Ψ and Ψ^+ obey the following relations:

$$[\Psi(\mathbf{r}), \Psi^+(\mathbf{r}')] = \delta(\mathbf{r} - \mathbf{r}') \tag{3.a}$$

$$[\Psi(\mathbf{r}), \Psi(\mathbf{r}')] = 0 \tag{3.b}$$

a) The Heisenberg equation for the operator Ψ is written

$$i\hbar \dot{\Psi}(\mathbf{r}) = [\Psi(\mathbf{r}), H]\,. \tag{4}$$

Find the commutator appearing in (4) using (1) for H and Equations (3), and show that Equation (4) coincides with the Schrödinger equation for the operator Ψ.

b) Show that the commutator $[\Psi(\mathbf{r}), \mathbf{P}]$, where $\mathbf{P}$ is defined by (2), is equal to $-i\hbar\nabla\Psi(r)$. What are the time and space translation operators for the field $\Psi(\mathbf{r})$?

c) Take $\rho(\mathbf{r}) = \Psi^+(\mathbf{r})\Psi(\mathbf{r})$. Show that $\rho(\mathbf{r})$ satisfies an equation of the form

$$\frac{d}{dt}\rho(\mathbf{r}) + \nabla \cdot \mathbf{j}(\mathbf{r}) = 0 \tag{5}$$

where $\mathbf{j}(\mathbf{r})$ is an operator which is to be found.

d) Denote by $\varphi_n(\mathbf{r})$ the eigenfunctions of $[-(\hbar^2/2m)\Delta + V(\mathbf{r})]$:

$$-\frac{\hbar^2}{2\,m} \Delta\varphi_n(\mathbf{r}) + V(r)\,\varphi_n(\mathbf{r}) = E_n\,\varphi_n(\mathbf{r})\,. \tag{6}$$

(*) See also Schiff, Chapter XIII.

The spectrum of eigenvalues E_n is assumed discrete. Let c_n and c_n^+ be operators defined by

$$c_n = \int d^3r\, \varphi_n^*(\mathbf{r})\, \Psi(\mathbf{r}) \tag{7}$$

and the adjoint relation. Find the commutators $[c_n, c_m]$ and $[c_n, c_m^+]$.

e) Show that the operator H defined in (1) can be put in the form

$$H = \sum_n E_n\, c_n^+\, c_n\,. \tag{8}$$

f) Assume that in the state space of the quantum field where the operators $\Psi(\mathbf{r})$, $\Psi^+(\mathbf{r})$, c_n, and c_n^+ act, there is a state $|0\rangle$, called the vacuum, which satisfies

$$c_n\,|\,0\,\rangle = 0 \tag{9}$$

for all n and which is normalizable ($\langle 0|0\rangle = 1$). Show that $(c_n^+)^p|0\rangle$, where p is a positive integer, is an eigenstate of H with eigenvalue pE_n. Give a particle interpretation for the elementary excitations of the quantum field.

g) Let the operator N be defined by

$$N = \int d^3r\, \Psi^+(\mathbf{r})\, \Psi(\mathbf{r})\,. \tag{10}$$

Show that N is equal to $\Sigma_m N_m$, where $N_m = c_m^+ c_m$. Find $\dot{N}$. How do you interpret this result?

(ii) Assume now that the fields Ψ and Ψ^+ obey the anticommutation relations

$$[\Psi(\mathbf{r}), \Psi^+(\mathbf{r}')]_+ = \delta(\mathbf{r} - \mathbf{r}') \tag{11.a}$$

$$[\Psi(\mathbf{r}), \Psi(\mathbf{r}')]_+ = 0\,. \tag{11.b}$$

a) Starting from the Heisenberg equation (4) for the operator Ψ and Equations (11), show that $\Psi(\mathbf{r})$ is always a solution of the Schrödinger equation. Derive the commutator $[\Psi(\mathbf{r}), \mathbf{P}]$ and show that it has the same value as in the preceding section. What can one conclude from this?

b) Show that Equation (5) remains valid. Derive the commutator $[\rho(\mathbf{r}), \rho(\mathbf{r}')]$. What can one conclude about the compatibility of the measurements of these variables? Do these results depend on the commutation or anticommutation relations between $\Psi(\mathbf{r})$ and $\Psi^+(\mathbf{r}')$ postulated to quantize the field?

c) Using (6) and the definition (7) of the operators c_n, find the anticommutators $[c_n, c_m]_+$ and $[c_n, c_m^+]_+$. Show that Equation (8) for H as a function of the operators $c_n^+ c_n$ remains valid.

d) Take [as in part (i)] $N_m = c_m^+ c_m$. Show that this operator satisfies the relationship

$$N_m^2 = N_m . \tag{12}$$

Show that the eigenvalues of N_m can only be 0 or 1.

e) Show that $c_m^+|0\rangle$ [the vacuum $|0\rangle$ being defined as in part (i)] is a state with one particle with energy E_m. Can one have states with more than one particle in state E_m? What can one conclude about the nature (boson or fermion) of the particles associated with the field?

Solution

(i) Field of bosons.

a) To find the commutator $[\Psi(\mathbf{r}), H]$, we use the identity

$$[A, BC] = [A, B]\, C + B[A, C] \tag{13}$$

which gives

$$[\Psi(\mathbf{r}), H] = \int d^3r' [\Psi(\mathbf{r}), \Psi^+(\mathbf{r}')] \left(-\frac{\hbar^2}{2m} \Delta_{r'} + V(r') \right) \Psi(\mathbf{r}') + $$

$$+ \int d^3r' \, \Psi^+(\mathbf{r}') \left[\Psi(\mathbf{r}), \left(-\frac{\hbar^2}{2m} \Delta_{r'} + V(r') \right) \Psi(\mathbf{r}') \right]. \tag{14}$$

Equation (3.b) implies that the second term of (14) is zero. The first term is found with the aid of (3.a), and so

$$[\Psi(\mathbf{r}), H] = \left(-\frac{\hbar^2}{2m} \Delta + V(r) \right) \Psi(\mathbf{r}) \tag{15}$$

Finally, using (4), one gets

$$i\hbar \dot{\Psi}(\mathbf{r}) = \left(-\frac{\hbar^2}{2m} \Delta + V(r) \right) \Psi(\mathbf{r}) . \tag{16}$$

The equation of evolution of the field operator $\Psi(\mathbf{r})$ has the same form as the equation of evolution of the classical field $\psi(\mathbf{r})$, that is, the Schrödinger equation. Note that here $\Psi(\mathbf{r})$ is a field operator and not a wave function. For this reason the procedure we are following here is sometimes called "second quantization".

b) The derivation of $[\Psi(\mathbf{r}), \mathbf{P}]$ is analogous to that of $[\Psi(\mathbf{r}), H]$:

$$[\Psi(\mathbf{r}), \mathbf{P}] = \int d^3r' [\Psi(\mathbf{r}), \Psi^+(\mathbf{r}')] \frac{\hbar}{i} \nabla_{r'} \Psi(\mathbf{r}') + \int d^3r' \, \Psi^+(\mathbf{r}') \left[\Psi(\mathbf{r}), \frac{\hbar}{i} \nabla_{r'} \Psi(\mathbf{r}') \right]. \tag{17}$$

The second term of (17) is zero from (3.b), and the first term is, using (3.a),

$$[\Psi(\mathbf{r}), \mathbf{P}] = - i\hbar \, \nabla \Psi(\mathbf{r}) . \tag{18}$$

This equation shows that $\mathbf{P}$ is the spatial translation generator for the field operator Ψ, just as H is the time translation operator. It follows that

$$\Psi(\mathbf{r} + \mathbf{a}, t) = e^{-i\mathbf{P}.\mathbf{a}/\hbar}\, \Psi(\mathbf{r}, t)\, e^{i\mathbf{P}.\mathbf{a}/\hbar} \tag{19.a}$$

$$\Psi(\mathbf{r}, t + \tau) = e^{iH\tau/\hbar}\, \Psi(\mathbf{r}, t)\, e^{-iH\tau/\hbar} \tag{19.b}$$

c) We calculate $\dot{\rho}(\mathbf{r})$:

$$\frac{d}{dt}\rho(\mathbf{r}) = \dot{\Psi}^+(\mathbf{r})\, \Psi(\mathbf{r}) + \Psi^+(\mathbf{r})\, \dot{\Psi}(\mathbf{r}). \tag{20}$$

Using (16) and the adjoint equation, we get

$$\frac{d}{dt}\rho(\mathbf{r}) = \frac{\hbar}{2\,mi}\left[(\Delta\Psi^+(\mathbf{r}))\, \Psi(\mathbf{r}) - \Psi^+(\mathbf{r})\, \Delta\Psi(\mathbf{r})\right] \tag{21}$$

since the term depending on $V(\mathbf{r})$ vanishes. One transforms the right-hand side of (21) by introducing

$$\mathbf{j}(\mathbf{r}) = \frac{\hbar}{2\,mi}\left[\Psi^+(\mathbf{r})\,(\nabla\Psi(\mathbf{r})) - (\nabla\Psi^+(\mathbf{r}))\,\Psi(\mathbf{r})\right]. \tag{22}$$

This becomes $\partial\rho/\partial t = -\nabla\cdot\mathbf{j}$, which coincides with Equation (5) as stated and which generalizes the continuity relation (conservation of probability) to quantum operators.

d) The commutator $[c_n, c_m]$ is equal to

$$[c_n, c_m] = \int d^3r\, d^3r'\, \varphi_n^*(\mathbf{r})\, \varphi_m^*(\mathbf{r}')\, [\Psi(\mathbf{r}), \Psi(\mathbf{r}')] \tag{23}$$

and thus vanishes on account of (3.b):

$$[c_n, c_m] = 0. \tag{24}$$

The commutator $[c_n, c_m^+]$ is found in an analogous fashion:

$$[c_n, c_m^+] = \int d^3r\, d^3r'\, \varphi_n^*(\mathbf{r})\, \varphi_m(\mathbf{r}')\, [\Psi(\mathbf{r}), \Psi^+(\mathbf{r}')] \tag{25}$$

which according to (3.a) is

$$[c_n, c_m^+] = \int d^3r\, d^3r'\, \varphi_n^*(\mathbf{r})\, \varphi_m(\mathbf{r}')\, \delta(\mathbf{r} - \mathbf{r}') \tag{26}$$

$$= \int d^3r\, \varphi_n^*(\mathbf{r})\, \varphi_m(\mathbf{r}).$$

Since the $\{\varphi_n(\mathbf{r})\}$ are orthonormal, one has

$$[c_n, c_m^+] = \delta_{n,m}. \tag{27}$$

e) Multiply both sides of (7) by $\varphi_n(\mathbf{r})$ and sum on n. Using the closure relation, one then gets

$$\Psi(\mathbf{r}) = \sum_n \varphi_n(\mathbf{r})\, c_n \tag{28}$$

which is analogous to the expansion of a wave function $\psi(\mathbf{r})$ in the orthonormal basis $\{\varphi_n(\mathbf{r})\}$. Note now that in (28), $\Psi(\mathbf{r})$ and c_n are operators. Substituting this expression in (1) and using (6), one finds

$$H = \sum_{n,m} \int d^3r\, \varphi_m^*(\mathbf{r})\, E_n\, \varphi_n(\mathbf{r})\, c_m^+\, c_n \tag{29}$$

so that using the orthonormalization of the wave functions $\varphi_n(\mathbf{r})$,

$$H = \sum_n E_n\, c_n^+\, c_n. \tag{30}$$

f) The form (8) for H and the commutation relations (24) and (27) are those of a set of independent harmonic oscillators. If we postulate the existence of the vacuum $|0\rangle$, the mathematical operations are identical to those used in the case of the harmonic oscillator. We know then that $[(c_n^+)^p/\sqrt{p!}]|0\rangle$ is a normalized eigenstate of H with eigenvalue pE_n. By making the creation operator c_n^+ act p times on $|0\rangle$ it is possible to obtain a state of the system whose energy is p times the energy E_n. This leads to the interpretation of this state of the field as a state with p particles in state E_n. The fact that p can be any integer, positive or zero, is characteristic of a system of bosons.

g) Using (28), one transforms (10) into

$$N = \sum_{n,m} \int \mathrm{d}^3r\, \varphi_m^*(\mathbf{r})\, \varphi_n(\mathbf{r})\, c_m^+\, c_n\,. \tag{31}$$

Orthonormalization of the wave functions then gives

$$N = \sum_m c_m^+\, c_m = \sum_m N_m\,. \tag{32}$$

To find $\dot{N}_m$, start with the Schrödinger equation (16) and express $\Psi(\mathbf{r})$ as in (28). One finds

$$\mathrm{i}\hbar \sum_m \varphi_m(\mathbf{r})\, \dot{c}_m = \sum_m \left(-\frac{\hbar^2}{2\,m} \Delta + V(r) \right) \varphi_m(\mathbf{r})\, c_m \tag{33}$$

so that, using (6) and the fact that the $\{\varphi_m(\mathbf{r})\}$ form a basis,

$$\mathrm{i}\hbar \dot{c}_m = E_m\, c_m\,. \tag{34}$$

The derivative of N_m with respect to time gives

$$\dot{N}_m = \dot{c}_m^+\, c_m + c_m^+\, \dot{c}_m \tag{35}$$

so that finally, using (34),

$$\dot{N}_m = 0\,. \tag{36}$$

We have seen in *f*) that N_m can be interpreted as the particle number operator for the energy state E_m [since the action of N_m on $(c_m^+)^p|0\rangle$ gives p times this same state]. It follows that $N = \Sigma_m N_m$ can be interpreted as the operator for the total number of particles. Equation (36) shows that for the quantized Schrödinger field, the number of particles in the energy state E_m remains constant. Note then that N can be written

$$N = \int \mathrm{d}^3r\, \rho(\mathbf{r})$$

and the conservation of N can be gotten directly from the equation of continuity (5).

(ii) Field of fermions.

a) To calculate $[\Psi(\mathbf{r}), H]$, one uses the following equations relating the commutator $[A, BC]$ to the anticommutators $[A, B]_+$ and $[A, C]_+$:

$$[A, BC] = [A, B]_+\, C - B[A, C]_+\,. \tag{37}$$

We get then

$$\begin{aligned} \mathrm{i}\hbar\dot{\Psi}(\mathbf{r}) &= [\Psi(\mathbf{r}), H] \\ &= \int \mathrm{d}^3r' [\Psi(\mathbf{r}), \Psi^+(\mathbf{r}')]_+ \left(-\frac{\hbar^2}{2\,m} \Delta_{\mathbf{r}'} + V(r') \right) \Psi(\mathbf{r}') - \\ &\quad - \int \mathrm{d}^3r'\, \Psi^+(\mathbf{r}') \left[\Psi(\mathbf{r}), \left(-\frac{\hbar^2}{2\,m} \Delta_{\mathbf{r}'} + V(r') \right) \Psi(\mathbf{r}') \right]_+ . \end{aligned} \tag{38}$$

The second term of the right-hand side of (38) is zero from (11.b), and the first one can be found using (11.a), so that

$$i\hbar\dot{\Psi}(\mathbf{r}) = [\Psi(\mathbf{r}), H] = \left(-\frac{\hbar^2}{2m}\Delta + V(r)\right)\Psi(\mathbf{r})\,. \tag{39}$$

It appears then that just as in part (i), the operator $\Psi(\mathbf{r})$ obeys the Schrödinger equation. The calculation of $[\Psi(\mathbf{r}), \mathbf{P}]$ is analogous:

$$[\Psi(\mathbf{r}), \mathbf{P}] = \int d^3r' [\Psi(\mathbf{r}), \Psi^+(\mathbf{r}')]_+ \frac{\hbar}{i}\nabla_{r'}\Psi(\mathbf{r}') - \\ - \int d^3r'\, \Psi^+(\mathbf{r}')\left[\Psi(\mathbf{r}), \frac{\hbar}{i}\nabla_{r'}\Psi(\mathbf{r}')\right]_+ \tag{40}$$

The second term on the right-hand side is zero as a result of (11.b); the first one is found using (11.a):

$$[\Psi(\mathbf{r}), \mathbf{P}] = -i\hbar\nabla\Psi(\mathbf{r})\,. \tag{41}$$

One finds then an expression identical to that in part (i). It is clear then that the quantization of the theory by means of anticommutators is compatible with the fact that the operators H and $\mathbf{P}$ introduced in (1) and (2) must be generators for the time and space translations.

b) The proof of (5) only requires Schrödinger's equation (39). It is identical to that in part (i). Now find the commutator $[\rho(\mathbf{r}), \rho(\mathbf{r}')]$:

$$[\rho(\mathbf{r}), \rho(\mathbf{r}')] = [\Psi^+(\mathbf{r})\,\Psi(\mathbf{r}), \Psi^+(\mathbf{r}')\,\Psi(\mathbf{r}')] \\ = \Psi^+(\mathbf{r})\,[\Psi(\mathbf{r}), \Psi^+(\mathbf{r}')\,\Psi(\mathbf{r}')] + [\Psi^+(\mathbf{r}), \Psi^+(\mathbf{r}')\,\Psi(\mathbf{r}')]\,\Psi(\mathbf{r})\,. \tag{42}$$

Using (37), one rewrites (42):

$$[\rho(\mathbf{r}), \rho(\mathbf{r}')] = \Psi^+(\mathbf{r})[\Psi(\mathbf{r}), \Psi^+(\mathbf{r}')]_+\,\Psi(\mathbf{r}') - \Psi^+(\mathbf{r})\,\Psi^+(\mathbf{r}')\,[\Psi(\mathbf{r}), \Psi(\mathbf{r}')]_+ + \\ + [\Psi^+(\mathbf{r}), \Psi^+(\mathbf{r}')]_+\,\Psi(\mathbf{r}')\,\Psi(\mathbf{r}) - \Psi^+(\mathbf{r}')\,[\Psi^+(\mathbf{r}), \Psi(\mathbf{r}')]_+\,\Psi(\mathbf{r}) \tag{43}$$

which simplifies using the anticommutation relations (11):

$$[\rho(\mathbf{r}), \rho(\mathbf{r}')] = \Psi^+(\mathbf{r})\,\Psi(\mathbf{r}')\,\delta(\mathbf{r}-\mathbf{r}') - \Psi^+(\mathbf{r}')\,\Psi(\mathbf{r})\,\delta(\mathbf{r}-\mathbf{r}') = 0\,. \tag{44}$$

One sees then that $\rho(\mathbf{r})$ and $\rho(\mathbf{r}')$ commute even if $\Psi(\mathbf{r})$ and $\Psi(\mathbf{r}')$ anticommute. This result about the operator associated with the particle density has a clear physical meaning. It is satisfying that two operators associated with local physical quantities taken at the same time but at different points commute (see the remark (i) at the end of §A.2.*g*). An identical result for $[\rho(\mathbf{r}), \rho(\mathbf{r}')]$ will be gotten if one postulates the commutation relations (3) rather than Equations (11).

c) The anticommutator $[c_n, c_m]_+$ is equal to

$$[c_n, c_m]_+ = \int d^3r\, d^3r'\, \varphi_n^*(\mathbf{r})\,\varphi_m^*(\mathbf{r}')\,[\Psi(\mathbf{r}), \Psi(\mathbf{r}')]_+ = 0\,. \tag{45}$$

as a result of (11.b). For the anticommutator $[c_n, c_m^+]_+$, an analogous calculation using (11.a) gives

$$[c_n, c_m^+]_+ = \delta_{n,m}\,. \tag{46}$$

The proof of (8) in *e*) of part (i) does not rely on the commutation relations and thus remains valid.

d) From the definition of N_m, one has

$$N_m^2 = c_m^+\,c_m\,c_m^+\,c_m\,. \tag{47}$$

Using the anticommutator $[c_m, c_m]_+ = 1$, one gets

$$N_m^2 = c_m^+ c_m - c_m^+ c_m^+ c_m c_m \tag{48}$$

Now the fact that the anticommutator $[c_m, c_m]_+$ must be zero shows that $c_m c_m = 0$, which demonstrates the identity between (48) and (12). The eigenvalues of the Hermitian operator N_m then must satisfy

$$\lambda^2 = \lambda \tag{49}$$

and can only have values 0 or 1.

e) One finds the action of N_m on $c_m^+ |0\rangle$:

$$N_m c_m^+ | 0 \rangle = c_m^+ c_m c_m^+ | 0 \rangle . \tag{50}$$

By using the anticommutation relation $[c_m, c_m^+]_+ = 1$ and the fact that $c_m |0\rangle = 0$ one gets

$$N_m c_m^+ | 0 \rangle = c_m^+ | 0 \rangle \tag{51}$$

$c_m^+ |0\rangle$ is thus an eigenstate of N_m with eigenvalue 1. Furthermore, it is clear that $c_m^+ |0\rangle$ is nonzero. Its norm $\langle 0 | c_m c_m^+ |0\rangle$ can be found using $[c_m, c_m^+]_+ = 1$ and is 1. In contrast, since the c_m^+ anticommute with themselves, we have $(c_m^+)^2 = 0$, and more generally $(c_m^+)^p = 0$ if $p > 1$. It is then impossible to construct, as was done in part (i), a state with more than one particle in the same state E_n. Actually we have already shown that the eigenvalues of N_m are restricted to 0 and 1. Thus the particles associated with the quantized field are fermions.

9. Schrödinger equation for a particle in an electromagnetic field: Arbitrariness of phase and gauge invariance

Consider the Lagrangian $L = \int \mathrm{d}^3 r \, \mathscr{L}$ where $\mathscr{L}$ is the real Lagrangian density

$$\mathscr{L} = \frac{i\hbar}{2}(\psi^* \dot{\psi} - \dot{\psi}^* \psi) - \frac{1}{2m}\left[\left(-\frac{\hbar}{i}\nabla - q\mathbf{A}\right)\psi^*\right]\cdot\left[\left(\frac{\hbar}{i}\nabla - q\mathbf{A}\right)\psi\right] - (V(r) + qU)\,\psi^* \psi . \tag{1}$$

In this Lagrangian, $\psi(\mathbf{r})$ is a classical complex field. The purpose of this exercise is to show first that the Lagrange equations associated with L are the Schrödinger equation of a particle in an electromagnetic field defined by the potentials $\mathbf{A}(\mathbf{r}, t)$ and $U(\mathbf{r}, t)$. One will see then that any phase modification of the matter field $\psi(\mathbf{r})$ is equivalent to a gauge transformation for the electromagnetic field.

a) Find the Lagrange equations associated with (1).

b) Make the change of variable

$$\psi(\mathbf{r}, t) = \psi'(\mathbf{r}, t)\, \mathrm{e}^{-iqF(\mathbf{r},t)/\hbar} \tag{2}$$

where $F(\mathbf{r}, t)$ is an arbitrary function of $\mathbf{r}$ and t. Give $\mathscr{L}$ as a function of ψ' and F, and show that a change in phase of ψ is mathematically equivalent to a change of gauge.

Solution

a) To get the Lagrange equation (A.52.b) one must take the following partial derivatives:

$$\frac{\partial \mathscr{L}}{\partial \dot{\psi}^*} = -\frac{i\hbar}{2}\psi \qquad (3.a)$$

$$\frac{\partial \mathscr{L}}{\partial \psi^*} = \frac{i\hbar}{2}\dot{\psi} + \sum_j \frac{qA_j}{2m}\left(\frac{\hbar}{i}\frac{\partial}{\partial x_j} - qA_j\right)\psi - (V + qU)\psi \qquad (3.b)$$

$$\frac{\partial \mathscr{L}}{\partial(\partial_j \psi^*)} = \frac{\hbar}{2mi}\left(\frac{\hbar}{i}\frac{\partial}{\partial x_j} - qA_j\right)\psi . \qquad (3.c)$$

One then gets the Lagrange equation

$$i\hbar\dot{\psi} - (V + qU)\psi - \frac{1}{2m}\left(\frac{\hbar}{i}\nabla - q\mathbf{A}\right)^2 \psi = 0 \qquad (4)$$

which is also the Schrödinger equation for a particle in an external electromagnetic field.

b) In the transformation (2), the time and space derivatives become

$$\dot{\psi} = \dot{\psi}' \, e^{-iqF/\hbar} - i\frac{q}{\hbar}\dot{F}\psi' \, e^{-iqF/\hbar} \qquad (5.a)$$

$$\partial_j \psi = (\partial_j \psi') \, e^{-iqF/\hbar} - i\frac{q}{\hbar}\frac{\partial F}{\partial x_j}\psi' \, e^{-iqF/\hbar} \qquad (5.b)$$

Putting these in the Lagrangian density (1) gives

$$\mathscr{L} = \frac{i\hbar}{2}(\psi'^* \dot{\psi}' - \dot{\psi}'^* \psi') - \frac{1}{2m}\left[\left(-\frac{\hbar}{i}\nabla - q(\mathbf{A} + \nabla F)\right)\psi'^*\right]\cdot\left[\left(\frac{\hbar}{i}\nabla - q(\mathbf{A} + \nabla F)\right)\psi'\right]$$
$$- [V + q(U - \dot{F})]\,\psi'^*\psi' . \qquad (6)$$

The same Lagrangian density could have been gotten by keeping the same field ψ and making a gauge change on the potentials $\mathbf{A}$ and U:

$$\mathbf{A}' = \mathbf{A} + \nabla F \qquad (7.a) \qquad\qquad U' = U - \dot{F} . \qquad (7.b)$$

CHAPTER III

Quantum Electrodynamics in the Coulomb Gauge

This chapter is devoted to a general presentation of quantum electrodynamics in the Coulomb gauge. Its purpose is to establish the basic elements of a *quantum* description of the interactions between nonrelativistic charged particles and photons and to emphasize several important physical aspects of this theory.

We begin in Part A by introducing the general framework for such a quantum theory. We review the fundamental *commutation relations* introduced in an elementary fashion at the end of Chapter I and derived in a more rigorous way in Chapter II. We make explicit the quantum operators associated with the various physical variables of the system and analyze the structure of the *state space* in which these operators act.

The problem of *temporal evolution* is then approached in Part B. In the *Schrödinger picture* the state vector of the global system obeys a Schrödinger equation and the matrix elements of the evolution operator between two states of the system are the transition amplitudes between these two states. The *Heisenberg picture* leads to equations of motion for the various variables of the system which are closely analogous to the classical equations. In particular, the Maxwell–Lorentz equations remain valid between operators. An additional advantage of this viewpoint is that it is well suited to the introduction of important statistical functions like the symmetric correlation functions or the linear susceptibilities.

Part C is devoted to a discussion of some important physical aspects of the quantized free field. Different types of measurements and various quantum states are reviewed and analyzed. We will see in particular that the elementary excitations of the quantum field can be analyzed in terms of particles having a well-defined energy and momentum, *viz.*, the *photons*. A particularly important state is the ground state of the field, called the *vacuum* (of photons). We also introduce the *coherent* or *quasi-classical* states, which are the quantum states of the field which correspond most closely to a given classical state. Let us mention here that the discussion of Part C is extended in Complement A_{III} with an analysis of *interference* phenomena (with one or more photons) and of the *wave–particle duality* in the framework of the quantum theory of radiation.

We consider finally, in Part D, the Hamiltonian describing the *interaction* between radiation and matter. The various terms of this Hamiltonian are written down and their orders of magnitude evaluated. The selection rules which the matrix elements obey are also examined.

We note finally that the global Hamiltonian of the system particles + field can be transformed and put into a more convenient form in the *long-wavelength limit*. Complements A_{IV} and B_{IV} of the next chapter introduce in an elementary way the unitary transformation which allows one to pass from the usual Hamiltonian ("in $\mathbf{A} \cdot \mathbf{p}$") to the electric dipole Hamiltonian ("in $\mathbf{E} \cdot \mathbf{r}$"). This is an example of a transformation for passing from electrodynamics in the Coulomb gauge to an equivalent formulation, a problem treated with more detail in Chapter IV.

A—THE GENERAL FRAMEWORK

1. Fundamental Dynamical Variables. Commutation Relations

The fundamental dynamical variables for each particle α are the position $\mathbf{r}_\alpha$ and the momentum $\mathbf{p}_\alpha$ of the particle, which satisfy the commutation relations

$$\begin{cases} [r_{\alpha i}, r_{\beta j}] = [p_{\alpha i}, p_{\beta j}] = 0 \\ [r_{\alpha i}, p_{\beta j}] = \mathrm{i}\hbar\, \delta_{\alpha\beta}\, \delta_{ij} \end{cases} \tag{A.1}$$

$$i, j = x, y, z\,.$$

As was indicated at the end of Chapter I and justified in Chapter II, the normal variables α_i and α_i^*, which characterize the state of the classical transverse field, become, following quantization, the destruction and creation operators a_i and a_i^+ for fictitious harmonic oscillators associated with the various modes of the field and satisfying the commutation relations

$$\begin{cases} [a_i, a_j] = [a_i^+, a_j^+] = 0 \\ [a_i, a_j^+] = \delta_{ij}\,. \end{cases} \tag{A.2}$$

Remark

We have explicitly used above the Schrödinger point of view where the operators are time independent. The commutation relations (A.1) and (A.2) remain valid in the Heisenberg formulation, provided that the two operators in the commutation relation are taken at the same instant.

2. The Operators Associated with the Various Physical Variables of the System

We begin with the various field observables. The operators associated with the transverse fields $\mathbf{E}_\perp(\mathbf{r})$, $\mathbf{B}(\mathbf{r})$, and $\mathbf{A}_\perp(\mathbf{r})$ at each point $\mathbf{r}$ of space are gotten by replacing α_i with a_i and α_i^* with a_i^+ in the expansions of the corresponding classical variables [see Equations (C.38), (C.39), and (C.37) in Chapter I]:

$$\mathbf{E}_\perp(\mathbf{r}) = \sum_i \mathrm{i}\mathscr{E}_{\omega_i}[a_i\, \boldsymbol{\varepsilon}_i\, \mathrm{e}^{\mathrm{i}\mathbf{k}_i\cdot\mathbf{r}} - a_i^+\, \boldsymbol{\varepsilon}_i\, \mathrm{e}^{-\mathrm{i}\mathbf{k}_i\cdot\mathbf{r}}] \tag{A.3}$$

$$\mathbf{B}(\mathbf{r}) = \sum_i \mathrm{i}\mathscr{B}_{\omega_i}[a_i(\boldsymbol{\kappa}_i \times \boldsymbol{\varepsilon}_i)\, \mathrm{e}^{\mathrm{i}\mathbf{k}_i\cdot\mathbf{r}} - a_i^+(\boldsymbol{\kappa}_i \times \boldsymbol{\varepsilon}_i)\, \mathrm{e}^{-\mathrm{i}\mathbf{k}_i\cdot\mathbf{r}}] \tag{A.4}$$

$$\mathbf{A}_\perp(\mathbf{r}) = \sum_i \mathscr{A}_{\omega_i}[a_i\, \boldsymbol{\varepsilon}_i\, \mathrm{e}^{\mathrm{i}\mathbf{k}_i\cdot\mathbf{r}} + a_i^+\, \boldsymbol{\varepsilon}_i\, \mathrm{e}^{-\mathrm{i}\mathbf{k}_i\cdot\mathbf{r}}] \tag{A.5}$$

with

$$\mathscr{E}_{\omega_i} = \left[\frac{\hbar\omega_i}{2\,\varepsilon_0\,L^3}\right]^{1/2} \qquad \mathscr{B}_{\omega_i} = \frac{\mathscr{E}_{\omega_i}}{c} \qquad \mathscr{A}_{\omega_i} = \frac{\mathscr{E}_{\omega_i}}{\omega_i}. \tag{A.6}$$

The total electric field $\mathbf{E}(\mathbf{r})$ is given by

$$\mathbf{E}(\mathbf{r}) = \mathbf{E}_\perp(\mathbf{r}) + \mathbf{E}_\parallel(\mathbf{r}) \tag{A.7}$$

where

$$\mathbf{E}_\parallel(\mathbf{r}) = \frac{1}{4\,\pi\varepsilon_0}\sum_\alpha q_\alpha \frac{\mathbf{r}-\mathbf{r}_\alpha}{|\mathbf{r}-\mathbf{r}_\alpha|^3}. \tag{A.8}$$

The position $\mathbf{r}_\alpha$ of particle α in (A.8) is now an operator.

Since we have respected the ordering of α_i and α_i^* in the calculations leading to Equations (C.16) and (C.17) in Chapter I for the Hamiltonian H_{trans} and the momentum $\mathbf{P}_{\text{trans}}$ of the transverse field, it is not necessary to redo these calculations in the quantum case, and we can write the operator H_{trans} in the form

$$\begin{aligned} H_{\text{trans}} &= \frac{\varepsilon_0}{2}\int \mathrm{d}^3r[\mathbf{E}_\perp^2(\mathbf{r}) + c^2\,\mathbf{B}^2(\mathbf{r})] \\ &= \sum_i \frac{\hbar\omega_i}{2}[a_i^+\,a_i + a_i\,a_i^+] = \sum_i \hbar\omega_i\left[a_i^+\,a_i + \frac{1}{2}\right] \end{aligned} \tag{A.9}$$

[where we have used (A.2) to replace $a_i a_i^+$ by $a_i^+ a_i + 1$]. Similarly,

$$\begin{aligned} \mathbf{P}_{\text{trans}} &= \varepsilon_0 \int \mathrm{d}^3r\,\mathbf{E}_\perp(\mathbf{r}) \times \mathbf{B}(\mathbf{r}) \\ &= \sum_i \frac{\hbar\mathbf{k}_i}{2}[a_i^+\,a_i + a_i\,a_i^+] = \sum_i \hbar\mathbf{k}_i\,a_i^+\,a_i \end{aligned} \tag{A.10}$$

(using the relationship $\sum_i \hbar\mathbf{k}_i/2 = \mathbf{0}$).

Finally, since we are in the Coulomb gauge (*) we have

$$\mathbf{A}_\parallel(\mathbf{r}) = \mathbf{0} \tag{A.11}$$

$$U(\mathbf{r}) = \frac{1}{4\,\pi\varepsilon_0}\sum_\alpha \frac{q_\alpha}{|\mathbf{r}-\mathbf{r}_\alpha|} \tag{A.12}$$

U is simply the electrostatic potential of the charge distribution.

(*) As in Chapter II (see §C.1), we will omit hereafter the index $\perp$ in $\mathbf{A}_\perp$, since in the Coulomb gauge $\mathbf{A}$ and $\mathbf{A}_\perp$ are the same.

Remarks

(i) Starting from the expansions (A.3) to (A.5) for the transverse fields in a_i and a_i^+ and the commutation relations (A.2), one can establish the following expressions for the commutators between two components of the fields taken at two different points $\mathbf{r}$ and $\mathbf{r}'$ (see Exercise 1):

$$[A_i(\mathbf{r}), A_j(\mathbf{r}')] = 0 \tag{A.13}$$

$$[A_i(\mathbf{r}), E_{\perp j}(\mathbf{r}')] = \frac{1}{\varepsilon_0} \frac{\hbar}{\mathrm{i}} \delta_{ij}^{\perp}(\mathbf{r} - \mathbf{r}') \tag{A.14}$$

where $i, j = x, y, z$ and where $\delta_{ij}^{\perp}$ is the "transverse delta function" defined by (B.17.b) of Chapter I. One gets also:

$$[E_{\perp x}(\mathbf{r}), B_y(\mathbf{r}')] = -\frac{\mathrm{i}\hbar}{\varepsilon_0} \frac{\partial}{\partial z} \delta(\mathbf{r} - \mathbf{r}'). \tag{A.15}$$

Since the derivative of $\delta(\mathbf{r} - \mathbf{r}')$ vanishes at $\mathbf{r} = \mathbf{r}'$ as a result of $\delta(\mathbf{r} - \mathbf{r}')$ being even, it clearly appears in (A.15) that $\mathbf{E}_\perp(\mathbf{r})$ and $\mathbf{B}(\mathbf{r})$ commute when they are taken at the same point. It is this property which makes the symmetrization of $\mathbf{E}_\perp(\mathbf{r}) \times \mathbf{B}(\mathbf{r})$ in the expressions for $\mathbf{P}_{\text{trans}}$ and $\mathbf{J}_{\text{trans}}$ useless.

Since $\mathbf{E}_{\parallel}(\mathbf{r})$ can be reexpressed as a function of $\mathbf{r}_\alpha$ [see (A.8)], which commutes with a_i and a_i^+, Equations (A.14) and (A.15) remain valid if $\mathbf{E}_\perp$ is replaced by the total field $\mathbf{E}$.

(ii) All the commutators (A.13) to (A.15) are evaluated from the Schrödinger point of view. They remain valid in the Heisenberg picture if the two fields in the commutator are taken at the same time.

(iii) In fact, the commutators (A.13) and (A.14) were derived directly in Chapter II before the commutation relations (A.2). The relations (A.2) are then a consequence of (A.13) and (A.14) (see §C.4, Chapter II). The reason for this is that in the Lagrangian formulation of electrodynamics in the Coulomb gauge, $\mathbf{A}(\mathbf{r})$ and $-\varepsilon_0 \mathbf{E}_\perp(\mathbf{r})$ appear as canonically conjugate variables, with the result that canonical quantization of the theory leads directly to (A.13) and (A.14) [the fact that it is $\delta_{ij}^{\perp}(\mathbf{r} - \mathbf{r}')$ and not $\delta_{ij}\delta(\mathbf{r} - \mathbf{r}')$ which appears in (A.14) is related to the transverse character of the fields (see §C.4.*b*, Chapter II)].

The Hamiltonian is a particularly important operator in this theory. We recall its expression in the Coulomb gauge:

$$\begin{aligned} H = {} & \sum_\alpha \frac{1}{2\,m_\alpha} [\mathbf{p}_\alpha - q_\alpha \mathbf{A}(\mathbf{r}_\alpha)]^2 + \\ & + \sum_\alpha \varepsilon_{\text{Coul}}^\alpha + \frac{1}{8\,\pi\varepsilon_0} \sum_{\alpha \neq \beta} \frac{q_\alpha q_\beta}{|\mathbf{r}_\alpha - \mathbf{r}_\beta|} + \\ & + \sum_i \hbar\omega_i \left(a_i^+ a_i + \frac{1}{2} \right). \end{aligned} \tag{A.16}$$

Physically, the operator (A.16) is associated with the total energy of the system, the first line of (A.16) giving the kinetic energy of the particles, the second line their Coulomb interaction energy [$\varepsilon_{\text{Coul}}^{\alpha}$ being the Coulomb self-energy of particle α given by (B.36) of Chapter I], and the third line the energy of the transverse field.

Remarks

(i) The expression (A.16) for the Hamiltonian in the Coulomb gauge, which was postulated at the end of Chapter I, has its justification in the Lagrangian and Hamiltonian approaches presented in Chapter II (see §C.3, Chapter II). For those who have not read Chapter II, the fact that the Hamiltonian (A.16), taking into account (A.1) and (A.2), leads to the quantized Maxwell–Lorentz equations (as will be shown in Part B below) will be considered as an a posteriori justification of (A.1), (A.2), and (A.16).

(ii) It will be useful for subsequent developments to give the expression for the Hamiltonian H in the presence of external fields—that is, static or time-dependent fields whose motion is imposed externally. Let $\mathbf{A}_e(\mathbf{r}, t)$ and $U_e(\mathbf{r}, t)$ be the vector and scalar potentials describing such external fields (the Coulomb gauge need not be used for $\mathbf{A}_e$ and U_e). We have shown in Complement C_{II} that the expression for H is then

$$H = \sum_{\alpha} \frac{1}{2m_\alpha}[\mathbf{p}_\alpha - q_\alpha \mathbf{A}(\mathbf{r}_\alpha) - q_\alpha \mathbf{A}_e(\mathbf{r}_\alpha, t)]^2 + \\ + \sum_{\alpha} \varepsilon_{\text{Coul}}^{\alpha} + \frac{1}{8\pi\varepsilon_0}\sum_{\alpha \neq \beta} \frac{q_\alpha q_\beta}{|\mathbf{r}_\alpha - \mathbf{r}_\beta|} + \sum_{\alpha} q_\alpha U_e(\mathbf{r}_\alpha, t) + \\ + \sum_i \hbar\omega_i \left(a_i^+ a_i + \frac{1}{2}\right). \tag{A.17}$$

We will show subsequently (see the remarks at the ends of §§B.2.*a* and B.2.*b*) that the Hamiltonian (A.17) with the commutation relations (A.1) and (A.2) leads to "good" equations of motion for the particles and for the fields.

We complete this review of the various quantum operators by giving the expression for the total momentum $\mathbf{P}$,

$$\mathbf{P} = \sum_{\alpha} \mathbf{p}_\alpha + \sum_i \hbar \mathbf{k}_i a_i^+ a_i \tag{A.18}$$

and that for the total angular momentum $\mathbf{J}$,

$$\mathbf{J} = \sum_{\alpha} \mathbf{r}_\alpha \times \mathbf{p}_\alpha + \mathbf{J}_{\text{trans}} \tag{A.19}$$

where $\mathbf{J}_{\text{trans}}$ is given by (C.18) of Chapter I (with α and α^* replaced by a and a^+).

3. State Space

The state space $\mathscr{E}$ of the system is the tensor product of the state space of the particles, $\mathscr{E}_P$, in which $\mathbf{r}_\alpha, \mathbf{p}_\alpha, \ldots$ act, with the state space $\mathscr{E}_R$ of the radiation field, in which $a_i, a_i^+, \ldots$ act:

$$\mathscr{E} = \mathscr{E}_P \otimes \mathscr{E}_R . \tag{A.20}$$

The space $\mathscr{E}_R$ is itself the product of the state spaces $\mathscr{E}_i$ of the various oscillators i associated with the modes of the field:

$$\mathscr{E}_R = \mathscr{E}_1 \otimes \mathscr{E}_2 \cdots \otimes \mathscr{E}_i \otimes \cdots \tag{A.21}$$

One possible orthonormal basis of $\mathscr{E}_i$ is $\{|n_i\rangle\}$, where $n_i = 0, 1, 2, 3, \ldots$ labels the energy levels of oscillator i. If $\{|s\rangle\}$ is an orthonormal basis of $\mathscr{E}_P$, we can take in the total space $\mathscr{E}$ the following basis:

$$\{|s\rangle|n_1\rangle|n_2\rangle \ldots |n_i\rangle \ldots\} = \{|s; n_1, n_2 \ldots n_i \ldots\rangle\} \tag{A.22}$$

and the most general state vector of $\mathscr{E}$ is a linear superposition of the basis vectors (A.22).

B—TIME EVOLUTION

1. The Schrödinger Picture

In this picture the various observables of the system particles + fields are fixed, and it is the state vector $|\psi(t)\rangle$ which evolves according to the Schrödinger equation:

$$i\hbar \frac{d}{dt} | \psi(t) \rangle = H | \psi(t) \rangle \tag{B.1}$$

where H is the global Hamiltonian given in (A.16).

If $|\psi(t)\rangle$ is expanded in the orthonormal basis (A.22),

$$| \psi(t) \rangle = \sum_{sn_1n_2...} c_{sn_1n_2...}(t) | s; n_1, n_2 ... \rangle \tag{B.2}$$

and if the expansion (B.2) is substituted in (B.1), one gets a linear system of differential equations for the coefficients $c_{sn_1n_2...}(t)$.

The Schrödinger point of view is convenient for introducing the transition amplitudes

$$\langle \psi_f | U(t) | \psi_i \rangle \tag{B.3}$$

where

$$U(t) = e^{-iHt/\hbar} \tag{B.4}$$

is the evolution operator. Physically, (B.3) represents the probability amplitude that the system starting from the initial state $|\psi_i\rangle$ will end up in the final state $|\psi_f\rangle$ after a time interval t. These amplitudes allow one to describe various manifestations of the interaction between matter and radiation (absorption, emission or scattering of photons by atoms, photoionization, radiative corrections, etc.). One of the important objectives of the theory will be to calculate these transition amplitudes, either by perturbation theory or by some more complex approach. (*)

In the present chapter, which is devoted to more general considerations, we will stress the advantage of the Heisenberg point of view, which is more appropriate for comparisons between the classical and quantum theories.

2. The Heisenberg Picture. The Quantized Maxwell–Lorentz Equations

In the Heisenberg picture, it is the state vector $|\psi\rangle$ of the global system which remains fixed, whereas the observables $G(t)$ of the system evolve according to the Heisenberg equation

$$\frac{d}{dt} G(t) = \frac{1}{i\hbar} [G(t), H(t)] \tag{B.5}$$

(*) These problems are treated for example in Cohen-Tannoudji, Dupont-Roc, and Grynberg (Chapters I and III).

where H is the Hamiltonian (A.16) in which all of the operators are taken at time t.

a) The Heisenberg Equations for Particles

The Heisenberg equation for $\mathbf{r}_\alpha$,

$$\dot{\mathbf{r}}_\alpha = \frac{1}{i\hbar}[\mathbf{r}_\alpha, H] = \frac{1}{i\hbar}\left[\mathbf{r}_\alpha, \frac{1}{2\,m_\alpha}(\mathbf{p}_\alpha - q_\alpha\,\mathbf{A}(\mathbf{r}_\alpha))^2\right]$$
$$= \frac{1}{m_\alpha}(\mathbf{p}_\alpha - q_\alpha\,\mathbf{A}(\mathbf{r}_\alpha)) \tag{B.6}$$

is just the well-known relation between the mechanical momentum $m_\alpha \mathbf{v}_\alpha$ of particle α and its canonical momentum $\mathbf{p}_\alpha$:

$$\mathbf{p}_\alpha = m_\alpha\,\mathbf{v}_\alpha + q_\alpha\,\mathbf{A}(\mathbf{r}_\alpha) \tag{B.7}$$

($\mathbf{v}_\alpha = \dot{\mathbf{r}}_\alpha$ is the velocity of particle α). Equation (B.44) of Chapter I thus appears here as an equation of motion.

For the following calculations, it will be useful to evaluate the commutator between the two components $v_{\alpha j}$ and $v_{\alpha l}$ of $\mathbf{v}_\alpha$ ($j, l = x, y, z$):

$$m_\alpha^2[v_{\alpha j}, v_{\alpha l}] = -\,q_\alpha[p_{\alpha j}, A_l(\mathbf{r}_\alpha)] - q_\alpha[A_j(\mathbf{r}_\alpha), p_{\alpha l}]$$
$$= i\hbar q_\alpha[\partial_j A_l(\mathbf{r}_\alpha) - \partial_l A_j(\mathbf{r}_\alpha)]$$
$$= i\hbar q_\alpha \sum_k \varepsilon_{jlk}\,B_k(\mathbf{r}_\alpha)\,. \tag{B.8}$$

Consider now the Heisenberg equation for $m_\alpha v_{\alpha j}$,

$$m_\alpha\,\dot{v}_{\alpha j} = m_\alpha\,\ddot{r}_{\alpha j} = \frac{m_\alpha}{i\hbar}[v_{\alpha j}, H] \tag{B.9}$$

and calculate the contributions of the three terms appearing in the expression (A.16) for H. The first term of H (kinetic energy) gives, using (B.8),

$$\frac{m_\alpha}{i\hbar}\left[v_{\alpha j}, \sum_l m_\alpha\,v_{\alpha l}^2/2\right] =$$
$$= \frac{m_\alpha^2}{2\,i\hbar}\sum_l \{\,v_{\alpha l}[v_{\alpha j}, v_{\alpha l}] + [v_{\alpha j}, v_{\alpha l}]\,v_{\alpha l}\,\}$$
$$= \frac{q_\alpha}{2}\sum_k \sum_l \varepsilon_{jlk}\{\,v_{\alpha l}\,B_k(\mathbf{r}_\alpha) + B_k(\mathbf{r}_\alpha)\,v_{\alpha l}\,\} \tag{B.10}$$

and can be considered as the j-axis component of the magnetic part of the symmetrized Lorentz force,

$$\frac{q_\alpha}{2}\{\mathbf{v}_\alpha \times \mathbf{B}(\mathbf{r}_\alpha) - \mathbf{B}(\mathbf{r}_\alpha) \times \mathbf{v}_\alpha\}. \tag{B.11}$$

The second term of H (Coulomb energy) gives

$$\frac{m_\alpha}{i\hbar}[v_{\alpha j}, V_{\text{Coul}}] = \frac{1}{i\hbar}[p_{\alpha j}, V_{\text{Coul}}] = -\frac{\partial}{\partial r_{\alpha j}} V_{\text{Coul}} \tag{B.12}$$

which is just the j-axis component of the longitudinal electric force $q_\alpha \mathbf{E}_\parallel(\mathbf{r}_\alpha)$. Finally, the third term of H (the energy of the transverse field) gives

$$\frac{m_\alpha}{i\hbar}\left[v_{\alpha j}, \sum_i \hbar\omega_i\left(a_i^+ a_i + \frac{1}{2}\right)\right] = iq_\alpha \sum_i \omega_i [A_j(\mathbf{r}_\alpha), a_i^+ a_i]. \tag{B.13}$$

By using on one hand the commutator $[a_i, a_i^+ a_i] = a_i$ and on the other hand the expansions (A.3) and (A.5) for $\mathbf{E}_\perp$ and $\mathbf{A}$, one can show that (B.13) is nothing more than the component of the transverse electric force $q_\alpha \mathbf{E}_\perp(\mathbf{r}_\alpha)$ on the j-axis.

By regrouping all the preceding results one gets finally

$$m_\alpha \ddot{\mathbf{r}}_\alpha = q_\alpha \mathbf{E}(\mathbf{r}_\alpha) + \frac{q_\alpha}{2}[\mathbf{v}_\alpha \times \mathbf{B}(\mathbf{r}_\alpha) - \mathbf{B}(\mathbf{r}_\alpha) \times \mathbf{v}_\alpha] \tag{B.14}$$

where $\mathbf{E} = \mathbf{E}_\parallel + \mathbf{E}_\perp$ is the total electric field. Equation (B.14) is the quantum form of the Newton–Lorentz equation

Remark

In the presence of external fields, it is necessary to use the Hamiltonian (A.17). Equation (B.6) then becomes

$$m_\alpha \dot{\mathbf{r}}_\alpha = m_\alpha \mathbf{v}_\alpha = \mathbf{p}_\alpha - q_\alpha \mathbf{A}(\mathbf{r}_\alpha) - q_\alpha \mathbf{A}_e(\mathbf{r}_\alpha, t) \tag{B.15}$$

and contains the external vector potential $\mathbf{A}_e$. With this new expression for $m_\alpha \mathbf{v}_\alpha$, the calculation of the commutator (B.8) involves the curl of $\mathbf{A} + \mathbf{A}_e$, that is to say, the total magnetic field

$$\mathbf{B}_t = \mathbf{B} + \mathbf{B}_e \tag{B.16}$$

which is the sum of the quantized magnetic field $\mathbf{B}$ and the external magnetic field $\mathbf{B}_e$:

$$m_\alpha^2[v_{\alpha j}, v_{\alpha l}] = i\hbar q_\alpha \sum_k \varepsilon_{jlk} B_{tk}(\mathbf{r}_\alpha). \tag{B.17}$$

Additionally, since the new expression (B.15) for $m_\alpha \mathbf{v}_\alpha$ contains a term *explicitly* dependent on time, $-q_\alpha \mathbf{A}_e(\mathbf{r}_\alpha, t)$, the Heisenberg equation (B.9) for $m_\alpha \mathbf{v}_\alpha$ is modified and becomes

$$\begin{aligned} m_\alpha \dot{\mathbf{v}}_\alpha &= m_\alpha \frac{\partial \mathbf{v}_\alpha}{\partial t} + \frac{m_\alpha}{i\hbar}[\mathbf{v}_\alpha, H] \\ &= -q_\alpha \frac{\partial}{\partial t}\mathbf{A}_e(\mathbf{r}_\alpha, t) + \frac{m_\alpha}{i\hbar}[\mathbf{v}_\alpha, H]. \end{aligned} \tag{B.18}$$

Finding the commutator of (B.18) is a similar procedure. Since it is necessary to use (B.17) in place of (B.8), it is the total field $\mathbf{B}_t$ which now appears in the Lorentz magnetic force (B.11). The other new term which arises in this calculation comes from the coupling of the charges with the external scalar potential [the term $\Sigma_\alpha q_\alpha U_e(\mathbf{r}_\alpha, t)$ of (A.17)]. It is written

$$\frac{m_\alpha}{i\hbar}[\mathbf{v}_\alpha, q_\alpha U_e(\mathbf{r}_\alpha, t)] = -q_\alpha \nabla U_e(\mathbf{r}_\alpha, t). \tag{B.19}$$

Regrouping (B.19) with the first term of (B.18), one gets

$$q_\alpha\left[-\frac{\partial}{\partial t}\mathbf{A}_e(\mathbf{r}_\alpha, t) - \nabla U_e(\mathbf{r}_\alpha, t)\right] = q_\alpha \mathbf{E}_e(\mathbf{r}_\alpha, t) \tag{B.20}$$

that is to say, the force due to the external electric field. Equations (B.12) and (B.13) remain unchanged and give rise to the force due to the quantum electric field $\mathbf{E}$, $q_\alpha \mathbf{E}(\mathbf{r}_\alpha, t)$, which is added to (B.20) to yield the total electric field

$$\mathbf{E}_t = \mathbf{E} + \mathbf{E}_e. \tag{B.21}$$

Finally, by regrouping the preceding results, one gets the new Newton–Lorentz equation

$$m_\alpha \ddot{\mathbf{r}}_\alpha = q_\alpha \mathbf{E}_t(\mathbf{r}_\alpha) + \frac{q_\alpha}{2}[\mathbf{v}_\alpha \times \mathbf{B}_t(\mathbf{r}_\alpha) - \mathbf{B}_t(\mathbf{r}_\alpha) \times \mathbf{v}_\alpha] \tag{B.22}$$

which now contains the total fields $\mathbf{B}_t$ and $\mathbf{E}_t$.

b) The Heisenberg Equations for Fields

The same linear relationships exist between the classical transverse fields and $\{\alpha_i, \alpha_i^*\}$ on one hand and the quantum transverse fields and $\{a_i, a_i^+\}$ on the other. If we show that α_i and a_i obey similar equations, we will have at the same time shown that the equations of motion of the fields are the same in the classical and quantum theories. Instead of writing the Heisenberg equations for $\mathbf{E}_\perp$, $\mathbf{B}$, and $\mathbf{A}$, it is therefore easier to consider one such equation for a_i:

$$\dot{a}_i = \frac{1}{i\hbar}[a_i, H]. \tag{B.23}$$

As with $m_\alpha \mathbf{v}_\alpha$, we will now calculate the contributions of the three terms of

H. The last term gives $-i\omega_i a_i$. The second commutes with a_i and gives zero. Finally, the first term gives

$$\frac{1}{i\hbar}\left[a_i, \sum_\alpha \frac{m_\alpha \mathbf{v}_\alpha^2}{2}\right] = \sum_\alpha \frac{m_\alpha}{2i\hbar}\left\{\mathbf{v}_\alpha \cdot \frac{\partial \mathbf{v}_\alpha}{\partial a_i^+} + \frac{\partial \mathbf{v}_\alpha}{\partial a_i^+}\cdot \mathbf{v}_\alpha\right\} =$$

$$= \sum_\alpha \frac{iq_\alpha}{2\hbar} \mathscr{A}_{\omega_i} \boldsymbol{\varepsilon}_i \cdot [\mathbf{v}_\alpha \mathrm{e}^{-i\mathbf{k}_i\cdot\mathbf{r}_\alpha} + \mathrm{e}^{-i\mathbf{k}_i\cdot\mathbf{r}_\alpha}\mathbf{v}_\alpha]\,. \quad \text{(B.24)}$$

We have used $[a_i, f(a_i^+)] = \partial f/\partial a_i^+$, which follows from (A.2). If one introduces the symmetrized current

$$\mathbf{j}(\mathbf{r}) = \frac{1}{2}\sum_\alpha q_\alpha[\mathbf{v}_\alpha\,\delta(\mathbf{r}-\mathbf{r}_\alpha) + \delta(\mathbf{r}-\mathbf{r}_\alpha)\,\mathbf{v}_\alpha] \quad \text{(B.25)}$$

one can then transform (B.24) into

$$\frac{i}{\sqrt{2\varepsilon_0\hbar\omega_i}}\, j_i \quad \text{(B.26)}$$

where

$$j_i = \frac{1}{\sqrt{L^3}}\int \mathrm{d}^3r\, \mathrm{e}^{-i\mathbf{k}_i\cdot\mathbf{r}}\, \boldsymbol{\varepsilon}_i \cdot \mathbf{j}(\mathbf{r}) \quad \text{(B.27)}$$

is the Fourier component of $\mathbf{j}$ on mode i. Finally, the equation of motion for a_i is written

$$\dot{a}_i + i\omega_i a_i = \frac{i}{\sqrt{2\varepsilon_0\hbar\omega_i}}\, j_i \quad \text{(B.28)}$$

and has exactly the same form as the equation of motion for α_i [Equation (C.41) of Chapter I]. The argument given at the beginning of this subsection then shows that Maxwell equations remain valid between operators.

Finally, all the basic equations of classical electrodynamics can be generalized into the quantum domain (provided however that one symmetrizes the products of noncommuting Hermitian operators such as the Lorentz magnetic force or the charged-particle current).

Remarks

(i) In the presence of external fields, the evolution equation for a_i keeps the same form (B.28). The third line of (A.17) always gives $-i\omega_i a_i$, the contribution of the second is zero, and that of the first is the same as above. [It must be noted however that the velocities $\mathbf{v}_\alpha$ appearing in (B.25) for the current are now related to the momenta $\mathbf{p}_\alpha$ by Equation (B.15) and no longer by (B.7).] Actually, the equation of motion of a_i only concerns the evolution of the quantized transverse field. This is expected, since the evolution of $\mathbf{A}_e$ and U_e is assumed to be externally imposed. Since the source term of (B.28) depends only on the particle current, the external field can only influence the evolution of the quantized transverse field through the current. More precisely, the presence of the external field modifies the motion of the particles [see (B.22)], and thus the current associated with them, and thus the field which they radiate.

(ii) In the absence of particles ($j_i = 0$), the solution of Equation (B.28) is quite simple:

$$a_i(t) = a_i(0)\, e^{-i\omega_i t} \tag{B.29}$$

The expressions for the *free* fields in the Heisenberg picture are then gotten by replacing, in the expansions (A.3), (A.4), and (A.5) for $\mathbf{E}_\perp$, $\mathbf{B}$, and $\mathbf{A}$, the quantity a_i by (B.29) and a_i^+ by the adjoint expression. For example,

$$\mathbf{E}_\perp^{\text{free}}(\mathbf{r}, t) = \sum_i i\,\mathscr{E}_{\omega_i}\left[a_i(0)\,\boldsymbol{\varepsilon}_i\,e^{i(\mathbf{k}_i\cdot\mathbf{r}-\omega_i t)} - a_i^+(0)\,\boldsymbol{\varepsilon}_i\,e^{-i(\mathbf{k}_i\cdot\mathbf{r}-\omega_i t)}\right]. \tag{B.30}$$

One thus gets an expansion of the free field in traveling plane waves whose coefficients $a_i(0)$ and $a_i^+(0)$ are now operators satisfying the commutation relations (A.2). Starting from (B.30) and the analogous equation for $\mathbf{B}^{\text{free}}(\mathbf{r}, t)$, one can find the commutators between components of the free fields taken at two different space–time points $\mathbf{r}, t$ and $\mathbf{r}', t'$ (see Complement C_{III}). In particular, one finds that the commutators of the electric and magnetic fields are always zero when $\mathbf{r}, t$ and $\mathbf{r}', t'$ are separated by a spacelike interval. This result means that two measurements of the fields at $\mathbf{r}, t$ and $\mathbf{r}', t'$ cannot perturb each other when the two events at $\mathbf{r}, t$ and $\mathbf{r}', t'$ cannot be connected by a physical signal.

c) The Advantages of the Heisenberg Point of View

A first advantage of this point of view is that it allows one to discuss easily the analogies and differences between the classical and quantum theories. As seen above, this leads to analogous equations of motion, but these equations now involve operators and not the classical variables.

A second advantage of the Heisenberg point of view is that it allows one to define "two-time averages", that is to say, mean values in the state $|\psi\rangle$ of the system (remember that $|\psi\rangle$ is time independent) of a product of two operators $F(t)$ and $G(t')$ taken at two different times t and t':

$$\langle \psi \,|\, F(t)\, G(t') \,|\, \psi \rangle. \tag{B.31}$$

Important examples of two-time averages are the symmetric correlation functions and the linear response functions. They describe respectively the dynamics of the fluctuations occurring in the system and the linear response of the system to a small external perturbation. We will return below (§C.3.c and Complement C_{III}) to the symmetric correlation functions and the linear susceptibilities of the quantized free field (see also Exercise 6, Chapter IV).

C—OBSERVABLES AND STATES OF THE QUANTIZED FREE FIELD

To discuss the physical properties of a quantum system, it is necessary to know the *observables* G associated with the various physical variables of the system and at the same time the *state vector* $|\psi\rangle$ (or the *density operator* ρ) describing the state of the system at the time the measurement is being effected. It is by using these two distinct mathematical quantities G and $|\psi\rangle$ (or G and ρ) that predictions can be made about the results of measurements performed on that system.

For instance, if $|\psi\rangle$ coincides with one of the eigenstates of G, the result of the measurement of G is certain; it is the corresponding eigenvalue of G. If $|\psi\rangle$ is not an eigenstate of G, the result of the measurement can a priori be any eigenvalue of G with well-defined probabilities, the mean value of the results gotten from a large number of identical measurements (repeated on the same state $|\psi\rangle$) being $\langle\psi|G|\psi\rangle$. If two physical variables are not compatible, that is, if the corresponding observables F and G do not commute, it is not possible to find a common basis of eigenvectors for F and G. There is no state $|\psi\rangle$ which is well adapted to both F and G. One can seek a compromise in this case, for example by looking for a state $|\psi\rangle$ such that the mean values of F and G in this state are equal to the corresponding classical values of these variables.

This general approach can be applied to the electromagnetic field and will be followed in this section to study the important properties of the quantized field. In order to concentrate the discussion on the field variables, we will consider only the free field, that is to say the field in the absence of sources.

1. Review of Various Observables of the Free Field

a) Total Energy and Total Momentum of the Field

In the absence of particles, H and $\mathbf{P}$ reduce to H_{trans} and $\mathbf{P}_{\text{trans}}$, whose expressions have been given above in (A.9) and (A.10). Note that these variables are *global* variables in the sense that they are given as integrals over all space (free space or quantization volume) of functions of the fields $\mathbf{E}$ and $\mathbf{B}$.

b) The Fields at a Given Point r of Space

Unlike H_{trans} and $\mathbf{P}_{\text{trans}}$, these variables are *local*. The measurement of the field at a point requires placing a *test charge* at that point. The expressions for $\mathbf{E}_\perp(\mathbf{r})$ [which is equal to $\mathbf{E}(\mathbf{r})$ in the absence of particles, since $\mathbf{E}_\parallel$ is then zero], $\mathbf{B}(\mathbf{r})$, and $\mathbf{A}(\mathbf{r})$ are given above by (A.3), (A.4), and (A.5).

Let us examine the electric field more precisely and consider the contribution of one mode to $\mathbf{E}(\mathbf{r})$. Rather than using the operators a and a^+, one can introduce their linear combinations $a_P = (a + a^+)/2$ and $a_Q = (a - a^+)/2i$, analogous to the position x and momentum p of a harmonic oscillator. It is easy to see that a_P and a_Q correspond to two quadrature components of the electric field (see Exercise 6). One can then use the analogy with the harmonic oscillator to establish results for the quantum radiation field. For example, the Heisenberg relation for $\Delta x \, \Delta p$ here becomes $\Delta a_P \, \Delta a_Q \geq \frac{1}{4}$ and means that measurements of two quadrature components of the field are not compatible. If one wishes to measure one component with great precision, this will introduce an increase in the uncertainty in the quadrature component. Another interesting result concerns the distribution of possible values of each component a_P and a_Q when the mode is in the ground state (no photon in the mode). This distribution is a Gaussian just like the distribution of possible values of x and p in the ground state of a harmonic oscillator.

c) Observables Corresponding to Photoelectric Measurements

In the optical domain, one most often uses detectors based on the photoelectric effect to make *local* field measurements. Schematically one puts an atom in the radiation field at point $\mathbf{r}$ and observes the photoelectrons produced by photoionization of this atom. Such measurements are *destructive* in the sense that the photons responsible for the photoelectric signal disappear.

We will use below the results of photodetection theory to relate the signals obtained to the local field observables (see the references at the end of the chapter). These results will be useful for the physical discussions of this chapter and Complement $\mathrm{A}_{\mathrm{III}}$.

i) *Single Counting Signals*

Suppose that a broad-band detector is placed at point $\mathbf{r}$ in a free radiation field described by the state $|\psi\rangle$. One can show that the probability of observing a photoionization in this detector between times t and $t + \mathrm{d}t$ is proportional to $w_{\mathrm{I}}(\mathbf{r}, t)\,\mathrm{d}t$, where

$$w_{\mathrm{I}}(\mathbf{r}, t) = \langle \psi \,|\, \mathbf{E}^{(-)}(\mathbf{r}, t) \cdot \mathbf{E}^{(+)}(\mathbf{r}, t) \,|\, \psi \rangle \tag{C.1}$$

$\mathbf{E}^{(+)}(\mathbf{r}, t)$ and $\mathbf{E}^{(-)}(\mathbf{r}, t)$ are the positive- and negative-frequency components of the free field (B.30), containing respectively only the destruction operators a_i and the creation operators a_i^+:

$$\mathbf{E}^{(+)}(\mathbf{r}, t) = \sum_i \mathrm{i}\mathscr{E}_{\omega_i}\, a_i\, \boldsymbol{\varepsilon}_i\, \mathrm{e}^{\mathrm{i}(\mathbf{k}_i \cdot \mathbf{r} - \omega_i t)}$$

$$\mathbf{E}^{(-)}(\mathbf{r}, t) = [\mathbf{E}^{(+)}(\mathbf{r}, t)]^+ = -\sum_i \mathrm{i}\mathscr{E}_{\omega_i}\, a_i^+\, \boldsymbol{\varepsilon}_i\, \mathrm{e}^{-\mathrm{i}(\mathbf{k}_i \cdot \mathbf{r} - \omega_i t)} \qquad (\mathrm{C}.2)$$

[For simplification we omit the index "free" and write a_i and a_i^+ for $a_i(0)$ and $a_i^+(0)$.]

The "single counting rate" w_{I} is the mean value in the state $|\psi\rangle$ of the field of the observable

$$I(\mathbf{r}, t) = \mathbf{E}^{(-)}(\mathbf{r}, t) \cdot \mathbf{E}^{(+)}(\mathbf{r}, t) \qquad (\mathrm{C}.3)$$

taken in the Heisenberg picture. $I(\mathbf{r}, t)$ is a Hermitian operator arranged in the *normal order* (that is, with all the destruction operators on the right and all the creation operators on the left), which one can call the "light intensity" at point $\mathbf{r}$ at time t.

Remark

It is also possible to give a semiclassical treatment of the photoelectric effect where only the detector is quantized and not the field (see the references at the end of the chapter). For the single counting rate one finds in place of (C.1)

$$w_{\mathrm{I}}^{\mathrm{cl}}(\mathbf{r}, t) = \mathbf{E}_{\mathrm{cl}}^{(-)}(\mathbf{r}, t) \cdot \mathbf{E}_{\mathrm{cl}}^{(+)}(\mathbf{r}, t) = I_{\mathrm{cl}}(\mathbf{r}, t) \qquad (\mathrm{C}.4)$$

where $\mathbf{E}_{\mathrm{cl}}^{(+)}$ and $\mathbf{E}_{\mathrm{cl}}^{(-)}$ are the positive- and negative-frequency components of the classical electric field and where $I_{\mathrm{cl}} = |\mathbf{E}_{\mathrm{cl}}^{(+)}|^2$ is the classical intensity.

ii) *Double Counting Signals*

Consider now two photodetectors at $\mathbf{r}$ and $\mathbf{r}'$. The probability of observing one photoionization at point $\mathbf{r}'$ between t' and $t' + \mathrm{d}t'$, *and* another one at $\mathbf{r}$ between t and $t + \mathrm{d}t$ is found to be proportional to $w_{\mathrm{II}}(\mathbf{r}, t; \mathbf{r}', t')\,\mathrm{d}t\,\mathrm{d}t'$, where

$$w_{\mathrm{II}}(\mathbf{r}, t; \mathbf{r}', t') = $$

$$= \sum_{m,n} \langle \psi \,|\, E_m^{(-)}(\mathbf{r}', t')\, E_n^{(-)}(\mathbf{r}, t)\, E_n^{(+)}(\mathbf{r}, t)\, E_m^{(+)}(\mathbf{r}', t') \,|\, \psi \rangle \qquad (\mathrm{C}.5)$$

with $m, n = x, y, z$. The "double counting rate" w_{II} is equal to the mean value of the observable arranged in normal order:

$$\sum_m \sum_n E_m^{(-)}(\mathbf{r}', t')\, E_n^{(-)}(\mathbf{r}, t)\, E_n^{(+)}(\mathbf{r}, t)\, E_m^{(+)}(\mathbf{r}', t'). \qquad (\mathrm{C}.6)$$

Since $\mathbf{E}^{(-)}(\mathbf{r}, t)$ and $\mathbf{E}^{(+)}(\mathbf{r}, t)$ do not commute, it is not possible to write such an observable in the form $I(\mathbf{r}, t)I(\mathbf{r}', t')$, that is to say, as the product of two light intensities at $\mathbf{r}, t$ and $\mathbf{r}', t'$.

Remark

The semiclassical expression for the double counting rate is

$$w_{\text{II}}^{\text{cl}}(\mathbf{r}, t; \mathbf{r}', t') = I_{\text{cl}}(\mathbf{r}, t)\, I_{\text{cl}}(\mathbf{r}', t') . \tag{C.7}$$

For a fluctuating classical field it is necessary to take the average of (C.7) over all possible realizations of the field. The double counting rate is then equal to the correlation function of the light intensity.

2. Elementary Excitations of the Quantized Free Field. Photons

a) Eigenstates of the Total Energy and the Total Momentum

Consider first of all the oscillator i (mode i). The eigenvalues of $a_i^+ a_i$ are the integers $n_i = 0, 1, 2, \dots,$

$$a_i^+ a_i | n_i \rangle = n_i | n_i \rangle \quad n_i = 0, 1, 2, \dots \tag{C.8}$$

and the eigenvectors $|n_i\rangle$ obey the well-known relations

$$\begin{cases} a_i^+ | n_i \rangle = \sqrt{n_i + 1}\, | n_i + 1 \rangle \\ a_i | n_i \rangle = \sqrt{n_i}\, | n_i - 1 \rangle \\ a_i | 0_i \rangle = 0 \end{cases} \tag{C.9}$$

$$| n_i \rangle = \frac{(a_i^+)^{n_i}}{\sqrt{n_i !}} | 0_i \rangle . \tag{C.10}$$

Since $a_i^+ a_i$ commutes with $a_j^+ a_j$, the eigenstates of H_{trans} and $\mathbf{P}_{\text{trans}}$ are the tensor products of the eigenstates $|n_i\rangle$ of $a_i^+ a_i$:

$$H_R | n_1 \dots n_i \dots \rangle = \sum_i \left(n_i + \frac{1}{2} \right) \hbar\omega_i | n_1 \dots n_i \dots \rangle \tag{C.11.a}$$

$$\mathbf{P}_R | n_1 \dots n_i \dots \rangle = \sum_i n_i \hbar \mathbf{k}_i | n_1 \dots n_i \dots \rangle . \tag{C.11.b}$$

The ground state of the field corresponds to all n_i equal to zero and is denoted $|0\rangle$:

$$| 0 \rangle = | 0_1 \dots 0_i \dots \rangle . \tag{C.12}$$

From (C.10) it is clear that all the eigenstates $|n_1, n_2, \dots, n_i \dots\rangle$ can be gotten by applying a certain number of creation operators to the ground state $|0\rangle$:

$$| n_1 \dots n_i \dots \rangle = \frac{(a_1^+)^{n_1}}{\sqrt{n_1 !}} \cdots \frac{(a_i^+)^{n_i}}{\sqrt{n_i !}} \cdots | 0 \rangle . \tag{C.13}$$

b) The Interpretation in Terms of Photons

With respect to the ground state $|0\rangle$, the state $|n_1 \dots n_i \dots\rangle$ has an energy $\sum_i n_i \hbar\omega_i$ and a momentum $\sum_i n_i \hbar \mathbf{k}_i$. The situation appears as if this state represented a set of n_1 particles with energy $\hbar\omega_1$ and momentum $\hbar\mathbf{k}_1, \dots$ n_i particles with energy $\hbar\omega_i$ and momentum $\hbar\mathbf{k}_i, \dots$. These particles are called *photons*. They describe the elementary excitations of the various modes of the quantized field.

The ground state, which has no photon (all n_i are zero), is called the *vacuum*. From (C.9) it appears that a_i^+ creates a photon i whereas a_i destroys a photon i. The total number of photons is described by the operator

$$N = \sum_i a_i^+ a_i . \tag{C.14}$$

Finally, since the field has been quantized with commutators, the photons are bosons. Thus, the total number of photons i, n_i, can be greater than 1.

Remarks

(i) The energies of the states have been evaluated with respect to the vacuum. However, the absolute energy of the vacuum, equal to $\sum_i \hbar\omega_i/2$, is infinite. We return to this point below when we study vacuum fluctuations.

(ii) Instead of using the transverse plane-wave expansion, one can expand the field in multipole waves (see Complement B_I). In that case one gets, after quantization, elementary excitations, or photons, characterized by well-defined values of the energy $[\hbar\omega]$, of the square of the angular momentum, $\mathbf{J}^2$ $[J(J+1)\hbar^2]$, of J_z $[M\hbar]$, and of the parity $[+$ or $-]$.

(iii) A relativistic wave equation like the Maxwell equations, the Klein–Gordon equation, or the Dirac equation has solutions in $e^{i\omega t}$ and $e^{-i\omega t}$, which one can interpret as solutions with positive or negative energies. After second quantization of such a theory, the coefficients of the field expansion in the negative-energy solutions become the destruction operators of a particle of negative energy, which one reinterprets as the *creation* operators of an *antiparticle* of positive energy. The field operator appears then as a linear superposition of a (the destruction operator of a particle) and b^+ (the creation operator of an antiparticle). For the Maxwell field only a and a^+ arise. This is due to the fact that the Maxwell field is real and the photon coincides with its antiparticle $(b^+ = a^+)$.

c) Single-Photon States. Propagation

The creation operator $a_{\mathbf{k}}^+$, acting on the vacuum $|0\rangle$, gives a state with one photon $\mathbf{k}$. Such states can be linearly superposed to give

$$|\psi\rangle = \sum_{\mathbf{k}} c_{\mathbf{k}} a_{\mathbf{k}}^+ |0\rangle . \tag{C.15}$$

One such linear combination is an eigenstate of the operator N given in (C.14),

$$N\,|\,\psi\,\rangle = |\,\psi\,\rangle \tag{C.16}$$

but *not* of H_{trans} and $\mathbf{P}_{\text{trans}}$ [since multiple values of $\mathbf{k}$ appear in (C.15)]. It follows that in general $|\psi\rangle$ describes a single photon nonstationary state.

To discuss the propagation of such a state, consider a simple one-dimensional model. All modes appearing in the expansion (C.15) are assumed to have their wave vectors parallel to the Ox axis and the same polarization, so that $\mathbf{E}^{(+)}(\mathbf{r}, t)$ will be denoted simply as $E^{(+)}(x, t)$:

$$E^{(+)}(x, t) = \sqrt{\frac{\hbar c}{2\,\varepsilon_0\,L^3}}\sum_k \sqrt{k}\,a_k\,\mathrm{e}^{i(kx-\omega t)}\,. \tag{C.17}$$

The single counting rate $w_{\mathrm{I}}(x, t)$ in the state (C.15) is then given by

$$w_{\mathrm{I}}(x, t) = \frac{\hbar c}{2\,\varepsilon_0\,L^3}\left|\sum_k \sqrt{k}\,c_k\,\mathrm{e}^{i(kx-\omega t)}\right|^2 \tag{C.18}$$

It clearly appears in (C.18) that $w_{\mathrm{I}}(x, t)$ depends only on $x - ct$ and thus propagates without distortion at velocity c.

Remarks

(i) A measurement of $\mathbf{P}_{\text{trans}}$ on the field in the state (C.15) gives (for the x-component) the value $\hbar k$ with probability $|c_k|^2$. (We assume that $\langle\psi|\psi\rangle = \sum_k |c_k|^2 = 1$.) The quantity $|c_k|^2$ can then be considered as a probability distribution for $\mathbf{P}_{\text{trans}}$.

(ii) The single counting rate $w_{\mathrm{I}}(x, t)$ is proportional to the probability of observing a photoelectron at point x. It is tempting to think of $w_{\mathrm{I}}(x, t)$ in the one-photon subspace, as the probability for a photon to be at point x. This would introduce the idea of a "position" for the photon. To confirm such an interpretation, it would be necessary to show that it is possible to construct a complete set of localized states for the photon, that is to say, a complete set of states for which $w_{\mathrm{I}}(x, t)$ is everywhere zero except at one point. In fact, this is impossible owing to the transverse character of the field. Assume, for example, that one wants to localize a photon with a polarization parallel to the z-axis. The transverse nature of the field requires that one use only plane waves having their wave vectors in the (x, y) plane, which implies that $w_{\mathrm{I}}(r, t)$ is completely delocalized along Oz. In fact, it is impossible to define a position operator for the photon, as has already been indicated in §C.5 of Chapter I.

3. Some Properties of the Vacuum

a) Qualitative Discussion

For a real harmonic oscillator, it is well known that the fundamental commutation relation $[x, p] = i\hbar$ prevents the *simultaneous* vanishing of the potential energy (proportional to x^2) and the kinetic energy (proportional to p^2). The lowest energy state results from a "compromise" between these two energies, which vary oppositely as functions of the width of the wave function. One understands then why the ground state has an absolute energy which is not zero ("the zero-point energy", $\hbar\omega/2$), and why in this state the variances Δx^2 and Δp^2 of x and p are not zero.

The same phenomenon arises for the quantized field. The fundamental commutation relations (A.2) between a_i and a_j^+ [see also (A.15)] prevent simultaneous vanishing of the electric and magnetic energies. It follows then that the ground state of the quantum field, that is, the vacuum $|0\rangle$, has a nonzero absolute energy, and that the variances of **E** and **B** in this state are nonzero. This is a purely *quantum* effect.

b) Mean Values and Variances of the Vacuum Field

Starting with the expansion (A.3) for **E** in a_i and a_i^+ and with

$$a_i \,|\, 0 \rangle = 0 \qquad \langle 0 \,|\, a_j \, a_i^+ \,|\, 0 \rangle = \delta_{ij} \tag{C.19}$$

one sees that

$$\langle 0 \,|\, \mathbf{E}(\mathbf{r}) \,|\, 0 \rangle = 0 \tag{C.20}$$

$$\langle 0 \,|\, [\mathbf{E}(\mathbf{r})]^2 \,|\, 0 \rangle = \sum_i \mathscr{E}_{\omega_i}^2 = \sum_i \frac{\hbar\omega_i}{2\,\varepsilon_0\,L^3} \tag{C.21}$$

and finally, by replacing the discrete sum with an integral [see (C.34) of Chapter I].

$$\Delta\mathbf{E}^2(\mathbf{r}) = \frac{\hbar c}{2\,\varepsilon_0\,\pi^2} \int_0^{k_M} k^3\, \mathrm{d}k\,. \tag{C.22}$$

The variance of the electric field at a given point **r** is then, in the vacuum, proportional to $\hbar$ (quantum effect) and diverges as k_M^4 as the upper limit k_M of the integral (C.22) tends to infinity. An analogous result can be gotten for $c\mathbf{B}(\mathbf{r})$, which has in the vacuum the same mean value and the same variance as $\mathbf{E}(\mathbf{r})$.

Thus the quantum theory of radiation predicts that even in the vacuum there is at every point in space an electromagnetic field with zero mean value and infinite variance.

Remark

Rather than considering the fields at a point $\mathbf{r}$, we can average the field over a finite volume surrounding the point $\mathbf{r}$ and with linear dimension r_0. More precisely, we introduce the mean field

$$\overline{\mathbf{E}}(\mathbf{r}) = \int d^3\rho \, f(\boldsymbol{\rho}) \, \mathbf{E}(\mathbf{r} + \boldsymbol{\rho}) \tag{C.23}$$

where $f(\boldsymbol{\rho})$ is a real function (Figure 1*a*), depending only on $|\boldsymbol{\rho}|$ and of width r_0, such that

$$\int d^3\rho \, f(\boldsymbol{\rho}) = 1 . \tag{C.24}$$

Using the expansion (A.3) for $\mathbf{E}$, one then transforms (C.23) into

$$\overline{\mathbf{E}}(\mathbf{r}) = \mathrm{i} \sum_i \mathscr{E}_{\omega_i} \, a_i \, g(\mathbf{k}_i) \, \boldsymbol{\varepsilon}_i \, e^{i\mathbf{k}_i \cdot \mathbf{r}} + \text{herm. conj.} \tag{C.25}$$

where $g(\mathbf{k})$ is the Fourier transform of $f(\boldsymbol{\rho})$:

$$g(\mathbf{k}) = \int d^3\rho \, e^{i\mathbf{k}\cdot\boldsymbol{\rho}} \, f(\boldsymbol{\rho}) \tag{C.26}$$

$g(\mathbf{k})$ depends only on $|\mathbf{k}|$ and tends to zero when $|\mathbf{k}| \gg 1/r_0$. Additionally, from (C.24) and (C.26)

$$g(\mathbf{0}) = 1 . \tag{C.27}$$

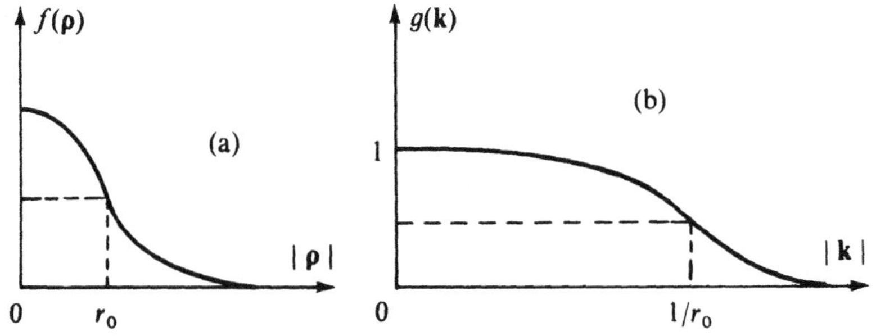

Figure 1. The shapes of the function $f(\boldsymbol{\rho})$ defining the mean field and its Fourier transform $g(\mathbf{k})$.

The shape of $g(\mathbf{k})$ is shown in Figure 1*b*. One sees then that averaging the field over a finite volume of linear extent r_0 about $\mathbf{r}$ is equivalent to introducing a "cutoff function" $g(\mathbf{k})$ in the mode expansion of the field. This cutoff suppresses the contributions of the modes with wave vectors greater than $1/r_0$. The same calculation as above then gives

$$\langle 0 | \overline{\mathbf{E}}(\mathbf{r}) | 0 \rangle = 0 \tag{C.28}$$

$$\Delta\overline{\mathbf{E}}^2(\mathbf{r}) = \frac{\hbar c}{2 \varepsilon_0 \pi^2} \int_0^\infty k^3 \, g(|\mathbf{k}|)^2 \, dk . \tag{C.29}$$

The integral of (C.29) converges if $g(|\mathbf{k}|)$ tends sufficiently rapidly toward zero when $|\mathbf{k}| \to \infty$. Thus, the variance of the mean field $\mathbf{E}$ in the vacuum can remain finite.

c) Vacuum Fluctuations

In the subsection above we have calculated the mean values of the fields and of the squares of the fields at a given instant. To study the *dynamics* of the vacuum field, it is necessary to further use the Heisenberg point of view and to calculate the *symmetric correlation functions*

$$C_{mn}(\mathbf{r}; t+\tau, t) = \\ = \frac{1}{2}\langle 0 | E_m(\mathbf{r}, t+\tau) E_n(\mathbf{r}, t) + E_n(\mathbf{r}, t) E_m(\mathbf{r}, t+\tau) | 0 \rangle \quad \text{(C.30)}$$

where $m, n = x, y, z$.

By using the expansion (B.30) for the free field $\mathbf{E}(\mathbf{r}, t)$ in the Heisenberg picture, one finds (see Complement C_{III})

$$C_{mn}(\mathbf{r}; t+\tau, t) = \delta_{mn} \frac{\hbar c}{12\,\varepsilon_0 \pi^2} \int_{-k_M}^{+k_M} |k|^3 \, e^{ick\tau} \, dk \,. \quad \text{(C.31)}$$

The correlation function C_{mn} is real and depends only on τ, because the vacuum is a stationary state whose properties are invariant under time translation. The magnitude of τ in $C_{mn}(\tau)$ is of the order of $1/ck_M$ and is thus quite small (recall that one should in principle let k_M go to infinity unless one averages the field over a finite volume of dimension r_0, which amounts to taking k_M of the order of $1/r_0$). It appears then that the vacuum fluctuations have a very short *correlation time*. It also appears from (C.31) that $C_{mn}(\tau)$ is the Fourier transform of a *spectral density* proportional to $|\omega|^3$. In Complement C_{III} we will discuss more precisely the shape of the variations of $C_{mn}(\tau)$ with τ for short τ (of the order of or less than $1/ck_M$) and for τ long with respect to $1/ck_M$ [in which case $C_{mn}(\tau)$ decreases as τ^{-4}].

The presence of a field fluctuating very rapidly about zero in the vacuum suggests interesting physical pictures for the interpretation of the spontaneous emission of radiation by an excited atom and of radiative corrections such as the Lamb shift. (*)

(*) See for example J. Dalibard, J. Dupont-Roc, and C. Cohen-Tannoudji, *J. Physique*, **43**, 1617 (1982) and the references therein.

4. Quasi-classical States

a) Introducing the Quasi-classical States

Consider the *classical* free field. Following the results of Chapter I, the state of this field is characterized by a set of normal variables $\{\alpha_i\}$. Once the set $\{\alpha_i\}$ is known, all the quantities relative to the field are known. For example:

$$H^{\text{cl}}_{\text{trans}}(\{\alpha_i\}) = \sum_i \hbar\omega_i \, \alpha_i^* \, \alpha_i \tag{C.32}$$

$$\mathbf{P}^{\text{cl}}_{\text{trans}}(\{\alpha_i\}) = \sum_i \hbar\mathbf{k}_i \, \alpha_i^* \, \alpha_i \tag{C.33}$$

$$\mathbf{E}_{\text{cl}}(\{\alpha_i\}; \mathbf{r}, t) = i\sum_i (\mathscr{E}_{\omega_i} \, \alpha_i \, \boldsymbol{\varepsilon}_i \, e^{i(\mathbf{k}_i\cdot\mathbf{r} - \omega_i t)} - \text{c.c}) \tag{C.34}$$

and so on.

For the quantized free field, the situation is more complex. Since the various observables of the field do not commute among themselves, it is impossible to find common eigenstates for these variables with eigenvalues equal to the values of the corresponding classical variables.

In this subsection, we try to find the quantum state $|\{\alpha_i\}\rangle$ which reproduces in the best possible fashion the properties of the classical state $\{\alpha_i\}$. The general idea is to seek a quantum state $|\{\alpha_i\}\rangle$ such that, for all the important observables, the mean values of these observables in the state $|\{\alpha_i\}\rangle$ coincide with the corresponding classical variables. More precisely we wish to have

$$\langle\{\alpha_i\}|H_{\text{trans}}|\{\alpha_i\}\rangle - E_{\text{vac}} = H^{\text{cl}}_{\text{trans}}(\{\alpha_i\}) \tag{C.35}$$

(we have subtracted the vacuum energy E_{vac}, since all the energies are taken with respect to the vacuum). Then

$$\langle\{\alpha_i\}|\mathbf{P}_{\text{trans}}|\{\alpha_i\}\rangle = \mathbf{P}^{\text{cl}}_{\text{trans}}(\{\alpha_i\}) \tag{C.36}$$

$$\langle\{\alpha_i\}|\mathbf{E}(\mathbf{r}, t)|\{\alpha_i\}\rangle = \mathbf{E}_{\text{cl}}(\{\alpha_i\}; \mathbf{r}, t) \tag{C.37}$$

for all $\mathbf{r}$ and all t, with analogous equations for $\mathbf{B}$ and $\mathbf{A}$.

b) Characterization of the Quasi-classical States

If the expansions (A.9), (A.10), and (B.30) for H_{trans}, $\mathbf{P}_{\text{trans}}$, and $\mathbf{E}(\mathbf{r}, t)$ in a_i and a_i^+ are substituted in the left-hand side of (C.35), (C.36), and (C.37) and compared with the expressions (C.32), (C.33), and (C.34) for $H^{\text{cl}}_{\text{trans}}$, $\mathbf{P}^{\text{cl}}_{\text{trans}}$, and $\mathbf{E}_{\text{cl}}(\{\alpha_i\}; \mathbf{r}, t)$, one finds that the conditions (C.35) to

(C.37) are equivalent to

$$\langle \{ \alpha_i \} | a_i | \{ \alpha_i \} \rangle = \alpha_i \qquad \forall i \tag{C.38}$$

$$\langle \{ \alpha_i \} | a_i^+ a_i | \{ \alpha_i \} \rangle = \alpha_i^* \alpha_i \quad \forall i. \tag{C.39}$$

We then introduce the operator

$$b_i = a_i - \alpha_i \mathbb{1} \tag{C.40}$$

where $\mathbb{1}$ is the unit operator. Equations (C.38) and (C.39) can be written

$$\langle \{ \alpha_i \} | b_i | \{ \alpha_i \} \rangle = 0 \qquad \forall i \tag{C.41}$$

$$\langle \{ \alpha_i \} | b_i^+ b_i | \{ \alpha_i \} \rangle = 0 \qquad \forall i. \tag{C.42}$$

Equation (C.42) shows that the norm of $b_i|\{\alpha_i\}\rangle$ is zero, so that the solution of (C.41) and (C.42) is

$$b_i | \{ \alpha_i \} \rangle = 0 \tag{C.43}$$

that is to say, finally,

$$a_i | \{ \alpha_i \} \rangle = \alpha_i | \{ \alpha_i \} \rangle . \tag{C.44}$$

It follows that the state $|\{\alpha_i\}\rangle$ is a product

$$| \{ \alpha_i \} \rangle = | \alpha_1 \rangle | \alpha_2 \rangle \dots | \alpha_i \rangle \dots \tag{C.45}$$

with

$$a_i | \alpha_i \rangle = \alpha_i | \alpha_i \rangle . \tag{C.46}$$

The quasi-classical state $|\{\alpha_i\}\rangle$, also called a *coherent state*, is thus the tensor product of the eigenstates of the various annihilation operators a_i with eigenvalues α_i which are precisely the corresponding classical normal variables.

From (C.46) it follows that

$$\langle \alpha_i | a_i^+ = \alpha_i^* \langle \alpha_i | \tag{C.47}$$

as well as

$$\mathbf{E}^{(+)}(\mathbf{r}, t) | \{ \alpha_i \} \rangle = \mathbf{E}_{\mathrm{cl}}^{(+)}(\{ \alpha_i \} ; \mathbf{r}, t) | \{ \alpha_i \} \rangle \tag{C.48}$$

$$\langle \{ \alpha_i \} | \mathbf{E}^{(-)}(\mathbf{r}, t) = \mathbf{E}_{\mathrm{cl}}^{(-)}(\{ \alpha_i \} ; \mathbf{r}, t) \langle \{ \alpha_i \} | \tag{C.49}$$

where $\mathbf{E}^{(+)}(\mathbf{r}, t)$ and $\mathbf{E}^{(-)}(\mathbf{r}, t)$ are the positive- and negative-frequency components of the free field in the Heisenberg picture, given in (C.2). More generally, all the observables arranged in the normal order have a mean value in the state $|\{\alpha_i\}\rangle$ equal to the value of the corresponding classical variable in the classical state $\{\alpha_i\}$.

c) Some Properties of the Quasi-classical States

We examine now several important properties of the states $|\{\alpha_i\}\rangle$ and more specifically of the eigenstate $|\alpha_i\rangle$ of the operator a_i. To simplify the notation, we will omit the index i. If additional detail is required, the reader is referred to the references at the end of the chapter.

By projecting (C.46) on the bra $\langle n-1|$, one gets the recurrence relation

$$\sqrt{n}\,\langle n \mid \alpha \rangle = \alpha \langle n-1 \mid \alpha \rangle \tag{C.50}$$

from which one can get

$$\mid \alpha \rangle = e^{-|\alpha|^2/2} \sum_{n=0}^{\infty} \frac{\alpha^n}{\sqrt{n\,!}} \mid n \rangle . \tag{C.51}$$

The probability $\mathcal{P}(n)$ of having n photons in a quasi-classical state $|\alpha\rangle$ is then given by a Poisson distribution

$$\mathcal{P}(n) = e^{-|\alpha|^2} \frac{|\alpha|^{2n}}{n\,!} \tag{C.52}$$

with mean

$$\langle n \rangle = |\alpha|^2 \tag{C.53}$$

and variance

$$\Delta n^2 = \langle n \rangle = |\alpha|^2 . \tag{C.54}$$

The orthonormalization and closure relations for the coherent states can be deduced from (C.51):

$$|\langle \beta \mid \alpha \rangle|^2 = e^{-|\beta-\alpha|^2} \tag{C.55}$$

$$\frac{1}{\pi}\int d^2\alpha \mid \alpha \rangle \langle \alpha \mid = \mathbb{1} \tag{C.56}$$

where $d^2\alpha = d\,\mathcal{Re}(\alpha)\, d\mathcal{Im}(\alpha)$.

Finally, it is possible to show that in the x-representation, a quasi-classical state is represented by a Gaussian wave packet which oscillates without deformation.

Remark

The quasi-classical states $|\alpha\rangle$ form a basis (overcomplete) in which it is possible to expand the operator density ρ of the mode. In many cases, ρ can be written in the form

$$\rho = \int d^2\alpha\, P(\alpha, \alpha^*) \mid \alpha \rangle \langle \alpha \mid \tag{C.57}$$

where $P(\alpha, \alpha^*)$ is a function of α and α^* which is *real* and *normalized*:

$$\int d^2\alpha \, P(\alpha, \alpha^*) = 1 \tag{C.58}$$

(this results from $\rho = \rho^+$ and $\mathrm{Tr}\,\rho = 1$). The function $P(\alpha, \alpha^*)$ is called the *P-representation* of the density operator (see the references at the end of the chapter as well as Exercise 5). The main interest in the P-representation is that it leads to simple expressions for the mean values of operators arranged in *normal order*, like $(a^+)^m(a)^l$. Thus the use of (C.57) in

$$\langle (a^+)^m (a)^l \rangle = \mathrm{Tr}\,\{ \rho(a^+)^m (a)^l \} \tag{C.59}$$

leads, taking account of (C.46) and (C.47), to

$$\begin{aligned} \langle (a^+)^m (a)^l \rangle &= \int d^2\alpha \, P(\alpha, \alpha^*) \, \mathrm{Tr}\,\{ | \alpha \rangle \langle \alpha | (a^+)^m (a)^l \} \\ &= \int d^2\alpha \, P(\alpha, \alpha^*) \langle \alpha | (a^+)^m (a)^l | \alpha \rangle \\ &= \int d^2\alpha \, P(\alpha, \alpha^*) \, (\alpha^*)^m (\alpha)^l \, . \end{aligned} \tag{C.60}$$

Through Equations (C.58) and (C.60), $P(\alpha, \alpha^*)$ appears as a probability density giving the distribution of possible values of α and α^*. Such an analogy is however misleading. First of all, simple results like (C.60) can only be gotten for operators arranged in normal order. They are not valid for $\langle (a)^l (a^+)^m \rangle$. Also, one can show that there are states ρ for which $P(\alpha, \alpha^*)$ can take on negative values. For this reason $P(\alpha, \alpha^*)$ is occasionally called a "*quasi-probability density*".

The P-representation allows a simple discussion of purely quantum effects. A true probability density is actually a positive definite function. As a consequence, some inequalities can be established. For some quantum states of the field, the negative values taken by $P(\alpha, \alpha^*)$ can lead to violations of these inequalities. Let us note finally that the master equation describing the damping of ρ under a relaxation process often takes the form of a Fokker–Planck equation for $P(\alpha, \alpha^*)$.

d) The Translation Operator for a and a^+

Consider the operator $T(\alpha)$ defined by

$$T(\alpha) = e^{\alpha^* a - \alpha a^+} \tag{C.61}$$

where α is a complex number. This operator is unitary, since

$$T^+(\alpha) = e^{\alpha a^+ - \alpha^* a} = T^{-1}(\alpha) \, . \tag{C.62}$$

A first interest in the operators $T(\alpha)$ and $T^+(\alpha)$ is that they allow one to relate the coherent state $|\alpha\rangle$ to the vacuum $|0\rangle$:

$$\left\{\begin{array}{l} |\,\alpha\,\rangle = T^+(\alpha)\,|\,0\,\rangle \\ |\,0\,\rangle = T(\alpha)\,|\,\alpha\,\rangle . \end{array}\right. \qquad \begin{array}{r}\text{(C.63.a)}\\ \text{(C.63.b)}\end{array}$$

To prove (C.63.a) it suffices to use the identity

$$e^{A+B} = e^A\, e^B\, e^{-\frac{1}{2}[A,B]} \tag{C.64}$$

valid if A and B commute with $[A, B]$. By taking $A = \alpha a^+$ and $B = -\alpha^* a$, and therefore $[A, B] = -|\alpha|^2[a^+, a] = |\alpha|^2$, which indeed commutes with a and a^+, we can transform (C.62) into

$$T^+(\alpha) = e^{\alpha a^+}\, e^{-\alpha^* a}\, e^{-|\alpha|^2/2} . \tag{C.65}$$

We then let (C.65) act on the vacuum after having expanded the first two exponentials of (C.65) in power series. By using $a|0\rangle = 0$ and $(a^+)^n|0\rangle = \sqrt{n!}\,|n\rangle$ we again get the expansion (C.51), which then proves (C.63).

Another interest of the unitary transformation T is that it leads to very simple results for the transforms of a and a^+:

$$\left\{\begin{array}{l} T(\alpha)\,a T^+(\alpha) = a + \alpha \\ T(\alpha)\,a^+\,T^+(\alpha) = a^+ + \alpha^* . \end{array}\right. \qquad \begin{array}{r}\text{(C.66.a)}\\ \text{(C.66.b)}\end{array}$$

To prove (C.66.a), one starts from

$$aT^+(\alpha) = T^+(\alpha)\,a + [a, T^+(\alpha)] \tag{C.67}$$

and uses (C.65) as well as $[a, f(a^+)] = \mathrm{d}f/\mathrm{d}a^+$ to calculate the commutator of (C.67). We get

$$aT^+(\alpha) = T^+(\alpha)\,a + \alpha T^+(\alpha) . \tag{C.68}$$

It suffices then to multiply both sides of (C.68) on the left by T and to use $TT^+ = 1$ to get (C.66.a). In what follows we will often use the equalities (C.66), which show that T is a translation operator for a and a^+.

D—THE HAMILTONIAN FOR THE INTERACTION BETWEEN PARTICLES AND FIELDS

1. Particle Hamiltonian, Radiation Field Hamiltonian, Interaction Hamiltonian

For what follows, it is useful to separate the Hamiltonian of the global system into three parts,

$$H = H_P + H_R + H_I \tag{D.1}$$

where H_P depends only on the variables $\mathbf{r}_\alpha$ and $\mathbf{p}_\alpha$ of the particles (the particle Hamiltonian), H_R depends only on the variables a_i and a_i^+ of the field (the radiation field Hamiltonian), and H_I depends simultaneously on $\mathbf{r}_\alpha$, $\mathbf{p}_\alpha$, a_i, and a_i^+ (the interaction Hamiltonian). Starting with (A.16) for H, one gets

$$H_P = \sum_\alpha \frac{\mathbf{p}_\alpha^2}{2\,m_\alpha} + V_{\text{Coul}} \tag{D.2}$$

$$H_R = H_{\text{trans}} = \sum_i \hbar\omega_i\left(a_i^+\,a_i + \frac{1}{2}\right) \tag{D.3}$$

$$H_I = H_{I1} + H_{I2} \tag{D.4}$$

where H_{I1} is linear with respect to the fields:

$$H_{I1} = -\sum_\alpha \frac{q_\alpha}{m_\alpha}\mathbf{p}_\alpha \cdot \mathbf{A}(\mathbf{r}_\alpha)\,. \tag{D.5}$$

[we have used the transversality of $\mathbf{A}$ in the Coulomb gauge, which implies $\mathbf{p}_\alpha \cdot \mathbf{A}(\mathbf{r}_\alpha) = \mathbf{A}(\mathbf{r}_\alpha) \cdot \mathbf{p}_\alpha$)], and H_{I2} quadratic:

$$H_{I2} = \sum_\alpha \frac{q_\alpha^2}{2\,m_\alpha}[\mathbf{A}(\mathbf{r}_\alpha)]^2\,. \tag{D.6}$$

Thus far, we have only considered charged particles without internal degrees of freedom. It is possible to remove this restriction and to add to the particle observables $\mathbf{r}_\alpha$ and $\mathbf{p}_\alpha$ the spin operators $\mathbf{S}_\alpha$. Because of the magnetic moment associated with such a spin, that is,

$$\mathbf{M}_\alpha^S = g_\alpha \frac{q_\alpha}{2\,m_\alpha}\mathbf{S}_\alpha \tag{D.7}$$

where g_α is the Lande factor of particle α, a new term must be added to H_I,

$$H_{I1}^S = -\sum_\alpha \mathbf{M}_\alpha^S \cdot \mathbf{B}(\mathbf{r}_\alpha) \tag{D.8}$$

representing the coupling of the magnetic spin moments of the various particles with the radiation magnetic field **B** evaluated at the points where the particles are located.

Remarks

(i) In the presence of external fields, it is necessary on one hand to start with the expression (A.17) for H and on the other to add the coupling of the spin magnetic moments with the external magnetic field. Since $\mathbf{A}_e(\mathbf{r}, t)$, $U_e(\mathbf{r}, t)$, and $\mathbf{B}_e(\mathbf{r}, t)$ are *classical* variables with a prescribed time dependence, the operators gotten by replacing $\mathbf{r}$ with the operator $\mathbf{r}_\alpha$ in these variables are atomic operators. One then gets for H_P and H_I the following new expressions:

$$H_P = \sum_\alpha (\mathbf{p}_\alpha^e)^2/2\,m_\alpha + V_{\text{Coul}} + \sum_\alpha q_\alpha\, U_e(\mathbf{r}_\alpha, t) - \sum_\alpha \mathbf{M}_\alpha^S \cdot \mathbf{B}_e(\mathbf{r}_\alpha, t) \tag{D.9}$$

where

$$\mathbf{p}_\alpha^e = \mathbf{p}_\alpha - q_\alpha\, \mathbf{A}_e(\mathbf{r}_\alpha, t) \tag{D.10}$$

$$H_I = H_{I1} + H_{I1}^S + H_{I2} \tag{D.11}$$

with

$$H_{I1} = -\sum_\alpha \frac{q_\alpha}{m_\alpha} \mathbf{p}_\alpha^e \cdot \mathbf{A}(\mathbf{r}_\alpha) \tag{D.12}$$

H_{I1}^S and H_{I2} are given by the same expressions as in (D.8) and (D.6).

(ii) All of the spin-dependent terms introduced in this section have been introduced heuristically. For electrons they can be justified by examining the nonrelativistic limit of the Dirac equation (see also Complement B_V). One then finds that the g-factor for an electron is equal to 2.

2. Orders of Magnitude of the Various Interaction Terms for Systems of Bound Particles

Consider first of all the ratio H_{I2}/H_{I1}. The orders of magnitude of A and p are taken equal to their root mean square values in the state considered:

$$\frac{H_{I2}}{H_{I1}} \simeq \frac{q^2 A^2/m}{qAp/m} = \frac{qAp/m}{p^2/m} \simeq \frac{H_{I1}}{H_P}. \tag{D.13}$$

For low-intensity radiation, the ratio H_{I1}/H_P is small, which implies that H_{I2}/H_{I1} is also small. On the other hand, at very high intensity, where the incident radiation field becomes of the order of the atomic field, H_{I2} can become of the same order as or larger than H_{I1}.

Note however that in certain scattering processes such as Rayleigh, Thomson, or Compton scattering, H_{I2} can arise in the first-order perturbation theory (since a single matrix element of H_{I2} is sufficient to describe the two-photon process corresponding to the destruction of the incident photon and the creation of the scattered photon), whereas H_{I1} only plays a role in second order (the matrix elements of H_{I1} correspond to one-photon processes, and two of them are necessary to describe a scattering process). Even if H_{I2} is much smaller than H_{I1}, the contribution of H_{I2} in first order can be of the same order of magnitude as that of H_{I1} in second order or even greater.

Remark

For a free particle or a weakly bound particle, H_{I2} (more precisely, the diagonal matrix elements of H_{I2}) can be interpreted as being the vibrational kinetic energy of the particle in the radiation field. For a field of frequency ω, we have $A \simeq E/\omega$, and

$$H_{I2} \simeq \frac{q^2 A^2}{2m} \simeq \frac{q^2 E^2}{2m\omega^2}. \tag{D.14}$$

One recognizes in (D.14) the kinetic energy of a particle vibrating in a field E of frequency ω with an amplitude $qE/m\omega^2$ and a velocity $qE/m\omega$. Note that such a picture remains valid when the particle interacts with the vacuum field. It is necessary in that case to replace E by $\mathscr{E}_\omega$, and H_{I2} then represents the vibrational kinetic energy of the particle in the vacuum fluctuations.

Consider now the ratio H_{I1}^S/H_{I1}. By using (D.7), (D.8), and the fact that $B \sim kA$ (since $\mathbf{B} = \nabla \times \mathbf{A}$), one gets

$$\frac{H_{I1}^S}{H_{I1}} \simeq \frac{q\hbar B/m}{qAp/m} \simeq \frac{\hbar k A}{pA} = \frac{\hbar k}{p} \tag{D.15}$$

—that is, the ratio between the momenta $\hbar k$ of the photon and p of the particle. For low-energy photons and a bound electron (for example, in the optical or microwave domains), such a ratio is very small compared to 1.

3. Selection Rules

In the absence of external fields (or in the presence of external fields invariant under spatial translation), one can show, starting with the commutation relations (A.1) and (A.2), that

$$[\mathbf{P}, H] = \mathbf{0} = [\mathbf{P}, H_P] = [\mathbf{P}, H_R] = [\mathbf{P}, H_I] \tag{D.16}$$

where $\mathbf{P}$ is the total momentum given in (A.18) (see Exercise 3). This result can be understood physically if one notes that $\mathbf{P}$ is the infinitesimal generator of the spatial translations of the global system field + particles. The fact that an operator commutes with $\mathbf{P}$ is a consequence of the invariance of the corresponding physical variable under a translation of the global system. Thus the velocity of each particle does not change under a translation, so that the total kinetic energy [first line of (A.16)] remains unchanged. Likewise, a global translation does not change the distances between particles, and thus their Coulomb energy [second line of (A.16)] remains unchanged, as does the integral over all space of $\mathbf{E}^2 + c^2\mathbf{B}^2$ [third line of (A.16)]. Finally, to understand why H_I commutes with $\mathbf{P}$, it suffices to note that $\mathbf{A}(\mathbf{r}_\alpha)$ and $\mathbf{B}(\mathbf{r}_\alpha)$ do not change when one translates by the same amount the fields $\mathbf{A}$ and $\mathbf{B}$ and the point $\mathbf{r}_\alpha$ where these fields are evaluated.

It follows from (D.16) that the total momentum is, as in the classical theory, a constant of the motion in the Heisenberg picture:

$$\frac{\mathrm{d}}{\mathrm{d}t}\mathbf{P}(t) = \mathbf{0}\,. \tag{D.17}$$

Another consequence of (D.16) is that H_I has nonzero matrix elements only between eigenstates of $H_P + H_R$ having the same total momentum. (Since $\mathbf{P}$ commutes with $H_P + H_R$, one can use a basis of eigenvectors common to $\mathbf{P}$ and $H_P + H_R$.) Such a selection rule implies the conservation of the total momentum during the absorption or emission of photons by a system of particles. Combined with the conservation of energy, which arises when one solves the Schrödinger equation over sufficiently long times, the conservation of the total momentum allows one to explain important physical effects such as the Doppler effect and the recoil shift.

One could in the same way show that the total angular momentum given in (A.19) commutes with H, H_P, H_R, and H_I in the absence of external fields or in the presence of external fields invariant under rotation. One consequence of this result is that H_I has nonzero matrix elements only between eigenstates of $H_P + H_R$ having the same total angular momentum (conservation of angular momentum during the absorption or emission of photons by systems of charged particles).

4. Introduction of a Cutoff (*)

The Hamiltonians (A.16) and (A.17) studied in this chapter are only valid for slow (nonrelativistic) particles. They do not correctly describe the interaction of such particles with the "relativistic modes" of the field, that is, modes for which $\hbar\omega \gtrsim mc^2$, since such interactions would impart high

(*) An analogous cutoff has already been introduced in §C.5.*d* of Chapter II for the standard interaction Lagrangian. For the reader who has not read Chapter II, we follow an analogous path here, but work directly with the interaction Hamiltonian without reference to the Lagrangian.

velocities to the particles or even create new particles as in electron–positron pair production.

Now, the mode expansion of the fields **A** and **B** appearing in the interaction Hamiltonian examined above contains arbitrarily high-frequency modes. Rather than retaining the coupling terms with the relativistic modes which are certainly inexact, we prefer here to eliminate them from the interaction Hamiltonian. This is accomplished by introducing a cutoff in all the field expansions appearing in H_I. More precisely, all the summations on $\mathbf{k}_i$ are limited to

$$|k_i| = k_i \leqslant k_c \tag{D.18}$$

with

$$\hbar c k_c = \hbar \omega_c \ll m_\alpha c^2 \tag{D.19}$$

The cutoff frequency ω_c is chosen large with respect to the characteristic frequencies ω_0 of the particle motion:

$$\omega_0 \ll \omega_c \ll m_\alpha c^2/\hbar \tag{D.20}$$

so as to keep in H_I a sufficiently large spectral interval to correctly describe the important electromagnetic interactions of the particles, in particular the resonant absorptions or emissions of photons. In fact, with the cutoff (D.18) we abandon for the time being the description of the effect on the particles of "virtual" emissions and reabsorptions of high-frequency photons.

At this point in the discussion, it is convenient to recall that by reexpressing the longitudinal field as a function of the coordinates of the particles, $\mathbf{r}_\alpha$, we have in fact included in the Hamiltonian of the particles a part of the electromagnetic interactions of the particles [the Coulomb interaction term V_{Coul} of (A.16)]. To be consistent with the above, we must also introduce the same cutoff in V_{Coul}. For this, we return to the calculation of V_{Coul} in reciprocal space, that is to say to Equation (B.35) of Chapter I, giving V_{Coul} in the form of an integral on k, and we introduce an upper limit k_c in this integral. The Coulomb self-energy of particle α, $\varepsilon^\alpha_{\text{Coul}}$, becomes finite and equal to

$$\varepsilon^\alpha_{\text{Coul}} = \frac{q_\alpha^2\, k_c}{4\,\pi^2\,\varepsilon_0}. \tag{D.21}$$

One gets in fact the Coulomb energy of a charge q_α distributed over a volume of linear extent $1/k_c$, which is not surprising, since a cutoff at k_c is equivalent to a spatial average on a volume k_c^{-3} (see the remark in §C.3.*b* above). As to the Coulomb interaction term between pairs of particles, it remains practically unchanged if $1/k_c$ is small compared to the distances between particles, which we assume to be the case here.

We illustrate the previous discussion on the simplest atomic system, the hydrogen atom. Figure 2, which is not to scale, gives several important characteristic energies $\hbar\omega$ and the corresponding wavelengths $\lambdabar = c/\omega = 1/k$.

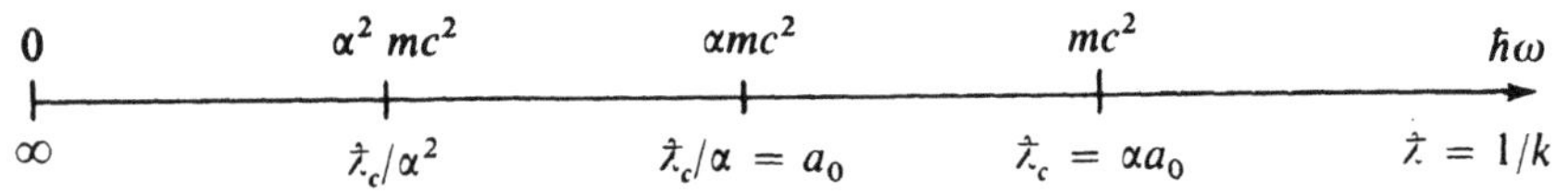

Figure 2. Various energies and characteristic wavelengths for the hydrogen atom. (Not to scale.)

A first important energy is the rest energy, mc^2, of the electron. From (D.19), the cutoff energy $\hbar\omega_c$ should be to the left of mc^2. The wavelength corresponding to mc^2 is the Compton wavelength $\lambdabar_c$, smaller than the Bohr radius a_0 by a factor $\alpha = \frac{1}{137}$ (α is the fine structure constant). For a cutoff energy $\hbar\omega_c$ of the order of mc^2, the charges are distributed over a volume of linear extent $1/k_c = \lambdabar_c$, small with respect to the mean distance a_0 between the two charges, and it is therefore legitimate to neglect the modifications of the Coulomb interaction energy between the electron and the proton.

The characteristic atomic energies, denoted $h\omega_0$ above, are here of the order of the ionization energy, that is, of the order of α^2mc^2. Such an energy is smaller than mc^2 by at least four orders of magnitude, and one thus has no difficulty in finding a cutoff frequency ω_c satisfying (D.20).

Finally, since the Bohr radius a_0 is equal to $\lambdabar_c/\alpha$, a wavelength of the order of a_0 corresponds to an energy $\hbar\omega$ of the order of αmc^2, much smaller than mc^2 but also much larger than α^2mc^2. The interval 0 to αmc^2 of Figure 2, which is very large compared to the typical atomic energy α^2mc^2, then corresponds to wavelengths large with respect to atomic dimensions. If one is interested only in the interaction of the hydrogen atom with photons of energy falling within this interval, it is possible to do the "long-wavelength approximation", which consists of neglecting the spatial variation of the electromagnetic field on the length scale of the atomic system (*) The consequences of such an approximation on the Hamiltonian of one or many localized systems of charges are treated in Complement A_{IV}.

(*) Of course, there are physical effects, such as the existence of quadrupole transitions, which are related to the first-order corrections to the long-wavelength approximation.

General references and supplementary reading

General works: Kroll, Heitler, Power, Loudon, Haken.

For photoelectric signals (quantum theory) see Glauber, Nussenzveig; see also Cohen-Tannoudji, Dupont-Roc, and Grynberg (Complement A_{II}). For the semiclassical theory of photoelectric signals see Lamb and Scully, (p. 363); L. Mandel, *Prog. Opt.*, **XIII**, 27 (1976); L. Mandel, E. C. G. Sudarshan, and E. Wolf, *Proc. Phys. Soc.*, **84**, 435 (1964).

For quasi-classical states and quasi-probability densities see Glauber; Cohen-Tannoudji, Diu, and Laloë (Complement G_V); Sargent, Scully, and Lamb (Chapters XI and XVI); Klauder and Skagerstam.

COMPLEMENT A_{III}

THE ANALYSIS OF INTERFERENCE PHENOMENA IN THE QUANTUM THEORY OF RADIATION

The wave aspect of light is seen experimentally through the ability of light to give rise to interference. A second equally essential aspect, the particle aspect, is observed through the discrete character of energy and momentum exchanges between matter and radiation.

It is indeed for light that the idea of wave–particle duality was first introduced by Einstein in 1909 and subsequently extended to all physical objects. It is therefore not surprising that the majority of textbooks on quantum physics begin with a discussion of the inseparability of the wave and particle aspects of light in order to introduce subsequently the simple idea that these two aspects can be integrated into a quantum description of phenomena where "the wave allows one to calculate the probability of finding the particle".

Although it is appealing to refer to this well-known example of light, such a discussion presents the drawback of suggesting (in spite of warnings) the false idea that the Maxwell waves are the wave functions of the photon. It is unfortunate that the interference phenomena most familiar to physicists occur for a system, light, for which there is no nonrelativistic approximation: one can introduce nonrelativistic wave functions $\psi(\mathbf{r})$ for a slow electron, but not for a photon, which is fundamentally relativistic.

It is therefore worthwhile to reconsider here the analysis of interference phenomena and wave–particle duality in the general framework of the quantum theory of radiation which has been established in Chapter III. We now have at our disposal all of the elements allowing a thorough discussion of several problems concerning interference phenomena. Can one construct states $|\psi\rangle$ of the quantum field such that local signals, like the photoelectric detection signal $w_I(\mathbf{r}, t)$, vary sinusoidally as a function of $\mathbf{r}$ with fixed t (or as a function of t with fixed $\mathbf{r}$)? Can one observe interference fringes on the double counting signal w_{II}? Can one observe fringes with two independent light beams? What happens if one has only a single photon in the field? Since the Maxwell wave is not the photon wave function, what are the quantities which interfere?

We begin (§A_{III}.1) by establishing the simple model used in this complement to discuss interference. We subsequently analyze the interference phenomena observable in single (§A_{III}.2) and double (§A_{III}.3) counting experiments. We can then interpret the results obtained in terms of interferences between transition amplitudes (§A_{III}. 4) and we conclude by summarizing how the quantum theory of radiation describes the wave-particle duality (§A_{III}.5).

1. A Simple Model

In order to simplify the calculations, we assume that the free field has been prepared in a state where only two modes, 1 and 2, contain photons, all the other modes $i \neq 1, 2$ being empty:

$$|\psi\rangle = |\psi_{12}\rangle \otimes \prod_{i \neq 1,2} |0_i\rangle . \tag{1}$$

Such a situation can be realized by reflecting a parallel beam of light from two plane mirrors with a small angle between them (Fresnel mirrors) or by using two independent laser beams. The two modes 1 and 2 are assumed to have the same polarization, so that we may ignore the vector character of the field in the following.

The most general form of the state vector $|\psi_{12}\rangle$ appearing in (1) is a linear superposition of basis states $|n_1, n_2\rangle$ relative to the set of modes 1 and 2:

$$|\psi_{12}\rangle = \sum_{n_1 n_2} c_{n_1 n_2} |n_1, n_2\rangle . \tag{2}$$

It can happen that $|\psi_{12}\rangle$ can be factored (particularly when one uses two independent laser sources) as

$$|\psi_{12}\rangle = |\psi_1\rangle |\psi_2\rangle . \tag{3}$$

A quasi-classical state is a special case of (3) (see §C.4 of this chapter):

$$|\psi_{12}\rangle = |\alpha_1\rangle |\alpha_2\rangle . \tag{4}$$

It will also be useful to consider the single-photon state

$$|\psi_{12}\rangle = c_1 |1_1, 0_2\rangle + c_2 |0_1, 1_2\rangle \tag{5}$$

in which the photon has a probability amplitude c_1 to be in mode 1 and c_2 to be in mode 2.

Remark

The state of the field is described in (1) by a state vector (pure state). More generally, one will rather use a density operator (a statistical mixture of states)

$$\rho = \rho(1, 2) \otimes \prod_{i \neq 1,2} (|0_i\rangle \langle 0_i|) . \tag{6}$$

If $\rho(1,2)$ has a P-representation (see the remark at the end of §C.4.c of this chapter), one has

$$\rho(1, 2) = \int d^2\alpha_1 \, d^2\alpha_2 \, P(\alpha_1, \alpha_2) \, | \, \alpha_1, \alpha_2 \rangle \langle \alpha_1, \alpha_2 \, | \tag{7}$$

where $|\alpha_1, \alpha_2\rangle$ is the quasi-classical state (4) and where $P(\alpha_1, \alpha_2)$ [a simplified notation for $P(\alpha_1, \alpha_1^*, \alpha_2, \alpha_2^*)$] is a quasi-probability density, real and normalized but not necessarily positive. The field state appears then as a "statistical mixture" of quasi-classical states with a "weighting function" $P(\alpha_1, \alpha_2)$ not necessarily positive.

The free field being in the state (1) [or (6)], we now assume that we place a photodetector at $\mathbf{r}$. How does the single counting rate $w_{\mathrm{I}}(\mathbf{r}, t)$ (see §C.1.c of this chapter) vary as a function of $\mathbf{r}$ for t fixed (or as a function of t for $\mathbf{r}$ fixed)? We can also set two photodetectors, one at $\mathbf{r}$ and one at $\mathbf{r}'$, and study the *correlations* between their signals. More precisely, how does the double counting rate $w_{\mathrm{II}}(\mathbf{r}, t; \mathbf{r}', t')$ (defined in §C.1.c above) vary as a function of $\mathbf{r} - \mathbf{r}'$ for $t = t'$ (or as a function of $t - t'$ for $\mathbf{r} = \mathbf{r}'$)?

Since w_{I} and w_{II} are the mean values of products of operator arranged in the normal order [see Equations (C.1) and (C.5) above], and since all modes $i \neq 1, 2$ are empty, the contributions of the modes $i \neq 1, 2$ in the expansions (in a_i and a_i^+) of the field operators $E^{(+)}$ and $E^{(-)}$ appearing in w_{I} and w_{II} vanish. This occurs because the operators a_i with $i \neq 1, 2$ appearing in $E^{(+)}$ give zero when they act on $|0_i\rangle$. The situation is similar for the operators a_i^+ appearing in $E^{(-)}$ and acting on $\langle 0_i|$. Thus, in all the following calculations it is sufficient to keep only modes 1 and 2 in the expansions of $E^{(+)}$ and $E^{(-)}$, and to write

$$E^{(+)}(\mathbf{r}, t) = E_1^{(+)}(\mathbf{r}, t) + E_2^{(+)}(\mathbf{r}, t) = \mathscr{E}_1 \, a_1 \, e^{i(\mathbf{k}_1 \cdot \mathbf{r} - \omega_1 t)} + \mathscr{E}_2 \, a_2 \, e^{i(\mathbf{k}_2 \cdot \mathbf{r} - \omega_2 t)} \tag{8}$$

where, using (B.30) and (A.6),

$$\mathscr{E}_1 = i \left[\frac{\hbar\omega_1}{2 \, \varepsilon_0 \, L^3} \right]^{1/2}, \qquad \mathscr{E}_2 = i \left[\frac{\hbar\omega_2}{2 \, \varepsilon_0 \, L^3} \right]^{1/2} \tag{9}$$

2. Interference Phenomena Observable with Single Photodetection Signals

a) The General Case

By substituting (8) and the adjoint expression for $E^{(-)}$ in the expression for w_{I} one gets

$$\begin{aligned} w_{\mathrm{I}} &= \langle \psi_{12} \, | (E_1^{(-)} + E_2^{(-)}) (E_1^{(+)} + E_2^{(+)}) \, | \, \psi_{12} \rangle = \\ &= \langle \psi_{12} \, | \, E_1^{(-)} E_1^{(+)} \, | \, \psi_{12} \rangle + \langle \psi_{12} \, | \, E_2^{(-)} E_2^{(+)} \, | \, \psi_{12} \rangle + \\ &\quad + \langle \psi_{12} \, | \, E_1^{(-)} E_2^{(+)} \, | \, \psi_{12} \rangle + \langle \psi_{12} \, | \, E_2^{(-)} E_1^{(+)} \, | \, \psi_{12} \rangle \end{aligned} \tag{10}$$

which can also be written

$$w_I(\mathbf{r}, t) = |\mathscr{E}_1|^2 \langle \psi_{12} | a_1^+ a_1 | \psi_{12} \rangle + 1 \rightleftarrows 2 + \\ + 2 \mathcal{R}e\, \mathscr{E}_1^* \mathscr{E}_2 \langle \psi_{12} | a_1^+ a_2 | \psi_{12} \rangle \times \\ \times e^{i[(\mathbf{k}_2 - \mathbf{k}_1).\mathbf{r} - (\omega_2 - \omega_1)t]} \quad (11)$$

If $\langle \psi_{12} | a_1^+ a_2 | \psi_{12} \rangle$ is nonzero, it appears from (11) that the single counting signal has a sinusoidal dependence on **r** for fixed t, so that an interference phenomenon can be observed.

b) Quasi-classical States

Assume that $|\psi_{12}\rangle$ is a quasi-classical state (4). Since $|\alpha_1, \alpha_2\rangle$ is an eigenket of $E_1^{(+)}(\mathbf{r}, t)$ and $E_2^{(+)}(\mathbf{r}, t)$ with eigenvalues equal to $E_{1\,\mathrm{cl}}^{(+)}(\{\alpha_1\}; \mathbf{r}, t)$ and $E_{2\,\mathrm{cl}}^{(+)}(\{\alpha_2\}; \mathbf{r}, t)$, and since $\langle \alpha_1 \alpha_2|$ is the eigenbra of the adjoint operators with the conjugate eigenvalues [see (8) and the equations (C.48) and (C.49) above], (11) becomes

$$w_I(\mathbf{r}, t) = | E_{1\,\mathrm{cl}}^{(+)}(\{ \alpha_1 \}; \mathbf{r}, t) + E_{2\,\mathrm{cl}}^{(+)}(\{ \alpha_2 \}; \mathbf{r}, t) |^2. \quad (12)$$

For a quasi-classical state, $w_I(\mathbf{r}, t)$ thus appears as the square of the modulus of the superposition of two classical Maxwell waves. In this particular case, it is possible to argue in terms of classical electromagnetic waves and to make them interfere to calculate the probability that the photon manifests its presence at **r**, t.

c) Factored States

Assume now that $|\psi_{12}\rangle$ is a factored state (3) (as is the case with two independent laser beams). The general expression (11) then becomes

$$w_I(\mathbf{r}, t) = |\mathscr{E}_1|^2 \langle \psi_1 | a_1^+ a_1 | \psi_1 \rangle + 1 \rightleftarrows 2 + \\ + 2 \mathcal{R}e\, \mathscr{E}_1^* \mathscr{E}_2 \langle \psi_1 | a_1^+ | \psi_1 \rangle \langle \psi_2 | a_2 | \psi_2 \rangle \times \\ \times e^{i[(\mathbf{k}_2 - \mathbf{k}_1).\mathbf{r} - (\omega_2 - \omega_1)t]}. \quad (13)$$

The interference fringes exist only if $\langle \psi_1 | a_1 | \psi_1 \rangle$ and $\langle \psi_2 | a_2 | \psi_2 \rangle$ are simultaneously nonzero, that is, if the mean values of the fields $\langle E_1 \rangle$ and $\langle E_2 \rangle$ are both nonzero.

In particular, if

$$|\psi_{12}\rangle = |n_1\rangle |n_2\rangle, \quad (14)$$

that is, if the number of photons in each mode has a well-defined value, then there are no fringes, since $\langle n_1|a_1|n_1\rangle = \langle n_2|a_2|n_2\rangle = 0$. A state $|n\rangle$ is in some way the quantum analogue of a single-mode classical field of well-defined energy but with a random phase equally distributed from 0 to 2π. Since the relative phases of the fields of the two beams are not well defined, interference cannot be observed.

The foregoing discussion shows that it is in principle possible to observe interference fringes out of w_I with two *independent* lasers. Actually, at a given instant, the phases of the fields of the two lasers have well-defined values. However, in practice, these phases evolve independently as a result of noise in the atomic amplifiers. After a time interval of the order of τ_d, the phase diffusion time, the relative phase between the two lasers has lost any memory of its initial value (*). To observe the fringes it is then necessary that the observation time be short with respect to τ_d in order to keep a well-defined phase and to prevent the washing out of the fringes.

Note that in general

$$\langle \psi_1 | a_1^+ a_1 | \psi_1 \rangle \neq \langle \psi_1 | a_1^+ | \psi_1 \rangle \langle \psi_1 | a_1 | \psi_1 \rangle \tag{15}$$

so that

$$w_I(\mathbf{r}, t) \neq |\langle E_1 \rangle + \langle E_2 \rangle|^2 . \tag{16}$$

The single counting rate cannot then be thought of as arising from the interference between two mean fields $\langle E_1\rangle$ and $\langle E_2\rangle$.

d) Single-Photon States

Consider the single-photon state (5). The action of $E_1^{(+)}(\mathbf{r}, t)$ on such a state gives

$$E_1^{(+)}(\mathbf{r}, t) | \psi_{12} \rangle = c_1 \, \mathscr{E}_1 \, e^{i[\mathbf{k}_1 \cdot \mathbf{r} - \omega_1 t]} | 0_1, 0_2 \rangle . \tag{17}$$

Destroying a photon in a state which contains only one photon gives the vacuum. An analogous expression can be gotten for $E_2^{(+)}(\mathbf{r}, t)$, so that finally

$$w_I(\mathbf{r}, t) = | c_1 \, \mathscr{E}_1 \, e^{i[\mathbf{k}_1 \cdot \mathbf{r} - \omega_1 t]} + c_2 \, \mathscr{E}_2 \, e^{i[\mathbf{k}_2 \cdot \mathbf{r} - \omega_2 t]} |^2 . \tag{18}$$

(*) In the Fresnel mirror experiment, the two beams which interfere are gotten from the same initial beam by wavefront division. The relative phase of the two beams then remains fixed even if the phase of the initial beam fluctuates. This explains why it is so much easier to observe fringes in this case.

It appears then that it is quite possible to observe interference with a quantized field which contains only one photon, provided that this photon has a nonzero amplitude to be in two different modes (in practice the experiment should be repeated several times with the same initial conditions, since each detection destroys the photon).

Remarks

(i) For a single-photon state (5), one can show that

$$\langle \psi_{12} \mid E_1 \mid \psi_{12} \rangle = \langle \psi_{12} \mid E_2 \mid \psi_{12} \rangle = 0 . \tag{19}$$

It follows that the wave $c_1 \mathscr{E}_1 e^{i(\mathbf{k}_1 \cdot \mathbf{r} - \omega_1 t)}$ appearing in (18) is *not* the mean electric field in mode 1, which is zero [this is also true for the other term in (18)]. The two waves which interfere in (18) are not related to mean fields.

(ii) Interference fringes would not arise in a state which was a statistical mixture of the states $|1_1, 0_2\rangle$ and $|0_1, 1_2\rangle$ with weights $|c_1|^2$ and $|c_2|^2$. The phase relationship between the two states appearing in the expansion (5) is essential for the appearance of fringes (this phase is fixed by the argument of the complex number $c_1^* c_2$).

3. Interference Phenomena Observable with Double Photodetection Signals.

If one replaces the four operators $E^{(\pm)}$ appearing in the expression for w_{II} by $E_1^{(\pm)} + E_2^{(\pm)}$, one gets $2^4 = 16$ terms with various sinusoidal dependences on the variables $(\mathbf{r}, t)$, $(\mathbf{r}', t')$, $(\mathbf{r} + \mathbf{r}', t + t')$, and $(\mathbf{r} - \mathbf{r}', t - t')$. It thus appears that interference phenomena can quite generally be observed with double counting signals. We now look at several cases in more detail.

a) Quasi-classical States

Starting from (4), (8) and Equations (C.48) (C.49) above, one gets for such states

$$w_{II} = I_{cl}(\{ \alpha_1, \alpha_2 \}; \mathbf{r}, t)\, I_{cl}(\{ \alpha_1, \alpha_2 \}; \mathbf{r}', t') \tag{20}$$

where I_{cl} $(\{\alpha_1, \alpha_2\}; \mathbf{r}t)$ is the classical intensity

$$I_{cl}(\{ \alpha_1, \alpha_2 \}; \mathbf{r}, t) = \left| E_{1\,cl}^{(+)}(\{ \alpha_1 \}; \mathbf{r}, t) + E_{2\,cl}^{(+)}(\{ \alpha_2 \}; \mathbf{r}, t) \right|^2 . \tag{21}$$

In this particular case, w_{II} is the product of two classical intensities in $\mathbf{r}, t$ and $\mathbf{r}', t'$. It appears then that when the field is in a quasi-classical state, the results of the quantum theory of radiation coincide with those of semiclassical theories (see the remark at the end of §C.1.*c*).

b) SINGLE-PHOTON STATES

For such states one finds that

$$w_{II}(\mathbf{r}, t; \mathbf{r}', t') = 0 \quad \forall \mathbf{r}, t; \mathbf{r}', t'. \tag{22}$$

The first of the two operators $E^{(+)}$ appearing in the expression for w_{II} gives the vacuum $|0_1, 0_2\rangle$ when it acts on a single-photon state. The second operator $E^{(+)}$, acting on the vacuum, then gives zero. This result expresses the fact that, physically, it is impossible to detect two photons in a state containing only one.

Whereas it is possible to observe a nonzero single counting signal in the state (5) (see §A_{III}.2.d above), the double counting signal w_{II} is identically zero for *all* values of $\mathbf{r}$, t and $\mathbf{r}'$, t'. Such a situation can never arise with a classical field. It is impossible to find a classical field E_{cl} such that $w_I^{cl} \neq 0$ and $w_{II}^{cl} \equiv 0$ for *all* $\mathbf{r}$, t and *all* $\mathbf{r}'$, t'.

We will give another example of a typical quantum situation. In the experiment shown in Figure 1, a single photon emitted by a single initially excited atom passes a half-transmitting mirror, and the signals registered by the photomultipliers A and B symmetrically placed with respect to the mirror are analyzed. The quantum theory predicts that one can observe a photoelectron at A *or* B but *never* at A *and* B. A semiclassical theory, in contrast, predicts possible coincidences between A and B, since the two photodetectors are simultaneously subjected to two wave packets resulting from the division of the initial wave packet into two packets by the half-transmitting mirror.

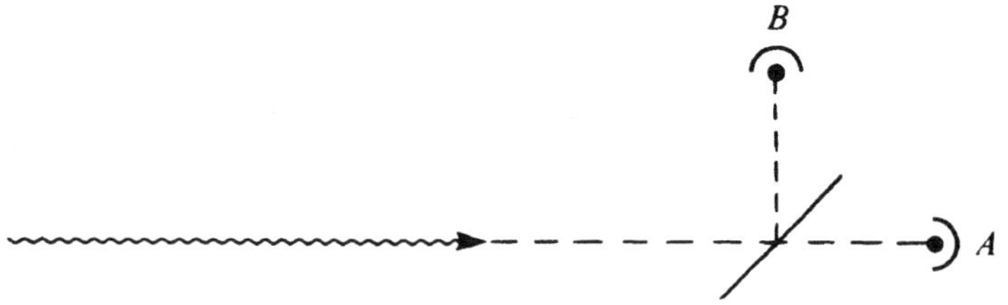

Figure 1. The scheme for a double counting experiment with a single photon.

Remarks

(i) An experiment closely related to that shown in Figure 1 has been done on the fluorescence light emitted by a very weak atomic beam excited by a resonant laser source (see the references at the end of the complement). Rather than having a single photon emitted by a single atom, one has a series of photons emitted by a single atom which is reexcited by a laser source after each emission. One can then observe at A and B a nonzero double counting rate. Since the atom can only emit a single photon at a time and since some time interval separates successive emissions, one finds that w_{II} is zero for $t = t'$ and is an increasing function of $|t - t'|$ near $t - t' = 0$ (recall that A and B are symmetric with respect to the half-transmitting mirror of Figure 1, so that we can set $\mathbf{r} = \mathbf{r}'$ in w_{II}). Such an "antibunching" of photoelectrons is a typical quantum effect, since one can show that it violates a semiclassical inequality predicting that w_{II}^{cl}, for $\mathbf{r} = \mathbf{r}'$, will be a nonincreasing function of $|t - t'|$.

(ii) At this stage of the discussion, it is useful to point out how the semiclassical and quantum calculations can lead to different predictions. Assume initially that the density operator ρ_{12} of the field allows a *P*-representation (see remark in §A_{III}.1 above). By substituting (7) for ρ_{12} in the expressions for w_I and w_{II}, one finds that these quantities are given by an "average" of the results (12) and (20) found above for the semiclassical states $|\alpha_1, \alpha_2\rangle$ "weighted" by the function $P(\alpha_1, \alpha_2)$ (which one should recall is real and normalized, since $\rho_{12} = \rho_{12}^+$ and $\operatorname{Tr} \rho_{12} = 1$):

$$w_I(\mathbf{r}, t) = \iint d^2\alpha_1 \, d^2\alpha_2 \, P(\alpha_1, \alpha_2) \, I_{cl}(\{\alpha_1, \alpha_2\}; \mathbf{r}, t) \tag{23}$$

$$w_{II}(\mathbf{r}, t; \mathbf{r}', t') = \iint d^2\alpha_1 \, d^2\alpha_2 \, P(\alpha_1, \alpha_2) \times \\ \times I_{cl}(\{\alpha_1, \alpha_2\}; \mathbf{r}, t) \, I_{cl}(\{\alpha_1, \alpha_2\}; \mathbf{r}', t') . \tag{24}$$

If the function $P(\alpha_1, \alpha_2)$ is positive, the signals (23) and (24) are identical with those given by a semiclassical theory (see §C.1.*c*) for a statistical mixture of semiclassical states $\{\alpha_1, \alpha_2\}$ with the *true weights* $P(\alpha_1, \alpha_2)$. There is then no difference between the quantum and semiclassical theories. The typically quantum effects arise when $P(\alpha_1, \alpha_2)$ does not exist or takes on *negative* values. The signals (23) and (24) can then violate the semiclassical inequalities derived from the positive definite character of the classical density $P_{cl}(\alpha_1, \alpha_2)$. One can then say in conclusion that the ensemble of quantum states of the radiation field is much larger than the ensemble of statistical mixtures of classical states. It follows in particular that it is not possible to interpret all the observable optical interference phenomena in terms of the superposition of classical Maxwell waves or of statistical mixtures of such superpositions.

c) Two-Photon States

We consider here a very simple example of a two-photon state

$$|\psi_{12}\rangle = |1_1\rangle |1_2\rangle \tag{25}$$

by taking a state with one photon in mode 1 and one photon in mode 2.

By substituting (25) in the expressions for w_I and w_{II} one gets, using (8),

$$w_I(\mathbf{r}, t) = |\mathscr{E}_1|^2 + |\mathscr{E}_2|^2 , \tag{26}$$

$$w_{II}(\mathbf{r}, t; \mathbf{r}', t') = 2 \mid \mathscr{E}_1 \mid^2 \mid \mathscr{E}_2 \mid^2 \times \\ \times \{ 1 + \mathcal{R}e\, e^{i[(\mathbf{k}_1 - \mathbf{k}_2).(\mathbf{r} - \mathbf{r}') - (\omega_1 - \omega_2)(t - t')]} \} . \quad (27)$$

It appears then that it is possible to observe interference effects on the double counting signals [associated with the last term of (27)] under conditions where they are not observable on single counting signals. The absence of interference terms in (26) is due to the fact mentioned above, that in a state like (25), where the number of photons in each mode is well defined, the mean fields $\langle E_1 \rangle$ and $\langle E_2 \rangle$ in each mode are zero (random phases). Detecting a photoelectron is equally probable everywhere: according to (26), $w_I(\mathbf{r}, t)$ is independent of $\mathbf{r}$. In contrast, once a photoelectron is detected at a point $\mathbf{r}'$, the probability of detecting a second one immediately after at another point $\mathbf{r}$ depends on $\mathbf{r} - \mathbf{r}'$. The detections of the two photons are not independent and give rise to an interference phenomenon. It is this that (27) illustrates.

The double counting signal w_{II} is also very useful when the phases of the fields in the two modes fluctuate independently with a characteristic diffusion time τ_d (as with two independent lasers). We have seen above (§A_{III}.2.c) that it is possible in that case to observe fringes with w_I provided that the observation time is short with respect to τ_d. However, the signal accumulated during such a period is generally quite weak, with the result that it is necessary to rerun the experiment many times. One encounters the difficulty then that, from one experiment to the next, it is not easy to control the relative phase between the two lasers. A signal like w_{II} which contains terms as in (27) independent of the phases of the two beams is much more suitable, since it can be accumulated over many experiments. This explains why interference phenomena between two independent lasers have been first observed experimentally, not from w_I, but from double and even multiple counting experiments (see the references at the end of this complement).

Another interesting example is that of observing the radiation coming from a star. The signal w_I is washed out by the phase fluctuations introduced by the atmosphere (*), whereas w_{II} contains terms independent of these phase fluctuations. The measurement of the variations of w_{II} with $|\mathbf{r} - \mathbf{r}'|$ allows the determination of the apparent diameter of the star.

(*) In speckle interferometry, the observation time is short compared to the correlation time of the phase fluctuations, and it is possible then to extract the information of interest from w_I.

4. Physical Interpretation in Terms of Interference between Transition Amplitudes

In §A_{III}.2 above, we have mentioned several times that w_I does not appear in general as the modulus squared of the sum of two classical waves. One can then ask what are the entities which in the general case interfere in w_I.

To see this we insert the closure relation

$$\sum_f |\psi_f\rangle\langle\psi_f| = 1 \tag{28}$$

for a complete orthonormal set $\{|\psi_f\rangle\}$ of states of the field between the two operators $E^{(-)}$ and $E^{(+)}$ appearing in the expression for w_I. This yields

$$\begin{aligned} w_I(\mathbf{r}, t) &= \sum_f \langle\psi|E^{(-)}(\mathbf{r}, t)|\psi_f\rangle\langle\psi_f|E^{(+)}(\mathbf{r}, t)|\psi\rangle \\ &= \sum_f |\langle\psi_f|E^{(+)}(\mathbf{r}, t)|\psi\rangle|^2 \\ &= \sum_f |\langle\psi_f|[E_1^{(+)}(\mathbf{r}, t) + E_2^{(+)}(\mathbf{r}, t)]|\psi\rangle|^2 . \end{aligned} \tag{29}$$

It is possible then to interpret (29) in the following manner. There are two different possible paths going from the "initial" state $|\psi\rangle$ to the "final" state $|\psi_f\rangle$ (Figure 2). The first path corresponds to the absorption of a photon from mode 1 at $\mathbf{r}$, t, and the second to the absorption of a photon from mode 2 at $\mathbf{r}$, t. The *amplitudes associated with these two paths* are

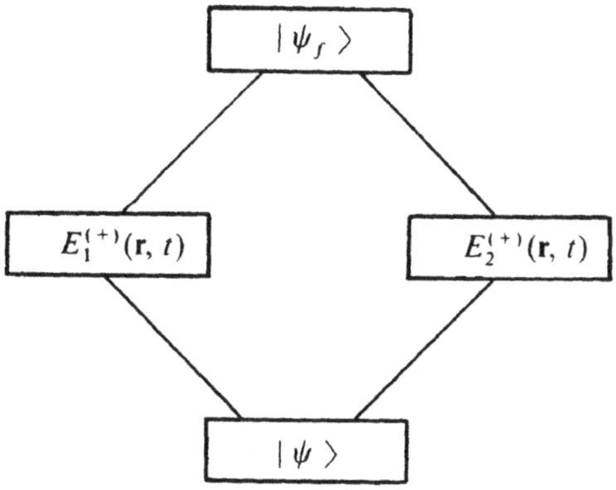

Figure 2. Schematic representation of the two amplitudes which interfere in a single counting experiment. $|\psi\rangle$ is the initial state, $|\psi_f\rangle$ the final state, and $E_i^{(+)}(\mathbf{r},t)$ represents the absorption of a photon in mode i at point $\mathbf{r}$ and time t.

$\langle \psi_f | E_1^{(+)}(\mathbf{r}, t) | \psi \rangle$ and $\langle \psi_f | E_2^{(+)}(\mathbf{r}, t) | \psi \rangle$, with the result that the *total amplitude* for going from $|\psi\rangle$ to $|\psi_f\rangle$ is written

$$\langle \psi_f \mid E_1^{(+)}(\mathbf{r}, t) \mid \psi \rangle + \langle \psi_f \mid E_2^{(+)}(\mathbf{r}, t) \mid \psi \rangle \tag{30}$$

The *transition probability* $|\psi\rangle \to |\psi_f\rangle$ is gotten by taking the square of the modulus of the sum appearing in (30). Since one does not observe the final state of the field, it is necessary to sum these probabilities over all possible states $|\psi_f\rangle$. [If the initial state is not a pure state $|\psi\rangle$ but a statistical mixture of states, it is necessary in addition to average (29) over all the possible states of this mixture.] The interference fringes observable from w_{I} are then due to interference between the transition amplitudes associated with the two paths of Figure 2.

An analogous interpretation can be given for w_{II}. Inserting the closure relation (28) between the two operators $E^{(-)}$ and the two operators $E^{(+)}$ appearing in the expression for w_{II}, one can show that the amplitude $|\psi\rangle \to |\psi_f\rangle$ in a double detection process is the sum of four amplitudes corresponding to the four different paths represented in Figure 3. For each of these paths two photons are absorbed, one at $\mathbf{r}'$, t' and the other at $\mathbf{r}$, t, each of these two photons belonging either to mode 1 or to mode 2. This gives $2^2 = 4$ possibilities. These four amplitudes interfere when the modulus of the global amplitude is squared in order to get the transition probability $|\psi\rangle \to |\psi_f\rangle$, then summed on the unobserved states $|\psi_f\rangle$. In particular, the fringes appearing for two-photon states [see Equation (27)] correspond to interference between two processes where the two photons in modes 1 and 2 are absorbed in different orders: 1 at $\mathbf{r}'$, t' and 2 at $\mathbf{r}$, t, or 2 at $\mathbf{r}'$, t' and 1 at $\mathbf{r}$, t.

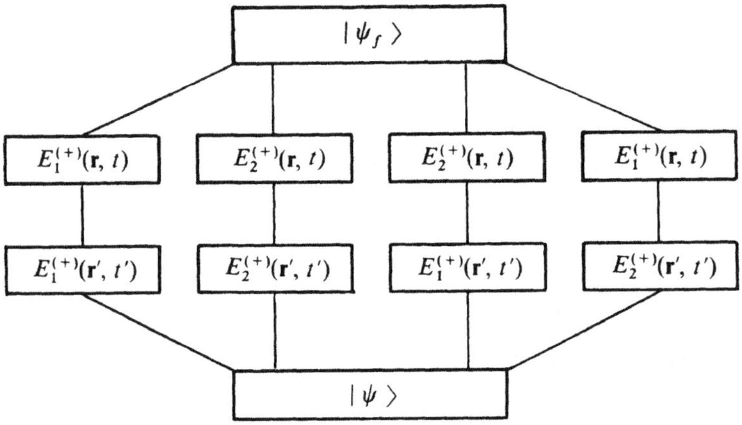

Figure 3. Schematic representation of the four amplitudes which interfere in the double counting experiment. The symbols have the same meanings as in Figure 2.

The general idea which emerges from the preceding discussion is that, in quantum theory, the interfering quantities are the *transition amplitudes*. To explain that the interference fringes for w_I are observable even when the photons arrive one by one, one often says that "each photon can only interfere with itself" and that "interference between two different photons is impossible". The discussion above shows that it is not the photons which interfere but rather the transition amplitudes, and these amplitudes can involve many photons. For instance, the fringes observable in w_{II} reveal the interference between two-photon amplitudes (Figure 3). One can easily generalize and show that amplitudes involving three, four,... photons can also interfere.

5. Conclusion: The Wave–Particle Duality in the Quantum Theory of Radiation

The discussion presented in this complement gives a better insight into the description of the wave–particle duality by the quantum theory of radiation.

The *wave* aspect is tied to the fact that the operators E of the various modes superpose linearly and each have a sinusoidal dependence in $\mathbf{r}$ and t. It is because $E^{(\pm)} = E_1^{(\pm)} + E_2^{(\pm)}$ that, in the quadratic detection signal $E^{(-)}E^{(+)}$, in addition to the squared terms $E_1^{(-)}E_1^{(+)}$ and $E_2^{(-)}E_2^{(+)}$, the terms $E_1^{(-)}E_2^{(+)}$ and $E_2^{(-)}E_1^{(+)}$ arise. The calculations closely resemble the classical calculations. However, it must be kept in mind that E is an operator and not a number.

The *particle* aspect is contained in the states $|\psi\rangle$ of the field, which indicate in some way what are the populated modes and how many photons they contain. It must be kept in mind that the states also are linearly superposable and that this property can be very important for the observation of interference effects. The states $|\psi\rangle$ and $|\psi_f\rangle$ in Figures 2 and 3 cannot be arbitrary if one wants to have more than one path for passing from $|\psi\rangle$ to $|\psi_f\rangle$. For example, a single-photon state which is a statistical mixture of $|1_1, 0_2\rangle$ and $|0_1, 1_2\rangle$ with the weights $|c_1|^2$ and $|c_2|^2$ does not give fringes in w_I whereas these are observable with the state $c_1|1_1, 0_2\rangle + c_2|0_1, 1_2\rangle$.

One sees then the richness of the quantum formalism which describes the physical systems with two different mathematical objects: the operators for the physical variables, and the vector $|\psi\rangle$ or the density operator ρ for the state of the system.

General references and further reading

For photon antibunching, see H. J. Kimble, M. Dagenais, and L. Mandel, *Phys. Rev. Lett.*, **39**, 691 (1977), and J. D. Cresser, J. Hager, G. Leuchs,

M. Rateike, and H. Walther, in *Dissipative Systems in Quantum Optics*, R. Bonifacio, ed., Springer, Berlin, 1982.

For single-photon interference see F. M. Pipkin, *Adv. At. Mol. Phys.*, **14**, 281 (1978), and P. Grangier, G. Roger, and A. Aspect, *Europhys. Lett.*, **1**, 173 (1986).

For interference with two independent beams, see G. Magyar and L. Mandel, *Nature*, **198**, 255 (1963), and R. L. Pfleegor and L. Mandel, *J. Opt. Soc. Am.*, **58**, 946 (1968).

For intensity correlations and interference between amplitudes involving several photons, see R. Hanbury Brown and R. Q. Twiss, *Nature*, **177**, 27 (1956); E. R. Pike, in *Quantum Optics*, S. M. Kay and A. Maitland, ed., Academic, New York, 1970; U. Fano, *Am. J. Phys.*, **29**, 539 (1961).

COMPLEMENT B_{III}

QUANTUM FIELD RADIATED BY CLASSICAL SOURCES

The purpose of this complement is to study a simple problem of electrodynamics showing the importance of the coherent states introduced in §C.4 of this chapter. We calculate the quantum field radiated by classical sources whose motion is not perturbed by the field and show that the state of such a field is a coherent state.

1. Assumptions about the Sources

We assume that before $t = 0$ no source is present. The radiation is initially in the vacuum state

$$|\psi(0)\rangle = |0\rangle . \tag{1}$$

At $t = 0$ the sources are "switched on". The currents $\mathbf{j}(\mathbf{r})$ associated with them are assumed not to be affected by the radiation they create (one can imagine that the radiation damping of the sources is compensated by the experimental setup which creates the sources). The preceding hypothesis implies that the sources have an externally imposed motion. In addition, we assume that the sources are macroscopic, that is, that the quantum fluctuations of the currents about their mean values are negligible. This set of hypotheses thus allows us to approximate the quantum currents $\mathbf{j}(\mathbf{r})$ by well-defined classical functions of $\mathbf{r}$ and t, $\mathbf{j}_{\text{cl}}(\mathbf{r}, t)$. Taking these approximations into account, what is then the state $|\psi(t)\rangle$ of the field for $t > 0$?

The preceding question favors the Schrödinger point of view, where the temporal evolution involves only the state vector $|\psi(t)\rangle$ of the system and not the operators. In fact, as a result of the approximations made above concerning the currents, the calculation of the temporal evolution is much simpler from the Heisenberg point of view, where it is the operators which evolve and where the state vector remains fixed and equal to (1). We will use such a point of view in §B_{III}.2 to calculate the evolution of the annihilation operator $a_i(t)$ of mode i. A unitary transformation will allow us in §B_{III}.3 to go from the Heisenberg point of view to that of Schrödinger and to determine what the state vector $|\psi(t)\rangle$ of the system is at time t.

2. Evolution of the Fields in the Heisenberg Picture

The equation of evolution for $a_i(t)$ [Equation (B.28)] can be written

$$\dot{a}_i + i\omega_i a_i = s_i \tag{2}$$

where

$$s_i = \frac{\mathrm{i}}{\sqrt{2\,\varepsilon_0\,\hbar\omega_i\,L^3}} \int \mathrm{d}^3r\;\mathrm{e}^{-\mathrm{i}\mathbf{k}_i\cdot\mathbf{r}}\;\boldsymbol{\varepsilon}_i\cdot\mathbf{j}(\mathbf{r})\,. \tag{3}$$

Strictly speaking, the current $\mathbf{j}(\mathbf{r})$ is given as a function of the variables of the particles forming the sources [see Equation (A.5.b), Chapter I] which themselves evolve under the influence of the forces exerted on them by the fields. The evolution of the right-hand side s_i of (2) is then in fact coupled to that of all the a_j, and the solution of (2) is not found easily in the general case. It is necessary to adjoin the equation of motion of the particles to Equation (2) and to solve these coupled equations.

The situation is much simpler if it is possible, as we will assume here, to replace the operator $\mathbf{j}(\mathbf{r})$ by a known function $\mathbf{j}_{\mathrm{cl}}(\mathbf{r}, t)$. Equation (2) then becomes

$$\dot{a}_i + \mathrm{i}\omega_i\,a_i = s_i^{\mathrm{cl}}(t) \tag{4}$$

where $s_i^{\mathrm{cl}}(t)$ is a known classical function of t,

$$s_i^{\mathrm{cl}}(t) = \frac{\mathrm{i}}{\sqrt{2\,\varepsilon_0\,\hbar\omega_i\,L^3}} \int \mathrm{d}^3r\,\mathrm{e}^{-\mathrm{i}\mathbf{k}_i\cdot\mathbf{r}}\;\boldsymbol{\varepsilon}_i\cdot\mathbf{j}_{\mathrm{cl}}(\mathbf{r}, t)\,. \tag{5}$$

Equation (4) is then easily integrated to give

$$a_i(t) = a_i(0)\,\mathrm{e}^{-\mathrm{i}\omega_i t} + \int_0^t \mathrm{d}t'\;s_i^{\mathrm{cl}}(t')\,\mathrm{e}^{-\mathrm{i}\omega_i(t-t')}\,. \tag{6}$$

The first term appearing on the right-hand side of (6) is the initial quantum field $a_i(0)$, which has evolved freely from 0 to t, and the second term is the field radiated by the sources.

Remarks

(i) Rather than introducing the simplifying hypotheses on $\mathbf{j}(\mathbf{r})$ in the equation of evolution (2) for a_i, we could make such an approximation directly on the Hamiltonian. The Hamiltonian of a quantized field coupled to known classical currents $\mathbf{j}_{\mathrm{cl}}(\mathbf{r}, t)$ is written in the Coulomb gauge as

$$H = H_R - \int \mathrm{d}^3r\;\mathbf{j}_{\mathrm{cl}}(\mathbf{r}, t)\cdot\mathbf{A}(\mathbf{r}) \tag{7}$$

where H_R is the Hamiltonian of the quantized free radiation field and $\mathbf{A}(\mathbf{r})$ the transverse vector-potential operator [see Equations (A.9) and (A.5)]. The operator (7) acts only on the radiation variables, and it is easy to show that the Heisenberg equation for a_i derived from this Hamiltonian is identical with (4).

(ii) The expression (6) for $a_i(t)$ allows one to get the quantized transverse field at time t. All that is necessary is to substitute (6) and the adjoint expression in the expansions of the transverse fields in a_i and a_i^+. To get the total field, it is necessary to add the longitudinal fields to these transverse fields, that is, the Coulomb field of the charge density $\rho_{cl}(\mathbf{r}, t)$ associated with the sources and assumed, like $\mathbf{j}_{cl}(\mathbf{r}, t)$, to be a known classical function of $\mathbf{r}$ and t.

To go further, we also write the equation of evolution of the classical fields coupled to the known classical currents $\mathbf{j}_{cl}(\mathbf{r}, t)$. For this, it is sufficient to replace the operator a_i in Equation (4) by the classical normal variable $\alpha_i(t)$ of mode i. One then gets

$$\dot{\alpha}_i + i\omega_i \alpha_i = s_i^{cl}(t) \tag{8}$$

whose solution corresponding to the initial condition $\alpha_i(0) = 0$ (no radiation at $t = 0$) is written

$$\alpha_i(t) = \int_0^t dt' \, s_i^{cl}(t') \, e^{-i\omega_i(t-t')} . \tag{9}$$

The last term on the right-hand side of (6) can then be interpreted as the classical field $\alpha_i(t)$ radiated by the known classical sources, and we rewrite (6) taking account of (9):

$$a_i(t) = a_i(0) \, e^{-i\omega_i t} + \alpha_i(t) . \tag{10}$$

Finally we apply the two terms of the operator equation (10) to state (1), which is the state of the system for all t in the Heinsenberg picture. Since

$$a_i(0) \mid \psi(0) \rangle = a_i \mid 0 \rangle = 0 \tag{11}$$

one gets

$$a_i(t) \mid \psi(0) \rangle = \alpha_i(t) \mid \psi(0) \rangle . \tag{12}$$

It appears then that the state $|\psi(0)\rangle$ is the eigenstate of the operator $a_i(t)$ with the normal classical variable $\alpha_i(t)$ of the classical field radiated by the sources as eigenvalue.

3. The Schrödinger Point of View. The Quantum State of the Field at Time t

Let $U(t,0)$ be the evolution operator from 0 to t in the Schrödinger picture. We will apply $U(t,0)$ to both sides of Equation (12) and in the first term insert $U^+(t,0)U(t,0) = 1$ between $a_i(t)$ and $|\psi(0)\rangle$. This leads to

$$U(t, 0) \, a_i(t) \, U^+(t, 0) \, U(t, 0) \mid \psi(0) \rangle = \alpha_i(t) \, U(t, 0) \mid \psi(0) \rangle . \tag{13}$$

The ket $U(t,0)|\psi(0)\rangle$ is the state vector $|\psi(t)\rangle$ of the system at time t in the Schrödinger picture. Since $U(t,0)$ is also the unitary operator allowing one to go from the Heisenberg point of view to that of Schrödinger at time t (note that the two points of view are identical at $t = 0$), $U(t,0)a_i(t)U^+(t,0)$ coincides with the annihilation operator a_i, from the Schrödinger point of view and is independent of time. Finally, Equation (13) becomes

$$a_i \mid \psi(t) \rangle = \alpha_i(t) \mid \psi(t) \rangle \tag{14}$$

and shows that the quantum state of the field at time t is the coherent state associated with the classical field radiated by the sources at the same time t.

Such a result is physically satisfying. When the sources are quasi-classical, the quantum state of the field which they radiate is the quantum state which most closely approximates the classical radiation field $\{\alpha_i(t)\}$, that is, the coherent state $|\{\alpha_i(t)\}\rangle$. The only deviation is due to quantum fluctuations, which in a coherent state are minimal and equal to those of the vacuum.

References

See Glauber.

COMPLEMENT C_{III}

COMMUTATION RELATIONS FOR FREE FIELDS AT DIFFERENT TIMES. SUSCEPTIBILITIES AND CORRELATION FUNCTIONS OF THE FIELDS IN THE VACUUM

Starting from the fundamental commutation relations between a_i and a_i^+, we have established in Part A (see also Exercise 1) the commutation relations between the observables $\mathbf{A}_\perp(\mathbf{r})$, $\mathbf{E}_\perp(\mathbf{r})$, and $\mathbf{B}(\mathbf{r})$ taken at different points but at the same instant. It is also interesting to consider the commutation relations for the same fields taken at different times from the Heisenberg point of view. For such a purpose, it is necessary to know the temporal evolution of the fields. The objective of this complement is to determine these commutation relations in the particular case of a free field for which the temporal evolution has been found in §B.2.*b*.

Following the introduction in §C_{III}.1 of singular functions which are useful in electrodynamics, we evaluate the various commutators in §C_{III}.2. Physically, these commutation relations allow us to find those observables of the field which can be measured independently of one another. Additionally, the commutators play a fundamental role in linear response theory; if the field, initially free, is coupled to an external source $j(t)$ by a coupling term $-j(t)M$, where M is an observable of the field, then the evolution of another observable of the field, N, in this perturbed state is given to first order in j by

$$\langle N(t) \rangle^{(1)} = \int_0^t \mathrm{d}t' \, \chi_{NM}(t, t') \, j(t') \tag{1}$$

where the *susceptibility* χ_{NM} is just the mean value, in the initial quantum state, of the commutator of the free fields $N(t)$ and $M(t')$:

$$\chi_{NM}(t, t') = \frac{\mathrm{i}}{\hbar} \langle [N(t), M(t')] \rangle . \tag{2}$$

The name free fields signifies that the temporal evolution of the operators $N(t)$ and $M(t')$ arising in (2) is free, that is to say, calculated in absence of sources j. Hence the free-field commutators at two different times give also the mean fields created in first order by an arbitrary source. An application of this procedure is given in Exercise 6, Chapter IV.

We have already encountered in §C.3.*c* another important function relative to free fields: the symmetric correlation function. For fundamental reasons (the fluctuation–dissipation theorem), the calculation of the correlation functions in the vacuum turns out to be quite similar to that of the commutators, and we will treat this in §C_{III}.3.

1. Preliminary Calculations

We will need later on the function $D_+(\boldsymbol{\rho}, \tau)$ defined by the Fourier transform

$$D_+(\boldsymbol{\rho}, \tau) = \mathrm{i}c \int \frac{\mathrm{d}^3k}{(2\pi)^3} \frac{\mathrm{e}^{\mathrm{i}(\mathbf{k}.\boldsymbol{\rho} - \omega\tau)}}{\omega} \tag{3}$$

where $\omega = ck$. One can note in (3) that D_+ is the solution of the wave equation

$$\left(\frac{1}{c^2} \frac{\partial^2}{\partial t^2} - \Delta_{\boldsymbol{\rho}}\right) D_+(\boldsymbol{\rho}, \tau) = 0 \tag{4}$$

and thus propagates with velocity c.

To calculate (3) we use spherical coordinates and denote by u the cosine of the angle between $\mathbf{k}$ and $\boldsymbol{\rho}$. The integral becomes

$$\begin{aligned} D_+(\boldsymbol{\rho}, \tau) &= \mathrm{i} \int_0^\infty \frac{k\,\mathrm{d}k}{(2\pi)^3} 2\pi \int_0^1 \mathrm{d}u\, \mathrm{e}^{\mathrm{i}(k\rho u - ck\tau)} \\ &= \frac{1}{(2\pi)^2 \rho} \int_0^\infty \mathrm{d}k [\mathrm{e}^{\mathrm{i}k(\rho - c\tau)} - \mathrm{e}^{-\mathrm{i}k(\rho + c\tau)}] . \end{aligned} \tag{5}$$

Regularizing the integrals by a convergence factor $\mathrm{e}^{-k\eta}$ (where η is a positive infinitesimal), one gets

$$\begin{aligned} D_+(\boldsymbol{\rho}, \tau) &= \frac{1}{(2\pi)^2 \rho} \left[\frac{1}{\eta - \mathrm{i}(\rho - c\tau)} - \frac{1}{\eta + \mathrm{i}(\rho + c\tau)}\right] \\ &= \frac{\mathrm{i}}{2\pi^2} \frac{1}{\rho^2 - (c\tau - \mathrm{i}\eta)^2} . \end{aligned} \tag{6}$$

It is of interest to separate the real and imaginary parts of D_+ in (6):

$$D_+(\boldsymbol{\rho}, \tau) = D(\boldsymbol{\rho}, \tau) + \mathrm{i}D_1(\boldsymbol{\rho}, \tau) , \tag{7}$$

$$D = \frac{1}{(2\pi)^2 \rho} \left[\frac{\eta}{\eta^2 + (\rho - c\tau)^2} - \frac{\eta}{\eta^2 + (\rho + c\tau)^2}\right] \tag{8}$$

$$D_1 = \frac{1}{(2\pi)^2 \rho} \left[\frac{\rho - c\tau}{\eta^2 + (\rho - c\tau)^2} + \frac{\rho + c\tau}{\eta^2 + (\rho + c\tau)^2}\right] . \tag{9}$$

The function D is infinitely small outside the light cone $\rho^2 = c^2\tau^2$ and is odd in τ. It is then 0 for $\tau = 0$, and its derivative with respect to τ at this point can be found directly from Equation (3) and has a value $c\delta(\boldsymbol{\rho})$:

$$D(\boldsymbol{\rho}, \tau = 0) = 0; \qquad \frac{\partial}{\partial \tau} D(\boldsymbol{\rho}, \tau)\bigg|_{\tau=0} = c\,\delta(\boldsymbol{\rho}). \tag{10}$$

From Equation (8) we see also that D can be found in the limit where $\eta \to 0$ as a sum of two δ-functions:

$$D = \frac{1}{4\pi\rho}[\delta(\rho - c\tau) - \delta(\rho + c\tau)] \tag{11}$$

D_1 is even in τ, diverges like a principal part in the neighborhood of the light cone, and decreases outside as $(\rho^2 - c^2\tau^2)^{-1}$. One can regroup the two terms of (9) to give D_1 as a function of ρ^2:

$$D_1 = \frac{1}{2\pi^2} \frac{\rho^2 - c^2\tau^2 + \eta^2}{(\rho^2 - c^2\tau^2)^2 + 2\eta^2(\rho^2 + c^2\tau^2) + \eta^4}. \tag{12}$$

2. Field Commutators

We start with the expression for the free fields from the Heisenberg point of view:

$$\mathbf{A}_\perp(\mathbf{r}, t) = \sum_{\mathbf{k},\boldsymbol{\varepsilon}} \mathscr{A}_\omega[a_{\mathbf{k}\boldsymbol{\varepsilon}}\, \boldsymbol{\varepsilon}\, e^{i(\mathbf{k}.\mathbf{r} - \omega t)} + a^+_{\mathbf{k}\boldsymbol{\varepsilon}}\, \boldsymbol{\varepsilon}\, e^{-i(\mathbf{k}.\mathbf{r} - \omega t)}] \tag{13.a}$$

$$\mathbf{E}_\perp(\mathbf{r}, t) = \sum_{\mathbf{k},\boldsymbol{\varepsilon}} i\, \mathscr{E}_\omega[a_{\mathbf{k}\boldsymbol{\varepsilon}}\, \boldsymbol{\varepsilon}\, e^{i(\mathbf{k}.\mathbf{r} - \omega t)} - a^+_{\mathbf{k}\boldsymbol{\varepsilon}}\, \boldsymbol{\varepsilon}\, e^{-i(\mathbf{k}.\mathbf{r} - \omega t)}] \tag{13.b}$$

$$\mathbf{B}(\mathbf{r}, t) = \sum_{\mathbf{k},\boldsymbol{\varepsilon}} i\, \mathscr{B}_\omega[a_{\mathbf{k}\boldsymbol{\varepsilon}}\, \boldsymbol{\kappa} \times \boldsymbol{\varepsilon}\, e^{i(\mathbf{k}.\mathbf{r} - \omega t)} - a^+_{\mathbf{k}\boldsymbol{\varepsilon}}\, \boldsymbol{\kappa} \times \boldsymbol{\varepsilon}\, e^{-i(\mathbf{k}.\mathbf{r} - \omega t)}] \tag{13.c}$$

where

$$\mathscr{E}_\omega = \left[\frac{\hbar\omega}{2\varepsilon_0 L^3}\right]^{1/2} \qquad \mathscr{B}_\omega = \frac{\mathscr{E}_\omega}{c} \qquad \mathscr{A}_\omega = \frac{\mathscr{E}_\omega}{\omega}. \tag{14}$$

The commutators of the components m and n ($m, n = x, y, z$) of any two of the fields are given in the form of a linear combination of the commutators $[a_{\mathbf{k}\boldsymbol{\varepsilon}}, a_{\mathbf{k}'\boldsymbol{\varepsilon}'}]$, $[a^+_{\mathbf{k}\boldsymbol{\varepsilon}}, a^+_{\mathbf{k}'\boldsymbol{\varepsilon}'}]$, and $[a_{\mathbf{k}\boldsymbol{\varepsilon}}, a^+_{\mathbf{k}'\boldsymbol{\varepsilon}'}]$. The first two are zero, and the last is $\delta_{\boldsymbol{\varepsilon}\boldsymbol{\varepsilon}'}\,\delta_{\mathbf{k}\mathbf{k}'}$. One then gets

$$[E_m(\mathbf{r}_1, t_1), E_n(\mathbf{r}_2, t_2)] = \sum_{\mathbf{k}\boldsymbol{\varepsilon}} \mathscr{E}_\omega^2\, \varepsilon_m\, \varepsilon_n\, e^{i[\mathbf{k}.(\mathbf{r}_1 - \mathbf{r}_2) - \omega(t_1 - t_2)]} - \text{c.c.} \tag{15.a}$$

$$[E_m(\mathbf{r}_1, t_1), B_n(\mathbf{r}_2, t_2)] = \sum_{\mathbf{k}\boldsymbol{\varepsilon}} \mathscr{E}_\omega\, \mathscr{B}_\omega\, \varepsilon_m(\boldsymbol{\kappa} \times \boldsymbol{\varepsilon})_n\, e^{i[\mathbf{k}.(\mathbf{r}_1 - \mathbf{r}_2) - \omega(t_1 - t_2)]} - \text{c.c.} \tag{15.b}$$

$$[B_m(\mathbf{r}_1, t_1), B_n(\mathbf{r}_2, t_2)] = \sum_{\mathbf{k}\boldsymbol{\varepsilon}} \mathscr{B}_\omega^2 (\boldsymbol{\kappa} \times \boldsymbol{\varepsilon})_m (\boldsymbol{\kappa} \times \boldsymbol{\varepsilon})_n \, e^{i[\mathbf{k}.(\mathbf{r}_1 - \mathbf{r}_2) - \omega(t_1 - t_2)]} - \text{c.c.} \tag{15.c}$$

$$[A_{\perp m}(\mathbf{r}_1, t_1), E_n(\mathbf{r}_2, t_2)] = \sum_{\mathbf{k}\boldsymbol{\varepsilon}} - i\mathscr{A}_\omega \mathscr{E}_\omega \varepsilon_m \varepsilon_n \, e^{i[\mathbf{k}.(\mathbf{r}_1 - \mathbf{r}_2) - \omega(t_1 - t_2)]} - \text{c.c.} \tag{15.d}$$

a) Reduction of the Expressions in Terms of D

In each of the expressions (15), the sum $\Sigma_{\mathbf{k}\boldsymbol{\varepsilon}}$ breaks up into a sum on the two transverse polarizations $\boldsymbol{\varepsilon}$ and $\boldsymbol{\varepsilon}'$, and then a sum over $\mathbf{k}$. The summations on the polarizations have already been made in §A_I.1.*a* and give

$$\sum_{\boldsymbol{\varepsilon} \perp \mathbf{k}} \varepsilon_m \varepsilon_n = \delta_{mn} - \kappa_m \kappa_n \tag{16.a}$$

$$\sum_{\boldsymbol{\varepsilon} \perp \mathbf{k}} \varepsilon_m (\boldsymbol{\kappa} \times \boldsymbol{\varepsilon})_n = \sum_l \varepsilon_{mnl} \kappa_l \tag{16.b}$$

$$\sum_{\boldsymbol{\varepsilon} \perp \mathbf{k}} (\boldsymbol{\kappa} \times \boldsymbol{\varepsilon})_m (\boldsymbol{\kappa} \times \boldsymbol{\varepsilon})_n = \delta_{mn} - \kappa_m \kappa_n \tag{16.c}$$

ε_{mnl} being the completely antisymmetric tensor.

To simplify the notation we denote $(\mathbf{r}_1, t_1)$ by (1) and $(\mathbf{r}_2, t_2)$ by (2). The differences between (1) and (2) will be denoted by

$$\mathbf{r}_1 - \mathbf{r}_2 = \boldsymbol{\rho}; \qquad t_1 - t_2 = \tau . \tag{17}$$

By using the sums (16), this notation, and the expression (14) for $\mathscr{A}_\omega$, $\mathscr{E}_\omega$, $\mathscr{B}_\omega$, one gets

$$[E_m(1), E_n(2)] = \frac{\hbar}{2\varepsilon_0} \sum_{\mathbf{k}} \frac{\omega}{L^3} \left(\delta_{mn} - \frac{k_m k_n}{k^2} \right) e^{i(\mathbf{k}.\boldsymbol{\rho} - \omega\tau)} - \text{c.c.} \tag{18.a}$$

$$[E_m(1), B_n(2)] = \frac{\hbar}{2\varepsilon_0 c} \sum_{\mathbf{k},l} \frac{\omega}{L^3} \varepsilon_{mnl} \frac{k_l}{k} e^{i(\mathbf{k}.\boldsymbol{\rho} - \omega\tau)} - \text{c.c.} \tag{18.b}$$

$$[B_m(1), B_n(2)] = \frac{1}{c^2} [E_m(1), E_n(2)] \tag{18.c}$$

$$[A_{\perp m}(1), E_n(2)] = \frac{\hbar}{2\varepsilon_0} \sum_{\mathbf{k}} \frac{-i}{L^3} \left(\delta_{mn} - \frac{k_m k_n}{k^2} \right) e^{i(\mathbf{k}.\boldsymbol{\rho} - \omega\tau)} - \text{c.c.} \tag{18.d}$$

The discrete sum on $\mathbf{k}$ can be replaced by an integral (see Equation (C.34), Chapter I). One can then insert $D_+(\boldsymbol{\rho}, \tau)$, or its derivatives and primitives with respect to $\boldsymbol{\rho}$ or τ, as follows:

$$\sum_{\mathbf{k}} \frac{\omega}{L^3}\left(\delta_{mn} - \frac{k_m k_n}{k^2}\right) e^{i(\mathbf{k}.\boldsymbol{\rho} - \omega\tau)} = \int \frac{d^3k}{(2\pi)^3}(\omega^2 \delta_{mn} - c^2 k_m k_n)\frac{e^{i(\mathbf{k}.\boldsymbol{\rho} - \omega\tau)}}{\omega} =$$

$$= -\left(\delta_{mn}\frac{\partial^2}{\partial\tau^2} - c^2 \frac{\partial}{\partial\rho_m}\frac{\partial}{\partial\rho_n}\right)\frac{D_+(\boldsymbol{\rho}, \tau)}{ic} \quad (19.a)$$

$$\sum_{\mathbf{k}} \frac{\omega}{L^3}\frac{k_l}{k} e^{i(\mathbf{k}.\boldsymbol{\rho} - \omega\tau)} = c\int \frac{d^3k}{(2\pi)^3}\omega k_l \frac{e^{i(\mathbf{k}.\boldsymbol{\rho} - \omega\tau)}}{\omega}$$

$$= \frac{1}{i}\frac{\partial}{\partial\tau}\frac{\partial}{\partial\rho_l} D_+(\boldsymbol{\rho}, \tau) \quad (19.b)$$

$$\sum_{\mathbf{k}} \frac{1}{L^3}\left(\delta_{mn} - \frac{k_m k_n}{k^2}\right) e^{i(\mathbf{k}.\boldsymbol{\rho} - \omega\tau)} = \int \frac{d^3k}{(2\pi)^3}\left[\omega\,\delta_{mn} - c^2\frac{k_m k_n}{\omega}\right]\frac{e^{i(\mathbf{k}.\boldsymbol{\rho} - \omega\tau)}}{\omega} =$$

$$= i\,\delta_{mn}\frac{\partial}{\partial\tau}\frac{D_+(\boldsymbol{\rho}, \tau)}{ic} + \frac{c^2}{i}\frac{\partial}{\partial\rho_m}\frac{\partial}{\partial\rho_n}\int_{-\infty}^{\tau} d\tau' \frac{D_+(\boldsymbol{\rho}, \tau')}{ic}. \quad (19.c)$$

In (19.c) we have used the fact that D_+ decreases like $1/\tau^2$ at infinity, so that its primitive gives a zero contribution at the lower limit of the integral. The contribution of the complex conjugate term appearing in (18) causes the imaginary part of D_+ to vanish, so that finally it is its real part $D(\boldsymbol{\rho}, \tau)$ which is present:

$$[E_m(1), E_n(2)] = c^2[B_m(1), B_n(2)] =$$

$$= \frac{i\hbar}{\varepsilon_0 c}\left[\delta_{mn}\frac{\partial^2}{\partial\tau^2} - c^2\frac{\partial}{\partial\rho_m}\frac{\partial}{\partial\rho_n}\right]D(\boldsymbol{\rho}, \tau) \quad (20.a)$$

$$[E_m(1), B_n(2)] = \frac{-i\hbar}{\varepsilon_0 c}\sum_l \varepsilon_{mnl}\frac{\partial}{\partial\tau}\frac{\partial}{\partial\rho_l}D(\boldsymbol{\rho}, \tau) \quad (20.b)$$

$$[A_{\perp m}(1), E_n(2)] = \frac{-i\hbar}{\varepsilon_0 c}\left[\delta_{mn}\frac{\partial}{\partial\tau}D(\boldsymbol{\rho}, \tau) - c^2\frac{\partial}{\partial\rho_m}\frac{\partial}{\partial\rho_n}\int_{-\infty}^{\tau} d\tau'\,D(\boldsymbol{\rho}, \tau')\right] \quad (20.c)$$

$\boldsymbol{\rho}$ and τ are defined in (17).

b) Explicit Expressions for the Commutators

Equations (20) allow one to discuss the essential properties of the commutators by relating them to the function D examined in §$C_{III}.1$. One can however go farther by substituting in (20.a, b, c) the expression (11) for D and by explicitly calculating the derivatives. The derivatives with respect to ρ_m or ρ_n are found using

$$\frac{\partial}{\partial \rho_m} f(\rho) = \frac{\rho_m}{\rho} f'(\rho) . \tag{21}$$

One then gets

$$\begin{aligned}
[E_m(1), E_n(2)] &= \frac{i\hbar c}{4\pi\varepsilon_0}\left(\delta_{mn}\frac{\partial^2}{c^2\,\partial\tau^2} - \frac{\partial}{\partial\rho_m}\frac{\partial}{\partial\rho_n}\right)\left[\frac{\delta(\rho - c\tau) - \delta(\rho + c\tau)}{\rho}\right] \\
&= \frac{i\hbar c}{4\pi\varepsilon_0}\left\{\left(\frac{3\,\rho_m\rho_n}{\rho^2} - \delta_{mn}\right)\left[\frac{\delta'(\rho - c\tau) - \delta'(\rho + c\tau)}{\rho^2} - \frac{\delta(\rho - c\tau) - \delta(\rho + c\tau)}{\rho^3}\right]\right. \\
&\qquad \left. - \left(\frac{\rho_m\rho_n}{\rho^2} - \delta_{mn}\right)\left[\frac{\delta''(\rho - c\tau) - \delta''(\rho + c\tau)}{\rho}\right]\right\}
\end{aligned} \tag{22}$$

$$\begin{aligned}
[E_m(1), B_n(2)] &= \frac{-i\hbar}{4\pi\varepsilon_0}\sum_l \varepsilon_{mnl}\frac{\partial}{c\,\partial\tau}\frac{\partial}{\partial\rho_l}\left[\frac{\delta(\rho - c\tau) - \delta(\rho + c\tau)}{\rho}\right] \\
&= \frac{i\hbar}{4\pi\varepsilon_0}\sum_l \varepsilon_{mnl}\frac{\rho_l}{\rho}\left[\frac{\delta''(\rho - c\tau) + \delta''(\rho + c\tau)}{\rho} - \frac{\delta'(\rho - c\tau) + \delta'(\rho + c\tau)}{\rho^2}\right]
\end{aligned} \tag{23}$$

$$\begin{aligned}
[A_{\perp m}(1), E_n(2)] &= \frac{-i\hbar}{4\pi\varepsilon_0}\left\{\delta_{mn}\frac{\partial}{c\,\partial\tau}\left[\frac{\delta(\rho - c\tau) - \delta(\rho + c\tau)}{\rho}\right] - \right. \\
&\qquad \left. - \frac{\partial}{\partial\rho_m}\frac{\partial}{\partial\rho_n}\int_{-\infty}^{\tau} c\,\mathrm{d}\tau'\left[\frac{\delta(c\tau' - \rho) - \delta(c\tau' + \rho)}{\rho}\right]\right\} \\
&= \frac{i\hbar}{4\pi\varepsilon_0}\left\{\delta_{mn}\frac{\delta'(\rho - c\tau) + \delta'(\rho + c\tau)}{\rho} - \frac{\partial}{\partial\rho_m}\frac{\partial}{\partial\rho_n}\frac{\theta(c\tau - \rho) - \theta(c\tau + \rho)}{\rho}\right\} \\
&= \frac{i\hbar}{4\pi\varepsilon_0}\left\{\left(\frac{3\,\rho_m\rho_n}{\rho^2} - \delta_{mn}\right)\left[\frac{\theta(c\tau - \rho) - \theta(c\tau + \rho)}{\rho^3} + \frac{\delta(\rho - c\tau) - \delta(\rho + c\tau)}{\rho^2}\right]\right. \\
&\qquad \left. - \left(\frac{\rho_m\rho_n}{\rho^2} - \delta_{mn}\right)\left[\frac{\delta'(\rho - c\tau) + \delta'(\rho + c\tau)}{\rho}\right]\right\}
\end{aligned} \tag{24}$$

$\theta(x)$ is the Heaviside function [$\theta(x) = 0$ for $x < 0$, $\theta(x) = 1$ for $x > 0$].

Remark

The equations (22), (23), and (24) are ambiguous when $\tau \to 0$ because they introduce in this case products of the function $\delta(\rho)$ with functions which diverge at $\rho = 0$. It is then necessary to return to expressions with η finite or else to Equations (18). For example, with η finite, Equation (20a) is no longer singular. Since $D(\boldsymbol{\rho}, \tau)$ is odd in τ, its value is zero at $\tau = 0$, as are its second derivatives

$$[E_m(\mathbf{r}_1, t), E_n(\mathbf{r}_2, t)] = [B_m(\mathbf{r}_1, t), B_n(\mathbf{r}_2, t)] = 0 \tag{25}$$

From (20b) and (10),

$$[E_m(\mathbf{r}_1, t), B_n(\mathbf{r}_2, t)] = \frac{-i\hbar}{\varepsilon_0} \sum_l \varepsilon_{mnl} \frac{\partial}{\partial \rho_l} \delta(\boldsymbol{\rho}) . \tag{26}$$

Finally, using (18d) for $\tau = 0$,

$$[A_{\perp m}(\mathbf{r}_1, t), E_n(\mathbf{r}_2, t)] = \frac{-i\hbar}{\varepsilon_0} \delta^{\perp}_{mn}(\boldsymbol{\rho}) . \tag{27}$$

This gives the result gotten in a more elementary fashion in Exercise 1.

c) Properties of the Commutators

All of the commutators (20) depend only on $\boldsymbol{\rho} = \mathbf{r}_1 - \mathbf{r}_2$ and $\tau = t_1 - t_2$. This property reflects the invariance of the theory under translations over space and time. With respect to rotations, the commutators are second-rank tensors.

All of the commutators are numbers and not operators. This property is directly related to the linearity of the field equations. One can also express this result by stating that the susceptibility of the field is independent of its state; the field created by a known source is independent of the previous field.

The commutators of the electric and magnetic fields are zero outside the light cone defined by $\rho^2 = c^2\tau^2$ [the only functions appearing in these commutators are the functions $\delta(\rho \pm c\tau)$ and their derivatives]. Outside the light cone, these physical variables can be measured independently of one another. Finally, even on the light cone, the components of **E** and **B** on a given axis always commute; they correspond to independent degrees of freedom of the field.

The commutators of $\mathbf{A}_\perp$ and **E** appear differently; in addition to a zero component outside the light cone, they contain a term made up of the product of a function of $\boldsymbol{\rho}$ with $[\theta(c\tau - \rho) - \theta(c\tau + \rho)]$. This factor is zero for the timelike intervals ($c^2\tau^2 > \rho^2$) and -1 for the spacelike intervals ($\rho^2 > c^2\tau^2$). The commutators are thus nonzero outside the light cone. This again demonstrates the existence in $\mathbf{A}_\perp$ of an instantaneous component.

3. Symmetric Correlation Functions of the Fields in the Vacuum

The product of two fields expressed in terms of the creation and annihilation operators generates four types of products,

$$a_{\mathbf{k}\varepsilon}\, a_{\mathbf{k}'\varepsilon'} \quad a_{\mathbf{k}\varepsilon}\, a^{+}_{\mathbf{k}'\varepsilon'} \quad a^{+}_{\mathbf{k}\varepsilon}\, a_{\mathbf{k}'\varepsilon'} \quad a^{+}_{\mathbf{k}\varepsilon}\, a^{+}_{\mathbf{k}'\varepsilon'} .$$

When one takes the mean value in the vacuum, all give a zero contribution except the second, which is $\delta_{\varepsilon\varepsilon'}\delta_{\mathbf{k}\mathbf{k}'}$:

$$\langle 0 \mid E_m(1)\, E_n(2) + E_n(2)\, E_m(1) \mid 0 \rangle = \sum_{\mathbf{k},\varepsilon} \mathscr{E}_\omega^2\, \varepsilon_m\, \varepsilon_n\, e^{i[\mathbf{k}.(\mathbf{r}_1 - \mathbf{r}_2) - \omega(t_1 - t_2)]} + \text{c.c.} \tag{28.a}$$

$$\langle 0 \mid E_m(1)\, B_n(2) + B_n(2)\, E_m(1) \mid 0 \rangle = \sum_{\mathbf{k},\varepsilon} \mathscr{E}_\omega\, \mathscr{B}_\omega\, \varepsilon_m (\boldsymbol{\kappa} \times \boldsymbol{\varepsilon})_n \times \\ \times e^{i[\mathbf{k}.(\mathbf{r}_1 - \mathbf{r}_2) - \omega(t_1 - t_2)]} + \text{c.c.} \tag{28.b}$$

$$\langle 0 \mid B_m(1)\, B_n(2) + B_n(2)\, B_m(1) \mid 0 \rangle = \sum_{\mathbf{k},\varepsilon} \mathscr{B}_\omega^2 (\boldsymbol{\kappa} \times \boldsymbol{\varepsilon})_m\, (\boldsymbol{\kappa} \times \boldsymbol{\varepsilon})_n \times \\ \times e^{i[\mathbf{k}.(\mathbf{r}_1 - \mathbf{r}_2) - \omega(t_1 - t_2)]} + \text{c.c.}\,. \tag{28.c}$$

The expressions on the right are identical to those of (15.a, b, c) up to replacement of $-$c.c. by $+$c.c. It is thus not necessary to repeat these calculations; it is sufficient to replace on the right of (20.a, b) the real part D of D_+ by its imaginary part iD_1. One gets

$$\langle 0 \mid E_m(1)\, E_n(2) + E_n(2)\, E_m(1) \mid 0 \rangle = c^2 \langle 0 \mid B_m(1)\, B_n(2) + B_n(2)\, B_m(1) \mid 0 \rangle \\ = -\frac{\hbar}{\varepsilon_0\, c} \left(\delta_{mn} \frac{\partial^2}{\partial \tau^2} - c^2 \frac{\partial}{\partial \rho_m} \frac{\partial}{\partial \rho_n} \right) D_1(\boldsymbol{\rho}, \tau) \tag{28.d}$$

$$\langle 0 \mid E_m(1)\, B_n(2) + B_n(2)\, E_m(1) \mid 0 \rangle = \frac{\hbar}{\varepsilon_0\, c} \sum_l \varepsilon_{mnl} \frac{\partial}{\partial \tau} \frac{\partial}{\partial \rho_l} D_1(\boldsymbol{\rho}, \tau) \tag{28.e}$$

where D_1 is given by Equations (9) and (12).

Off the light cone ($\rho^2 \neq c^2\tau^2$), D_1 is finite. Taking $\eta = 0$ in (12) gives immediately

$$D_1(\boldsymbol{\rho}, \tau) = \frac{1}{2\,\pi^2} \frac{1}{\rho^2 - c^2\,\tau^2} \qquad (\rho^2 \neq c^2\,\tau^2)\,. \tag{29}$$

The calculation of the derivatives presents no difficulty, and one gets

$$\langle 0 \mid E_m(1)\, E_n(2) + E_n(2)\, E_m(1) \mid 0 \rangle = \frac{2\,\hbar c}{\pi^2\, \varepsilon_0} \frac{\rho_m\, \rho_n - \delta_{mn}(\rho^2 + c^2\,\tau^2)}{(\rho^2 - c^2\,\tau^2)^3} \tag{30.a}$$

$$\langle 0 \mid E_m(1)\, B_n(2) + B_n(2)\, E_m(1) \mid 0 \rangle = -\frac{2\,\hbar c}{\pi^2\, \varepsilon_0} \sum_l \varepsilon_{mnl} \frac{2\,\rho_l\, c\tau}{(\rho^2 - c^2\,\tau^2)^3}\,. \tag{30.b}$$

The autocorrelation functions of **E** and of **B** decrease as $1/\rho^4$ and $1/\tau^4$ at infinity; between different components ($m \neq n$), the decrease is more rapid: as $1/\tau^6$. The correlation functions between **E** and **B** decrease as $1/\rho^5$ and $1/\tau^5$.

The explicit form of the correlation functions near the light cone is very complex. To get them one starts from Equation (12), being careful to keep η finite, and takes the derivative twice with respect to ρ and τ. The result is a rational fraction in ρ and $c\tau$ whose numerator is of 8th degree and whose denominator is of 12th degree in the general case. We will give here the result in the special case of the time correlation functions of the field at a given point ($\boldsymbol{\rho} = \mathbf{0}$). One can find them starting with (28.d) and (28.e) or else with the sums of the type (28.a) and (28.b), which simplify significantly for $\boldsymbol{\rho} = \mathbf{0}$. One gets

$$\langle 0 | E_m(\mathbf{r}, t_1)\, E_n(\mathbf{r}, t_2) + E_n(\mathbf{r}, t_2)\, E_m(\mathbf{r}, t_1) | 0 \rangle = \frac{2\,\hbar c}{\pi^2\,\varepsilon_0}\,\delta_{mn}\,\frac{c^4\,\tau^4 - 6\,\eta^2\,c^2\,\tau^2 + \eta^4}{(\eta^2 + c^2\,\tau^2)^4} \tag{31.a}$$

$$\langle 0 | E_m(\mathbf{r}, t_1)\, B_n(\mathbf{r}, t_2) + B_n(\mathbf{r}, t_2)\, E_m(\mathbf{r}, t_1) | 0 \rangle = 0\,. \tag{31.b}$$

The second result is evident; since D_1 is even in $\boldsymbol{\rho}$, its derivative with respect to ρ_l is odd in ρ_l and therefore zero at $\boldsymbol{\rho} = \mathbf{0}$. The autocorrelation function of the fields (31.a) is clearly even in τ. It decreases as $1/\tau^4$ at infinity, and its general shape is given in Figure 1.

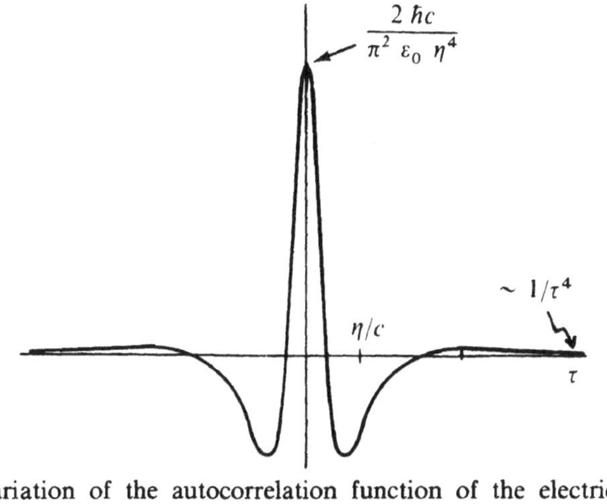

Figure 1. Variation of the autocorrelation function of the electric field in the vacuum with τ [Equation (31.a)]. η is vanishingly small. For $\tau \gg \eta/c$, the function is positive and decreases as $|\tau|^{-4}$.

General references and additional reading

Commutators between the components of the free field and the properties of the special functions like D, D_1, etc. are studied in most works on relativistic quantum electrodynamics, such as Akhiezer and Berestetski (§13) and Heitler (Chapter II, §§8, 9).

COMPLEMENT D_{III}

EXERCISES

Exercise 1. Commutators of **A**, $\mathbf{E}_\perp$, and **B** in the Coulomb gauge
Exercise 2. Hamiltonian of a system of two particles with opposite charges coupled to the electromagnetic field
Exercise 3. Commutation relations for the total momentum **P** with H_P, H_R, and H_I
Exercise 4. Bose–Einstein distribution
Exercise 5. Quasi-probability densities and characteristic functions
Exercise 6. Quadrature components of a single-mode field. Graphical representation of the state of the field
Exercise 7. Squeezed states of the radiation field
Exercise 8. Generation of squeezed states by two-photon interactions
Exercise 9. Quasi-probability density of a squeezed state

1. COMMUTATORS OF A, $\mathbf{E}_\perp$, AND B IN THE COULOMB GAUGE

Express the commutators $[A_m(\mathbf{r}), A_n(\mathbf{r}')]$, $[A_m(\mathbf{r}), E_{\perp n}(\mathbf{r}')]$, $[E_{\perp m}(\mathbf{r}), E_{\perp n}(\mathbf{r}')]$, and $[E_{\perp m}(\mathbf{r}), B_n(\mathbf{r}')]$ $(m, n = x, y, z)$ as a function of the creation and annihilation operators a_i^+ and a_i of the different modes. Find their values.

Solution

Let V_m and W_n be two field components. These can be expressed as linear combinations of a_i and a_i^+ [cf. (A.3), (A.4), and (A.5)]:

$$V_m = \sum_i v_{mi}\, a_i + v_{mi}^*\, a_i^+ \tag{1.a}$$

$$W_n = \sum_j w_{nj}\, a_j + w_{nj}^*\, a_j^+ \tag{1.b}$$

$$[V_m, W_n] = \sum_{i,j} v_{mi}\, w_{mj}[a_i, a_j] + v_{mi}\, w_{nj}^*[a_i, a_j^+] + v_{mi}^*\, w_{nj}[a_i^+, a_j] + v_{mi}^*\, w_{nj}^*[a_i^+, a_j^+]\,. \tag{2}$$

From the fundamental commutation relations (A.2) this last expression reduces to

$$[V_m, W_n] = \sum_i (v_{mi}\, w_{ni}^* - v_{mi}^*\, w_{ni})\,. \tag{3}$$

We apply this equation in the different cases:

$$V_m = A_m(\mathbf{r}) \to v_{mi} = \mathscr{A}_{\omega_i}\, \varepsilon_{im}\, e^{i\mathbf{k}_i\cdot\mathbf{r}} \tag{4.a}$$

$$W_n = A_n(\mathbf{r}') \to w_{ni} = \mathscr{A}_{\omega_i}\, \varepsilon_{in}\, e^{i\mathbf{k}_i\cdot\mathbf{r}'} \tag{4.b}$$

$$[A_m(\mathbf{r}), A_n(\mathbf{r}')] = \sum_{\mathbf{k}\varepsilon} \mathscr{A}_\omega^2 \, \varepsilon_m \, \varepsilon_n (e^{i\mathbf{k}.\mathbf{r}} \, e^{-i\mathbf{k}.\mathbf{r}'} - e^{-i\mathbf{k}.\mathbf{r}} \, e^{i\mathbf{k}.\mathbf{r}'}) = 0 \tag{5}$$

since the contributions of the modes $\mathbf{k}$ and $-\mathbf{k}$ are opposite in the summation which arises in (5). The calculation is identical for the commutator between components of the electric field to within the multiplicative factor of $i\omega$ for $\mathscr{A}_\omega$.

$$[E_{\perp m}(\mathbf{r}), E_{\perp n}(\mathbf{r}')] = 0 \, . \tag{6}$$

For the commutator between A_m and $E_{\perp n}$, (4.a) remains unchanged and

$$w_{ni} = i\omega_i \, \mathscr{A}_{\omega_i} \, \varepsilon_{in} \, e^{i\mathbf{k}_i.\mathbf{r}'} \tag{7}$$

We get

$$\begin{aligned} [A_m(\mathbf{r}), E_{\perp n}(\mathbf{r}')] &= \sum_{\mathbf{k},\varepsilon} - i\omega \, \mathscr{A}_\omega^2 \, \varepsilon_m \, \varepsilon_n (e^{i\mathbf{k}.\mathbf{r}} \, e^{-i\mathbf{k}.\mathbf{r}'} + e^{-i\mathbf{k}.\mathbf{r}} \, e^{i\mathbf{k}.\mathbf{r}'}) \\ &= \sum_{\mathbf{k},\varepsilon} - 2\,i \frac{\hbar}{2\,\varepsilon_0 \, L^3} \, \varepsilon_m \, \varepsilon_n \, e^{i\mathbf{k}.(\mathbf{r}-\mathbf{r}')} \end{aligned} \tag{8}$$

We have used the expression (A.6) for $\mathscr{A}_\omega$ and the fact that the contributions of the two terms in the parentheses of the right-hand side of (8) are equal (to see this, change $\mathbf{k}$ to $-\mathbf{k}$ in the summation over the modes). We now replace this discrete summation on the modes by an integral [see (C.34) of Chapter I], and use (1) from Complement A_I. Then we have

$$\begin{aligned} [A_m(\mathbf{r}), E_{\perp n}(\mathbf{r}')] &= \frac{\hbar}{i\varepsilon_0} \frac{1}{(2\,\pi)^3} \int d^3k \, e^{i\mathbf{k}.(\mathbf{r}-\mathbf{r}')} \sum_{\varepsilon} \varepsilon_m \, \varepsilon_n \\ &= \frac{\hbar}{i\varepsilon_0} \frac{1}{(2\,\pi)^3} \int d^3k \left(\delta_{mn} - \frac{k_m \, k_n}{k^2} \right) e^{i\mathbf{k}.(\mathbf{r}-\mathbf{r}')} \, . \end{aligned} \tag{9}$$

The function $\delta_{mn}^{\perp}(\mathbf{r} - \mathbf{r}')$ defined by (B.17), Chapter I, appears and leads to

$$[A_m(\mathbf{r}), E_{\perp n}(\mathbf{r}')] = \frac{\hbar}{i\varepsilon_0} \, \delta_{mn}^{\perp}(\mathbf{r} - \mathbf{r}') \, . \tag{10}$$

Finally we calculate the last commutator,

$$\begin{aligned} [E_{\perp m}(\mathbf{r}), B_n(\mathbf{r}')] &= [E_{\perp m}(\mathbf{r}), (\nabla' \times \mathbf{A}(\mathbf{r}'))_n] \\ &= \sum_{pq} \varepsilon_{npq} \, \nabla'_p [E_{\perp m}(\mathbf{r}), A_q(\mathbf{r}')] \\ &= \frac{-\hbar}{i\varepsilon_0} \frac{1}{(2\,\pi)^3} \int d^3k \sum_{\varepsilon} \left(\sum_{pq} \varepsilon_{npq} \, ik_p \, \varepsilon_q \right) \varepsilon_m \, e^{i\mathbf{k}.(\mathbf{r}'-\mathbf{r})} \, . \end{aligned} \tag{11}$$

The sum on ε of $\varepsilon_m i(\mathbf{k} \times \varepsilon)_n$ becomes, following (4) of Complement A_I, $i \sum_l \varepsilon_{mnl} k_l$. The factor ik_l can be replaced by the action of the operator $-\nabla_l$ on the function of $\mathbf{r}$ resulting from integration on $\mathbf{k}$. We get then

$$\begin{aligned} [E_{\perp m}(\mathbf{r}), B_n(\mathbf{r}')] &= \frac{\hbar}{i\varepsilon_0} \sum_l \varepsilon_{mnl} \, \nabla_l \frac{1}{(2\,\pi)^3} \int d^3k \, e^{i\mathbf{k}.(\mathbf{r}'-\mathbf{r})} \\ &= \frac{-i\hbar}{\varepsilon_0} \sum_l \varepsilon_{mnl} \, \nabla_l \, \delta(\mathbf{r} - \mathbf{r}') \end{aligned} \tag{12}$$

which gives (A.15).

Note finally that (5), (6), (10), and (12) are special cases of the commutation relations between fields at two different times discussed in Complement C_{III}.

2. Hamiltonian of a system of two particles with opposite charges coupled to the electromagnetic field

Consider a system of two particles 1 and 2 with masses m_1 and m_2, opposite charges $q_1 = -q_2 = q$, positions $\mathbf{r}_1$ and $\mathbf{r}_2$, and momenta $\mathbf{p}_1$ and $\mathbf{p}_2$. These two particles are assumed to form a bound state whose dimensions, of the order of a_o, are small compared to the wavelength λ of the modes of the radiation field which are taken into account in the interaction Hamiltonian. All calculations are made to order 0 in a_o/λ (long-wavelength approximation), and the magnetic spin couplings are neglected.

a) Write the Hamiltonian in the Coulomb gauge for such a system of two particles interacting with the electromagnetic field described by the transverse vector potential **A**. The results will be expressed as a function of the center-of-mass variables (external variables)

$$\mathbf{R} = \frac{m_1 \mathbf{r}_1 + m_2 \mathbf{r}_2}{m_1 + m_2} \quad (1.a) \qquad \mathbf{P} = \mathbf{p}_1 + \mathbf{p}_2 \quad (1.b)$$

and the relative variables (internal variables)

$$\mathbf{r} = \mathbf{r}_1 - \mathbf{r}_2 \quad (2.a) \qquad \frac{\mathbf{p}}{m} = \frac{\mathbf{p}_1}{m_1} - \frac{\mathbf{p}_2}{m_2} \quad (2.b)$$

where m is the reduced mass

$$m = \frac{m_1 m_2}{m_1 + m_2}. \tag{3}$$

We call the total mass $M = m_1 + m_2$.

b) Calculate the matrix element of the interaction Hamiltonian between the state $|\mathbf{K}, b; 0\rangle$ (center of mass in a momentum state $\hbar\mathbf{K}$, internal atomic state b, 0 photon) and the state $|\mathbf{K}', a; \mathbf{k}\boldsymbol{\varepsilon}\rangle$ (center of mass in a momentum state $\hbar\mathbf{K}'$, internal atomic state a, one photon $\mathbf{k}\boldsymbol{\varepsilon}$). How does the conservation of total momentum manifest itself in this matrix element?

Solution

a) The Hamiltonian of the particles is

$$H_P = \frac{\mathbf{p}_1^2}{2\,m_1} + \frac{\mathbf{p}_2^2}{2\,m_2} - \frac{q^2}{4\,\pi\varepsilon_0\,|\mathbf{r}_1 - \mathbf{r}_2|} \tag{4}$$

(one omits the Coulomb self-energies of the particles, which are constant). It can be written (*) as a function of the variables (1) and (2):

$$H_P = \frac{\mathbf{P}^2}{2\,M} + \frac{\mathbf{p}^2}{2\,m} - \frac{q^2}{4\,\pi\varepsilon_0\,r}. \tag{5}$$

(*) See for instance Cohen-Tannoudji, Diu, and Laloë, Chapter VII, §B.

In the interaction Hamiltonian $H_{I1} + H_{I2}$ with

$$H_{I1} = -\frac{q_1}{m_1}\mathbf{p}_1 . \mathbf{A}(\mathbf{r}_1) - \frac{q_2}{m_2}\mathbf{p}_2 . \mathbf{A}(\mathbf{r}_2) \tag{6}$$

$$H_{I2} = \frac{q_1^2}{2\,m_1}\mathbf{A}^2(\mathbf{r}_1) + \frac{q_2^2}{2\,m_2}\mathbf{A}^2(\mathbf{r}_2), \tag{7}$$

one can, to order 0 in a_o/λ, replace $\mathbf{A}(\mathbf{r}_1)$ and $\mathbf{A}(\mathbf{r}_2)$ by $\mathbf{A}(\mathbf{R})$, since $|\mathbf{r}_1 - \mathbf{R}|$ and $|\mathbf{r}_2 - \mathbf{R}|$ are of the order of a_o and $\mathbf{A}$ varies over distances of the order of λ. Since $q_1 = -q_2 = q$, this then becomes

$$H_{I1} = -q\left(\frac{\mathbf{p}_1}{m_1} - \frac{\mathbf{p}_2}{m_2}\right)\cdot \mathbf{A}(\mathbf{R}) = -q\frac{\mathbf{p}}{m}\cdot \mathbf{A}(\mathbf{R}) \tag{8}$$

$$H_{I2} = \frac{q^2}{2}\left(\frac{1}{m_1} + \frac{1}{m_2}\right)\mathbf{A}^2(\mathbf{R}) = \frac{q^2}{2\,m}\mathbf{A}^2(\mathbf{R}). \tag{9}$$

The external variables only arise in $\mathbf{A}$, which is evaluated at the center of mass of the system. The internal variables appear only in H_{I1} through the relative velocity $\mathbf{p}/m$.

b) In the matrix element

$$\langle\, \mathbf{K}', a; \mathbf{k}\boldsymbol{\varepsilon} \,|\, H_I \,|\, \mathbf{K}, b; 0 \,\rangle, \tag{10}$$

only H_{I1} contributes (the term linear in a and a^+). Additionally, for the creation of a photon $\mathbf{k}\boldsymbol{\varepsilon}$, it is necessary to take the coefficient of $a^+_{\mathbf{k}\boldsymbol{\varepsilon}}$ in the expansion of $\mathbf{A}(\mathbf{R})$. One then gets for (10)

$$\sqrt{\frac{\hbar}{2\,\varepsilon_0\,\omega L^3}}\;\langle\, \mathbf{K}', a; \mathbf{k}\boldsymbol{\varepsilon} \left| -\frac{q}{m}\mathbf{p}.\boldsymbol{\varepsilon}\, \mathrm{e}^{-i\mathbf{k}.\mathbf{R}}\, a^+_{\mathbf{k}\boldsymbol{\varepsilon}} \right| \mathbf{K}, b; 0 \,\rangle. \tag{11}$$

The matrix elements of $\mathbf{p}$, $\mathrm{e}^{-i\mathbf{k}\cdot\mathbf{R}}$, and $a^+_{\mathbf{k}\boldsymbol{\varepsilon}}$ are factored to give

$$-\frac{q}{m}\sqrt{\frac{\hbar}{2\varepsilon_0\,\omega L^3}}\;\langle\, a \,|\, \boldsymbol{\varepsilon}.\mathbf{p} \,|\, b \,\rangle \langle\, \mathbf{K}' \,|\, \mathrm{e}^{-i\mathbf{k}.\mathbf{R}} \,|\, \mathbf{K} \,\rangle. \tag{12}$$

The last matrix element is proportional to

$$\int \mathrm{d}^3 R\, \mathrm{e}^{-i\mathbf{K}'.\mathbf{R}}\, \mathrm{e}^{-i\mathbf{k}.\mathbf{R}}\, \mathrm{e}^{i\mathbf{K}.\mathbf{R}},$$

that is to say, finally, to $\delta(\mathbf{K} - \mathbf{k} - \mathbf{K}')$, which implies that the total momentum of the initial state, $\hbar\mathbf{K}$, must be equal to the total momentum of the final state, $\hbar\mathbf{K}' + \hbar\mathbf{k}$.

3. Commutation relations for the total momentum P with H_P, H_R, and H_I

Let $\mathbf{P}$ be the total momentum of the system (charged particles + quantized electromagnetic field). Let H be its Hamiltonian. Recall that $\mathbf{P}$ and H are given respectively by

$$\mathbf{P} = \sum_\alpha \mathbf{p}_\alpha + \sum_i \hbar\mathbf{k}_i\, a_i^+ a_i \tag{1}$$

$$H = H_P + H_R + H_{I1} + H_{I2} \tag{2}$$

with

$$H_P = \sum_\alpha \frac{\mathbf{p}_\alpha^2}{2\,m_\alpha} + \sum_{\alpha>\beta} \frac{q_\alpha q_\beta}{4\,\pi\varepsilon_0\,|\,\mathbf{r}_\alpha - \mathbf{r}_\beta\,|} + \sum_\alpha \varepsilon_\alpha^{\text{coul}} \tag{3.a}$$

$$H_R = \sum_i \hbar\omega_i\left(a_i^+ a_i + \frac{1}{2}\right) \tag{3.b}$$

$$H_{I1} = -\sum_\alpha \frac{q_\alpha}{m_\alpha} \mathbf{p}_\alpha \cdot \mathbf{A}(\mathbf{r}_\alpha) \tag{3.c}$$

$$H_{I2} = \sum_\alpha \frac{q_\alpha^2}{2\,m_\alpha^2} \mathbf{A}^2(\mathbf{r}_\alpha) \tag{3.d}$$

Show by using the fundamental commutation relations that $\mathbf{P}$ commutes with H_P, H_R, H_{I1}, and H_{I2}.

Solution

To solve this exercise we need the following commutation relations between $\mathbf{p}_\alpha$, a_i, a_i^+, and a general operator M:

$$[(\mathbf{p}_\alpha)_j, M] = -i\hbar \frac{\partial M}{\partial (\mathbf{x}_\alpha)_j} \qquad j = x, y, z \tag{4.a}$$

$$[a_i, M] = \frac{\partial M}{\partial a_i^+} \quad (4.b) \qquad\qquad [a_i^+, M] = -\frac{\partial M}{\partial a_i}. \quad (4.c)$$

Let us begin with $[\mathbf{P}, H_P]$. Since a_i and a_i^+ commute with the particle operators, we have

$$[\mathbf{P}, H_p] = \left[\sum_\alpha \mathbf{p}_\alpha, H_p\right] = \left[\sum_\alpha \mathbf{p}_\alpha, \sum_{\beta > \gamma} \frac{q_\beta\, q_\gamma}{4\,\pi\varepsilon_0\, |\mathbf{r}_\beta - \mathbf{r}_\gamma|}\right]. \tag{5}$$

For a given pair of particles β, γ in V_{Coul}, the commutators relative to $\mathbf{p}_\beta$ and $\mathbf{p}_\gamma$ have opposite signs, so that (5) is zero.

The commutator $[\mathbf{P}, H_R]$ is clearly zero, since the operator $N_P = a_i^+ a_i$ arises both in $\mathbf{P}$ and in H_R.

To study the commutators $[\mathbf{P}, H_{I1}]$ and $[\mathbf{P}, H_{I2}]$, we show first of all that $[\mathbf{P}, A_j(\mathbf{r}_\alpha)] = \mathbf{0}$. For this we examine the two terms of (1) separately and use the expansion (A.5) of $\mathbf{A}(\mathbf{r}_\alpha)$:

$$\left[\sum_\beta \mathbf{p}_\beta, A_j(\mathbf{r}_\alpha)\right] = -i\hbar \nabla_{\mathbf{r}_\alpha} A_j(\mathbf{r}_\alpha) = \sum_i \hbar \mathbf{k}_i\, \mathscr{A}_{\omega_i} (\boldsymbol{\varepsilon}_i)_j (a_i\, e^{i\mathbf{k}_i \cdot \mathbf{r}_\alpha} - a_i^+\, e^{-i\mathbf{k}_i \cdot \mathbf{r}_\alpha}) \tag{6}$$

$$\left[\sum_i \hbar \mathbf{k}_i\, a_i^+ a_i, A_j(\mathbf{r}_\alpha)\right] = \sum_i \hbar \mathbf{k}_i \left(a_i^+ \frac{\partial A_j}{\partial a_i^+} - \frac{\partial A_j}{\partial a_i} a_i\right) =$$

$$= \sum_i \hbar \mathbf{k}_i\, \mathscr{A}_{\omega_i} (\boldsymbol{\varepsilon}_i)_j (a_i^+\, e^{-i\mathbf{k}_i \cdot \mathbf{r}_\alpha} - a_i\, e^{i\mathbf{k}_i \cdot \mathbf{r}_\alpha}). \tag{7}$$

By adding (6) and (7) it is clear that $[\mathbf{P}, A_j(\mathbf{r}_\alpha)]$ is zero. The expressions (3.c) and (3.d) show that H_{I1} and H_{I2} are constructed starting from the operators $(p_\alpha)_i$ and $A_j(\mathbf{r}_\alpha)$, which both commute with $\mathbf{P}$. As a result, they themselves commute with $\mathbf{P}$.

4. Bose–Einstein distribution

Consider a mode of the electromagnetic field in thermodynamic equilibrium at temperature T. The density matrix of the field is

$$\rho = \frac{1}{Z} e^{-\beta H} \tag{1}$$

with $\beta = 1/k_B T$ and

$$H = \hbar\omega\left(a^+ a + \frac{1}{2}\right). \tag{2}$$

a) Calculate Z by using the property that $\mathrm{Tr}\,\rho = 1$.

b) Calculate the mean value $\langle N \rangle$ and the variance $(\Delta N)^2 = \langle (N - \langle N \rangle)^2 \rangle$ of the number of photons in the mode.

c) Derive the probability $\mathcal{P}(n)$ of having n photons in the mode as a function of n and $\langle N \rangle$ (it is called the Bose distribution law).

Solution

a) The normalization condition for ρ,

$$\mathrm{Tr}\,\rho = 1 \tag{3}$$

is written

$$\frac{1}{Z}\sum_n e^{-\beta(n+\frac{1}{2})\hbar\omega} = 1 \tag{4.a}$$

so that by carrying out the summation over n, we have

$$\frac{1}{Z}\left(\frac{e^{-\beta\hbar\omega/2}}{1 - e^{-\beta\hbar\omega}}\right) = 1 \tag{4.b}$$

from which

$$Z = \frac{e^{-\beta\hbar\omega/2}}{1 - e^{-\beta\hbar\omega}}. \tag{5}$$

b) The mean number of photons in mode n is given by

$$\langle N \rangle = \sum_n n\rho_{nn} \tag{6}$$

where ρ_{nn} is evaluated using (5):

$$\rho_{nn} = e^{-\beta n\hbar\omega}(1 - e^{-\beta\hbar\omega}). \tag{7}$$

To calculate $\langle N \rangle$, start with the following expression used in going from (4.a) to (4.b):

$$\sum_n e^{-nx} = \frac{1}{1 - e^{-x}} \tag{8}$$

where $x = \beta\hbar\omega$, and take the derivative of both sides with respect to x. One gets

$$\sum_n n\,e^{-nx} = \frac{e^{-x}}{(1 - e^{-x})^2}. \tag{9}$$

By comparing this expression with that gotten in (6) and (7) we find that the mean number of photons in the mode being considered is

$$\langle N \rangle = \frac{e^{-\beta\hbar\omega}}{1 - e^{-\beta\hbar\omega}} = \frac{1}{e^{\beta\hbar\omega} - 1}. \tag{10}$$

To get the variance $(\Delta N)^2$, it is necessary to first calculate

$$\langle N^2 \rangle = \sum_n n^2 \rho_{nn} = \sum_n n^2 e^{-n\beta\hbar\omega}(1 - e^{-\beta\hbar\omega}). \tag{11}$$

For this it is sufficient to differentiate (9) with respect to x:

$$\sum_n n^2 \, e^{-nx} = \frac{e^{-x}(1 + e^{-x})}{(1 - e^{-x})^3} \tag{12}$$

which gives

$$\langle N^2 \rangle = \frac{e^{-\beta\hbar\omega}(1 + e^{-\beta\hbar\omega})}{(1 - e^{-\beta\hbar\omega})^2} = \frac{1}{(e^{\beta\hbar\omega} - 1)^2} + \frac{e^{\beta\hbar\omega}}{(e^{\beta\hbar\omega} - 1)^2}, \tag{13}$$

The first term of (13) is just $\langle N \rangle^2$, so that

$$(\Delta N)^2 = \frac{e^{\beta\hbar\omega}}{(e^{\beta\hbar\omega} - 1)^2} = \frac{(e^{\beta\hbar\omega} - 1) + 1}{(e^{\beta\hbar\omega} - 1)^2} = \frac{1}{(e^{\beta\hbar\omega} - 1)} + \frac{1}{(e^{\beta\hbar\omega} - 1)^2} \tag{14}$$

which can be transformed by means of (10) into

$$\Delta N = [\langle N \rangle^2 + \langle N \rangle]^{1/2}. \tag{15}$$

We can conclude that ΔN is of the order of $\langle N \rangle$ for $\langle N \rangle$ large and of the order of $\sqrt{\langle N \rangle}$ for $\langle N \rangle$ small.

c) The probability $\mathcal{P}(n)$ is given in (7). To express it as a function of $\langle N \rangle$, we use (10) to eliminate $e^{-\beta\hbar\omega}$ and we get

$$\mathcal{P}(n) = \frac{1}{1 + \langle N \rangle} \left(\frac{\langle N \rangle}{1 + \langle N \rangle} \right)^n . \tag{16}$$

One notes that, contrary to the Poisson distribution (C.52) gotten for a quasi-classical state, $\mathcal{P}(n)$ always decreases with increasing n.

5. Quasi-probability densities and characteristic functions

Throughout this exercise one considers only a single mode of the electromagnetic field, whose creation and annihilation operators are a^+ and a. Let ρ be the density operator of this mode of the field. The purpose of the exercise is to present various properties of two quasi-probability densities $P_A(\alpha, \alpha^*)$ and $P_N(\alpha, \alpha^*)$ and of the associated "characteristic functions".

a) λ being a complex number, one defines the functions $C_N(\lambda, \lambda^*)$ and $C_A(\lambda, \lambda^*)$ by

$$C_N(\lambda, \lambda^*) = \mathrm{Tr}\,(\rho \, e^{\lambda a^+} \, e^{-\lambda^* a}) \tag{1}$$

$$C_A(\lambda, \lambda^*) = \mathrm{Tr}\,(\rho \, e^{-\lambda^* a} \, e^{\lambda a^+}). \tag{2}$$

By using the Glauber relations

$$e^{(\lambda a^+ - \lambda^* a)} = e^{\lambda a^+} \, e^{-\lambda^* a} \, e^{-|\lambda|^2/2} = e^{-\lambda^* a} \, e^{\lambda a^+} \, e^{|\lambda|^2/2} \tag{3}$$

establish the relationship between $C_N(\lambda, \lambda^*)$ and $C_A(\lambda, \lambda^*)$.

b) Consider the classical functions $f(\lambda, \lambda^*)$ and $g(\alpha, \alpha^*)$ such that

$$f(\lambda, \lambda^*) = \int d^2\alpha \; g(\alpha, \alpha^*) \, e^{(\lambda\alpha^* - \lambda^*\alpha)} \tag{4}$$

where $d^2\alpha = d\mathcal{Re}\,\alpha\, d\mathcal{Im}\,\alpha$. By taking

$$\alpha = \alpha_P + i\alpha_Q \tag{5.a}$$

$$\lambda = \frac{1}{2}(v - iu) \tag{5.b}$$

where α_P, α_Q, u, and v are real, show that f and g are related by a two-dimensional Fourier transformation and that

$$g(\alpha, \alpha^*) = \frac{1}{\pi^2}\int d^2\lambda\, f(\lambda, \lambda^*)\, e^{-(\lambda\alpha^* - \lambda^*\alpha)}. \tag{6}$$

c) Consider the function of α

$$P_A(\alpha, \alpha^*) = \frac{1}{\pi}\langle \alpha \,|\, \rho \,|\, \alpha \rangle \tag{7}$$

where $|\alpha\rangle$ is a quasi-classical state. Prove the following two relations:

$$\int d^2\alpha\, P_A(\alpha, \alpha^*) = 1 \tag{8.a}$$

$$\langle (a)^m (a^+)^l \rangle = \int d^2\alpha\, P_A(\alpha, \alpha^*)\, (\alpha)^m (\alpha^*)^l \tag{8.b}$$

Infer from these relations that $P_A(\alpha, \alpha^*)$ is a quasi-probability density suited to antinormal order.

d) By using the definition (2) for $C_A(\lambda, \lambda^*)$, show that there exist between $C_A(\lambda, \lambda^*)$ and $P_A(\alpha, \alpha^*)$ Fourier transform relations of the type of (4) and (6).

One calls $C_A(\lambda, \lambda^*)$ the characteristic function of $P_A(\alpha, \alpha^*)$. How can one express $\langle a^m (a^+)^l \rangle$ as a function of the derivatives of $C_A(\lambda, \lambda^*)$ with respect to λ and λ^* (considered as independent variables) evaluated at $\lambda = \lambda^* = 0$?

e) Assume that ρ has a *P*-representation, that is, that there exists a real function $P_N(\alpha, \alpha^*)$ such that

$$\rho = \int d^2\alpha\, P_N(\alpha, \alpha^*)\, |\, \alpha \rangle \langle \alpha \,| \tag{9}$$

where $|\alpha\rangle$ is a quasi-classical state. Show that

$$\int d^2\alpha\, P_N(\alpha, \alpha^*) = 1 \tag{10.a}$$

$$\langle (a^+)^l (a)^m \rangle = \int d^2\alpha\, P_N(\alpha, \alpha^*)\, (\alpha^*)^l (\alpha)^m . \tag{10.b}$$

Infer from these relations that $P_N(\alpha, \alpha^*)$ is a quasi-probability density distribution well suited to normal order.

f) By using (1) and (9), show that $C_N(\lambda, \lambda^*)$ and $P_N(\alpha, \alpha^*)$ are related by Fourier transform relations of the type of (4) and (6). For this reason, $C_N(\lambda, \lambda^*)$ is called the characteristic function for normal order.

Give $\langle (a^+)^l a^m \rangle$ as a function of the partial derivatives of $C_N(\lambda, \lambda^*)$ with respect to λ and λ^* (taken as independent variables) evaluated at $\lambda = \lambda^* = 0$.

g) Calculate $\langle \beta | \rho | \beta \rangle$ with the help of Equation (9) ($|\beta\rangle$ being a quasi-classical state), and derive the relationship between $P_A(\beta, \beta^*)$ and $P_N(\alpha, \alpha^*)$. Can one establish this result directly beginning with the results of *a*) on the Fourier transforms relating $C_N(\lambda, \lambda^*)$ and $C_A(\lambda, \lambda^*)$ to $P_N(\alpha, \alpha^*)$ and $P_A(\alpha, \alpha^*)$?

h) To illustrate these results with a physical example, find the functions $P_A(\alpha, \alpha^*)$ and $P_N(\alpha, \alpha^*)$ for radiation in thermodynamic equilibrium at temperature T, for which (see Exercise 4)

$$\rho = \frac{1}{Z} \mathrm{e}^{-\beta\hbar\omega(a^+a + \frac{1}{2})} \tag{11}$$

where $\beta = 1/k_B T$, Z being such that $\mathrm{Tr}\,\rho = 1$. For this particular example find $P_A(\alpha, \alpha^*)$ and $P_N(\alpha, \alpha^*)$ using the results of the preceding questions.

Solution

a) Equation (3) immediately gives the equality

$$\mathrm{e}^{\lambda a^+} \mathrm{e}^{-\lambda^* a} = \mathrm{e}^{-\lambda^* a} \mathrm{e}^{\lambda a^+} \mathrm{e}^{|\lambda|^2} \tag{12}$$

which substituted in (1) leads to

$$C_N(\lambda, \lambda^*) = C_A(\lambda, \lambda^*)\, \mathrm{e}^{|\lambda|^2}. \tag{13}$$

b) Using Equations (5.a) and (5.b), we rewrite (4) in the form

$$f(\lambda, \lambda^*) = \frac{1}{2\pi} \iint \mathrm{d}\alpha_P\, \mathrm{d}\alpha_Q (2\pi g(\alpha, \alpha^*))\, \mathrm{e}^{-\mathrm{i}\alpha_P u}\, \mathrm{e}^{-\mathrm{i}\alpha_Q v}. \tag{14}$$

We see that $f(\lambda, \lambda^*)$ and $2\pi g(\alpha, \alpha^*)$ are related by a Fourier transformation. The inverse transformation of (14) is written

$$2\pi g(\alpha, \alpha^*) = \frac{1}{2\pi} \int \mathrm{d}u\, \mathrm{d}v\, f(\lambda, \lambda^*)\, \mathrm{e}^{\mathrm{i}\alpha_P u}\, \mathrm{e}^{\mathrm{i}\alpha_Q v} \tag{15}$$

which coincides with (6), since $\mathrm{d}^2\lambda = (\mathrm{d}u\, \mathrm{d}v)/4$.

c) To derive (8.a), we start from

$$\mathrm{Tr}\,\rho = \sum_n \langle n | \rho | n \rangle = 1 \tag{16}$$

and introduce the closure relation (C.56) between $\langle n|$ and ρ:

$$\operatorname{Tr}\rho = \frac{1}{\pi}\int d^2\alpha\Big(\sum_n \langle n \mid \alpha\rangle\langle \alpha \mid \rho \mid n\rangle\Big) = \frac{1}{\pi}\int d^2\alpha\Big(\sum_n \langle \alpha \mid \rho \mid n\rangle\langle n \mid \alpha\rangle\Big) =$$

$$= \frac{1}{\pi}\int d^2\alpha \langle \alpha \mid \rho \mid \alpha\rangle \tag{17}$$

which corresponds with (8.a). We now calculate $\langle a^m(a^+)^l\rangle$:

$$\langle (a)^m (a^+)^l \rangle = \operatorname{Tr}\rho(a)^m(a^+)^l = \sum_n \langle n \mid \rho(a)^m(a^+)^l \mid n\rangle \tag{18}$$

so that, by introducing the closure relation (C.56) between a^m and $(a^+)^l$, we get

$$\begin{aligned}\langle (a)^m (a^+)^l \rangle &= \frac{1}{\pi}\int d^2\alpha\Big(\sum_n \langle n \mid \rho(a)^m \mid \alpha\rangle\langle \alpha \mid (a^+)^l \mid n\rangle\Big)\\ &= \frac{1}{\pi}\int d^2\alpha(\alpha)^m(\alpha^*)^l\Big(\sum_n \langle n \mid \rho \mid \alpha\rangle\langle \alpha \mid n\rangle\Big)\\ &= \frac{1}{\pi}\int d^2\alpha(\alpha)^m(\alpha^*)^l \langle \alpha \mid \rho \mid \alpha\rangle\end{aligned} \tag{19}$$

which agrees with (8.b).

Equations (8.a) and (8.b), together with the fact that $P_A(\alpha, \alpha^*)$ is real and positive (as a result of the general properties of the density matrix), show that $P_A(\alpha, \alpha^*)$ has properties similar to those of a probability density. In fact, it doesn't behave like a true probability density, but rather a quasi-probability density. Actually, for a system in the quasi-classical state $|\beta\rangle$, one would expect to find a zero probability density when α is different from β. This is not true for $P_A(\alpha, \alpha^*)$, which is equal to $(1/\pi)e^{-|\beta-\alpha|^2}$ when $\rho = |\beta\rangle\langle\beta|$, according to (C.55). $P_A(\alpha, \alpha^*)$ is a quasi-probability density well suited to antinormal order, since the mean value of the operator $a^m(a^+)^l$ where a and a^+ are arranged in antinormal order is easily expressed [see (8.b)] as a function of $P_A(\alpha, \alpha^*)$.

d) Introducing the closure relation (C.56) between $e^{-\lambda^* a}$ and $e^{\lambda a^+}$ in Equation (2), we get

$$\begin{aligned}C_A(\lambda, \lambda^*) &= \sum_n \frac{1}{\pi}\int d^2\alpha \langle n \mid \rho\, e^{-\lambda^* a} \mid \alpha\rangle\langle \alpha \mid e^{\lambda a^+} \mid n\rangle\\ &= \frac{1}{\pi}\int d^2\alpha\Big(\sum_n \langle n \mid \rho \mid \alpha\rangle\langle \alpha \mid n\rangle\Big)e^{-\lambda^*\alpha}\, e^{\lambda\alpha^*}\\ &= \int d^2\alpha\, P_A(\alpha, \alpha^*)\, e^{-\lambda^*\alpha}\, e^{\lambda\alpha^*}\end{aligned} \tag{20}$$

This shows that $C_A(\lambda, \lambda^*)$ is related to $P_A(\alpha, \alpha^*)$ by an expression of the form (4).

We calculate $\partial^{m+l}C_A/(\partial\lambda^*)^m(\partial\lambda)^l$ using (2):

$$\frac{\partial^{m+l}C_A}{(\partial\lambda^*)^m(\partial\lambda)^l} = \operatorname{Tr}(\rho\, e^{-\lambda^* a}(-a)^m e^{\lambda a^+}(a^+)^l) \tag{21}$$

whose value taken at $\lambda = \lambda^* = 0$ is

$$\operatorname{Tr}(\rho(-a)^m(a^+)^l) = (-1)^m \langle (a)^m(a^+)^l\rangle \tag{22}$$

which shows the relationship existing between the derivatives of the characteristic function and the mean values of the operators a and a^+ arranged in antinormal order.

e) Equations (10.a) and (10.b) have been already established in the remark of §C.4.*c*. $P_N(\alpha, \alpha^*)$ is a quasi-probability density well suited to normal order, since the mean value of a product of operators a and a^+ arranged in normal order is easily expressed by means of (10.b).

f) We calculate $C_N(\lambda, \lambda^*)$ with ρ defined by (9). Substituting this relation in (1), we find

$$C_N(\lambda, \lambda^*) = \int d^2\alpha \, P_N(\alpha, \alpha^*) \, \mathrm{Tr}\left(| \alpha \rangle \langle \alpha | e^{\lambda a^+} e^{-\lambda^* a}\right). \tag{23}$$

The cyclic property of the trace allows us to write

$$\mathrm{Tr}\left(| \alpha \rangle \langle \alpha | e^{\lambda a^+} e^{-\lambda^* a}\right) = \mathrm{Tr}\left(e^{-\lambda^* a} | \alpha \rangle \langle \alpha | e^{\lambda a^+}\right) = e^{-\lambda^* \alpha} e^{\lambda \alpha^*} \mathrm{Tr}\left(| \alpha \rangle \langle \alpha |\right) = e^{-\lambda^* \alpha} e^{\lambda \alpha^*} \tag{24}$$

which gives for $C_N(\lambda, \lambda^*)$

$$C_N(\lambda, \lambda^*) = \int d^2\alpha \, P_N(\alpha, \alpha^*) \, e^{\lambda \alpha^*} e^{-\lambda^* \alpha} \tag{25}$$

which coincides with (4).

By proceeding as in part *d*) we find

$$\frac{\partial^{l+m} C_N(\lambda, \lambda^*)}{(\partial \lambda)^l \, (\partial \lambda^*)^m} = \mathrm{Tr}\left(\rho \, e^{\lambda a^+} (a^+)^l \, e^{-\lambda^* a} (-a)^m\right) \tag{26}$$

whose value for $\lambda = \lambda^* = 0$ is equal to $(-1)^m \langle (a^+)^l a^m \rangle$.

g) We find the diagonal matrix element $\langle \beta | \rho | \beta \rangle$ by using (9) and (C.55):

$$\langle \beta | \rho | \beta \rangle = \int d^2\alpha \, P_N(\alpha, \alpha^*) \, | \langle \beta | \alpha \rangle |^2 = \int d^2\alpha \, P_N(\alpha, \alpha^*) \, e^{-|\beta - \alpha|^2} \tag{27}$$

which shows, taking into account (7), that $P_A(\beta, \beta^*)$ is the convolution of $P_N(\alpha, \alpha^*)$ and a Gaussian function

$$P_A(\beta, \beta^*) = \frac{1}{\pi} \int d^2\alpha \, P_N(\alpha, \alpha^*) \, e^{-|\beta - \alpha|^2}. \tag{28}$$

This result can be found by other methods. Equation (13) shows that $C_A(\lambda, \lambda^*)$ is the product of $C_N(\lambda, \lambda^*)$ with $e^{-|\lambda|^2}$. By taking the Fourier transform of both sides, one finds on one hand P_A and on the other the convolution of the Fourier transforms of C_N and of $e^{-|\lambda|^2}$, that is, as in (28), the convolution of P_N and the Gaussian distribution.

h) Starting with the form

$$\rho = \frac{1}{1 + \langle N \rangle} \sum_n \left(\frac{\langle N \rangle}{1 + \langle N \rangle} \right)^n | n \rangle \langle n | \tag{29}$$

gotten from (16) of Exercise 4, we calculate $\langle \alpha | \rho | \alpha \rangle$:

$$\begin{aligned} \langle \alpha | \rho | \alpha \rangle &= \frac{1}{1 + \langle N \rangle} \sum_n \left(\frac{\langle N \rangle}{1 + \langle N \rangle} \right)^n | \langle n | \alpha \rangle |^2 \\ &= \frac{e^{-|\alpha|^2}}{1 + \langle N \rangle} \sum_n \left(\frac{\langle N \rangle}{1 + \langle N \rangle} \right)^n \frac{| \alpha |^{2n}}{n!} \\ &= \frac{e^{-|\alpha|^2}}{1 + \langle N \rangle} \exp \frac{\langle N \rangle | \alpha |^2}{1 + \langle N \rangle} \end{aligned} \tag{30}$$

from which we get

$$P_A(\alpha, \alpha^*) = \frac{1}{\pi(1 + \langle N \rangle)} \exp \frac{-| \alpha |^2}{1 + \langle N \rangle} \tag{31}$$

By using (20) and the fact that the Fourier transform of a Gaussian is a Gaussian, we find

$$C_A(\lambda, \lambda^*) = \exp[-(1 + \langle N \rangle) | \lambda |^2] \qquad (32)$$

$C_N(\lambda, \lambda^*)$ is then gotten using (13):

$$C_N(\lambda, \lambda^*) = e^{-\langle N \rangle |\lambda|^2} \qquad (33)$$

and one finds $P_N(\alpha, \alpha^*)$ by the Fourier transform

$$P_N(\alpha, \alpha^*) = \frac{1}{\pi^2} \int d^2\lambda \; C_N(\lambda, \lambda^*) \, e^{(\lambda^* \alpha - \lambda \alpha^*)}$$

$$= \frac{1}{\pi \langle N \rangle} e^{-|\alpha|^2/\langle N \rangle} . \qquad (34)$$

The representation $P_N(\alpha, \alpha^*)$ of radiation at thermal equilibrium is thus a Gaussian centered about the origin and whose width is of the order of $\sqrt{\langle N \rangle}$.

6. Quadrature components of a single-mode field. Graphical representation of the state of the field

Throughout this exercise, assume that only one mode $\mathbf{k}\boldsymbol{\varepsilon}$ of the field is populated, all other modes being empty (single-mode field). One is only interested in the contribution of this mode to the free electric field, which is written from the Heisenberg point of view:

$$\mathbf{E}(\mathbf{r}, t) = i\mathscr{E}_\omega \, \boldsymbol{\varepsilon}[a \, e^{i(\mathbf{k}.\mathbf{r} - \omega t)} - a^+ \, e^{-i(\mathbf{k}.\mathbf{r} - \omega t)}] \qquad (1)$$

with

$$\mathscr{E}_\omega = [\hbar\omega / 2 \, \varepsilon_0 \, L^3]^{1/2} .$$

a) One introduces the two Hermitian linear combinations of the creation and annihilation operators a and a^+ defined by

$$a_P = \frac{1}{2}(a + a^+) \qquad (2.a)$$

$$a_Q = \frac{1}{2i}(a - a^+) . \qquad (2.b)$$

Find the commutator $[a_P, a_Q]$, and show that $\Delta a_P \, \Delta a_Q \geq \frac{1}{4}$, where Δa_P and Δa_Q are the dispersions (root mean square deviations) of a_P and a_Q in the state of the mode under consideration.

Show that the Hamiltonian H of the mode can be written

$$\frac{H}{\hbar\omega} = \frac{1}{2}(a^+ a + a a^+) = a_P^2 + a_Q^2 . \qquad (3)$$

b) Give the electric field $\mathbf{E}(\mathbf{r}, t)$ as a function of a_P and a_Q. Show that the operators a_P and a_Q represent two quadrature components of the field. What is the physical consequence of the nonzero value of the commutator $[a_P, a_Q]$?

c) Show that

$$a_\theta = a_P \cos\theta + a_Q \sin\theta \tag{4}$$

represents physically the component of the field out of phase by θ with respect to that described by a_P. Show that $H/\hbar\omega$ can also be written

$$\frac{H}{\hbar\omega} = a_\theta^2 + a_{\theta+\pi/2}^2 \,. \tag{5}$$

d) Assume that the state of the field is a quasi-classical state $|\alpha\rangle$, and take

$$\begin{cases} \alpha_P = \frac{1}{2}(\alpha + \alpha^*) & (6.a) \\ \alpha_Q = \frac{1}{2\mathrm{i}}(\alpha - \alpha^*) & (6.b) \\ \alpha_\theta = \alpha_P \cos\theta + \alpha_Q \sin\theta\,. & (6.c) \end{cases}$$

Calculate the means values $\langle a_P\rangle$, $\langle a_Q\rangle$, and $\langle a_\theta\rangle$ of the operators a_P, a_Q, and a_θ in the state $|\alpha\rangle$ as well as the dispersions Δa_P, Δa_Q, and Δa_θ. Show that these three dispersions are equal, and find their common value δ. By using the well-known results for the harmonic oscillator, show that the distributions of the possible values of a_P, a_Q, anda_θ in the state $|\alpha\rangle$ are Gaussian.

e) Consider a classical single-mode field whose state is described by the normal variable α. It is convenient to represent such a state by a point M, of abscissa α_P and ordinate α_Q, in a plane with two orthogonal axes Ox and Oy. This plane can equally well be thought of as the complex plane, α_P and α_Q being the real and imaginary parts of the complex number α. Show that α_θ is gotten by projecting M on an axis $O\theta$ passing through the origin and making an angle θ with Ox. How can one characterize graphically the phase φ and the energy $H_{\text{cl}}/\hbar\omega$ (in units of $\hbar\omega$) of the classical field in state α?

f) Consider now a quantum field in the quasi-classical state $|\alpha\rangle$. The incompatibility of a_P and a_Q suggests no longer representing the state of the field by a point, but rather by a small area centered about the point M representing the classical state α. Show that if one takes a disk of radius δ, where δ has been found above in *d*), then one gets, by projecting all of the points of this surface onto the Ox, Oy, and $O\theta$ axes, three segments correctly centered on $\langle a_P\rangle$, $\langle a_Q\rangle$, and $\langle a_\theta\rangle$ and whose width 2δ gives a good indication of the dispersions Δa_P, Δa_Q, and Δa_θ.

g) The quantum field is still in state $|\alpha\rangle$ but assume now in addition that $|\alpha| \gg 1$. By using results relating to quasi-classical states, show that the dispersion ΔN of the number of photons in the state $|\alpha\rangle$ is simply related to the length of the segment OM.

Let $2\,\Delta\varphi$ be the angle subtended at the origin by the disk centered at M and associated with the state $|\alpha\rangle$. Explain qualitatively, without trying to define a phase operator precisely, why $\Delta\varphi$ gives the order of magnitude of the phase dispersion in the state $|\alpha\rangle$. Give $\Delta\varphi$ as a function of δ and OM in the limit $|\alpha| \gg 1$. What relationship exists between $\Delta\varphi$ and ΔN?

h) There are states of the field which, while being minimal like the state $|\alpha\rangle$ (that is, such that $\Delta a_P\, \Delta a_Q = \frac{1}{4}$), and while leading to the same mean values of a_P, a_Q, and a_θ as the state $|\alpha\rangle$, do not have equivalent dispersions for these various observables. These states can be represented qualitatively by ellipses centered on M with the same area as the disks considered in *f* and *g*. They are called squeezed states. Examples of such states are examined in Exercise 7.

For a given ratio between the major and minor axes of the ellipse, sketch the squeezed states corresponding to each of the following cases:

i) the dispersion is minimal on a_P,
ii) the dispersion is minimal on a_Q,
iii) the dispersion is minimal on a_θ,
iv) the dispersion is minimal on φ,
v) the dispersion is minimal on the amplitude.

Solution

a) From $[a, a^+] = 1$ and Equations (2), it follows that

$$[a_P, a_Q] = \frac{i}{2}. \tag{7}$$

The general relation

$$\Delta a_P\, \Delta a_Q \geqslant \frac{1}{2}\,|\langle [a_P, a_Q] \rangle| \tag{8}$$

then gives

$$\Delta a_P\, \Delta a_Q \geqslant 1/4. \tag{9}$$

Finally, Equation (3) is a consequence of (2).

b) By replacing $e^{\pm i(\mathbf{k}\cdot\mathbf{r}-\omega t)}$ with $\cos(\mathbf{k}\cdot\mathbf{r}-\omega t) \pm i\sin(\mathbf{k}\cdot\mathbf{r}-\omega t)$ in (1) and using (2), we can write $\mathbf{E}(\mathbf{r}, t)$ in the form

$$\mathbf{E}(\mathbf{r}, t) = -\,2\,\mathscr{E}_\omega\,\boldsymbol{\varepsilon}[a_P \sin(\mathbf{k}\cdot\mathbf{r}-\omega t) + a_Q \cos(\mathbf{k}\cdot\mathbf{r}-\omega t)] \tag{10}$$

which shows that a_P and a_Q describe two quadrature components of the field.

Equation (7) shows that these physical variables are incompatible. Any increase in the precision of one component of the field is accompanied by an increase in the uncertainty of the quadrature component.

c) Changing from θ to $\theta + \pi/2$ in (4) yields

$$a_{\theta+\pi/2} = -a_P \sin\theta + a_Q \cos\theta . \tag{11}$$

Starting from (4) and (11), it is possible to give a_P and a_Q as functions of a_θ and $a_{\theta+\pi/2}$. Substitution of these expressions in (10) then gives

$$\mathbf{E}(\mathbf{r}, t) = -2\,\mathscr{E}_\omega\,\boldsymbol{\varepsilon}[a_\theta \sin(\mathbf{k}\cdot\mathbf{r} - \omega t + \theta) + a_{\theta+\pi/2}\cos(\mathbf{k}\cdot\mathbf{r} - \omega t + \theta)] \tag{12}$$

which shows that the components of the field associated with a_P and a_θ are out of phase by θ.

It follows also from (4) and (11) that

$$a_\theta^2 + a_{\theta+\pi/2}^2 = a_P^2 + a_Q^2 \tag{13}$$

which, taking account of (3), gives (5).

d) The calculation of the mean values of the operators (2.a), (2.b), and (4) in the state $|\alpha\rangle$ gives, when taken with the relations $a|\alpha\rangle = \alpha|\alpha\rangle$ and $\langle\alpha|a^+ = \alpha^*\langle\alpha|$ and the definitions (6),

$$\langle a_P \rangle = \alpha_P \tag{14.a}$$

$$\langle a_Q \rangle = \alpha_Q \tag{14.b}$$

$$\langle a_\theta \rangle = \alpha_\theta . \tag{14.c}$$

Now we calculate Δa_P^2:

$$\begin{aligned}\Delta a_P^2 &= \langle \alpha | a_P^2 | \alpha \rangle - (\langle \alpha | a_P | \alpha \rangle)^2 \\ &= \frac{1}{4}\langle \alpha | a^2 + a^{+2} + aa^+ + a^+ a | \alpha \rangle - \alpha_P^2 \\ &= \frac{1}{4}\langle \alpha | a^2 + a^{+2} + 2a^+ a + 1 | \alpha \rangle - \alpha_P^2 \\ &= \frac{1}{4}(\alpha^2 + \alpha^{*2} + 2\alpha^*\alpha + 1) - \alpha_P^2 \\ &= \alpha_P^2 + \frac{1}{4} - \alpha_P^2 = \frac{1}{4}\end{aligned} \tag{15}$$

Analogous calculations can be made for Δa_Q^2 and Δa_θ^2. One finds then that

$$\Delta a_P = \Delta a_Q = \Delta a_\theta = \delta = \frac{1}{2} . \tag{16}$$

Equations (3) and (5) along with Equation (7) and an analogous equation for $[a_\theta, a_{\theta+\pi/2}]$, show that the Hermitian operators a_P and a_Q or a_θ and $a_{\theta+\pi/2}$ are analogous to the position x and the momentum p of a one-dimensional harmonic oscillator. Since the distributions of x and p in a quasi-classical state are Gaussians and since this state is a minimal state, it follows that the distributions of a_P, a_Q, and a_θ are Gaussians whose dispersion has a minimal value (16) compatible with the uncertainty relation (9).

e) The definition (6.c) of α_θ implies that α_θ is the projection of $\mathbf{OM} = \alpha_P \mathbf{e}_x + \alpha_Q \mathbf{e}_y$ on the axis $O\theta$ (Figure 1). The phase φ of the field is described by the angle φ between Ox and OM. Finally, following (3), $H_{cl}/\hbar\omega$ is equal to OM^2, the square of the distance from O to M.

f) The projection of the circle centered on M with radius $\delta = \frac{1}{2}$ on Ox, Oy, and $O\theta$ gives three segments of length $2\delta = 1$ centered on α_P, α_Q, and α_θ. One then gets a good representation of the mean values and dispersions of a_P, a_Q, and a_θ in the state $|\alpha\rangle$.

g) In a quasi-classical state (see §C.4.*c*),

$$\Delta N = \sqrt{\langle N \rangle} = \sqrt{|\alpha|^2} = |\alpha| . \tag{17}$$

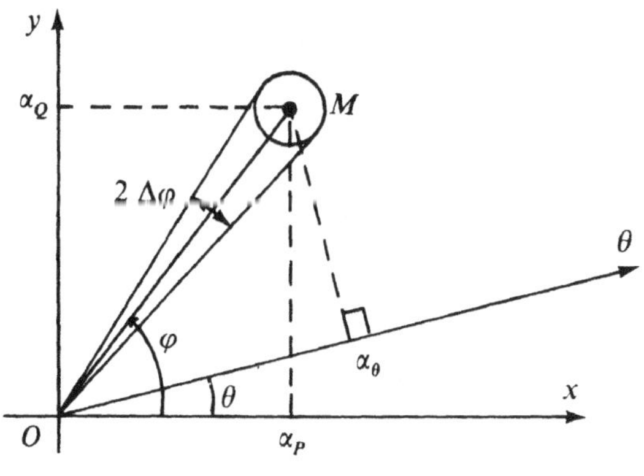

Figure 1

Furthermore, from (6.a) and (6.b), it follows that $|\alpha^2| = \alpha_P^2 + \alpha_Q^2$, so that $|\alpha| = OM$. One has then

$$\Delta N = OM. \tag{18}$$

On the other hand, if $|\alpha| \gg 1$,

$$\Delta\varphi = \frac{\delta}{OM} = \frac{1}{2} \times \frac{1}{OM} \tag{19}$$

$\Delta\varphi$ corresponds to the dispersion of the values of φ for the various points inside the circle in the figure. $\Delta\varphi$ thus clearly represents a phase dispersion. Finally, the combination of (18) and (19) gives

$$\Delta N\,\Delta\varphi = \frac{1}{2}. \tag{20}$$

The number of photons in the mode and the phase of the field are therefore incompatible variables.

h) The major axis of the ellipse centered on M is

— parallel to Oy in case i),
— parallel to Ox in case ii),
— perpendicular to $O\theta$ in case iii),
— parallel to OM in case iv),
— perpendicular to OM in case v).

Consider case iv) in more detail. Figure 2 represents the circle centered on M associated with the state $|\alpha\rangle$ and the ellipse with equal area and whose major axis is parallel to OM. Since the areas of the circle and the ellipse are the same, the minor axis, perpendicular to OM, is smaller than the radius δ of the circle. The angle which the ellipse subtends at O is smaller than the angle subtended by the circle. The phase uncertainty $\Delta\varphi$ is thus smaller for this squeezed state than for the coherent state. In contrast, the major axis of the ellipse is larger than the radius of the circle. The dispersion of distances from O, that is, the amplitude dispersion, is greater for the ellipse than for the circle. The gain in precision in specifying the phase is accompanied by an increase in the amplitude uncertainty. Similar arguments can be developed for the other cases.

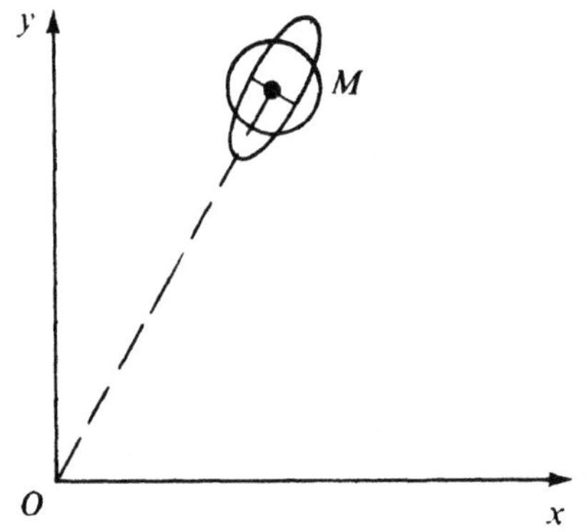

Figure 2

7. Squeezed states of the radiation field

The purpose of this exercise is to introduce states of the field which have unequal dispersions on two quadrature components of the field, the product of these two dispersions having the minimal value compatible with the uncertainty relation (*).

Consider the following operator acting on a mode of the field:

$$B = \frac{r}{2}(a^2 - a^{+2}) \tag{1}$$

where r is a real number and a^+ and a are the creation and annihilation operators for that mode.

a) Find the commutators $[a, B]$ and $[a^+, B]$.

b) Let $T = e^B$. Show that T is a unitary operator.

c) Using the operator equation

$$e^B X e^{-B} = X + [B, X] + \frac{1}{2!}[B, [B, X]] + \cdots + \frac{1}{n!}[B, [B, \ldots [B, X] \ldots]] + \cdots \tag{2}$$

calculate TaT^+ and Ta^+T^+.

d) One introduces the operators a_P and a_Q characterizing two quadrature components of the field using the relationships (see the previous exercise)

$$a_P = \frac{1}{2}(a + a^+) \quad (3.\text{a}) \qquad a_Q = \frac{1}{2\text{i}}(a - a^+). \quad (3.\text{b})$$

(*) See for example D. F. Walls, *Nature*, **306**, 141 (1983).

Calculate $b_P = Ta_PT^+$, $b_Q = Ta_QT^+$ as well as $Ta_P^2T^+$ and $Ta_Q^2T^+$.

e) Assume that the field is in the state $|c\rangle = T^+|\alpha\rangle$ where $|\alpha\rangle$ is a quasi-classical state. Find the mean values of a_P and a_Q in state $|c\rangle$ as well as the corresponding dispersions Δa_P and Δa_Q.

Solution

a) First find $[a, B]$:

$$\left[a, \frac{r}{2}(a^2 - a^{+2})\right] = -\frac{r}{2}[a, a^{+2}] = -ra^+ . \tag{4}$$

Consider now $[a^+, B]$:

$$\left[a^+, \frac{r}{2}(a^2 - a^{+2})\right] = \frac{r}{2}[a^+, a^2] = -ra . \tag{5}$$

b) The expression for T^+,

$$T^+ = e^{r(a^{+2} - a^2)/2} = e^{-B} \tag{6}$$

shows that $TT^+ = T^+T = 1$.

c) By using Equation (2), we find that TaT^+ is equal to

$$e^B a e^{-B} = a + [B, a] + \frac{1}{2!}[B, [B, a]] + \frac{1}{3!}[B, [B, [B, a]]] + \cdots \tag{7}$$

which one can transform using (4) and (5):

$$\begin{aligned} e^B a e^{-B} &= a + ra^+ + \frac{r^2}{2!}a + \frac{r^3}{3!}a^+ + \cdots \\ &= a\left(1 + \frac{r^2}{2!} + \cdots\right) + a^+\left(r + \frac{r^3}{3!} + \cdots\right). \end{aligned} \tag{8}$$

It is easy to check, by examining the next terms of the expansion (8) that the coefficient of a is $\cosh r$ and that of a^+ is $\sinh r$:

$$e^B a e^{-B} = a \cosh r + a^+ \sinh r. \tag{9}$$

The adjoint expression to (9) is written

$$e^B a^+ e^{-B} = a^+ \cosh r + a \sinh r. \tag{10}$$

d) By combining (9) and (10) we get

$$\begin{aligned} b_P &= \tfrac{1}{2}(TaT^+ + Ta^+T^+), \\ &= (a + a^+)\frac{\cosh r + \sinh r}{2} = a_P e^r \end{aligned} \tag{11}$$

and analogously

$$b_Q = \frac{1}{2i}(TaT^+ - Ta^+T^+) = a_Q e^{-r}. \tag{12}$$

The relations (11) and (12) define changes of scale for the observables a_P and a_Q. We now find $Ta_P^2T^+$:

$$Ta_P^2T^+ = Ta_PT^+Ta_PT^+ = b_P^2 = a_P^2 e^{2r} \tag{13}$$

and likewise

$$T a_Q^2 T^+ = b_Q^2 = a_Q^2 \, e^{-2r} . \tag{14}$$

e) The mean values of a_P, a_Q, a_P^2, and a_Q^2 in the state $|c\rangle$ are respectively identical to those of b_P, b_Q, b_P^2, and b_Q^2 in the state $|\alpha\rangle$. Using (11) and (3.a) as well as (C.46) and (C.47), we find

$$\langle c \,|\, a_P \,|\, c \rangle = \langle \alpha \,|\, b_P \,|\, \alpha \rangle = \frac{\alpha + \alpha^*}{2} \, e^r \tag{15}$$

and by using (12) and (3.b)

$$\langle c \,|\, a_Q \,|\, c \rangle = \langle \alpha \,|\, b_Q \,|\, \alpha \rangle = \frac{\alpha - \alpha^*}{2\,i} \, e^{-r} . \tag{16}$$

For $\langle b_P^2 \rangle$, we find using (13) and (3.a)

$$\langle c \,|\, a_P^2 \,|\, c \rangle = \langle \alpha \,|\, b_P^2 \,|\, \alpha \rangle = \frac{1}{4} \langle \alpha \,|\, a^2 + a^{+2} + 2\, a^+ a + 1 \,|\, \alpha \rangle \, e^{2r}$$

$$= \frac{\alpha^2 + \alpha^{*2} + 2\, \alpha^* \alpha + 1}{4} \, e^{2r} \tag{17}$$

which gives finally, using (15) and (17),

$$\Delta a_P = \frac{e^r}{2} . \tag{18}$$

An analogous calculation made for b_Q and using (3.b), (14), and (16) gives

$$\Delta a_Q = \frac{e^{-r}}{2} . \tag{19}$$

We conclude that the state $T^+ |\alpha\rangle$ is a minimal state, since $\Delta a_P \, \Delta a_Q = \frac{1}{4}$, but for which the uncertainties in a_P and a_Q are different. In the graphical representation in Exercise 8, one gets the region representing the state $T^+ |\alpha\rangle$ by effecting on the disk associated with $|\alpha\rangle$ an affinity transformation on the x-axis with amplitude e^r, and an affinity transformation on the y-axis with amplitude e^{-r}. The circle associated with the state $|\alpha\rangle$ is thus transformed into an ellipse with major axis parallel to Ox and minor axis parallel to Oy.

8. Generation of Squeezed States by Two-Photon Interactions

The purpose of this exercise is to show how it is theoretically possible to generate squeezed states of the radiation field. For these states, the dispersions of two quadrature components of the field are different, and their product is equal to the minimal value compatible with the Heisenberg relations. (It is recommended that the reader have previously worked through Exercises 6 and 7.)

Consider a mode $\mathbf{k}\boldsymbol{\varepsilon}$ of the electromagnetic field with frequency ω whose Hamiltonian H is given by

$$H = \hbar\omega a^+ a + i\hbar\Lambda(a^{+2} \, e^{-2i\omega t} - a^2 \, e^{2i\omega t}) \tag{1}$$

where a^+ and a are the creation and annihilation operators of the mode.

The first term of (1) is the energy of the mode for the free field. The second describes a two-photon interaction process such as parametric amplification (a classical wave of frequency 2ω generating two photons with frequency ω). Λ is a real quantity characterizing the strength of the interaction.

a) Write, using the Heisenberg point of view, the equation of motion for $a(t)$. Take

$$a(t) = b(t)\,e^{-i\omega t}. \qquad (2)$$

What are the equations of motion for $b(t)$ and $b^+(t)$?

b) Using the Heisenberg picture, the contribution of the mode $\mathbf{k}\varepsilon$ to the electric field is written

$$\mathbf{E}(\mathbf{r}, t) = i\mathscr{E}_\omega\, \boldsymbol{\varepsilon}\left[a(t)\,e^{i\mathbf{k}.\mathbf{r}} - a^+(t)\,e^{-i\mathbf{k}.\mathbf{r}}\right] \qquad (3)$$

where $a(t)$ is the solution of the equation studied in *a*). Show that

$$b_P(t) = \frac{b(t) + b^+(t)}{2} \quad (4.a); \qquad b_Q(t) = \frac{b(t) - b^+(t)}{2i} \quad (4.b)$$

[where $b(t)$ is defined in (2)] represent physically two quadrature components of the field. Find the equations of motion of $b_P(t)$ and $b_Q(t)$ and give their solutions, assuming that $b_P(0)$ and $b_Q(0)$ are known.

c) Assume that at $t = 0$, the electromagnetic field is in the vacuum state. Calculate at time t the mean number of photons, $\langle N \rangle$, in the mode $\mathbf{k}\varepsilon$ as well as the dispersions $\Delta b_P(t)$ and $\Delta b_Q(t)$ on the two quadrature components of the field. Explain the result.

Solution

a) The Heisenberg equation for a,

$$i\hbar\dot{a} = [a, H] \qquad (5.a)$$

is written using (1) as

$$i\hbar\dot{a} = \hbar\omega a + 2\,i\hbar\,\Lambda a^+\,e^{-2i\omega t} \qquad (5.b)$$

$$\dot{a} = -\,i\omega a + 2\,\Lambda a^+\,e^{-2i\omega t}. \qquad (5.c)$$

With the change of variables (2), equation (5.c) becomes

$$\dot{b} = 2\,\Lambda b^+ \qquad (6.a)$$

and the adjoint equation is written

$$\dot{b}^+ = 2\,\Lambda b. \qquad (6.b)$$

b) We get $a(t)$ and $a^+(t)$ as functions of $b(t)$ and $b^+(t)$ using (2) and the adjoint equation, and then as functions of $b_P(t)$ and $b_Q(t)$ by means of (4.a) and (4.b). Finally we substitute these values into Equation (3) for $\mathbf{E}(\mathbf{r}, t)$. A calculation similar to that in Exercise 6 gives

$$\mathbf{E}(\mathbf{r}, t) = -\,2\,\mathscr{E}_\omega\,\boldsymbol{\varepsilon}\left[b_P(t)\sin(\mathbf{k}\cdot\mathbf{r} - \omega t) + b_Q(t)\cos(\mathbf{k}\cdot\mathbf{r} - \omega t)\right] \qquad (3')$$

which shows that $b_P(t)$ and $b_Q(t)$ represent two quadrature components of the field whose amplitude varies over time as a result of the parametric interaction described by the last term of (1).

Using the equations of motion (6) for $b(t)$ and $b^+(t)$ and the definitions (4), we get

$$\dot{b}_P(t) = 2\,\Lambda b_P(t) \qquad (7.\text{a}); \qquad \dot{b}_Q(t) = -\,2\,\Lambda b_Q(t) \qquad (7.\text{b})$$

whose solution is

$$b_P(t) = b_P(0)\,\mathrm{e}^{2\Lambda t} \qquad (8.\text{a}); \qquad b_Q(t) = b_Q(0)\,\mathrm{e}^{-2\Lambda t}. \qquad (8.\text{b})$$

c) The number of photons in the mode $\mathbf{k}\varepsilon$ is given by the operator

$$N(t) = a^+(t)\,a(t) = b^+(t)\,b(t)\,. \qquad (9)$$

By inverting the relations (4) one can get $b(t)$ and $b^+(t)$ as functions of $b_P(t)$ and $b_Q(t)$, so that using (8.a) and (8.b), we have

$$b(t) = b_P(0)\,\mathrm{e}^{2\Lambda t} + \mathrm{i}b_Q(0)\,\mathrm{e}^{-2\Lambda t} \qquad (10.\text{a})$$

$$b^+(t) = b_P(0)\,\mathrm{e}^{2\Lambda t} - \mathrm{i}b_Q(0)\,\mathrm{e}^{-2\Lambda t} \qquad (10.\text{b})$$

$$b(t) = b(0)\cosh \Lambda t + b^+(0)\sinh \Lambda t \qquad (11.\text{a})$$

$$b^+(t) = b(0)\sinh \Lambda t + b^+(0)\cosh \Lambda t. \qquad (11.\text{b})$$

We see then, since $b(0) = a(0)$, that

$$N(t) = [a^+(0)\cosh \Lambda t + a(0)\sinh \Lambda t][a(0)\cosh \Lambda t + a^+(0)\sinh \Lambda t] \qquad (12)$$

whose mean value in the vacuum is

$$\langle N(t)\rangle = \sinh^2 \Lambda t. \qquad (13)$$

The equation for the electric field $\mathbf{E}$ at t is derived from (3′) and (8):

$$\mathbf{E}(\mathbf{r}, t) = -\,2\,\mathscr{E}_\omega\,\varepsilon[b_P(0)\,\mathrm{e}^{2\Lambda t}\sin(\mathbf{k}\cdot\mathbf{r} - \omega t) + b_Q(0)\,\mathrm{e}^{-2\Lambda t}\cos(\mathbf{k}\cdot\mathbf{r} - \omega t)]\,. \qquad (14)$$

It appears then that the effect of the parametric interaction is to increase one component of the electric field and to decrease the quadrature component. More precisely, from (8.a) and (8.b),

$$(\Delta b_P(t))^2 = \langle 0 \mid b_P^2(0) \mid 0\rangle\,\mathrm{e}^{4\Lambda t} - \langle 0 \mid b_P(0) \mid 0\rangle^2\,\mathrm{e}^{4\Lambda t} \qquad (15.\text{a})$$

$$(\Delta b_Q(t))^2 = \langle 0 \mid b_Q^2(0) \mid 0\rangle\,\mathrm{e}^{-4\Lambda t} - \langle 0 \mid b_Q(0) \mid 0\rangle^2\,\mathrm{e}^{-4\Lambda t} \qquad (15.\text{b})$$

Using (4), we find then

$$\Delta b_P(t) = \frac{\mathrm{e}^{2\Lambda t}}{2} \qquad (16.\text{a}); \qquad \Delta b_Q(t) = \frac{\mathrm{e}^{-2\Lambda t}}{2}. \qquad (16.\text{b})$$

This shows that the uncertainty in one of the field components, here b_Q, can be made very small. This result is achieved at the expense of the quadrature component, in agreement with the Heisenberg relationship $\Delta b_P\,\Delta b_Q \geq \frac{1}{4}$ (see Exercise 6). Note that the result corresponds to the minimum of this uncertainty. Note also that this compression of uncertainty in one of the field components is accompanied by an increase in the mean number of photons (13).

9. Quasi-probability density of a squeezed state

The purpose of this exercise is to calculate, for a squeezed state of the radiation field, the characteristic functions and the quasi-probability densities for both normal and antinormal order. To do so here one must have previously worked Exercises 5 and 7.

a) Consider a mode of the radiation field, and let $|0\rangle$ be the vacuum state of this mode. One can construct a squeezed state by causing the operator e^{-B} to act on $|0\rangle$, where $B = r(a^2 - a^{+2})/2$, r being a real number and a and a^+ the annihilation and creation operators of the mode. The density operator for the field is then

$$\rho = e^{-B}\,|\,0\,\rangle\langle\,0\,|\,e^{B} \tag{1}$$

Find for this field the characteristic functions $C_N(\lambda, \lambda^*)$ and $C_A(\lambda, \lambda^*)$ respectively related to the normal and antinormal order and introduced in Exercise 5. One can use the Glauber equation

$$e^X e^Y = e^{(X+Y)}\, e^{[X,Y]/2} \tag{2}$$

valid when X and Y commute with $[X, Y]$.

b) Calculate the quasi-probability density $P_A(\alpha, \alpha^*)$ associated with such a radiation state. Give $P_A(\alpha, \alpha^*)$ as a function of the real and imaginary parts, α_P and α_Q, of α.

Can one find a quasi-probability density $P_N(\alpha, \alpha^*)$ for this state?

Solution

a) Starting with the definition of $C_N(\lambda, \lambda^*)$ given in Exercise 5 (Equation 1) and introducing the density operator ρ as given in (1), we get

$$C_N(\lambda, \lambda^*) = \mathrm{Tr}\,(e^{-B}\,|\,0\,\rangle\langle\,0\,|\,e^{B}\,e^{\lambda a^+}\,e^{-\lambda^* a}). \tag{3}$$

Using the cyclic property of the trace and introducing $e^{-B}e^{B} = 1$ between $e^{\lambda a^+}$ and $e^{-\lambda^* a}$, we obtain

$$\begin{aligned} C_N(\lambda, \lambda^*) &= \mathrm{Tr}\,(|\,0\,\rangle\langle\,0\,|\,e^{B}\,e^{\lambda a^+}\,e^{-B}\,e^{B}\,e^{-\lambda^* a}\,e^{-B}) \\ &= \langle\,0\,|\,e^{B}\,e^{\lambda a^+}\,e^{-B}\,e^{B}\,e^{-\lambda^* a}\,e^{-B}\,|\,0\,\rangle. \end{aligned} \tag{4}$$

Equations (9) and (10) from Exercise 7 give then

$$e^{B}e^{\lambda a^+}e^{-B} = e^{\lambda(a^+\cosh r + a\sinh r)} \tag{5.a}$$

$$e^{B}e^{-\lambda^* a}e^{-B} = e^{-\lambda^*(a\cosh r + a^+\sinh r)} \tag{5.b}$$

so that (4) can be written

$$C_N(\lambda, \lambda^*) = \langle 0|\, e^{\lambda(a^+\cosh r + a\sinh r)}\, e^{-\lambda^*(a\cosh r + a^+\sinh r)}\,|0\rangle. \tag{6}$$

Denote by e^X and e^Y the operators on the right-hand sides of (5.a) and (5.b). The commutator $[X, Y]$ is a c-number equal to

$$[X, Y] = -|\lambda|^2\big(\cosh^2 r[a^+, a] + \sinh^2 r[a, a^+]\big) = |\lambda|^2 \tag{7}$$

which allows one to use the Glauber relation and to transform (6) into

$$C_N(\lambda, \lambda^*) = e^{|\lambda|^2/2}\langle 0|\, e^{[a^+(\lambda\cosh r - \lambda^*\sinh r) + a(\lambda\sinh r - \lambda^*\cosh r)]}\,|0\rangle. \tag{8}$$

Using the Glauber relation again to separate the exponentials containing a^+ from those containing a, we have

$$\begin{aligned} e^{[a^+(\lambda\cosh r - \lambda^*\sinh r) + a(\lambda\sinh r - \lambda^*\cosh r)]} &= e^{a^+(\lambda\cosh r - \lambda^*\sinh r)} \\ &\quad\times e^{a(\lambda\sinh r - \lambda^*\cosh r)}e^{-|\lambda|^2(\cosh^2 r + \sinh^2 r)/2} \\ &\quad\times e^{(\lambda^2 + \lambda^{*2})(\cosh r\sinh r)/2} \end{aligned} \tag{9}$$

and substituting this in (8), we find then

$$C_N(\lambda, \lambda^*) = e^{-|\lambda|^2 \sinh^2 r} e^{(\lambda^2 + \lambda^{*2})(\cosh r \sinh r)/2} \tag{10}$$

since

$$e^{a(\lambda \sinh r - \lambda^* \cosh r)}|0\rangle = |0\rangle \tag{11.a}$$

$$\langle 0| e^{a^+(\lambda \cosh r - \lambda^* \sinh r)} = \langle 0| . \tag{11.b}$$

Equation (13) of Exercise (5) allows one to obtain $C_A(\lambda, \lambda^*)$ as

$$\begin{aligned} C_A(\lambda, \lambda^*) &= C_N(\lambda, \lambda^*) e^{-|\lambda|^2} \\ &= e^{-|\lambda|^2 \cosh^2 r} e^{(\lambda^2 + \lambda^{*2})(\cosh r \sinh r)/2} \end{aligned} \tag{12}$$

b) We have shown in Exercise 5 that P_A and C_A are related by a Fourier transform

$$P_A(\alpha, \alpha^*) = \frac{1}{\pi^2} \int d^2\lambda \, C_A(\lambda, \lambda^*) \, e^{(\lambda^* \alpha - \lambda \alpha^*)} . \tag{13.a}$$

The notation (5.a) and (5.b) of Exercise 5 allows us to rewrite (13.a) in the form

$$P_A(\alpha, \alpha^*) = \frac{1}{4\pi^2} \int du \, dv \, C_A(\lambda, \lambda^*) \, e^{i\alpha_P u} \, e^{i\alpha_Q v} \tag{13.b}$$

where $C_A(\lambda, \lambda^*)$ comes from (12):

$$\begin{aligned} C_A(\lambda, \lambda^*) &= \exp\left(-\frac{u^2 + v^2}{4} \cosh^2 r \right) \exp\left(-\frac{u^2 - v^2}{4} \cosh r \sinh r \right) \\ &= \exp\left(-\frac{u^2}{4} \cosh r \, (\cosh r + \sinh r) \right) \exp\left(-\frac{v^2}{4} \cosh r \, (\cosh r - \sinh r) \right) \\ &= \exp\left(-\frac{u^2}{8} (1 + e^{2r}) \right) \exp\left(-\frac{v^2}{8} (1 + e^{-2r}) \right) . \end{aligned} \tag{14}$$

It is then necessary to find the Fourier transform of two Gaussian functions of different widths. We get

$$P_A(\alpha, \alpha^*) = \frac{2}{\pi[(1 + e^{2r})(1 + e^{-2r})]^{1/2}} \exp[-2\,\alpha_P^2/(1 + e^{2r})] \exp[-2\,\alpha_Q^2/(1 + e^{-2r})] \tag{15}$$

that is, a product of two Gaussians of very different widths in α_P and α_Q, which reflects the different behavior of the two quadrature components of the field (see Exercise 7).

The same calculation in the case of $C_N(\lambda, \lambda^*)$ leads to

$$C_N(\lambda, \lambda^*) = \exp\left[\frac{u^2}{8} (1 - e^{2r}) \right] \exp\left[\frac{v^2}{8} (1 - e^{-2r}) \right] . \tag{16}$$

It is clear that, for $r \neq 0$, one of the components of $C_N(\lambda, \lambda^*)$ diverges exponentially at infinity. Finding the Fourier transform is then impossible. This state of the field does not have a *P*-representation.

To understand this point, one can note that the state $|0\rangle$ is represented by a quasi-probability density $P_N(\alpha, \alpha^*)$ which is a Dirac distribution $\delta^{(2)}(\alpha)$. A state for which $\Delta a_P = \Delta a_Q = \frac{1}{2}$ is thus represented by a point distribution. For the squeezed state considered here, one reduces the uncertainty on one of the field components. Along this direction, the quasi-probability distribution should be narrower than for the state $|0\rangle$. This is clearly impossible starting from a point distribution. On the other hand, in the case corresponding to the antinormal order, the state $|0\rangle$ is represented by a Gaussian distribution whose width can be reduced.

CHAPTER IV

Other Equivalent Formulations of Electrodynamics

In the preceding chapters, and especially in Chapter II, we have followed the simplest possible procedure to construct quantum electrodynamics. Starting with the standard Lagrangian, we eliminated the scalar potential by expressing it as a function of the other dynamical variables; then we chose the Coulomb gauge, which sets the longitudinal vector potential $\mathbf{A}_{\parallel}$ equal to zero. This formalism can then be used in any situation provided that the particle velocities and the frequencies of the fields remain in the nonrelativistic domain.

It is clear however that such a description is not the only one possible. One can formulate others, equivalent with respect to their physical predictions but formally different. Generally, the aim is to obtain a new formulation in which a given problem can be treated more easily than in the standard formulation. The purpose of this chapter is to present and examine examples of such developments (*).

We begin in Part A by reviewing several methods which can be utilized to construct alternative descriptions of electrodynamics. A first possibility is to choose a different gauge and take $\mathbf{A}_{\parallel}$ different from zero. One can also add to the standard Lagrangian in the Coulomb gauge the derivative with respect to time of a function of the generalized coordinates of the system. The new Lagrangian is equivalent to the old and gives the same equations of motion. However, the new conjugate momenta are in general different from the old, and when one applies the canonical quantization procedure, the same mathematical operators in the new description represent in general different physical variables. The relationship between the

(*) Here we will not consider the formulation of electrodynamics in the Lorentz gauge, since that is the subject of Chapter V.

two points of view will be studied in detail, and we will show that the two quantum representations can be derived from one another by a *unitary transformation*. The gauge change discussed above is, in fact, only one special case of this type of transformation. One can finally get a new description of electrodynamics by applying to the standard representation in the Coulomb gauge a unitary transformation more general than those leading to an equivalent Lagrangian. Such a method is more powerful than that above, in the same way that, in classical mechanics, canonical transformations are more general than those changing the Lagrangian.

We then treat, in Part B, several simple examples of such transformations applied to the case of a system of localized charges, like an atom or molecule, interacting with a long-wavelength external electromagnetic wave. We first consider a simple gauge change, and then introduce the Göppert-Mayer transformation which allows one to go from the standard point of view, where the interaction Hamiltonian between the system of charges and the field is proportional to $\mathbf{A} \cdot \mathbf{p}$, to the electric dipole point of view where the interaction Hamiltonian is proportional to $\mathbf{E} \cdot \mathbf{r}$. The equivalence between the two descriptions will be discussed once more, in view of the practical importance of this transformation. We present also in Part B an example of a transformation of the Hamiltonian which is not equivalent to a change in the Lagrangian: the Henneberger transformation, which is used to study some interaction processes with intense electromagnetic waves.

In Part C, we return to the general case where the electromagnetic field is considered, not as an external field with a given time dependence, but as a dynamical system coupled to the particles. We present the Power–Zienau–Woolley transformation, which generalizes the Göppert-Mayer transformation and gives, for the interaction Hamiltonian between the system of charges and the field, the complete multipole expansion in a compact form. The system of charges is described by a polarization and magnetization density, and the displacement appears naturally as the momentum conjugate with the transverse vector potential.

We finally study, in Part D, how the equivalence between two points of view derived from one another by a unitary transformation manifests itself in the scattering S-matrix. We show that the equality of the transition amplitudes takes, in this case, a simpler form, since the same kets can be used to describe the initial and final states in the two approaches. This property of the S-matrix explains the current practice of attributing (wrongly) the same physical meaning to the "unperturbed" Hamiltonian whatever the point of view used.

A—HOW TO GET OTHER EQUIVALENT FORMULATIONS OF ELECTRODYNAMICS

In this part we review different methods which can be used to get formulations of electrodynamics equivalent to those of Chapters II and III. The emphasis here will be on the correspondence between the different descriptions and on the equivalence of the associated quantum theories.

1. Change of Gauge and of Lagrangian

We briefly review the arguments followed in Chapter II. Since the standard Lagrangian does not depend on the velocity $\dot{\mathscr{U}}$ relative to the scalar potential $\mathscr{U}$, Lagrange's equation relative to $\mathscr{U}$ leads to a relationship [(C.1) of Chapter II]

$$\mathscr{U} = \frac{1}{k^2}\left[ik\dot{\mathscr{A}}_{\parallel} + \frac{\rho}{\varepsilon_0}\right] \tag{A.1}$$

which allows one to reformulate $\mathscr{U}$ as a function of the other dynamical variables. When $\mathscr{U}$ has been eliminated, the standard Lagrangian will only depend on $\mathbf{r}_\alpha$, $\dot{\mathbf{r}}_\alpha$, $\mathscr{A}_\perp$, $\dot{\mathscr{A}}_\perp$, $\mathscr{A}_\parallel$, and $\dot{\mathscr{A}}_\parallel$ and can be expressed in the form

$$\begin{aligned} L = \sum_\alpha \frac{1}{2} m_\alpha \dot{\mathbf{r}}_\alpha^2 - \int \mathrm{d}^3k\, \frac{\rho^*\rho}{\varepsilon_0 k^2} + \varepsilon_0 \int \mathrm{d}^3k [\dot{\mathscr{A}}_\perp^* \cdot \dot{\mathscr{A}}_\perp - c^2 k^2 \mathscr{A}_\perp^* \cdot \mathscr{A}_\perp] + \\ + \int \mathrm{d}^3k [\mathscr{j}^* \cdot \mathscr{A}_\perp + \mathscr{j} \cdot \mathscr{A}_\perp^*] + \\ + \int \mathrm{d}^3k \left[\mathscr{j}_\parallel^* \mathscr{A}_\parallel + \mathscr{j}_\parallel \mathscr{A}_\parallel^* - \frac{\mathrm{i}}{k}(\rho^* \dot{\mathscr{A}}_\parallel - \rho \dot{\mathscr{A}}_\parallel^*)\right]. \end{aligned} \tag{A.2}$$

We then found that $\mathscr{A}_\parallel$ is not a true dynamical variable, in the sense that the associated Lagrange equation is not an equation of motion for $\mathscr{A}_\parallel$ (*), which can then take any value. The choice of the Coulomb gauge, $\mathscr{A}_\parallel = 0$, appears then to be the simplest possible, but it is not necessary. It is always possible at this stage to set $\mathscr{A}_\parallel \neq 0$ [this also changes the scalar potential according to (A.1)] and to continue the procedure of canonical quantization. One can, for example, take $\mathscr{A}_\parallel$ as an arbitrary function of $\mathbf{k}$ and t, which can also eventually depend on the other true dynamical

(*) In fact, this Lagrange equation is nothing more than the equation for the conservation of charge, which adds nothing new, since it follows directly from the definition of ρ and $\mathbf{j}$ as functions of the dynamical variables $\mathbf{r}_\alpha$ and $\dot{\mathbf{r}}_\alpha$ of the particles.

variables of the system, $\mathbf{r}_\alpha$ and $\mathscr{A}_\perp$ (*). The consequences of this choice can now be foreseen: since $\dot{\mathbf{r}}_\alpha$ appears in the last term of (A.2) through $\mathscr{j}_\parallel$ and possibly $\dot{\mathscr{A}}_\parallel$, the conjugate momentum of $\mathbf{r}_\alpha$ will be modified. If $\mathscr{A}_\parallel$ depends on $\mathscr{A}_\perp$, so will the momentum conjugate with $\mathscr{A}_\perp(\mathbf{k})$.

Another way of understanding why $\mathscr{A}_\parallel$ is not a true dynamical variable involves noting, as in §C.1.*b* of Chapter II, that the last term of (A.2) where $\mathscr{A}_\parallel$ and $\dot{\mathscr{A}}_\parallel$ appear can also be written as the total time derivative of the function

$$F = \int d^3k \left[\frac{\rho^*}{ik}\mathscr{A}_\parallel - \frac{\rho}{ik}\mathscr{A}_\parallel^*\right]. \tag{A.3}$$

Thus going from the Coulomb gauge to another gauge is an illustration of the general law which allows one to go from one Lagrangian to an equivalent Lagrangian by adding the total time derivative of a function of the generalized coordinates. We are now going to examine how such transformations are expressed in quantum theory and to show that they amount to applying a unitary transformation to the initial quantum representation.

2. Changes of Lagrangian and the Associated Unitary Transformation

Such a problem is not specific to the system (electromagnetic field + charged particles). We are going to study a much simpler system with one degree of freedom on the x-axis and later generalize the results gotten there. The interest of such a simple case is that we can use unambiguous notation. Later we will return to more compact notation.

Let $L(x, \dot{x})$ be a Lagrangian describing the dynamics of the system. Since we will introduce other Lagrangians hereafter, we take the precaution of denoting by p_L the momentum conjugate with x with respect to the Lagrangian L:

$$p_L = \frac{\partial L}{\partial \dot{x}}. \tag{A.4}$$

It is a quantity which is expressed as a function of x and $\dot{x}$. The equation of motion is Lagrange's equation

$$\dot{p}_L \equiv \frac{d}{dt}\left(\frac{\partial L}{\partial \dot{x}}\right) = \frac{\partial L}{\partial x}. \tag{A.5}$$

(*) In contrast, $\mathscr{A}_\parallel$ cannot depend on the velocities $\dot{\mathbf{r}}_\alpha$ and $\dot{\mathscr{A}}_\perp$, since that would cause the accelerations to appear in the Lagrangian (A.2).

a) Changing the Lagrangian

By adding to L the time derivative of a function $F(x, t)$, one gets a new Lagrangian

$$L'(x, \dot{x}) = L(x, \dot{x}) + \frac{\mathrm{d}}{\mathrm{d}t} F(x, t)$$

$$= L(x, \dot{x}) + \dot{x} \frac{\partial F}{\partial x} + \frac{\partial F}{\partial t} \tag{A.6}$$

equivalent to L in the sense that it gives for x the same equation of motion as (A.5) (see §A.1.*c* of Chapter II). The momentum conjugate with x with respect to the new Lagrangian L' is

$$p_{L'} = \frac{\partial L'}{\partial \dot{x}} = \frac{\partial L}{\partial \dot{x}} + \frac{\partial F}{\partial x}. \tag{A.7}$$

The expression for $p_{L'}$ as a function of x and $\dot{x}$ is different from that for p_L. Going from a Lagrangian L to an equivalent Lagrangian L' has as a consequence a change in the momentum conjugate with x. The old and new conjugate momenta are two functions of x and $\dot{x}$ which are related by

$$p_{L'} = p_L + \frac{\partial F}{\partial x}. \tag{A.8}$$

A single dynamical state of the system, characterized by given values of x and $\dot{x}$, will then be described by different values of the conjugate momenta p_L and $p_{L'}$ related by (A.8). One can verify that, if this relationship is true at one time, the dynamical equations assure that it will remain so.

The physical variables of the system are functions of x and $\dot{x}$. Their values depend only on the dynamical state of the system characterized by x and $\dot{x}$. Consider one such variable, for example the kinetic energy, described by the function $\mathscr{G}(x, \dot{x})$. In the Hamiltonian formalism, one uses x and the conjugate momentum p_L as dynamical variables. The value of the physical variable $\mathscr{G}$ will be a function $G_L(x, p_L)$. The index L reminds us that this function depends on the Lagrangian which has been used to define p_L. The value of G_L when one replaces p_L by its expression (A.4) should coincide with $\mathscr{G}(x, \dot{x})$:

$$G_L\left(x, \frac{\partial L}{\partial \dot{x}}\right) = \mathscr{G}(x, \dot{x}) \tag{A.9}$$

which completely fixes G_L. For the new Lagrangian L', $G_{L'}$ is different from G_L and (A.9) is written

$$G_{L'}\left(x, \frac{\partial L'}{\partial \dot{x}}\right) = \mathscr{G}(x, \dot{x}). \tag{A.10}$$

Equations (A.8), (A.9), and (A.10) then give quite simply the relationship between G_L and $G_{L'}$,

$$G_{L'}\left(x, p_L + \frac{\partial F}{\partial x}\right) = G_L(x, p_L) \,. \tag{A.11}$$

This equation assures that for the same dynamical state, the values of $G_{L'}(x, p_{L'})$ and $G_L(x, p_L)$ are identical, p_L and $p_{L'}$ being related by (A.8). The physical predictions with regard to $\mathscr{G}$ are the same from both points of view.

Consider now the Hamiltonian. Its form, as a function of x and $p_{L'}$, depends on the Lagrangian chosen. Furthermore, its value for a given dynamical state is not necessarily identical in the two descriptions, since

$$\begin{aligned} H_{L'}(x, p_{L'}) &= \dot{x} p_{L'} - L' = \dot{x}\left(p_L + \frac{\partial F}{\partial x}\right) - \left(L + \dot{x}\frac{\partial F}{\partial x} + \frac{\partial F}{\partial t}\right) \\ &= H_L(x, p_L) - \frac{\partial F}{\partial t} \end{aligned} \tag{A.12}$$

where $p_{L'}$ and p_L are related by (A.8). It appears then that the Hamiltonian behaves like a physical quantity only when F is time independent [compare (A.12) and (A.11)].

b) The Two Quantum Descriptions

When one applies the standard canonical quantization procedure starting with L on the one hand and L' on the other, one gets two quantum descriptions for the system. To eliminate all ambiguity we will use superscript indices (1) for all the elements in the first case and (2) in the second. Thus $X^{(1)}$ is the operator representing the "position" variable in the first description, and $|\psi^{(2)}\rangle$ is the state vector representing the dynamical state of the system in the second one. Finally, we introduce the two fundamental mathematical operators, the operator "multiplication by x" which we call X and the operator $(\hbar/i)\,\partial/\partial x$ which we call P:

$$X = x \tag{A.13}$$

$$P = \frac{\hbar}{i}\frac{\partial}{\partial x} \,. \tag{A.14}$$

If one applies the standard canonical quantization procedure starting with the Lagrangian L, one gets a first quantum description. The coordinate x is represented by the operator "multiplication by x", and its conjugate momentum p_L by the operator $(\hbar/i)\,\partial/\partial x$. One has then

$$X^{(1)} = X \tag{A.15}$$

$$P_L^{(1)} = P\,. \tag{A.16}$$

The variable $\mathcal{G}$ is represented by the observable

$$G^{(1)} = G_L(X, P) \tag{A.17}$$

where G_L is the function defined by (A.9). The Hamiltonian is

$$H_L^{(1)} = H_L(X, P)\,. \tag{A.18}$$

we have kept the index L for $H_L^{(1)}$ and $P_L^{(1)}$ to remind ourselves that the momentum conjugate with x and the Hamiltonian depend in the general case on the Lagrangian from which they have been defined. The state of the system at a given instant is represented by the state vector $|\psi^{(1)}\rangle$.

Applying the same quantization procedure starting with L', one gets a second description of the system in which the coordinate is always represented by the operator X, and the new conjugate momentum $P_{L'}$ by the operator P:

$$X^{(2)} = X \tag{A.19}$$

$$P_{L'}^{(2)} = P\,. \tag{A.20}$$

The physical variable $\mathcal{G}$ is represented by

$$G^{(2)} = G_{L'}(X, P) \tag{A.21}$$

where the function $G_{L'}$ is defined by (A.10). The new Hamiltonian $H_{L'}$ is represented by

$$H_{L'}^{(2)} = H_{L'}(X, P)\,. \tag{A.22}$$

In general, the functions G_L and $G_{L'}$ are different, with the result that *the same physical quantity $\mathcal{G}$ is represented by different mathematical operators in the two representations*:

$$G^{(1)} \neq G^{(2)}\,. \tag{A.23}$$

Starting with the Lagrangian L or L', one is then led to descriptions of the system observables which are different. In the same way, the state vectors which describe a given state are not the same:

$$|\psi^{(1)}\rangle \neq |\psi^{(2)}\rangle\,. \tag{A.24}$$

Recall that it is the same classically: a given dynamical state of the system is represented by different values of p_L and $p_{L'}$.

c) The Correspondence between the Two Quantum Descriptions

We are now going to examine the correspondence between the two quantum descriptions. Clearly, the coordinate is represented by the same operator [cf. (A.15) and (A.19)]. For the conjugate momenta, it is necessary to recall that two have been introduced, p_L and $p_{L'}$, and that they represent different physical variables, as can be seen from their functional dependence on x and $\dot{x}$. Thus we have to consider four operators: $P_L^{(1)}$, $P_L^{(2)}$, $P_{L'}^{(1)}$, and $P_{L'}^{(2)}$. The first and the last are already known [Equations (A.16) and (A.20)]. The other two can be derived from Equation (A.8) between $p_{L'}$ and p_L. If one expresses this in representation (1), for example, one gets

$$P_{L'}^{(1)} = P_L^{(1)} + \frac{\partial F}{\partial X} = P + \frac{\partial F}{\partial X} \tag{A.25.a}$$

by using (A.16). In the same way, expressing (A.8) in representation (2) and using (A.20), one gets the following equation for $P_L^{(2)}$:

$$P_L^{(2)} = P_{L'}^{(2)} - \frac{\partial F}{\partial X} = P - \frac{\partial F}{\partial X}\,. \tag{A.25.b}$$

We can summarize the expressions for the operators representing the different variables in the two representations as follows:

— Representation (1):

$$\begin{cases} X^{(1)} = X & \text{(A.26.a)} \\ P_L^{(1)} = P & \text{(A.26.b)} \\ P_{L'}^{(1)} = P + \dfrac{\partial F}{\partial X} & \text{(A.26.c)} \end{cases}$$

— Representation (2):

$$\begin{cases} X^{(2)} = X & \text{(A.27.a)} \\ P_L^{(2)} = P - \dfrac{\partial F}{\partial X} & \text{(A.27.b)} \\ P_{L'}^{(2)} = P & \text{(A.27.c)} \end{cases}$$

Comparison of the two representations suggests that passing from one to the other involves the unitary transformation

$$T = \exp \frac{i}{\hbar} F(X) \tag{A.28}$$

which amounts to translating P by an amount $-\partial F/\partial X$. It is indeed possible to go from the expressions (A.26) for $X^{(1)}$, $P_L^{(1)}$, and $P_{L'}^{(1)}$ to the expressions (A.27) for $X^{(2)}$, $P_L^{(2)}$, and $P_{L'}^{(2)}$ through the following relations:

$$X^{(2)} = TX^{(1)}T^{+} \tag{A.29.a}$$

$$P_L^{(2)} = TP_L^{(1)}T^{+} \tag{A.29.b}$$

$$P_{L'}^{(2)} = TP_{L'}^{(1)}T^{+} \tag{A.29.c}$$

where T is the unitary operator (A.28).

More generally, we can check that the transformation T establishes the correspondence between the two representations $G^{(1)}$ and $G^{(2)}$ of the same physical variable $\mathcal{G}$

$$G^{(2)} = TG^{(1)}T^{+} \tag{A.30}$$

$G^{(1)}$ and $G^{(2)}$ being defined by (A.17) and (A.21). The expression (A.11) which relates the two functions G_L and $G_{L'}$, can be written by replacing the classical variables x and p_L with the operators X and P. One gets then

$$G_{L'}\left(X, P + \frac{\partial F}{\partial X}\right) = G_L(X, P) . \qquad \text{(A.31)}$$

Transforming both sides of this equation by T and applying the transformation on the operators within the function appearing on the left-hand side, one finds that X is unchanged and $P + \partial F/\partial X$ is transformed into P, so that

$$G_{L'}(X, P) = T G_L(X, P) T^+ . \qquad \text{(A.32)}$$

Using (A.17) and (A.21), this reduces to (A.30). The representation (2) of the system observables is gotten simply by applying the transformation T to representation (1).

The state vectors are related by the same transformation

$$| \psi^{(2)} \rangle = T | \psi^{(1)} \rangle . \qquad \text{(A.33)}$$

Indeed, if one imagines that the state of the system results from the measurement of a variable $\mathscr{G}$ (or a set of such variables), then $|\psi^{(1)}\rangle$ and $|\psi^{(2)}\rangle$ are the eigenvectors respectively of $G^{(1)}$ and $G^{(2)}$ corresponding to the same eigenvalue. From (A.30), this implies Equation (A.33). The expressions (A.30) and (A.33) are then sufficient to assure the equality of the mean values and of the measurement results in both approaches.

The temporal evolution of $|\psi^{(1)}\rangle$ and $|\psi^{(2)}\rangle$ is governed respectively by $H_L^{(1)}$ and $H_{L'}^{(2)}$; the relationship between the two operators is not of the type (A.30). Starting with (A.12) between the classical functions and proceeding as for G, one finds that

$$H_{L'}^{(2)} = T H_L^{(1)} T^+ - \frac{\partial F}{\partial t} . \qquad \text{(A.34)}$$

This is precisely the relationship which assures that equation (A.33) between $|\psi^{(1)}\rangle$ and $|\psi^{(2)}\rangle$ continues over time. Indeed, if one compares the rate of change of $|\psi^{(2)}\rangle$ on one hand and of $T|\psi^{(1)}\rangle$ on the other, one finds

$$i\hbar \frac{\mathrm{d}}{\mathrm{d}t} | \psi^{(2)} \rangle = H_{L'}^{(2)} | \psi^{(2)} \rangle \qquad \text{(A.35)}$$

and

$$\begin{aligned} i\hbar \frac{\mathrm{d}}{\mathrm{d}t} T | \psi^{(1)} \rangle &= T H_L^{(1)} | \psi^{(1)} \rangle + \left(i\hbar \frac{\mathrm{d}T}{\mathrm{d}t} \right) | \psi^{(1)} \rangle \\ &= \left(T H_L^{(1)} T^+ - \frac{\partial F}{\partial t} \right) T | \psi^{(1)} \rangle . \end{aligned} \qquad \text{(A.36)}$$

The two rates are equal on account of (A.33) and (A.34). Thus the correspondence (A.33) between the two points of view is preserved over time.

d) Application to the Electromagnetic Field

The preceding considerations are easily generalized for a more complex system and in particular for the case of the electromagnetic field interacting with an ensemble of particles.

Now, each change in gauge ($\mathbf{A}_{\parallel} \neq \mathbf{0}$) or each transformation from the Coulomb Lagrangian to an equivalent one is accompanied by a corresponding *change in quantum representation*. With each of these transformations is associated a unitary transformation T of the form

$$T = \exp \frac{i}{\hbar} F(\ldots, \mathbf{r}_\alpha, \ldots; \ldots, \mathscr{A}_\perp(\mathbf{k}), \ldots; t) \tag{A.37}$$

where F is a function of the generalized coordinates of the system, $\mathbf{r}_\alpha$ and $\mathscr{A}_\perp(\mathbf{k})$, and possibly of time. To simplify the notation, we no longer use, as in the preceding paragraph, different symbols for the generalized coordinates (position $\mathbf{r}_\alpha$, transverse vector potential $\mathscr{A}_\perp(\mathbf{k})$, etc.) and the corresponding operators.

All of the equations established in the preceding subsection and relating the system descriptions in one or the other points of view [Equations (A.30), (A.33), and (A.34) relating $G^{(2)}$ to $G^{(1)}$, $|\psi^{(2)}\rangle$ to $|\psi^{(1)}\rangle$, and $H_L^{(2)}$ to $H_L^{(1)}$] remain valid provided that (A.28) is replaced by (A.37). We will examine in detail hereafter specific examples of the transformations (A.37). Beforehand, we introduce unitary transformations more general than (A.37) and prove the equivalence between the transition amplitudes calculated from two viewpoints related by a unitary transformation.

3. The General Unitary Transformation. The Equivalence between the Different Formulations of Quantum Electrodynamics

The unitary transformation T defined by (A.37) depends only on the generalized coordinates. One can consider more general transformations depending also on the conjugate momenta of the type (*)

$$T(t) = \exp \frac{i}{\hbar} F(\ldots, \mathbf{r}_\alpha, \mathbf{p}_\alpha, \ldots; \ldots, \mathscr{A}_\perp(\mathbf{k}), \boldsymbol{\pi}(\mathbf{k}), \ldots; t). \tag{A.38}$$

The presence of the conjugate momenta in (A.38) implies that such a transformation is not associated with a change of Lagrangian. It is applied

(*) For these calculations which follow, it is useful to use the notation $T(t)$, showing that T depends explicitly on t, if this is also the case for F.

directly to the states and observables of the system in accordance with

$$| \psi^{(2)}(t) \rangle = T(t) \, | \psi^{(1)}(t) \rangle \tag{A.39}$$

$$G^{(2)}(t) = T(t) \, G^{(1)}(t) \, T^{+}(t) \tag{A.40}$$

which are analogous to (A.33) and (A.30). Note incidentally that the presence of $\mathbf{p}_\alpha$ in (A.38) implies that the transformation $T(t)$ does not generally leave the coordinates invariant: from the new point of view, the operator "multiplication by $\mathbf{r}_\alpha$" no longer represents the position of the particle, but some other physical variable.

Let $H^{(1)}(t)$ be the Hamiltonian describing the temporal evolution from point of view (1) (*). To get the Schrödinger equation satisfied by the vector $|\psi^{(2)}(t)\rangle$ related at each instant t to $|\psi^{(1)}(t)\rangle$ by (A.39), it suffices to take the derivative of each side of (A.39) with respect to t and to use the Schrödinger equation for $|\psi^{(1)}(t)\rangle$. We get

$$\begin{aligned} i\hbar \frac{\mathrm{d}}{\mathrm{d}t} | \psi^{(2)}(t) \rangle &= \left[i\hbar \frac{\mathrm{d}T(t)}{\mathrm{d}t} + T(t) \, H^{(1)}(t) \right] | \psi^{(1)}(t) \rangle \\ &= \left[i\hbar \frac{\mathrm{d}T(t)}{\mathrm{d}t} T^{+}(t) + T(t) \, H^{(1)}(t) \, T^{+}(t) \right] | \psi^{(2)}(t) \rangle . \end{aligned} \tag{A.41}$$

The Hamiltonian $H^{(2)}$ in approach (2) is thus written

$$H^{(2)}(t) = T(t) \, H^{(1)}(t) \, T^{+}(t) + i\hbar \frac{\mathrm{d}T(t)}{\mathrm{d}t} T^{+}(t) . \tag{A.42}$$

The simultaneous presence of the generalized coordinates and conjugate momenta in the operator F of (A.38) implies that $\partial F/\partial t$ does not commute with F in general. That explains why the last term of (A.42) cannot be written $-\partial F/\partial t$ as in (A.34).

We will now establish an important relationship between the evolution operators $U^{(1)}(t, t_0)$ and $U^{(2)}(t, t_0)$. Since (A.39) is valid regardless of t, we can write

$$\begin{aligned} | \psi^{(2)}(t) \rangle &= T(t) \, | \psi^{(1)}(t) \rangle = T(t) \, U^{(1)}(t, t_0) \, | \psi^{(1)}(t_0) \rangle \\ &= T(t) \, U^{(1)}(t, t_0) \, T^{+}(t_0) \, | \psi^{(2)}(t_0) \rangle \\ &= U^{(2)}(t, t_0) \, | \psi^{(2)}(t_0) \rangle . \end{aligned} \tag{A.43}$$

It follows that

$$U^{(2)}(t, t_0) = T(t) \, U^{(1)}(t, t_0) \, T^{+}(t_0) . \tag{A.44}$$

Such an expression together with (A.39) assures the identity of the physical predictions in the two approaches. To show this, consider the

(*) Since the two descriptions (1) and (2) are no longer associated with two Lagrangians L and L', there is no longer justification for using H_L.

probability amplitudes that the system, starting at t_0 in an initial state described respectively by

$$| \varphi^{(1)}(t_0) \rangle \tag{A.45.a}$$

$$| \varphi^{(2)}(t_0) \rangle = T(t_0) \, | \varphi^{(1)}(t_0) \rangle \tag{A.45.b}$$

in the two representations, ends at time t in a final state

$$| \chi^{(1)}(t) \rangle \tag{A.46.a}$$

$$| \chi^{(2)}(t) \rangle = T(t) \, | \chi^{(1)}(t) \rangle . \tag{A.46.b}$$

The equations (A.44), (A.45), and (A.46) insure the identity of the transition amplitudes calculated in the two representations:

$$\langle \chi^{(2)}(t) \, | \, U^{(2)}(t, t_0) \, | \, \varphi^{(2)}(t_0) \rangle = \langle \chi^{(1)}(t) \, | \, U^{(1)}(t, t_0) \, | \, \varphi^{(1)}(t_0) \rangle \tag{A.47}$$

and thus the equality of the physical predictions. The identity (A.47) established for the general unitary transformation (A.38) is of course also valid for (A.37).

Thus, the equivalence of all the different descriptions of quantum electrodynamics that one can construct using the procedures described in this Part A is assured in a fundamental way by the existence of a unitary transformation relating the various descriptions.

The foregoing considerations concerning the equivalence of the various descriptions can seem at first blush elementary and even superfluous. In fact this is not the case. Their translation into a specific case is often far from easy and can give rise to incorrect interpretations or to subtle errors. It behooves one in each case to be certain what operators represent the different physical variables in one or the other of the representations and to ascertain that the state vectors used to represent the system from the two points of view describe the same physical state—for example, are eigenstates of the same physical observable with the same eigenvalue (*).

Remarks

(i) The correspondence between the description of electrodynamics in the Coulomb gauge and that which one gets by the covariant quantization described in Chapter V is not so simple as that of the preceding unitary transformations. In the covariant description, the electromagnetic field is described with a larger number of degrees of freedom and the state space does not have the same structure. We will return to this problem in Complement B_V.

(ii) A subtle error can be introduced, for example, when one causes the system to interact suddenly at $t = 0$ with a field described by a potential. Causing a

(*) See for example the discussion presented in Y. Aharonov and C. K. Au, *Phys. Lett.*, **86A**, 269 (1981).

vector potential to change at $t = 0$ to a finite value is accompanied in fact by a pulse of electric field. This pulse will be absent if one describes the electric field by a scalar potential. Thus, one can have the impression that the calculation carried out in two different gauges gives different results, when in fact the reality of the situation is that the system was interacting with two *different* electric fields in the two cases. (An example of this is discussed in Exercise 1.)

B—SIMPLE EXAMPLES DEALING WITH CHARGES COUPLED TO AN EXTERNAL FIELD

Before considering transformations involving the ensemble field + particles as a dynamical system, we will first examine the simplest situation relative to a system of charged particles localized about the coordinate origin and interacting with an external electromagnetic wave. Such a system can be, for example, an atom or a molecule. We begin (§B.1) by recalling the expressions for the Lagrangian and the Hamiltonian for such a system. We then illustrate the general considerations of Part A by means of three examples: a simple gauge change which will allow us to state what is meant by gauge invariance (§B.2), the Göppert-Mayer transformation, which gives rise to the electric dipole interaction at the long-wavelength limit (§B.3), and the Henneberger transformation, which is a unitary transformation depending on the conjugate momenta (§B.4).

1. The Lagrangian and Hamiltonian of the System

The Lagrangian and Hamiltonian of a system of charges in the presence of an external field have been introduced in Complement C_{II}. Assume that the particles are sufficiently near to one another that the Coulomb interaction is a very good approximation to their real interaction. In the Lagrangian and Hamiltonian of the problem, it is then possible to neglect the terms relative to the transverse free field and to the interaction between the particles and the transverse field. In contrast, we retain the coupling terms with the external field described by the potentials

$$\mathbf{A}_e(\mathbf{r}, t), \quad U_e(\mathbf{r}, t). \tag{B.1}$$

The Lagrangian L and the Hamiltonian H, which are now functions of the dynamical variables of the particles only, are written

$$L = \sum_\alpha \frac{1}{2} m_\alpha \dot{\mathbf{r}}_\alpha^2 - V_{\text{Coul}} + \sum_\alpha [q_\alpha \dot{\mathbf{r}}_\alpha \cdot \mathbf{A}_e(\mathbf{r}_\alpha, t) - q_\alpha U_e(\mathbf{r}_\alpha, t)] \tag{B.2}$$

$$H_L = \sum_\alpha \frac{1}{2 m_\alpha} [\mathbf{p}_{\alpha L} - q_\alpha \mathbf{A}_e(\mathbf{r}_\alpha, t)]^2 + V_{\text{Coul}} + \sum_\alpha q_\alpha U_e(\mathbf{r}_\alpha, t). \tag{B.3}$$

The conjugate momentum $\mathbf{p}_{\alpha L}$ of $\mathbf{r}_\alpha$ is given by

$$\mathbf{p}_{\alpha L} = m_\alpha \dot{\mathbf{r}}_\alpha + q_\alpha \mathbf{A}_e(\mathbf{r}_\alpha, t). \tag{B.4}$$

In quantum theory, the fundamental operators satisfying the canonical commutation relations are $\mathbf{r}_\alpha$ and $\mathbf{p}_\alpha = -i\hbar \nabla_\alpha$. In the standard quantum representation developed starting with L, which we call (1) and which will serve as our reference, the operator representing the position of particle α is

$$\mathbf{r}_\alpha^{(1)} = \mathbf{r}_\alpha \tag{B.5}$$

and the one representing its conjugate momentum $\mathbf{p}_{\alpha L}$ is

$$\mathbf{p}_{\alpha L}^{(1)} = \mathbf{p}_\alpha = \frac{\hbar}{\mathrm{i}} \boldsymbol{\nabla}_\alpha . \tag{B.6}$$

The Hamiltonian operator is gotten by replacing in (B.3) the quantities $\mathbf{r}_\alpha$ and $\mathbf{p}_{\alpha L}$ with the corresponding operators (B.5) and (B.6). By separating the "particle Hamiltonian" and "interaction Hamiltonian", one gets

$$H_L^{(1)} = H_{PL}^{(1)} + h_{IL}^{(1)} \tag{B.7}$$

$$H_{PL}^{(1)} = \sum_\alpha \frac{\mathbf{p}_\alpha^2}{2\,m_\alpha} + V_{\text{Coul}}(\dots, \mathbf{r}_\alpha, \dots) \tag{B.8}$$

$$h_{IL}^{(1)} = \sum_\alpha \left[-\frac{q_\alpha}{m_\alpha} \mathbf{p}_\alpha \cdot \mathbf{A}_e(\mathbf{r}_\alpha, t) + \frac{q_\alpha^2}{2\,m_\alpha} \mathbf{A}_e^2(\mathbf{r}_\alpha, t) + q_\alpha\, U_e(\mathbf{r}_\alpha, t) \right] . \tag{B.9}$$

We are now going to consider other descriptions of the same system constructed using the methods of Part A.

2. Simple Gauge Change; Gauge Invariance

Consider initially the gauge transformation defined by the function $\chi(\mathbf{r},t)$ which depends only on $\mathbf{r}$ and t. The external field is now described by

$$\mathbf{A}_e'(\mathbf{r}, t) = \mathbf{A}_e(\mathbf{r}, t) + \boldsymbol{\nabla}\chi(\mathbf{r}, t) \tag{B.10a}$$

$$U_e'(\mathbf{r}, t) = U_e(\mathbf{r}, t) - \frac{\partial}{\partial t} \chi(\mathbf{r}, t) . \tag{B.10b}$$

a) The New Description

The new Lagrangian is written

$$L' = \sum_\alpha \frac{1}{2} m_\alpha \dot{\mathbf{r}}_\alpha^2 - V_{\text{Coul}} + \sum_\alpha \left[q_\alpha \dot{\mathbf{r}}_\alpha \cdot \mathbf{A}_e'(\mathbf{r}_\alpha, t) - q_\alpha\, U_e'(\mathbf{r}_\alpha, t) \right]$$

$$= L + \frac{\mathrm{d}}{\mathrm{d}t} \left[\sum_\alpha q_\alpha\, \chi(\mathbf{r}_\alpha, t) \right] . \tag{B.11}$$

The new conjugate momenta defined by

$$\mathbf{p}_{\alpha L'} = m_\alpha \dot{\mathbf{r}}_\alpha + q_\alpha \mathbf{A}_e'(\mathbf{r}_\alpha, t) = \mathbf{p}_{\alpha L} + q_\alpha \boldsymbol{\nabla}\chi(\mathbf{r}_\alpha, t) \tag{B.12}$$

are different from the old. The Hamiltonian $H_{L'}$ has a form similar to H_L

in which $\mathbf{p}_{\alpha L}$, $\mathbf{A}_e$, and U_e are replaced respectively by $\mathbf{p}_{\alpha L'}$, $\mathbf{A}'_e$, and U'_e.

In the new quantum representation, the position of the particle is always represented by $\mathbf{r}_\alpha$, and $\mathbf{p}_{\alpha L'}$ by $\mathbf{p}_\alpha = -i\hbar\nabla_\alpha$:

$$\mathbf{r}_\alpha^{(2)} = \mathbf{r}_\alpha \tag{B.13.a}$$

$$\mathbf{p}_{\alpha L'}^{(2)} = \mathbf{p}_\alpha . \tag{B.13.b}$$

The new Hamiltonian operator is then

$$H_{L'}^{(2)} = H_{PL'}^{(2)} + h_{IL'}^{(2)} \tag{B.14.a}$$

$$H_{PL'}^{(2)} = \sum_\alpha \frac{\mathbf{p}_\alpha^2}{2m_\alpha} + V_{\text{Coul}}(\ldots, \mathbf{r}_\alpha, \ldots) \tag{B.14.b}$$

$$h_{IL'}^{(2)} = \sum_\alpha \left[-\frac{q_\alpha}{m_\alpha}\mathbf{p}_\alpha \cdot \mathbf{A}'_e(\mathbf{r}_\alpha, t) + \frac{q_\alpha^2}{2m_\alpha}\mathbf{A}'^2_e(\mathbf{r}_\alpha, t) + q_\alpha U'_e(\mathbf{r}_\alpha, t)\right]. \tag{B.14.c}$$

b) The Unitary Transformation Relating the Two Descriptions—Gauge Invariance (*)

From (B.11), the new Lagrangian L' differs from the old by a total derivative. The treatment in §A.2 above shows then that one passes from representation (1) to representation (2) by a unitary transformation

$$T = \exp\frac{i}{\hbar}\sum_\alpha q_\alpha \chi(\mathbf{r}_\alpha, t). \tag{B.15}$$

Thus, if $\psi^{(1)}(\ldots, \mathbf{r}_\alpha, \ldots)$ is the wave function representing the particle state in the first representation, this same state is represented in the new description by

$$\psi^{(2)}(\ldots, \mathbf{r}_\alpha, \ldots) = \psi^{(1)}(\ldots, \mathbf{r}_\alpha, \ldots)\exp\frac{i}{\hbar}\sum_\alpha q_\alpha \chi(\mathbf{r}_\alpha, t). \tag{B.16}$$

The operators representing the same physical variable in the two descriptions are also connected by T. Consider for example the velocity of particle α. Following (B.4) and (B.6),

$$m_\alpha \mathbf{v}_\alpha^{(1)} = \mathbf{p}_{\alpha L}^{(1)} - q_\alpha \mathbf{A}_e(\mathbf{r}_\alpha, t) = \mathbf{p}_\alpha - q_\alpha \mathbf{A}_e(\mathbf{r}_\alpha, t). \tag{B.17}$$

Similarly

$$m_\alpha \mathbf{v}_\alpha^{(2)} = \mathbf{p}_{\alpha L'}^{(2)} - q_\alpha \mathbf{A}'_e(\mathbf{r}_\alpha, t) = \mathbf{p}_\alpha - q_\alpha \mathbf{A}_e(\mathbf{r}_\alpha, t) - q_\alpha \nabla\chi(\mathbf{r}_\alpha, t) \tag{B.18}$$

and one sees then that

$$\mathbf{v}_\alpha^{(2)} = T\mathbf{v}_\alpha^{(1)}T^+ . \tag{B.19}$$

(*) Cohen-Tannoudji, Diu, and Laloe, Complement H_{III}.

In contrast, the situation is different for the Hamiltonians

$$H_{L'}^{(2)} = TH_L^{(1)}T^+ - \sum_\alpha q_\alpha \frac{\partial}{\partial t}\chi(\mathbf{r}_\alpha, t). \tag{B.20}$$

In the same way, $H_{PL}^{(1)}$ and $H_{PL'}^{(2)}$ or $h_{IL}^{(1)}$ and $h_{IL'}^{(2)}$ are not related by T. Thus, the operator

$$\sum_\alpha \frac{\mathbf{p}_\alpha^2}{2\,m_\alpha} + V_{\text{Coul}}(\ldots, \mathbf{r}_\alpha, \ldots) \tag{B.21}$$

called the particle Hamiltonian $H_{PL}^{(1)}$ or $H_{PL'}^{(2)}$, as in (B.8) or (B.14.b), does not represent the same physical quantity in the two descriptions. An eigenstate of this operator does not describe the same physical state in the two representations. Furthermore, the particle Hamiltonian (B.21) generally does not coincide with the proper energy of the particles defined as the sum of their kinetic and mutual potential energy, since the operator $\mathbf{p}_\alpha/m_\alpha$ in general does not describe the velocity of the particle [see (B.17) and (B.18)].

Note finally that the general demonstration of §A.3 above concerning the equivalence of physical predictions applies here. Since the gauge change is described in quantum theory by a unitary transformation, the transition amplitudes calculated using the gauge $(\mathbf{A}_e, U_e)$ and the gauge $(\mathbf{A}'_e, U'_e)$ are identical [see (A.47)]. It is of course necessary that the kets describing the initial and final states in either gauge correspond to one another by the unitary transformation (B.15) [see (A.45) and (A.46)]. The fact that the transition amplitudes, and as a result all the physical predictions, are the same regardless of gauge reflects the *gauge invariance* of quantum electrodynamics.

3. The Göppert-Mayer Transformation (*)

a) The Long-Wavelength Approximation

Assume that the charges q_α, localized near the origin, form a globally neutral system

$$\sum_\alpha q_\alpha = 0 \tag{B.22}$$

whose spatial extent a is small with respect to the distance characterizing the spatial variations of $\mathbf{A}_e$ and U_e (for example, the wavelength for incident radiation). In Equation (B.2) for the Lagrangian one can then expand the potentials $\mathbf{A}_e(\mathbf{r}_\alpha, t)$ and $U_e(\mathbf{r}_\alpha, t)$ in powers of $\mathbf{r}_\alpha$, which gives rise to the multipole moments of increasing order for the system of

(*) M. Göppert-Mayer, *Ann. Phys.*, (Leipzig) **9**, 273 (1931).

charges with respect to the origin. The electric dipole approximation consists of retaining only the lowest-order terms, which can be expressed in terms of the electric dipole moment with respect to the origin:

$$\mathbf{d} = \sum_\alpha q_\alpha \mathbf{r}_\alpha . \tag{B.23}$$

With this approximation, the Lagrangian (B.2) is written using (B.22) as

$$L = \sum_\alpha \frac{1}{2} m_\alpha \dot{\mathbf{r}}_\alpha^2 - V_{\text{Coul}} + \dot{\mathbf{d}} \cdot \mathbf{A}_e(\mathbf{0}, t) - \mathbf{d} \cdot \nabla U_e(\mathbf{0}, t) . \tag{B.24}$$

One then gets for the momentum conjugate with $\mathbf{r}_\alpha$

$$\mathbf{p}_{\alpha L} = \frac{\partial L}{\partial \dot{\mathbf{r}}_\alpha} = m_\alpha \dot{\mathbf{r}}_\alpha + q_\alpha \mathbf{A}_e(\mathbf{0}, t) \tag{B.25}$$

and for the Hamiltonian

$$\begin{aligned} H_L &= \sum_\alpha \mathbf{p}_{\alpha L} \cdot \dot{\mathbf{r}}_\alpha - L \\ &= \sum_\alpha \frac{1}{2 m_\alpha} [\mathbf{p}_{\alpha L} - q_\alpha \mathbf{A}_e(\mathbf{0}, t)]^2 + V_{\text{Coul}} + \mathbf{d} \cdot \nabla U_e(\mathbf{0}, t) . \end{aligned} \tag{B.26}$$

b) Gauge Change Giving Rise to the Electric Dipole Interaction

We select now a new gauge $\mathbf{A}'_e$, U'_e . The new expressions for L' , $\mathbf{p}_{\alpha L'}$, and $H_{L'}$ are gotten by replacing $\mathbf{A}_e(\mathbf{0}, t)$ with $\mathbf{A}'_e(\mathbf{0}, t)$ and $\nabla U_e(\mathbf{0}, t)$ with $\nabla U'_e(\mathbf{0}, t)$ in (B.24), (B.25), and (B.26). The Göppert-Mayer transformation seeks to get $\mathbf{A}'_e(\mathbf{0}, t) = \mathbf{0}$, so as to simplify as far as possible the expression for $\mathbf{p}_{\alpha L'}$ and thereby that of the first term of $H_{L'}$. For this, one uses the gauge change defined by

$$\chi(\mathbf{r}, t) = - \mathbf{r} \cdot \mathbf{A}_e(\mathbf{0}, t) \tag{B.27}$$

or, which amounts to the same thing, the change in Lagrangian gotten by adding to (B.24) the total derivative

$$\frac{\mathrm{d}}{\mathrm{d}t}\left[- \sum_\alpha q_\alpha \mathbf{r}_\alpha \cdot \mathbf{A}_e(\mathbf{0}, t) \right] = \frac{\mathrm{d}}{\mathrm{d}t}[- \mathbf{d} \cdot \mathbf{A}_e(\mathbf{0}, t)] . \tag{B.28}$$

The new potentials $\mathbf{A}'_e$ and U'_e, using (B.10) and (B.27), are

$$\begin{cases} \mathbf{A}'_e(\mathbf{r}, t) = \mathbf{A}_e(\mathbf{r}, t) - \mathbf{A}_e(\mathbf{0}, t) & \text{(B.29.a)} \\ U'_e(\mathbf{r}, t) = U_e(\mathbf{r}, t) + \mathbf{r} \cdot \dot{\mathbf{A}}_e(\mathbf{0}, t) & \text{(B.29.b)} \end{cases}$$

which yields

$$\begin{cases} \mathbf{A}'_e(\mathbf{0}, t) = \mathbf{0} & \text{(B.30.a)} \\ \nabla U'_e(\mathbf{0}, t) = \nabla U_e(\mathbf{0}, t) + \dot{\mathbf{A}}_e(\mathbf{0}, t) = -\mathbf{E}_e(\mathbf{0}, t) & \text{(B.30.b)} \end{cases}$$

where $\mathbf{E}_e(\mathbf{0}, t)$ is the total external field at $\mathbf{0}$. It is sufficient then to substitute (B.30) in Equations (B.24), (B.25), and (B.26), written in the new gauge, to get

$$L' = \sum_\alpha \frac{1}{2} m_\alpha \dot{\mathbf{r}}_\alpha^2 - V_{\text{Coul}} + \mathbf{d} \cdot \mathbf{E}_e(\mathbf{0}, t) \tag{B.31}$$

$$\mathbf{p}_{\alpha L'} = m_\alpha \dot{\mathbf{r}}_\alpha \tag{B.32}$$

$$H_{L'} = \sum_\alpha \frac{\mathbf{p}_{\alpha L'}^2}{2 m_\alpha} + V_{\text{Coul}} - \mathbf{d} \cdot \mathbf{E}_e(\mathbf{0}, t) . \tag{B.33}$$

Beyond their simpler form, Equations (B.31) and (B.33) have the advantage of making explicit the electric dipole interaction between $\mathbf{d}$ and $\mathbf{E}_e$.

In quantum theory, the transition from the usual description (1) to the Göppert-Mayer description (2) is realized, according to (A.28) and (B.28), by the unitary transformation

$$T(t) = \exp\left\{ -\frac{\mathrm{i}}{\hbar} \mathbf{d} \cdot \mathbf{A}_e(\mathbf{0}, t) \right\} = \exp\left\{ -\frac{\mathrm{i}}{\hbar} \sum_\alpha q_\alpha \mathbf{r}_\alpha \cdot \mathbf{A}_e(\mathbf{0}, t) \right\}. \tag{B.34}$$

It is indeed possible to study directly the effect of the unitary transformation (B.34) on the initial representation. This is done in Complement A_{IV} in order to introduce the electric dipole Hamiltonian in an elementary way without using the Lagrangian formalism.

c) The Advantages of the New Point of View

In the new representation, the operator $\mathbf{p}_\alpha = -i\hbar \nabla_\alpha$ describes the variable $\mathbf{p}_{\alpha L'}$ given in (B.32), that is, the *mechanical momentum* of the particle. It follows then that the particle Hamiltonian $H_{PL'}^{(2)}$ given in (B.14.b) truly represents in this new description the energy of the particle system, that is, *the sum of the kinetic and Coulomb energies*. As a consequence, the eigenstates $|\varphi_a\rangle$ of $H_{PL'}^{(2)}$ with eigenvalue E_a are now the physical states where the energy of the particles has a well-defined value E_a. The amplitudes

$$\langle \varphi_b \mid U^{(2)}(t_f, t_0) \mid \varphi_a \rangle \tag{B.35}$$

are the transition amplitudes between an initial state with energy E_a at time t_0 and a final state with energy E_b at t_f. Note finally that the calculation of $U^{(2)}(t, t_0)$ uses the interaction Hamiltonian in representation (2),

$$h_{IL'}^{(2)} = - \mathbf{d} \cdot \mathbf{E}_e(\mathbf{0}, t) \tag{B.36.a}$$

which is much simpler than the one in representation (1), which, using (B.26), is written

$$h_{IL}^{(1)} = - \sum_\alpha \frac{q_\alpha}{m_\alpha} \mathbf{p}_\alpha \cdot \mathbf{A}_e(\mathbf{0}, t) + \sum_\alpha \frac{q_\alpha^2}{2\, m_\alpha} \mathbf{A}_e^2(\mathbf{0}, t) + \mathbf{d} \cdot \nabla U_e(\mathbf{0}, t) \tag{B.36.b}$$

$h_{IL'}^{(2)}$, which reduces to a single term, is linear in the fields; it depends only on the field $\mathbf{E}_e$ and not on the potentials.

It clearly appears then that the calculations are much simpler and more direct in the new representation. It should not be forgotten however that, if they are approached correctly, the calculations should lead to the same results in both representations. Given the practical importance of the interaction Hamiltonians $\mathbf{A} \cdot \mathbf{p}$ and $\mathbf{E} \cdot \mathbf{r}$, we will discuss their equivalence in detail.

d) The Equivalence between the Interaction Hamiltonians $\mathbf{A} \cdot \mathbf{p}$ and $\mathbf{E} \cdot \mathbf{r}$

i) *The Simple Case Where the Potentials are Zero at the Initial and Final Times*

Assume initially that the potentials $\mathbf{A}_e(\mathbf{0}, t)$ and $U_e(\mathbf{0}, t)$ are zero at the initial time t_0 and final time t_f. This must also be the case for the field $\mathbf{E}_e(\mathbf{0}, t)$. Such a situation arises for example when a wave packet impinges on an atom, the times t_0 and t_f being respectively before and after the packet passes the origin.

Since $\mathbf{A}_e(\mathbf{0}, t)$ is zero for $t = t_0$ and $t = t_f$, it follows from (B.34) that

$$T(t_0) = T(t_f) = 1\,. \tag{B.37}$$

The vanishing of $\mathbf{A}_e(\mathbf{0}, t_0)$ and $\mathbf{A}_e(\mathbf{0}, t_f)$ entails likewise that $\mathbf{p}_{\alpha L}$ coincides at $t = t_0$ and $t = t_f$ with the mechanical momentum of particle α [see (B.25)], so that the eigenstates $|\varphi_a\rangle$ and $|\varphi_b\rangle$ of $H_{PL}^{(1)}$ [see (B.8)] represent at $t = t_0$ and $t = t_f$ states with well-defined total energies E_a and E_b in representation (1). It follows that the transition amplitude between an initial energy state E_a at $t = t_0$ and a final energy state E_b at $t = t_f$ can be written in representation (1) as

$$\langle\, \varphi_b \mid U^{(1)}(t_f, t_0) \mid \varphi_a \,\rangle\,. \tag{B.38}$$

Indeed, the equality between the two amplitudes (B.35) and (B.38) follows directly from (A.44) and (B.37).

It is also possible to verify the equality of the amplitudes (B.35) and (B.38) directly through an explicit calculation. This is done in Complement $\mathrm{B_{IV}}$ for one- and two-photon processes induced by a nearly monochromatic wave packet whose time of passage T tends to infinity. The amplitudes (B.35) and (B.38), evaluated in the interaction representation with respect to the particle Hamiltonian, are then elements of the scattering S-matrix. The problem of the equivalence of the scattering S-matrices in two representations related by a unitary transformation is reexamined in Part D of this chapter, the radiation field being no longer treated as an external field but as a quantum-dynamical system.

ii) *The General Case*

If $\mathbf{A}_e(\mathbf{0}, t)$ is not zero at t_0 and t_f, the equation (B.38) for the transition amplitude is no longer correct, since the states $|\varphi_a\rangle$ and $|\varphi_b\rangle$ no longer represent well-defined energy states in representation (1), and $\mathbf{p}_{\alpha L}$ differs from the mechanical momentum $m_\alpha \dot{\mathbf{r}}_\alpha$ when $\mathbf{A}_e$ is nonzero. In contrast, since $|\varphi_a\rangle$ and $|\varphi_b\rangle$ always represent the energy states E_a and E_b in representation (2), it is always possible, whether $\mathbf{A}_e$ is zero or not [see (B.32) and (B.33)], to get the corresponding states in representation (1) by means of the transformations $T^+(t_0)$ and $T^+(t_f)$ which transform from (2) to (1). The initial and final states are then written in (1) as

$$\begin{cases} |\psi^{(1)}(t_0)\rangle = T^+(t_0)\,|\varphi_a\rangle & \text{(B.39.a)} \\ |\psi^{(1)}(t_f)\rangle = T^+(t_f)\,|\varphi_b\rangle & \text{(B.39.b)} \end{cases}$$

and the transition amplitude in this description becomes

$$\langle \varphi_b \,|\, T(t_f)\, U^{(1)}(t_f, t_0)\, T^+(t_0) \,|\, \varphi_a \rangle =$$

$$= \left\langle \varphi_b \left| \exp\left\{-\frac{\mathrm{i}}{\hbar}\mathbf{d}\cdot\mathbf{A}_e(\mathbf{0}, t_f)\right\} U^{(1)}(t_f, t_0) \exp\left\{\frac{\mathrm{i}}{\hbar}\mathbf{d}\cdot\mathbf{A}_e(\mathbf{0}, t_0)\right\} \right| \varphi_a \right\rangle. \tag{B.40}$$

One should not make the mistake of omitting the two exponentials in (B.40).

The equality between the amplitudes (B.35) and (B.40) follows from the general property (A.47) established above. It is also possible to verify this equality directly by explicitly calculating the amplitudes, for example to a given order in q_α. The interested reader will find an example of such a calculation in §B.3.*d*.iii below.

iii) *Direct Verification of the Equality of the Two Transition Amplitudes to First Order*

We will expand the two exponentials of (B.40) in powers of q_α and use the well-known perturbation expansion of the operator U associated with $H_L^{(1)}$,

$$U(t_f, t_0) = U_0(t_f, t_0) + \frac{1}{\mathrm{i}\hbar}\int_{t_0}^{t_f} U_0(t_f, \tau)\, V(\tau)\, U_0(\tau, t_0)\, \mathrm{d}\tau + \cdots \tag{B.41}$$

where U_0 is the unperturbed evolution operator associated with $H_{PL}^{(1)}$ and $V(t)$ is the "perturbation" $h_{IL}^{(1)}$ given by (B.36.b). To order 1 in q_α, (B.40) is written

$$\langle \psi^{(1)}(t_f) \mid U^{(1)}(t_f, t_0) \mid \psi^{(1)}(t_0) \rangle \simeq \delta_{ba}\, e^{-i\omega_a(t_f - t_0)} +$$

$$+ \frac{1}{i\hbar} \langle \varphi_b \mid \mathbf{d} \cdot \mathbf{A}_e(\mathbf{0}, t_f)\, e^{-i\omega_a(t_f - t_0)} - e^{-i\omega_b(t_f - t_0)}\, \mathbf{d} \cdot \mathbf{A}_e(\mathbf{0}, t_0) \mid \varphi_a \rangle +$$

$$+ \frac{1}{i\hbar} e^{-i(\omega_b t_f - \omega_a t_0)} \int_{t_0}^{t_f} \mathrm{d}\tau\, e^{-i\omega_{ab}\tau} \times$$

$$\times \left\langle \varphi_b \left| \sum_\alpha \frac{-q_\alpha}{m_\alpha} \mathbf{p}_\alpha \cdot \mathbf{A}_e(\mathbf{0}, \tau) + \mathbf{d} \cdot \nabla U_e(\mathbf{0}, \tau) \right| \varphi_a \right\rangle + \cdots \qquad (\text{B.42.a})$$

where $\omega_a = E_a/\hbar$, $\omega_b = E_b/\hbar$, and $\omega_{ab} = \omega_a - \omega_b$. An analogous calculation gives for the amplitude (B.35), with $V(t)$ being now replaced in (B.41) by the Hamiltonian $h_{IL}^{(2)}$, given in (B.36.a),

$$\langle \varphi_b \mid U^{(2)}(t_f, t_0) \mid \varphi_a \rangle \simeq \delta_{ba}\, e^{-i\omega_a(t_f - t_0)} +$$

$$+ \frac{1}{i\hbar} e^{-i(\omega_b t_f - \omega_a t_0)} \int_{t_0}^{t_f} \mathrm{d}\tau\, e^{-i\omega_{ab}\tau} \langle \varphi_b \mid -\mathbf{d} \cdot \mathbf{E}_e(\mathbf{0}, \tau) \mid \varphi_a \rangle . \qquad (\text{B.42.b})$$

We can now replace the matrix element for $\mathbf{p}_\alpha$ by that of $\mathbf{r}_\alpha$ in (B.42.a), thanks to the algebraic relationship

$$\left\langle \varphi_b \left| \frac{\mathbf{p}_\alpha}{m_\alpha} \right| \varphi_a \right\rangle = i\omega_{ba} \langle \varphi_b \mid \mathbf{r}_\alpha \mid \varphi_a \rangle \qquad (\text{B.43})$$

obtained by taking the matrix elements between $|\varphi_a\rangle$ and $|\varphi_b\rangle$ of the identity

$$[\mathbf{r}_\alpha, H_{PL}^{(1)}] = i\hbar \frac{\partial H_{PL}^{(1)}}{\partial \mathbf{p}_\alpha} = i\hbar \frac{\mathbf{p}_\alpha}{m_\alpha} . \qquad (\text{B.44})$$

We also integrate by parts the term in $\mathbf{p}_\alpha \cdot \mathbf{A}_e(\mathbf{0}, \tau)$ of (B.42.a) to get $\dot{\mathbf{A}}_e$. The integrated terms cancel exactly the second line of (B.42.a) and the remaining terms coincide exactly with (B.42.b). Thus we have demonstrated the identity of the amplitudes (B.35) and (B.40) to first order in q_α. Similar relationships exist between the matrix elements to all orders.

e) Generalizations

The transformation presented in §B.3.*a* can be generalized in two ways. It is possible first of all to introduce an analogous transformation for the long-wavelength modes of the transverse field taken together as a dynamical system; the function χ depends then on the field variables, and the transformation modifies the momenta conjugate with the field. One such approach is presented in detail in Complement A_{IV}.

One can also try to get rid of the long-wavelength approximation and seek a gauge transformation which eliminates **A** in favor of the electric and magnetic fields. We will see in Part C of this chapter that such an objective can be attained by means of a change in the Lagrangian. We will also see in Complement D_{IV} that such a change in the Lagrangian is in certain cases equivalent to a change of gauge characterized by the function

$$\chi(\mathbf{r}) = -\int_0^1 \mathbf{r} \cdot \mathbf{A}_\perp(\mathbf{r}u)\, \mathrm{d}u \tag{B.45}$$

which generalizes (B.27).

4. A Transformation Which Does Not Reduce to a Change of Lagrangian: The Henneberger Transformation (*)

a) Motivation

One of the advantages of the Göppert-Mayer transformation is to establish between the velocity of a particle and the momentum conjugate with its coordinate a simpler relationship than in the standard description, as the result of a change in the conjugate momentum. One can reach this same objective by changing the particle "coordinate". More precisely, we will look for the quantity $\mathbf{R}_\alpha$ whose velocity is $\mathbf{p}_\alpha/m_\alpha$ in the standard classical description. We still assume that the particles are localized near the origin and that the long-wavelength approximation applies. Equation (B.4) then becomes

$$m_\alpha \dot{\mathbf{R}}_\alpha = m_\alpha \dot{\mathbf{r}}_\alpha + q_\alpha \mathbf{A}_e(\mathbf{0}, t)\,. \tag{B.46}$$

If one introduces the new potential $\mathbf{Z}_e(\mathbf{r}, t)$ defined by

$$\mathbf{Z}_e(\mathbf{r}, t) = -\int^t \mathrm{d}\tau\, \mathbf{A}_e(\mathbf{r}, \tau) \tag{B.47}$$

the expression for $\mathbf{R}_\alpha$ then becomes

$$\mathbf{R}_\alpha = \mathbf{r}_\alpha - \boldsymbol{\xi}_\alpha(t) \tag{B.48.a}$$

with

$$\boldsymbol{\xi}_\alpha(t) = \frac{q_\alpha}{m_\alpha}\mathbf{Z}_e(\mathbf{0}, t)\,. \tag{B.48.b}$$

(*) W. C. Henneberger, *Phys. Rev. Lett.*, **21**, 838 (1968).

The physical meaning of $\mathbf{R}_\alpha$ is simple: if the particle is subject only to the action of the electric field derived from $\mathbf{A}_e(\mathbf{0}, t)$, its equation of motion will be

$$m_\alpha \ddot{\mathbf{r}}_\alpha = - q_\alpha \dot{\mathbf{A}}_e(\mathbf{0}, t) \tag{B.49}$$

and one has then

$$m_\alpha \ddot{\mathbf{R}}_\alpha = 0 . \tag{B.50}$$

The quantity $\mathbf{R}_\alpha$ will then describe the motion of a free particle. If $\mathbf{A}_e$ corresponds to an oscillating field, $\mathbf{R}_\alpha$ represents a kind of "mean position" about which the particle executes a forced oscillatory motion described by $\boldsymbol{\xi}_\alpha(t)$. In the presence of other forces created by U_e and the interaction V_{Coul} between the particles, the motion of $\mathbf{R}_\alpha$ will not be so simple. However, if the action of $\mathbf{A}_e$ is dominant, the choice of $\mathbf{R}_\alpha$ as a dynamical variable is advantageous, since it already takes account of the particle's dynamics under $\mathbf{A}_e$.

b) Determination of the Unitary Transformation. Transforms of the Various Operators

In the subsection above, we have demonstrated the interest which one has in considering the physical quantity $\mathbf{R}_\alpha$ defined as the "mean position" of particle α. We now seek a unitary transformation T such that in the new representation (2), the operator $\mathbf{R}_\alpha^{(2)}$ representing this mean position $\mathbf{R}_\alpha$ is simply the operator of multiplication by $\mathbf{r}_\alpha$:

$$\mathbf{R}_\alpha^{(2)} = \mathbf{r}_\alpha \tag{B.51}$$

whereas in (1), this same operator represents the instantaneous position $\mathbf{r}_\alpha^{(1)}$:

$$\mathbf{r}_\alpha^{(1)} = \mathbf{r}_\alpha . \tag{B.52}$$

Using (B.48), (B.51), and (B.52), we get for T the equation

$$\mathbf{r}_\alpha = T[\mathbf{r}_\alpha - \boldsymbol{\xi}_\alpha(t)]\, T^+ = T\left[\mathbf{r}_\alpha - \frac{q_\alpha}{m_\alpha}\mathbf{Z}_e(\mathbf{0}, t)\right] T^+ \tag{B.53}$$

which shows that T is a spatial translation operator

$$T = \exp \sum_\alpha \frac{\mathrm{i}}{\hbar}\frac{q_\alpha}{m_\alpha}\mathbf{p}_\alpha \cdot \mathbf{Z}_e(\mathbf{0}, t) . \tag{B.54}$$

The transformation leaves the momentum $\mathbf{p}_\alpha$ unchanged:

$$\mathbf{p}_\alpha^{(2)} = \mathbf{p}_\alpha = \mathbf{p}_\alpha^{(1)} . \tag{B.55}$$

Following (A.42), the Hamiltonian of the new representation is

$$H^{(2)} = TH^{(1)}T^+ + i\hbar \frac{\partial T}{\partial t} T^+ \tag{B.56}$$

Starting with Equations (B.8) and (B.36.b) for $H_{PL}^{(1)}$ and $h_{IL}^{(1)}$, and using the fact that T does not change $\mathbf{p}_\alpha$ and changes $\mathbf{r}_\alpha$ into $\mathbf{r}_\alpha + (q_\alpha/m_\alpha)\mathbf{Z}_e(\mathbf{0}, t)$, one gets

$$TH_{PL}^{(1)}T^+ = \sum_\alpha \frac{\mathbf{p}_\alpha^2}{2m_\alpha} + V_{\text{Coul}}\left(\dots, \mathbf{r}_\alpha + \frac{q_\alpha}{m_\alpha}\mathbf{Z}_e(\mathbf{0}, t), \dots\right) \tag{B.57.a}$$

$$Th_{IL}^{(1)}T^+ = \sum_\alpha \left[-\frac{q_\alpha}{m_\alpha}\mathbf{p}_\alpha \cdot \mathbf{A}_e(\mathbf{0}, t) + \frac{q_\alpha^2}{2m_\alpha}\mathbf{A}_e^2(\mathbf{0}, t) + \right.$$
$$\left. + q_\alpha\left(\mathbf{r}_\alpha + \frac{q_\alpha}{m_\alpha}\mathbf{Z}_e(\mathbf{0}, t)\right) \cdot \nabla U_e(\mathbf{0}, t) \right]. \tag{B.57.b}$$

From (B.54) and (B.47), it follows that

$$i\hbar \frac{\partial T}{\partial t} T^+ = \sum_\alpha \frac{q_\alpha}{m_\alpha}\mathbf{p}_\alpha \cdot \mathbf{A}_e(\mathbf{0}, t). \tag{B.58}$$

Finally, (B.56) becomes

$$H^{(2)} = \sum_\alpha \frac{\mathbf{p}_\alpha^2}{2m_\alpha} + \sum_\alpha \frac{q_\alpha^2}{2m_\alpha}\mathbf{A}_e^2(\mathbf{0}, t) + V_{\text{Coul}}\left(\dots, \mathbf{r}_\alpha + \frac{q_\alpha}{m_\alpha}\mathbf{Z}_e(\mathbf{0}, t), \dots\right) +$$
$$+ \sum_\alpha \left[q_\alpha \mathbf{r}_\alpha \cdot \nabla U_e(\mathbf{0}, t) + \frac{q_\alpha^2}{m_\alpha}\mathbf{Z}_e(\mathbf{0}, t) \cdot \nabla U_e(\mathbf{0}, t) \right]. \tag{B.59}$$

c) Physical Interpretation

The physical interpretation of $H^{(2)}$ is simple. The first term represents the kinetic energy associated with the motion of the "mean position". To see this, it suffices to note that the Heisenberg equation for $\mathbf{r}_\alpha$ is written

$$\dot{\mathbf{r}}_\alpha = \frac{1}{i\hbar}[\mathbf{r}_\alpha, H^{(2)}] = \frac{\partial H^{(2)}}{\partial \mathbf{p}_\alpha} = \frac{\mathbf{p}_\alpha}{m_\alpha} \tag{B.60}$$

and to remember that $\mathbf{r}_\alpha$ represents the mean position in representation (2) [see (B.51)]. The second term, which is a number and which has the same form in either representation, can be written using (B.47) and (B.48.b) as $\sum_\alpha m_\alpha \dot{\xi}_\alpha^2(t)/2$. It represents the kinetic energy of the oscillatory forced motion under the effect of $\mathbf{A}_e$. The third term represents the Coulomb energy of the system of charges depending on the mean position $\mathbf{r}_\alpha$ of the particles and their deviation $q_\alpha \mathbf{Z}_e(\mathbf{0}, t)/m_\alpha$ from the mean position. The fourth and fifth terms represent the coupling of the external potential with the mean position and the oscillatory motion respectively. (We always use the long-wavelength approximation.)

Note finally that in the new description, the only terms containing both types of variables, $\mathbf{r}_\alpha$ and the fields deriving from $\mathbf{A}_e$, involve V_{Coul} (third term of B.59). The motion of $\mathbf{r}_\alpha$ is only coupled to the field through V_{Coul}.

This is precisely what makes the Henneberger transformation interesting. Consider for example an electron subject to incident radiation with frequency ω. If this electron is free, it cannot absorb really (i.e. with simultaneous conservation of energy and momentum) one or more impinging photons. On the other hand, if this electron is also experiencing the Coulomb potential of other charges, such real transitions can occur, since the Coulomb potential can now give (or absorb) the corresponding momentum. The Henneberger representation, which introduces a clear separation between the oscillatory motion of the electron in the incident wave (which can be also interpreted in terms of "virtual" absorptions and reemissions of incident photons) and the mean motion (which changes only as a result of real transitions), is particularly well suited to the analysis of such processes. Exercise 4 shows indeed how the third term of (B.59) allows a simple calculation (to all orders in the coupling with the transverse field and to order 1 in V_{Coul}) of the scattering cross section of an electron in a Coulomb field in the presence of an intense laser radiation.

d) Generalization to a Quantized Field: The Pauli–Fierz–Kramers Transformation

Consider now the electromagnetic field as a quantized system having its own dynamics, and assume that one is interested in the interaction of a system of charges localized about the origin with only the long-wavelength modes of the field. Under these conditions, we can, in the Hamiltonian in Coulomb gauge [(A.16) of Chapter III], replace $\mathbf{A}(\mathbf{r}_\alpha)$ by $\mathbf{A}(\mathbf{0})$ and thus get

$$H = \sum_\alpha \frac{1}{2\,m_\alpha}\left[\mathbf{p}_\alpha - q_\alpha\,\mathbf{A}(\mathbf{0})\right]^2 + V_{\text{Coul}} + \sum_i \hbar\omega_i\left(a_i^+\,a_i + \frac{1}{2}\right). \quad \text{(B.61)}$$

The generalization to the quantum case of the transformation (B.54) is (*)

$$T = \exp\left\{\sum_\alpha \frac{i}{\hbar}\,\frac{q_\alpha}{m_\alpha}\,\mathbf{p}_\alpha\cdot\mathbf{Z}(\mathbf{0})\right\} \quad \text{(B.62)}$$

(*) This transformation was in fact introduced by Pauli and Fierz before that of Henneberger and with quite different motivations: W. Pauli and M. Fierz, *Nuovo Cimento*, **15**, 167 (1938).

where $\mathbf{Z}(\mathbf{r})$ is the quantized field

$$\mathbf{Z}(\mathbf{r}) = \int d^3k \sum_{\varepsilon} \sqrt{\frac{\hbar}{2\,\varepsilon_0\,\omega(2\,\pi)^3}}\,\boldsymbol{\varepsilon}\left[\frac{a_\varepsilon(\mathbf{k})}{i\omega}\,e^{i\mathbf{k}.\mathbf{r}} - \frac{a_\varepsilon^+(\mathbf{k})}{i\omega}\,e^{-i\mathbf{k}.\mathbf{r}}\right]. \tag{B.63}$$

This field is the quantum analogue to the external potential introduced in (B.47). Indeed, for free fields in the Heisenberg approach, one finds the following expression:

$$\mathbf{Z}(\mathbf{r}, t) = -\int^t d\tau\,\mathbf{A}(\mathbf{r}, \tau) \tag{B.64}$$

which generalizes (B.47).

The operators relative to the particles transform according to laws which simply generalize (B.53) and (B.55). However, it is the way in which the field operators are transformed which makes the Pauli–Fierz transformation particularly interesting. Indeed, it is possible to show that, in the long-wavelength approximation, the Pauli–Fierz transformation removes from the transverse field a part of the field which is "tied" to the particles. According to the results of Chapter III, the transverse field does not describe only the free radiation (in particular the vacuum fluctuations) and the radiation emitted by the particles. It also contains the part of the field depending on the velocity of the particles, for example, the magnetic field produced by their motion. This field is in some way bound to the particles as long as their velocity is unchanged. It is this last contribution to the transverse field which can, in a first approximation, be removed by the transformation (B.62). This explains why such a separation has been tried as a first approach to *renormalization*. One such approach, initiated by Kramers, allows one to understand in a qualitative way certain properties of radiative corrections (*), although it cannot actually be carried to its conclusion.

(*) A detailed study of the Pauli–Fierz transformation is presented in Cohen-Tannoudji, Dupont-Roc, and Grynberg in Complement B_{II}.

C—THE POWER-ZIENAU-WOOLLEY TRANSFORMATION: THE MULTIPOLE FORM OF THE INTERACTION BETWEEN CHARGES AND FIELD

We return to the case where the field is considered as a dynamical system, and construct a generalization of the Göppert-Mayer transformation which no longer describes the system of charges through their dipole moment but takes into account the precise distribution of charges and currents. One such transformation, which has been introduced by Power and Zienau, and using an alternative approach by Woolley, leads to a novel description of electrodynamics, rigorously equivalent to the standard description (see the references at the end of the chapter).

This new description has two advantages: first of all, the coupling between field and charges is expressed as a function of the electric and magnetic fields themselves and no longer as a function of the vector potential. Also, the system of charges is described by polarization and magnetization densities which are given directly as functions of the microscopic observables, the position and velocity of the particles. This provides then a rigorous basis for the electrodynamics of material media; in particular, the displacement **D** is introduced naturally as the momentum conjugate with the vector potential **A**.

One can also use this new approach to introduce the different electric and magnetic multipole moments of the system of charges and thus get a multipole expansion of the interaction between the system of charges and the electric and magnetic fields. On the other hand, it is equally clear that the magnitudes of these different multipole moments, and thus of the coupling Hamiltonian, increase rapidly as the system is extended. As a result, even though valid in principle in all circumstances, the approach described above is only useful in practice for localized systems of charges or for ensembles of such systems.

We begin (§C.1) by describing the localized system of charges by polarization and magnetization densities, which will allow us to introduce simply the Power–Zienau–Woolley transformation and the corresponding new Lagrangian (§C.2). We then derive the expression for the new conjugate momenta and the new Hamiltonian (§C.3). We analyze finally the new description of quantum electrodynamics which results (§C.4) and its equivalence with the description in the Coulomb gauge (§C.5).

1. Description of the Sources in Terms of a Polarization and a Magnetization Density (*)

In §B.3 above, the system of charges was approximated by a point dipole as far as its interactions with the long-wavelength modes of the

(*) The microscopic definition of these notions is discussed in detail by S. R. de Groot, *The Maxwell Equations*, North Holland, 1969.

field were concerned. To completely describe a distribution of charges with finite extension a, it is necessary to introduce a polarization density $\mathbf{P}(\mathbf{r})$ strictly equivalent to the charge distribution.

a) The Polarization Density Associated with a System of Charges

We initially construct the polarization field associated with a single charge q_α at $\mathbf{r}_\alpha$. The idea of polarization assumes that one considers the deviation of the charge distribution with respect to a reference distribution. We take as the origin O the point with respect to which the deviation of the charge q_α is referred. The reference distribution is thus the charge q_α at O. The real charge distribution (Figure 1a) can be gotten by adding to the reference distribution a line of n nonpoint electric dipoles, each made up of charges $-q_\alpha$ and $+q_\alpha$ separated by $\mathbf{r}_\alpha/n$ and disposed so that the charge $-q_\alpha$ of one dipole is superimposed on the charge $+q_\alpha$ of the preceding charge (Figure 1b). By letting n go to infinity, one gets a continuous distribution of point dipoles whose polarization density is given by

$$\mathbf{P}(\mathbf{r}) = \lim_{n\to\infty} \sum_{p=0}^{n-1} q_\alpha \frac{\mathbf{r}_\alpha}{n} \delta\left(\mathbf{r} - \frac{p+\frac{1}{2}}{n}\mathbf{r}_\alpha\right) = \int_0^1 du\, q_\alpha\, \mathbf{r}_\alpha\, \delta(\mathbf{r} - u\mathbf{r}_\alpha). \quad (\text{C}.1)$$

By construction, the charge density corresponding to q_α at O plus the polarization density $\mathbf{P}$ is strictly identical with that which corresponds to q_α at $\mathbf{r}_\alpha$. The polarization density is uniformly distributed on the line segment between O and $\mathbf{r}_\alpha$. It appears then immediately that the description of the charge q_α at $\mathbf{r}_\alpha$ with the help of a polarization density is only truly interesting if the charge is not displaced too far from the origin O, that is, if the charge is bound.

Figure 1. The real charge distribution q_α at $\mathbf{r}_\alpha$, represented in (a), is equivalent to the reference distribution (q_α at O) plus n dipoles $q_\alpha \mathbf{r}_\alpha / n$ (b).

Remark

Rather than distributing the dipoles along the line from O to $\mathbf{r}_\alpha$, one can also arrange them tangentially to any curve having the same endpoints. One can also add an arbitrary number of closed curves of the same type. All these polarization distributions correspond to the same charge density. One thereby has a certain freedom in the definition of the polarization field corresponding to a given charge density, beyond the definition of the reference distribution itself with respect to which the deviation is measured, just as the same electromagnetic field can be described by different potentials. The choice consisting of aligning the dipoles on a straight line gives the minimum polarization field.

The same procedure permits the description of a system of charges in terms of its total charge $\Sigma_\alpha q_\alpha$ at the origin O and the polarization distribution

$$\mathbf{P}(\mathbf{r}) = \sum_\alpha \int_0^1 \mathrm{d}u \, q_\alpha \, \mathbf{r}_\alpha \, \delta(\mathbf{r} - u\mathbf{r}_\alpha) \tag{C.2}$$

distributed on all the segments $O\mathbf{r}_\alpha$ and thus localized in a region of extent a. We also use the Fourier transform of $\mathbf{P}(\mathbf{r})$,

$$\mathscr{P}(\mathbf{k}) = \sum_\alpha \int_0^1 \mathrm{d}u \, \frac{q_\alpha \, \mathbf{r}_\alpha}{(2\pi)^{3/2}} \, \mathrm{e}^{-i\mathbf{k}.\mathbf{r}_\alpha u} \, . \tag{C.3}$$

Although this integral can be evaluated easily, it is more convenient for what follows to retain it in the form (C.3).

Remark

When k is small with respect to $1/a$, one can take $\exp(-i\mathbf{k}\cdot\mathbf{r}_\alpha u) = 1$ in (C.3). One then gets the approximate expression of the polarization density used in Complement A_{IV} [Equation (24b)].

b) THE DISPLACEMENT

The divergence of $\mathbf{P}$ is directly related to the charge density. Actually, it follows from (C.3) that

$$\begin{aligned} i\mathbf{k}\cdot\mathscr{P} &= \sum_\alpha \int_0^1 \mathrm{d}u \, \frac{q_\alpha}{(2\pi)^{3/2}} (i\mathbf{k}\cdot\mathbf{r}_\alpha) \, \mathrm{e}^{-i\mathbf{k}.\mathbf{r}_\alpha u} \\ &= \sum_\alpha \frac{-q_\alpha}{(2\pi)^{3/2}} \, \mathrm{e}^{-i\mathbf{k}.\mathbf{r}_\alpha u} \Big|_0^1 \\ &= -\sum_\alpha q_\alpha \frac{\mathrm{e}^{-i\mathbf{k}.\mathbf{r}_\alpha}}{(2\pi)^{3/2}} + \sum_\alpha q_\alpha \frac{1}{(2\pi)^{3/2}} \end{aligned} \tag{C.4}$$

which gives, in real space,

$$\nabla \cdot \mathbf{P}(\mathbf{r}) = -\rho(\mathbf{r}) + \rho_0(\mathbf{r}) \tag{C.5}$$

where $\rho_0(\mathbf{r})$ is the reference charge density

$$\rho_0(\mathbf{r}) = \left(\sum_\alpha{}' q_\alpha\right)\delta(\mathbf{r}) \tag{C.6}$$

and $\rho(\mathbf{r})$ the real charge density given by (A.5.a) of Chapter I. Combining (C.5) with the equation

$$\nabla \cdot \mathbf{E}(\mathbf{r}) = \frac{\rho(\mathbf{r})}{\varepsilon_0} \tag{C.7}$$

one sees that one can construct a field, the *displacement*

$$\mathbf{D}(\mathbf{r}) = \varepsilon_0\, \mathbf{E}(\mathbf{r}) + \mathbf{P}(\mathbf{r}) \tag{C.8}$$

whose divergence is

$$\nabla \cdot \mathbf{D}(\mathbf{r}) = \rho_0(\mathbf{r})\,. \tag{C.9.a}$$

In contrast to **E**, whose divergence is related to the real charge density ρ which depends on the dynamical variables $\mathbf{r}_\alpha$, the divergence of **D** is related to the density ρ_0, which has the two fold advantage of being known and being static. If $\mathbf{E}_0$ is the static electric field produced by ρ_0, then

$$\mathbf{D}_{\|}(\mathbf{r}) = \varepsilon_0\, \mathbf{E}_0(\mathbf{r})\,. \tag{C.9.b}$$

Remark

If certain charges of the system are unbound and can be displaced far from the origin, one cannot include them in the definition of **P**. One includes them in ρ_0 on the right-hand side of (C.9.a). This is the "free charge" density of the electrodynamics of material media. We assume here that all the charges are bound.

For a globally *neutral* system, ρ_0 is identically zero, and **D** is then a transverse field:

$$\begin{cases} \mathbf{D}_{\|} = \mathbf{0} \\ \mathbf{D} = \mathbf{D}_{\perp}\,. \end{cases} \tag{C.10}$$

As a result then, following (C.8),

$$\varepsilon_0\, \mathbf{E}_{\|}(\mathbf{r}) = -\,\mathbf{P}_{\|}(\mathbf{r})\,. \tag{C.11}$$

One can take advantage of this expression to give the Coulomb energy of the system of charges, which is also the energy of the longitudinal field [see (B.31.a), Chapter I], in a form which will be used later:

$$V_{\text{Coul}} = \frac{1}{2\,\varepsilon_0}\int \mathrm{d}^3r\, \mathbf{P}_{\|}^2(\mathbf{r})\,. \tag{C.12}$$

c) Polarization Current and Magnetization Current

The motion of the charges q_α, which is the origin of the currents described by **j**, is accompanied by a motion of the polarization density **P**. There is then necessarily a relationship between the current density **j** and the rate of variation $\dot{\mathbf{P}}$ of **P**.

To get one such relationship, take the derivative of (C.5) with respect to time. Since the reference density ρ_0 is time independent, this gives

$$\dot{\rho} + \nabla \cdot \dot{\mathbf{P}} = 0 . \tag{C.13}$$

Comparison of (C.13) and the equation for the conservation of charge, $\dot{\rho} + \nabla \cdot \mathbf{j} = 0$, shows that $\mathbf{j} - \dot{\mathbf{P}}$ is a vector with zero divergence. We can then write the current **j** as the sum of two terms

$$\mathbf{j}(\mathbf{r}) = \mathbf{j}_p(\mathbf{r}) + \mathbf{j}_m(\mathbf{r}) \tag{C.14}$$

where the first term

$$\mathbf{j}_p(\mathbf{r}) = \dot{\mathbf{P}}(\mathbf{r}) \tag{C.15}$$

which is related to the motion of **P**, is called the *polarization current*, and where the other term, $\mathbf{j}_m$, with zero divergence, is called the *magnetization current* for reasons which will become apparent later on.

To get the expression for $\mathbf{j}_m$, rewrite (C.14) in reciprocal space by using (C.3) and the expression for the Fourier transform of **j**:

$$\mathscr{J}_m(\mathbf{k}) = \mathscr{J}(\mathbf{k}) - \dot{\mathscr{P}}(\mathbf{k}) =$$

$$= \sum_\alpha q_\alpha \dot{\mathbf{r}}_\alpha \frac{\mathrm{e}^{-i\mathbf{k}.\mathbf{r}_\alpha}}{(2\pi)^{3/2}} - \sum_\alpha \int_0^1 du\, q_\alpha \dot{\mathbf{r}}_\alpha \frac{\mathrm{e}^{-i\mathbf{k}.\mathbf{r}_\alpha u}}{(2\pi)^{3/2}} -$$

$$- \sum_\alpha \int_0^1 u\, du\, q_\alpha \mathbf{r}_\alpha (-i\mathbf{k} \cdot \dot{\mathbf{r}}_\alpha) \frac{\mathrm{e}^{-i\mathbf{k}.\mathbf{r}_\alpha u}}{(2\pi)^{3/2}} . \tag{C.16}$$

An integration by parts in the second term gives for the fully integrated term the negative of the first term. The remaining integral is grouped with the third term to give

$$\mathscr{J}_m(\mathbf{k}) = i\mathbf{k} \times \sum_\alpha \int_0^1 u\, du\, q_\alpha (\mathbf{r}_\alpha \times \dot{\mathbf{r}}_\alpha) \frac{\mathrm{e}^{-i\mathbf{k}.\mathbf{r}_\alpha u}}{(2\pi)^{3/2}} \tag{C.17}$$

so that, in real space,

$$\mathbf{j}_m(\mathbf{r}) = \nabla \times \mathbf{M}(\mathbf{r}) \tag{C.18}$$

where

$$\mathbf{M}(\mathbf{r}) = \sum_\alpha \int_0^1 u\, du\, q_\alpha \mathbf{r}_\alpha \times \dot{\mathbf{r}}_\alpha\, \delta(\mathbf{r} - u\mathbf{r}_\alpha) . \tag{C.19}$$

The current $\mathbf{j}_m$ appears then as the curl of a vector field $\mathbf{M}(\mathbf{r})$ which can be interpreted as a *magnetization density*. To see the physical origin of this density, return to the definition of the polarization $\mathbf{P}$. It is made up of a set of elementary dipoles $q_\alpha \mathbf{r}_\alpha \mathrm{d}u$ localized at $u\mathbf{r}_\alpha$. Such an elementary dipole is represented by the small arrow in Figure 2.

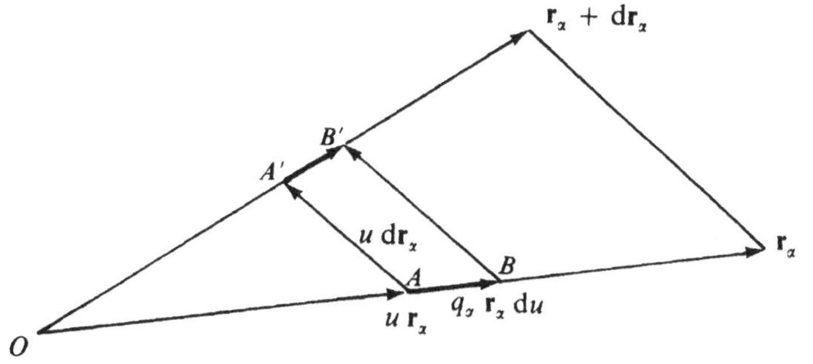

Figure 2. When particle α is displaced from $\mathbf{r}_\alpha$ to $\mathbf{r}_\alpha + \mathrm{d}\mathbf{r}_\alpha$, each elementary dipole $q_\alpha, -q_\alpha$ is displaced from AB to $A'B'$. Such a transportation is equivalent to a current $q_\alpha/\mathrm{d}t$ flowing along $ABB'A'$ and giving rise to a magnetic dipole moment, together with two radial elements of current BA and $A'B'$ which give rise to the polarization current.

When particle α is displaced from $\mathbf{r}_\alpha$ to $\mathbf{r}_\alpha + \mathrm{d}\mathbf{r}_\alpha$ during the time $\mathrm{d}t$, the charges $-q_\alpha$ and $+q_\alpha$ of the dipole are displaced respectively from A to A' and from B to B'. The currents associated with these displacements flow in opposite directions and are equal to $q_\alpha/\mathrm{d}t$. One can close the circuit by introducing currents with the same intensity along $\mathbf{r}_\alpha \mathrm{d}u$ and $(\mathbf{r}_\alpha + \mathrm{d}\mathbf{r}_\alpha)\,\mathrm{d}u$. Such a current loop gives rise to a magnetic moment $(q_\alpha/\mathrm{d}t)(\mathbf{r}_\alpha \mathrm{d}u) \times (u\,\mathrm{d}\mathbf{r}_\alpha)$ localized at $u\mathbf{r}_\alpha$. This is just the elementary magnetic moment of the magnetization density (C.19). The radial elements of current in AB and $B'A'$ must be compensated by two current densities localized at $u\mathbf{r}_\alpha$ and $u(\mathbf{r}_\alpha + \mathrm{d}\mathbf{r}_\alpha)$ and respectively proportional to $-q_\alpha \mathbf{r}_\alpha \mathrm{d}u/\mathrm{d}t$ and $q_\alpha(\mathbf{r}_\alpha + \mathrm{d}\mathbf{r}_\alpha)\,\mathrm{d}u/\mathrm{d}t$. The corresponding current density

$$\begin{aligned}\mathrm{d}\mathbf{j}_P &= q_\alpha[(\mathbf{r}_\alpha + \mathrm{d}\mathbf{r}_\alpha)\,\mathrm{d}u\,\delta(\mathbf{r} - u(\mathbf{r}_\alpha + \mathrm{d}\mathbf{r}_\alpha)) - \mathbf{r}_\alpha\,\mathrm{d}u\,\delta(\mathbf{r} - u\mathbf{r}_\alpha)]/\mathrm{d}t \\ &= q_\alpha\,\mathrm{d}u\,\frac{\mathrm{d}}{\mathrm{d}t}[\mathbf{r}_\alpha\,\delta(\mathbf{r} - u\mathbf{r}_\alpha)]\end{aligned}$$

is just the elementary polarization current $\mathrm{d}\dot{\mathbf{P}}$.

Remarks

(i) We have assumed the reference charge distribution fixed. One can generalize the foregoing expressions when this is not the case. Beside the polarization and magnetization current densities, other current densities then arise, corresponding to the transport of the total charge and of the polarization density (*).

(ii) If the particles have magnetic moments $\boldsymbol{\mu}_\alpha$, this causes $\Sigma_\alpha \boldsymbol{\mu}_\alpha \delta(\mathbf{r} - \mathbf{r}_\alpha)$ to be added to the magnetization density (C.19).

(iii) Following Power and Woolley, we have defined the polarization and magnetization densities as integrals over the dummy variable u. In order to get the expressions given by de Groot, one should expand the functions $\delta(\mathbf{r} - u\mathbf{r}_\alpha)$ in powers of the components of $u\mathbf{r}_\alpha$, and perform the integrations over u. This gives an expansion in powers of $\mathbf{r}_\alpha$, containing all the derivative of $\delta(\mathbf{r})$.

2. Changing the Lagrangian

a) The Power–Zienau–Woolley Transformation

This transformation consists of adding to the the standard Lagrangian L in the Coulomb gauge given by (C.15) of Chapter II the derivative with respect to time of a function F of the coordinates $\dots, \mathbf{r}_\alpha, \dots, \mathbf{A}(\mathbf{r}), \dots,$ which generalizes (B.28) to the extent that the system of charges is no longer approximated by an electric dipole.

The function F introduced by Power, Zienau, and Woolley is given by

$$F = -\int \mathrm{d}^3 r \, \mathbf{P}(\mathbf{r}) \cdot \mathbf{A}(\mathbf{r}) \tag{C.20}$$

$$F = -\int \mathrm{d}^3 k \, \boldsymbol{\mathscr{P}}^*(\mathbf{k}) \cdot \boldsymbol{\mathscr{A}}(\mathbf{k}) . \tag{C.21}$$

Except for the sign, it is the scalar product of $\mathbf{P}$ and $\mathbf{A}$. The vector potential $\mathbf{A}$ being transverse, only the transverse part $\mathbf{P}_\perp$ of $\mathbf{P}$ contributes to the integral.

Remark

In the case where the positions of all the particles are referred to the same point (such as the origin as we do here), one can show (Complement $\mathrm{D_{IV}}$) that the preceding transformation is equivalent to a gauge change.

b) The New Lagrangian

In the expression for the new Lagrangian

$$L' = L + \frac{\mathrm{d}F}{\mathrm{d}t}, \tag{C.22}$$

(*) E. A. Power and T. Thirunamachandran, *Proc. Roy. Soc. Lond.*, **A372**, 265 (1980).

it is interesting to group $\mathrm{d}F/\mathrm{d}t$ with the interaction Lagrangian L_I of the particles with the transverse field. The new interaction Lagrangian L_I' is written

$$L_I' = L_I + \frac{\mathrm{d}F}{\mathrm{d}t}$$
$$= \int \mathrm{d}^3 r\, \mathbf{j} \cdot \mathbf{A} - \int \mathrm{d}^3 r\, (\dot{\mathbf{P}} \cdot \mathbf{A} + \mathbf{P} \cdot \dot{\mathbf{A}}) . \qquad \text{(C.23)}$$

One gets the scalar product of $\mathbf{A}$ with $\mathbf{j} - \dot{\mathbf{P}}$, which from (C.14), (C.15), and (C.18) is just $\nabla \times \mathbf{M}$:

$$L_I' = \int \mathrm{d}^3 r (\nabla \times \mathbf{M}) \cdot \mathbf{A} - \int \mathrm{d}^3 r\, \mathbf{P} \cdot \dot{\mathbf{A}} . \qquad \text{(C.24)}$$

Integration by parts of the first term gives rise to the curl of $\mathbf{A}$, i.e. to the magnetic field $\mathbf{B}(\mathbf{r})$. Since $\dot{\mathbf{A}}$ is nothing more than $-\mathbf{E}_\perp$, we get finally

$$L_I' = \int \mathrm{d}^3 r\, \mathbf{M} \cdot \mathbf{B} + \int \mathrm{d}^3 r\, \mathbf{P} \cdot \mathbf{E}_\perp \qquad \text{(C.25)}$$

which expresses the interaction of the charges with the transverse field as the interaction of the magnetization density with the magnetic field and that of the polarization density with the transverse electric field.

The complete expression for the new Lagrangian L' is

$$L' = \sum_\alpha \frac{m_\alpha \dot{\mathbf{r}}_\alpha^2}{2} - \sum_{\alpha < \beta} \frac{q_\alpha q_\beta}{4 \pi \varepsilon_0 | \mathbf{r}_\alpha - \mathbf{r}_\beta |} - \sum_\alpha \varepsilon_{\mathrm{Coul}}^\alpha +$$
$$+ \frac{\varepsilon_0}{2} \int \mathrm{d}^3 r\, [\mathbf{E}_\perp^2 - c^2 \mathbf{B}^2] + \int \mathrm{d}^3 r\, \mathbf{M} \cdot \mathbf{B} + \int \mathrm{d}^3 r\, \mathbf{P} \cdot \mathbf{E}_\perp . \qquad \text{(C.26)}$$

c) Multipole Expansion of the Interaction between the Charged Particles and the Field

With a view to the multipole expansions in powers of a/λ which will be introduced hereafter, it is useful to limit the transformation of the Lagrangian to the long-wavelength modes, that is, those having a wave vector $\mathbf{k}$ with a modulus smaller than a limit k_M such that $k_M a \ll 1$.

For this we distinguish in the expansion of the transverse field $\mathbf{A}(\mathbf{r})$ the contributions of the long-wavelength modes and the others:

$$\mathbf{A}(\mathbf{r}) = \mathbf{A}^<(\mathbf{r}) + \mathbf{A}^>(\mathbf{r}) \qquad \text{(C.27)}$$

where

$$\mathbf{A}^{<}(\mathbf{r}) = \frac{1}{(2\pi)^{3/2}} \int_{<} \mathrm{d}^3k \, \mathscr{A}(\mathbf{k}) \, \mathrm{e}^{\mathrm{i}\mathbf{k}\cdot\mathbf{r}} \tag{C.28.a}$$

$$\mathbf{A}^{>}(\mathbf{r}) = \frac{1}{(2\pi)^{3/2}} \int_{>} \mathrm{d}^3k \, \mathscr{A}(\mathbf{k}) \, \mathrm{e}^{\mathrm{i}\mathbf{k}\cdot\mathbf{r}} \tag{C.28.b}$$

the symbol $<$ (or $>$) means that the sum over $\mathbf{k}$ involves all the values of $\mathbf{k}$ with modulus less than (or greater than) k_M. The function F is then chosen so as to contain only the long-wavelength modes:

$$F = -\int \mathrm{d}^3r \, \mathbf{P} \cdot \mathbf{A}^{<} \tag{C.29}$$

and one gets

$$L_I' = L_I'^{<} + L_I'^{>} \tag{C.30.a}$$

$$L_I'^{>} = \int \mathrm{d}^3r \, \mathbf{j} \cdot \mathbf{A}^{>} \tag{C.30.b}$$

$$L_I'^{<} = \int \mathrm{d}^3r \, \mathbf{M} \cdot \mathbf{B}^{<} + \int \mathrm{d}^3r \, \mathbf{P} \cdot \mathbf{E}_\perp^{<} \, . \tag{C.30.c}$$

Since by definition $\mathbf{B}^{<}(\mathbf{r})$ and $\mathbf{E}^{<}(\mathbf{r})$ vary only slightly over a distance of the order of a, one can expand these fields in a Taylor series near the origin. The integrals over $\mathbf{r}$ in (C.30.c) amount to projecting the magnetization and polarization densities $\mathbf{M}(\mathbf{r})$ and $\mathbf{P}(\mathbf{r})$ on monomials of increasing power in x, y, z. These projections are the multipole moments of the system of charges and currents. Taking into account the functions $\delta(\mathbf{r} - u\mathbf{r}_\alpha)$ appearing in the definitions (C.2) and (C.19) for $\mathbf{P}(\mathbf{r})$ and $\mathbf{M}(\mathbf{r})$, the integral is immediately evaluated. It remains only to integrate over u to get the expression for these multipole moments as a function of $\mathbf{r}_\alpha$ and $\dot{\mathbf{r}}_\alpha$. The interaction Lagrangian is then written in the form of a multipole expansion, whose first three terms are

$$L_I'^{<} = \mathbf{d} \cdot \mathbf{E}_\perp^{<}(\mathbf{0}) + \mathbf{m} \cdot \mathbf{B}^{<}(\mathbf{0}) + \sum_{ij} q_{ij} \frac{\partial}{\partial x_i} E_{\perp j}^{<}(\mathbf{0}) + \cdots \tag{C.31}$$

with

$$\mathbf{d} = \sum_\alpha q_\alpha \, \mathbf{r}_\alpha \tag{C.32.a}$$

$$\mathbf{m} = \sum_\alpha \frac{1}{2} q_\alpha \, \mathbf{r}_\alpha \times \dot{\mathbf{r}}_\alpha \tag{C.32.b}$$

$$q_{ij} = \sum_\alpha \frac{1}{2} q_\alpha \left(r_{\alpha i} \, r_{\alpha j} - \frac{1}{3} \delta_{ij} \, \mathbf{r}_\alpha^2 \right) . \tag{C.32.c}$$

One obviously gets the interaction of the electric dipole **d** with the transverse electric field at the origin as in §B.3, but to second order in a new terms also arise: the interaction of the magnetic moment **m** with the magnetic field at the origin, and that of the electric quadrupole moment q_{ij} with the gradient of the transverse electric field at the origin. [One removes the trace from the tensor $r_{\alpha i} r_{\alpha j}$, since it does not contribute to (C.31), $\mathbf{E}_\perp$ having a zero divergence.]

3. The New Conjugate Momenta and the New Hamiltonian

a) The Expressions for These Quantities

In (C.26) for L', the velocity $\dot{\mathbf{r}}_\alpha$ appears in the kinetic energy and in the magnetic interaction term. This latter quantity is given by

$$\begin{aligned}\int \mathrm{d}^3r\, \mathbf{M} \cdot \mathbf{B} &= \sum_\alpha \int \mathrm{d}^3r \int_0^1 u\, \mathrm{d}u\, q_\alpha(\mathbf{r}_\alpha \times \dot{\mathbf{r}}_\alpha) \cdot \mathbf{B}(\mathbf{r})\, \delta(\mathbf{r} - \mathbf{r}_\alpha u) \\ &= \sum_\alpha \dot{\mathbf{r}}_\alpha \cdot \int_0^1 u\, \mathrm{d}u\, q_\alpha[\mathbf{B}(\mathbf{r}_\alpha u) \times \mathbf{r}_\alpha] . \end{aligned} \tag{C.33}$$

Taking the derivative of L' with respect to $\dot{\mathbf{r}}_\alpha$ then gives

$$\mathbf{p}_{\alpha L'} = m_\alpha \dot{\mathbf{r}}_\alpha + \int_0^1 u\, \mathrm{d}u\, q_\alpha\, \mathbf{B}(\mathbf{r}_\alpha u) \times \mathbf{r}_\alpha . \tag{C.34}$$

In that same expression (C.26) for L', the vector $\dot{\mathbf{A}}$, which is just $-\mathbf{E}_\perp$, appears in the field energy and in the interaction Lagrangian, where it is multiplied by **P**. As has been done in §C.3 of Chapter II, it is useful, before taking the derivative, to express L' as a function of the independent dynamical variables of the transverse field, that is, $\mathscr{A}(\mathbf{k})$ and $\mathscr{A}^*(\mathbf{k})$ taken is one reciprocal half space. One has then

$$\pi_{L'}(\mathbf{k}) = \varepsilon_0\, \dot{\mathscr{A}}(\mathbf{k}) - \mathscr{P}_\perp(\mathbf{k}) . \tag{C.35}$$

Since $\mathscr{A}$ is transverse, only $\mathscr{P}_\perp$ contributes to the interaction and thus arises in the derivative.

The new Hamiltonian is given by

$$H_{L'} = \sum_\alpha \dot{\mathbf{r}}_\alpha \cdot \mathbf{p}_{\alpha L'} + \int \mathrm{d}^3k(\dot{\mathscr{A}} \cdot \pi^*_{L'} + \dot{\mathscr{A}}^* \cdot \pi_{L'}) - L' . \tag{C.36}$$

One can check that $H_{L'}$ is the same as H_L when it is reexpressed as a function of the velocities [see (A.12) with $\partial F/\partial t = 0$]. It is the sum of the kinetic energy of the particles, the energy of the transverse field, and the energy of the longitudinal field (Coulomb energy). As a function of the new canonical variables, $H_{L'}$ is given by

$$H_{L'} = \sum_\alpha \frac{1}{2\,m_\alpha}\left[\mathbf{p}_{\alpha L'} - \int_0^1 u\,\mathrm{d}u\,q_\alpha\,\mathbf{B}(\mathbf{r}_\alpha u) \times \mathbf{r}_\alpha\right]^2 + \\ + \int \mathrm{d}^3k \left[\frac{(\boldsymbol{\pi}_{L'} + \mathcal{P}_\perp)\cdot(\boldsymbol{\pi}_{L'} + \mathcal{P}_\perp)^*}{\varepsilon_0} + \varepsilon_0\,c^2\,k^2\,\mathcal{A}\cdot\mathcal{A}^*\right] + \\ + \sum_\alpha \varepsilon^\alpha_{\text{Coul}} + \sum_{\alpha<\alpha'} \frac{q_\alpha\,q_{\alpha'}}{4\,\pi\varepsilon_0\,|\mathbf{r}_\alpha - \mathbf{r}_{\alpha'}|} \tag{C.37}$$

$\varepsilon^\alpha_{\text{Coul}}$ is defined by (B.36) of Chapter I. One can separate $H_{L'}$ into a first part $H_{PL'}$ which depends only on the particle variables $\mathbf{r}_\alpha$ and $\mathbf{p}_{\alpha L'}$, a second part $H_{RL'}$ which depends only on the new variables $\mathcal{A}$ and $\boldsymbol{\pi}_{L'}$ of the transverse field, and an interaction Hamiltonian $H_{IL'}$ which depends on both:

$$H_{PL'} = \sum_\alpha \frac{\mathbf{p}^2_{\alpha L'}}{2\,m_\alpha} + \sum_\alpha \varepsilon^\alpha_{\text{Coul}} + \sum_{\alpha<\alpha'} \frac{q_\alpha\,q_{\alpha'}}{4\,\pi\varepsilon_0\,|\mathbf{r}_\alpha - \mathbf{r}_{\alpha'}|} + \int \mathrm{d}^3k\,\frac{|\mathcal{P}_\perp|^2}{\varepsilon_0} \tag{C.38}$$

$$H_{RL'} = \int \mathrm{d}^3k \left(\frac{\boldsymbol{\pi}_{L'}\cdot\boldsymbol{\pi}^*_{L'}}{\varepsilon_0} + \varepsilon_0\,\omega^2\,\mathcal{A}\cdot\mathcal{A}^*\right) = \\ = \int \mathrm{d}^3k\,\frac{\varepsilon_0}{2}\left[\left|\frac{\boldsymbol{\pi}_{L'}}{\varepsilon_0}\right|^2 + \omega^2\,|\mathcal{A}|^2\right] \tag{C.39}$$

$$H_{IL'} = \int \mathrm{d}^3k \left(\frac{\boldsymbol{\pi}_{L'}}{\varepsilon_0}\cdot\mathcal{P}^*_\perp + \frac{\boldsymbol{\pi}^*_{L'}}{\varepsilon_0}\cdot\mathcal{P}_\perp\right) - \\ - \sum_\alpha \int_0^1 u\,\mathrm{d}u\,q_\alpha\left(\mathbf{r}_\alpha \times \frac{\mathbf{p}_{\alpha L'}}{m_\alpha}\right)\cdot\mathbf{B}(\mathbf{r}_\alpha u) + \\ + \sum_\alpha \frac{q^2_\alpha}{2\,m_\alpha}\left[\int_0^1 u\,\mathrm{d}u\,\mathbf{r}_\alpha \times \mathbf{B}(\mathbf{r}_\alpha u)\right]^2 \tag{C.40}$$

b) The Physical Significance of the New Conjugate Momenta

In the approximation where the magnetic effects are neglected ($\mathbf{B} = \mathbf{0}$), $\mathbf{p}_{\alpha L'}$ is the mechanical momentum $m_\alpha\dot{\mathbf{r}}_\alpha$ as in the electric dipole approximation. More generally, the difference between $\mathbf{p}_{\alpha L'}$ and $m_\alpha\dot{\mathbf{r}}_\alpha$ is proportional to $|\mathbf{r}_\alpha|$, and can thus be very small if the system of charges is highly localized. Finally, this deviation is expressed directly as a function of the magnetic field and no longer as a function of the vector potential.

Remark

In fact, Equation (C.34) can be rewritten in the form

$$\mathbf{p}_{\alpha L'} = m_\alpha \dot{\mathbf{r}}_\alpha + q_\alpha \mathbf{A}'(\mathbf{r}_\alpha) \quad \text{(C.41.a)}$$

where

$$\mathbf{A}'(\mathbf{r}) = \int_0^1 u \, du \, \mathbf{B}(\mathbf{r}u) \times \mathbf{r} \,. \quad \text{(C.41.b)}$$

One can show by taking its curl that $\mathbf{A}'(\mathbf{r})$ is a special vector potential of the magnetic field $\mathbf{B}(\mathbf{r})$. It appears as a generalization of the vector potential $\mathbf{B} \times \mathbf{r}/2$ of a constant field. It has the property of being zero at the origin, of being expressible as a function of $\mathbf{B}$ itself, and of being at every point perpendicular to $\mathbf{r}$ (Poincaré gauge; see Complement D_{IV}).

The conjugate momentum $\boldsymbol{\pi}_{L'}(\mathbf{k})$ of the vector potential is transverse from (C.35). Since $\dot{\mathbf{A}}$ is equal to $-\mathbf{E}_\perp$, the equation (C.35) can be written

$$\begin{aligned} \mathbf{\Pi}_{L'}(\mathbf{r}) &= -\left[\varepsilon_0 \mathbf{E}_\perp(\mathbf{r}) + \mathbf{P}_\perp(\mathbf{r})\right] \\ &= -\mathbf{D}_\perp(\mathbf{r}) \end{aligned} \quad \text{(C.42)}$$

where $\mathbf{D}$ is the displacement defined by (C.8). The momentum conjugate with $\mathbf{A}$ then arises as the transverse part of $\mathbf{D}$, whereas in the standard description it corresponds with the transverse part of the electric field $\mathbf{E}$. Since from (C.9) $\mathbf{D}_\parallel$ is a time-independent Coulomb field (created by the reference charge distribution ρ_0), one can say that to within a constant field, $\mathbf{\Pi}_{L'}$ represents the displacement $\mathbf{D}$ in its entirety. The displacement $\mathbf{D}(\mathbf{r})$ being related *locally* to the *total* electric field by (C.8), the same is true for $\mathbf{\Pi}_{L'}$, whereas in the standard representation, $\mathbf{\Pi}_L$ is related to $\mathbf{E}_\perp$ and not to the total electric field. The tie between $\mathbf{\Pi}_{L'}(\mathbf{r})$ and $\mathbf{E}(\mathbf{r})$ is even clearer for a globally neutral system ($\rho_0 = 0$). One has then

$$\text{Neutral system} \quad \Rightarrow \quad \mathbf{\Pi}_{L'}(\mathbf{r}) = -\left[\varepsilon_0 \mathbf{E}(\mathbf{r}) + \mathbf{P}(\mathbf{r})\right]. \quad \text{(C.43)}$$

Away from the system of charges, $\mathbf{P}$ vanishes and $\mathbf{\Pi}_{L'}$ even coincides with $\mathbf{E}$ (to within a factor of $-\varepsilon_0$). One can say finally that $\mathbf{\Pi}_{L'}(\mathbf{r})$ represents the best possible description of the total electric field, which has a longitudinal part, by a transverse field. This description is exact throughout all space except in the finite volume occupied by the system of charges.

This local relation between $\mathbf{\Pi}_{L'}$ and $\mathbf{E}$ is the most interesting aspect of this new point of view; away from the system of charges, $\mathbf{P}$ is identically zero and $\mathbf{\Pi}_{L'}$, like $\mathbf{E}$, propagates with velocity c. It contains no instantaneous part, as $\mathbf{\Pi}_L$ does in the standard description. This property has important consequences for the study of the interaction between two distant charge systems (see Complement C_{IV}).

Remark

Rigorously, inside the system of charges there are instantaneous propagation effects for $\mathbf{\Pi}_{L'}$; in effect, $\mathbf{P}(\mathbf{r})$ is related instantaneously and nonlocally to the position of the charges. As a result, since $\mathbf{E}$ propagates with velocity c, $\mathbf{\Pi}_{L'}$ also has an instantaneous part. However, except in this small region of space where the retardation effects are everywhere small, the new dynamical variable $\mathbf{\Pi}_{L'}$ propagates with velocity c.

c) The Structure of the New Hamiltonian

The Hamiltonian of the transverse field, $H_{RL'}$, is (as in the standard description) the sum of independent harmonic-oscillator Hamiltonians corresponding to each of the modes. The dynamics of the field, described in terms of the conjugate variables, is then the same as in the standard description. However, the electric field itself is not expressed in the same way as a function of $\mathbf{\Pi}_{L'}$ and $\mathbf{\Pi}_L$, so that different motions of the electric field correspond to the same motion of $\mathbf{\Pi}_{L'}$ and $\mathbf{\Pi}_L$.

The Hamiltonian of the particles, $H_{PL'}$ is given in terms of the canonical variables and is little changed. Only the last term of (C.38) is added and appears as a correction to the Coulomb energy. It must be regrouped with the radiative corrections resulting from the coupling of the system of charges with the free field. Note however that this term plays an important role when one considers the case of two separated systems of charges $\mathscr{S}_A$ and $\mathscr{S}_B$. It can be shown then (see Complement $\mathrm{C_{IV}}$) that the contribution to the integral over $\mathbf{k}$ of $\mathscr{P}^*_{\perp A} \cdot \mathscr{P}_{\perp B}$ + c.c. cancels the instantaneous Coulomb interaction between the two systems of charges. There are no more direct interaction terms between $\mathscr{S}_A$ and $\mathscr{S}_B$. All interactions take place through the fields $\mathbf{B}$ and $\mathbf{\Pi}_{L'}$, which propagate with velocity c.

The interaction Hamiltonian $H_{IL'}$ between the particles and the fields $\mathbf{B}$ and $\mathbf{\Pi}_{L'}$ is given by (C.40). It contains three terms.

The first term can be written in real space in the form

$$\int \mathrm{d}^3r \, \frac{\mathbf{\Pi}_{L'}}{\varepsilon_0} \cdot \mathbf{P} = - \int \mathrm{d}^3r \, \frac{\mathbf{D}_\perp}{\varepsilon_0} \cdot \mathbf{P} \qquad \text{(C.44)}$$

$\mathbf{\Pi}_{L'}$ itself being transverse, one can replace $\mathbf{P}_\perp$ by $\mathbf{P}$ in the integral. This first term describes the interaction of the polarization $\mathbf{P}$ of the system of charges with $\mathbf{D}_\perp$, or $\mathbf{D}$ for a globally neutral system.

The second term can be written with the help of the magnetization density

$$\mathbf{M}'(\mathbf{r}) = \sum_\alpha \int_0^1 u \, \mathrm{d}u \, q_\alpha \left(\mathbf{r}_\alpha \times \frac{\mathbf{p}_{\alpha L'}}{m_\alpha} \right) \delta(\mathbf{r} - \mathbf{r}_\alpha u) \qquad \text{(C.45)}$$

in the form

$$-\int \mathrm{d}^3r\, \mathbf{M}'(\mathbf{r}) \cdot \mathbf{B}(\mathbf{r})\,. \tag{C.46}$$

It describes the paramagnetic interaction between the magnetization density $\mathbf{M}'(\mathbf{r})$ of the system of charges and the magnetic field $\mathbf{B}(\mathbf{r})$. It should be noted that since $\mathbf{p}_{\alpha L'}$ is different from $m_\alpha \dot{\mathbf{r}}_\alpha$ [cf. (C.34)], $\mathbf{M}'(\mathbf{r})$ is different from the true magnetization density $\mathbf{M}(\mathbf{r})$ introduced in (C.19). However, the density $\mathbf{M}'(\mathbf{r})$ is easier to use in the Hamiltonian formalism than $\mathbf{M}(\mathbf{r})$ in that it is given directly as a function of the dynamical variables $\mathbf{r}_\alpha$ and $\mathbf{p}_{\alpha L'}$.

The last term, quadratic in $\mathbf{B}$, represents the diamagnetic energy of the system of charges in the magnetic field $\mathbf{B}$, that is, the energy associated with the variation of current density when one switches on the field. We call this last term $H_{IL'}(\text{dia})$.

The total Hamiltonian is thus

$$H_{IL'} = \int \mathrm{d}^3r\, \frac{\mathbf{P} \cdot \mathbf{\Pi}_{L'}}{\varepsilon_0} - \int \mathrm{d}^3r\, \mathbf{M}' \cdot \mathbf{B} + H_{IL'}(\text{dia}) \tag{C.47}$$

$$= \int \mathrm{d}^3k\, \frac{\mathcal{P}^* \cdot \pi_{L'}}{\varepsilon_0} - \int \mathrm{d}^3k\, \mathcal{M}'^* \cdot \mathcal{B} + H_{IL'}(\text{dia})\,. \tag{C.48}$$

The first two terms of $H_{IL'}$ could be put in the form of a sum of terms describing the interaction of the multipole electric and magnetic moments of the system of charges with the corresponding electromagnetic multipole waves.

4. Quantum Electrodynamics from the New Point of View

One can now proceed to the canonical quantization of the preceding theory in exactly the same way as in Chapter II (§C.4).

a) Quantization

For particle α, the position $\mathbf{r}_\alpha$ and its conjugate momentum $\mathbf{p}_{\alpha L'}$ become operators $\mathbf{r}_\alpha^{(2)}$ and $\mathbf{p}_{\alpha L'}^{(2)}$ satisfying the fundamental commutation relations. One easy way to satisfy these commutation relations is to take for $\mathbf{r}_\alpha^{(2)}$ and $\mathbf{p}_{\alpha L'}^{(2)}$ the same operators $\mathbf{r}_\alpha$ (multiplication by $\mathbf{r}_\alpha$) and $\mathbf{p}_\alpha = -i\hbar \nabla_{\mathbf{r}_\alpha}$ as those used in the initial description to represent the conjugate variables $\mathbf{r}_\alpha^{(1)}$ and $\mathbf{p}_{\alpha L}^{(1)}$:

$$\begin{cases} \mathbf{r}_\alpha^{(2)} = \mathbf{r}_\alpha & \text{(C.49.a)} \\ \mathbf{p}_{\alpha L'}^{(2)} = \mathbf{p}_\alpha\,. & \text{(C.49.b)} \end{cases}$$

In a general way, where we later use a mathematical operator G without superscript, it will always coincide with the operator representing the variable $G^{(1)}$ in the earlier description:

$$G \equiv G^{(1)} . \tag{C.50}$$

Likewise, the electromagnetic field is described in this new representation by the two transverse conjugate fields $\mathscr{A}$ and $\pi_{L'}$, which become the operators $\mathscr{A}^{(2)}$ and $\pi_{L'}^{(2)}$ satisfying the canonical commutation relations (C.44) of Chapter II. As above, it is convenient to take for $\mathscr{A}^{(2)}$ and $\pi_{L'}^{(2)}$ the same operators $\mathscr{A}$ and π as those representing $\mathscr{A}^{(1)}$ and $\pi_L^{(1)}$ in the first representation:

$$\begin{cases} \mathscr{A}^{(2)}(\mathbf{k}) = \mathscr{A}(\mathbf{k}) & \text{(C.51.a)} \\ \pi_{L'}^{(2)}(\mathbf{k}) = \pi(\mathbf{k}) . & \text{(C.51.b)} \end{cases}$$

As in §C.4.c of Chapter II, the annihilation operator $a_\varepsilon(\mathbf{k})$ is introduced as a linear combination of $\mathscr{A}_\varepsilon(\mathbf{k})$ and $\pi_\varepsilon(\mathbf{k})$. The Fourier transforms $\mathbf{A}(\mathbf{r})$ and $\mathbf{\Pi}(\mathbf{r})$ of $\mathscr{A}(\mathbf{k})$ and $\pi(\mathbf{k})$ are then written

$$\begin{cases} \mathbf{A}(\mathbf{r}) = \displaystyle\int \mathrm{d}^3k \sum_{\boldsymbol{\varepsilon}} \sqrt{\frac{\hbar}{2\,\varepsilon_0\,\omega(2\,\pi)^3}} \\ \qquad\qquad \times \left[\boldsymbol{\varepsilon} a_\varepsilon(\mathbf{k})\,\mathrm{e}^{\mathrm{i}\mathbf{k}.\mathbf{r}} + \boldsymbol{\varepsilon} a_\varepsilon^+(\mathbf{k})\,\mathrm{e}^{-\mathrm{i}\mathbf{k}.\mathbf{r}}\right] & \text{(C.52.a)} \\ \mathbf{\Pi}(\mathbf{r}) = -\,\mathrm{i}\varepsilon_0 \displaystyle\int \mathrm{d}^3k \sum_{\boldsymbol{\varepsilon}} \sqrt{\frac{\hbar\omega}{2\,\varepsilon_0(2\,\pi)^3}} \times \\ \qquad\qquad \times \left[\boldsymbol{\varepsilon} a_\varepsilon(\mathbf{k})\,\mathrm{e}^{\mathrm{i}\mathbf{k}.\mathbf{r}} - \boldsymbol{\varepsilon} a_\varepsilon^+(\mathbf{k})\,\mathrm{e}^{-\mathrm{i}\mathbf{k}.\mathbf{r}}\right] . & \text{(C.52.b)} \end{cases}$$

These various mathematical operators, as well as $\mathbf{B} = \boldsymbol{\nabla} \times \mathbf{A}(\mathbf{r})$, are used in what follows to give the various observables in the new representation. Thus, the equations (C.51) become in real space

$$\begin{cases} \mathbf{A}^{(2)}(\mathbf{r}) = \mathbf{A}(\mathbf{r}) & \text{(C.53.a)} \\ \mathbf{\Pi}_{L'}^{(2)}(\mathbf{r}) = \mathbf{\Pi}(\mathbf{r}) . & \text{(C.53.b)} \end{cases}$$

b) The Expressions for the Various Physical Variables

As has been explained in §A.2 above, a given physical quantity is generally represented by two different operators $G^{(1)}$ and $G^{(2)}$, related by $G^{(2)} = TG^{(1)}T^+$, in the new representation (2) and in the original one (1).

Consider for example the velocity of particle α. In the standard

Coulomb description this variable is represented by the operator

$$\mathbf{v}_\alpha^{(1)} = \frac{1}{m_\alpha}[\mathbf{p}_\alpha - q_\alpha \mathbf{A}(\mathbf{r}_\alpha)] \tag{C.54.a}$$

whereas in the Power–Zienau–Woolley description one has, from (C.34),

$$\mathbf{v}_\alpha^{(2)} = \frac{1}{m_\alpha}\left[\mathbf{p}_\alpha - q_\alpha \int_0^1 \mathrm{d}u\, u\mathbf{B}(\mathbf{r}_\alpha u) \times \mathbf{r}_\alpha\right]. \tag{C.54.b}$$

Another example of a variable is the total electric field, which is written in representation (1) as

$$\mathbf{E}^{(1)}(\mathbf{r}) = -\frac{1}{\varepsilon_0}\mathbf{\Pi}(\mathbf{r}) + \frac{1}{4\pi\varepsilon_0}\sum_\alpha q_\alpha \frac{\mathbf{r} - \mathbf{r}_\alpha}{|\mathbf{r} - \mathbf{r}_\alpha|^3} \tag{C.55.a}$$

and in representation (2), using (C.8), (C.9.b), and (C.42), as

$$\mathbf{E}^{(2)}(\mathbf{r}) = -\frac{1}{\varepsilon_0}\mathbf{\Pi}(\mathbf{r}) - \frac{1}{\varepsilon_0}\mathbf{P}(\mathbf{r}) + \mathbf{E}_0(\mathbf{r}) \tag{C.55.b}$$

where $\mathbf{E}_0(\mathbf{r})$ is the static field created by the reference charge distribution $\rho_0(\mathbf{r})$.

Conversely, the physical meaning of a given mathematical operator depends on whether one is using representation (1) or (2). Thus, the operator $\mathbf{\Pi}(\mathbf{r})$ given in (C.52.b) is associated either with the transverse electric field or with the transverse displacement [see (C.42)]:

$$\mathbf{\Pi}(\mathbf{r}) = \mathbf{\Pi}_L^{(1)}(\mathbf{r}) = -\varepsilon_0 \mathbf{E}_\perp^{(1)}(\mathbf{r}) \tag{C.56.a}$$

$$\mathbf{\Pi}(\mathbf{r}) = \mathbf{\Pi}_{L'}^{(2)}(\mathbf{r}) = -\mathbf{D}_\perp^{(2)}(\mathbf{r}). \tag{C.56.b}$$

Finally we give the expression for $H^{(2)}$, the Hamiltonian in the new representation. It suffices to replace the various variables and conjugate momenta $\mathbf{r}_\alpha$, $\mathbf{p}_{\alpha L'}$, $\mathbf{A}(\mathbf{r})$, $\mathbf{\Pi}_{L'}(\mathbf{r})$ in (C.38), (C.39), and (C.40) by the corresponding operators $\mathbf{r}_\alpha$, $\mathbf{p}_\alpha$, $\mathbf{A}(\mathbf{r})$, and $\mathbf{\Pi}(\mathbf{r})$. One gets then

$$H^{(2)} = H'_P + H_R + H'_I \tag{C.57}$$

with

$$H'_P = H_{PL'}^{(2)} = \sum_\alpha \frac{\mathbf{p}_\alpha^2}{2m_\alpha} + \sum_\alpha \varepsilon_{\text{Coul}}^\alpha + \sum_{\alpha<\alpha'} \frac{q_\alpha q_{\alpha'}}{4\pi\varepsilon_0 |\mathbf{r}_\alpha - \mathbf{r}_{\alpha'}|} + \int \mathrm{d}^3k \frac{|\mathscr{P}_\perp|^2}{\varepsilon_0} \tag{C.58.a}$$

$$H_R = H_{RL'}^{(2)} = \frac{\varepsilon_0}{2} \int d^3r \left[\frac{\Pi^2(\mathbf{r})}{\varepsilon_0} + c^2 (\nabla \times \mathbf{A}(\mathbf{r}))^2 \right]$$

$$= \int d^3k \sum \hbar\omega \left[a_\varepsilon^+(\mathbf{k})\, a_\varepsilon(\mathbf{k}) + \frac{1}{2} \right] \tag{C.58.b}$$

$$H_I' = H_{IL'}^{(2)} = \frac{1}{\varepsilon_0} \int d^3r\, \mathbf{P}(\mathbf{r}) \cdot \Pi(\mathbf{r})$$

$$- \frac{1}{2} \sum_\alpha \int_0^1 u\, du\, q_\alpha \left[\left(\mathbf{r}_\alpha \times \frac{\mathbf{p}_\alpha}{m_\alpha} \right) \cdot \mathbf{B}(\mathbf{r}_\alpha u) + \mathbf{B}(\mathbf{r}_\alpha u) \cdot \left(\mathbf{r}_\alpha \times \frac{\mathbf{p}_\alpha}{m_\alpha} \right) \right]$$

$$+ \sum_\alpha \frac{q_\alpha^2}{2\, m_\alpha} \left[\int_0^1 u\, du\, \mathbf{r}_\alpha \times \mathbf{B}(\mathbf{r}_\alpha u) \right]^2 . \tag{C.58.c}$$

The new particle Hamiltonian $H_P' = H_{PL'}^{(2)}$ differs for the original $H_P = H_{PL}^{(1)}$ by the last term of (C.58.a). In contrast, the radiation Hamiltonians are the same: $H_{RL'}^{(2)} = H_{RL}^{(1)} = H_R$. Finally, in the new interaction Hamiltonian $H_I' = H_{IL'}^{(2)}$, which differs from the original one $H_I = H_{IL}^{(1)}$, we have correctly symmetrized the product of $\mathbf{p}_\alpha$ by the function of $\mathbf{r}_\alpha$ which appears in the paramagnetic coupling term.

5. The Equivalence of the Two Points of View. A Few Traps to Avoid

The two formulations of quantum electrodynamics based on the standard Lagrangian in the Coulomb gauge and on the Power–Zienau–Woolley Lagrangian are certainly equivalent, since, as a result of the general considerations of Part A above, they are related by a unitary transformation T whose expression is written, using (A.37) and (C.21),

$$T = \exp\left[-\frac{i}{\hbar} \int d^3k\, \mathscr{P}^* \cdot \mathscr{A} \right]. \tag{C.59}$$

It happens however that the general character of this equivalence is often forgotten and that errors of calculation or misinterpretations give the illusion that one of these two points of view is more valid than the other.

To avoid such errors, it is appropriate first of all to identify the physical states and the variables which arise in the process under study. For example, if the initial state of the process is a state where the particles have a well-defined velocity, then it is necessary to identify the operator associated with the velocity, since the initial state is an eigenstate of this operator.

Once the problem is correctly stated in physical terms, one must keep in mind the fact that its mathematical formulation generally depends on the representation. For example, the velocity operator does not have the same form in the Coulomb and in the Power–Zienau–Woolley representations [see (C.54.a) and (C.54.b)]. More generally, the operators $G^{(1)}$ and $G^{(2)}$ associated with the same variable in representations (1) and (2) must correspond through T and are most often different. It can happen that the operator associated with a variable has a simpler form in one description than in the other. The eigenvectors associated with this variable are simpler to find in this description, and it is always possible by use of T or T^+ to get their expressions in the other description. It can also happen that the exact diagonalization of the operator associated with a variable is not possible in any representation, and then it is necessary to use perturbation expansions in powers of a coupling parameter, such as the charge q_α of the particles. It is necessary then, if one wants to compare the predictions from the two points of view, to extend the expansions to the same perturbation order in the two representations. For example, the transition amplitudes, which are the matrix elements of the evolution operator between an initial and a final physical state, only have the same value in the two descriptions to a given order s in q_α if the expansions of the initial state, the final state, and the evolution operator are extended throughout to the same order s.

Finally, we return to the decomposition of the total Hamiltonian into three parts relative to the particles, the field, and the interaction. Since T does not depend explicitly on time, the total Hamiltonians $H^{(2)}$ and $H^{(1)}$ are related by the unitary transformation (see A.34)

$$H^{(2)} = TH^{(1)}T^+ \tag{C.60}$$

On the other hand, analogous relationships do not exist between $H_{PL'}^{(2)}$ and $H_{PL}^{(1)}$, $H_{IL'}^{(2)}$ and $H_{IL}^{(1)}$, and $H_{RL'}^{(2)}$ and $H_{RL}^{(1)}$. In other words, the unitary transformation redistributes in different ways certain terms of the Hamiltonian between the three parts which we distinguished in (C.58.a, b, c). One important consequence is that the eigenstates of $H_{PL}^{(1)}$ and $H_{PL'}^{(2)}$ are not put in correspondence by T; the ground state of $H_{PL}^{(1)}$ does not represent the same physical state as the ground state of $H_{PL'}^{(2)}$. One can then ask the question: what is the "true" ground state of an atom or a molecule? Is it that of $H_{PL}^{(1)}$ or $H_{PL'}^{(2)}$? It is in fact neither one or the other. One cannot really remove the interaction of the charges with the transverse field and observe the ground state of $H_{PL}^{(1)}$ or $H_{PL'}^{(2)}$. What we call the ground state of a system of charges is in fact the ground state of the system charges + transverse field, which must be an eigenstate of $H^{(1)}$ or $H^{(2)}$. These operators are really related by T from (C.60), so that the ground state from either point of view describes the same physical state. The eigenstates of $H_{PL}^{(1)}$ and of $H_{PL'}^{(2)}$ are different approximations of the real state, involving the neglect of this or that part of the effects of the transverse field on the system of charges.

D—SIMPLIFIED FORM OF EQUIVALENCE FOR THE SCATTERING S-MATRIX

The rigorous approach to the problem of equivalence between two points of view which we have treated in §C.5 above is valid in every case. As we have seen in Part A, it applies in a general way to every transformation appearing finally as a unitary transformation in state space. It can however lead to complex calculations. We will see in this part that, if the physical problem can be stated in terms of collisions, important simplifications appear. The transition amplitudes are equal in either approach, even if one does not take care to transform the state vectors representing the initial and final states.

To show this, we are going to follow a procedure which generalizes in a way that of §B.3.d.i above, where the external potentials are taken to be zero at the initial time t_i and at the final time t_f and are only "switched on" between t_i and t_f, with the result that the unitary transformation $T(t)$ connecting the two representations reduces to the identity for $t < t_i$ and $t > t_f$. Such a method is not directly applicable to quantum electrodynamics, which we are treating here, since the quantum fields are operators and not given functions of time. One can however draw inspiration from it for adiabatically "switching on" and "switching off" the coupling between the particles and the field with the help of a parameter $\lambda(t)$ which is formally introduced into the interaction Hamiltonian and which has a time variation corresponding to the temporal evolution of a collision process.

1. Introduction of the S-Matrix

We begin by explaining how it is possible to analyze in this way a collision process in the first representation. More generally, consider a Hamiltonian $H^{(1)}$ which can be separated in the form

$$H^{(1)} = H_0 + H_I \tag{D.1}$$

where H_0 is an unperturbed Hamiltonian describing the proper energy of the physical systems whose interaction is described by H_I. We then introduce formally a new Hamiltonian $H^{(1)}(\lambda)$ defined by

$$H^{(1)}(\lambda) = H_0 + \lambda H_I \tag{D.2}$$

where λ is a real parameter lying between 0 and 1. For $\lambda = 0$, $H^{(1)}(\lambda)$ reduces to H_0, whereas for $\lambda = 1$, $H^{(1)}(\lambda)$ is equal to $H^{(1)}$. To describe a collision process, it is convenient then to take for λ a function $\lambda(t)$ which increases slowly from 0 to 1 between t_i' and t_i, and then decreases slowly from 1 to 0 between t_f and t_f' (Figure 3). Indeed, assume that the global

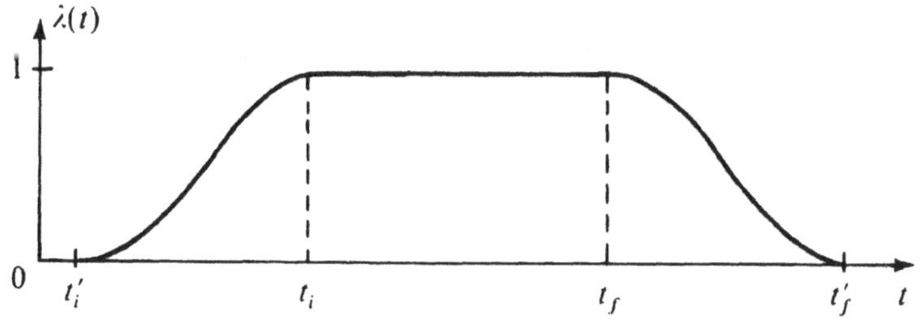

Figure 3. The temporal evolution of the coupling parameter.

system starts at time t_i' in an eigenstate $|\varphi_a\rangle$ of H_0 with energy E_a, describing the unperturbed system. The transition of $\lambda(t)$ from 0 to 1 between t_i' and t_i is an easy way to simulate the approach and overlapping of two "quasi-monochromatic" wave packets which initially do not interact because they are too remote. Similarly, the transition of $\lambda(t)$ from 1 to 0 between t_f and t_f' simulates the end of the collision, the wave packets separating and no longer interacting. The interaction of the systems, established between t_i' and t_i, acts between t_i and t_f and vanishes between t_f and t_f'. At time t_f' the system has evolved to a state $|\psi_f\rangle$ which, $\lambda(t)$ being slowly varying, is found on the same "energy shell" as $|\varphi_a\rangle$. It follows that the scalar product of $|\psi_f\rangle$ with another eigenstate $|\varphi_b\rangle$ of H_0 is different from zero only if $E_b = E_a$. This scalar product represents the scattering amplitude from $|\varphi_a\rangle$ to $|\varphi_b\rangle$.

The switching on and off of the coupling by the function $\lambda(t)$ in Figure 3 is, in fact, only one convenient and intuitive way to introduce the S-matrix by taking as asymptotic states the eigenstates $|\varphi_a\rangle$ and $|\varphi_b\rangle$ of H_0. (*) The calculation of the matrix element between $|\varphi_a\rangle$ and $|\varphi_b\rangle$ of the evolution operator (**) $U^{(1)}(t_f', t_i')$ associated with the Hamiltonian $H_0 + \lambda(t)H_I$ gives the matrix element of S between $|\varphi_a\rangle$ and $|\varphi_b\rangle$, which one can show may be written

$$S_{ba}^{(1)} = \delta_{ba} - 2\pi i\, \delta(E_b - E_a)\, \mathcal{T}_{ba}^{(1)}. \tag{D.3}$$

The function $\delta(E_b - E_a)$ expresses the fact mentioned above that the final state $|\varphi_b\rangle$ must be on the same energy shell as the initial state $|\varphi_a\rangle$. $\mathcal{T}_{ba}^{(1)}$

(*) There are of course more rigorous methods available, such as the formal theory of scattering. See the references at the end of the chapter.

(**) In fact, this operator must be taken in the interaction representation with respect to H_0, so as to eliminate free-evolution exponentials, which lack a well-defined limit when $t_f' - t_i'$ goes to infinity.

is the transition matrix, which can be calculated perturbatively as an expansion in powers of H_I (Born expansion),

$$\mathcal{T}_{ba}^{(1)} = \langle \varphi_b | H_I | \varphi_a \rangle + \lim_{\varepsilon \to 0_+} \sum_c \frac{\langle \varphi_b | H_I | \varphi_c \rangle \langle \varphi_c | H_I | \varphi_a \rangle}{E_a - E_c + i\varepsilon} + \cdots \tag{D.4}$$

2. The S-Matrix from Another Point of View. An Examination of the Equivalence

Consider now a second point of view derived from the first by the unitary transformation $e^{iF/\hbar}$, where F is time independent. The new Hamiltonian $H^{(2)}$ is related to the original $H^{(1)}$ by

$$H^{(2)} = T H^{(1)} T^{+} . \tag{D.5}$$

It is quite convenient for what follows to introduce formally the same parameter λ which was used to define $H^{(1)}(\lambda)$ in (D.2). We take then

$$T(\lambda) = \exp\left(\frac{i}{\hbar} \lambda F\right) \tag{D.6}$$

and we define the Hamiltonian $H^{(2)}(\lambda)$ as the transform of $H^{(1)}(\lambda)$ by $T(\lambda)$:

$$H^{(2)}(\lambda) = T(\lambda)\, H^{(1)}(\lambda)\, T^{+}(\lambda) = T(\lambda)\,[H_0 + \lambda H_I]\, T^{+}(\lambda) . \tag{D.7}$$

For $\lambda = 0$, $T(\lambda)$ is equal to the unit operator, and $H^{(2)}(\lambda)$ and $H^{(1)}(\lambda)$ reduce to H_0. For $\lambda = 1$, one gets (D.5). It is then possible to write

$$H^{(2)}(\lambda) = H_0 + \lambda H_I'(\lambda) \tag{D.8}$$

where we have regrouped all of the terms depending on λ in $\lambda H_I'(\lambda)$. This last term (taken for $\lambda = 1$) thus forms the interaction Hamiltonian in the second description, which we now denote by H_I'. Note that this Hamiltonian is not in general the transform of H_I by T.

Having thus shown the correspondence between the two descriptions for each value of λ, we can now analyze how the collision process associated with the function $\lambda(t)$ in Figure 3 appears from the second point of view. At the initial time t_i', $\lambda(t_i') = 0$ and $T(\lambda(t_i'))$ reduces to the unit operator. The initial state is thus described by the same ket $|\varphi_a\rangle$ in both representations, since $|\varphi_a\rangle$ is identical to its transform. The same is true at the final time t_f', since $\lambda(t_f') = 0$. The final state is thus also described by the same vector $|\varphi_b\rangle$ in both representations. Between t_i' and t_f' the state vector evolves under the influence of the Hamiltonian $H_0 + \lambda(t) H_I$ in the first representation. Since the unitary transformation

$T(\lambda(t))$ depends on time via $\lambda(t)$, the evolution in the second representation is governed using (A.42) and (D.6), by the Hamiltonian

$$H_0 + \lambda(t)\, H_I'(\lambda(t)) + \mathrm{i}\hbar\left[\frac{\mathrm{d}}{\mathrm{d}t}\, T(\lambda(t))\right] T^+(\lambda(t)) =$$
$$= H_0 + \lambda(t)\, H_I'((\lambda(t)) - \dot{\lambda}(t)\, F\,. \quad \text{(D.9)}$$

The matrix element between $\langle \varphi_b|$ and $|\varphi_a\rangle$ of the evolution operator associated with the Hamiltonian $H_0 + \lambda(t) H_I$ is then identical to the matrix element between the same states $\langle \varphi_b|$ and $|\varphi_a\rangle$ of the evolution operator associated with (D.9):

$$U_{ba}^{(1)}(t_f', t_i') = U_{ba}^{(2)}(t_f', t_i')\,. \quad \text{(D.10)}$$

In fact, since $\lambda(t)$ varies very slowly with t, the last term of (D.9), $\dot{\lambda}(t)F$, is very much smaller than the second. Its contribution to the evolution operator associated with (D.9) can thus in general be neglected (*). The operator $U^{(2)}$ can then be calculated by considering only the effect of the perturbation $\lambda(t)H_I'(\lambda(t))$, which increases slowly from 0 to H_I' between t_i' and t_i, remains equal to H_I' between t_i and t_f, and decreases slowly then from H_I' to 0 between t_f and t_f'. One then recovers, for the interaction Hamiltonian which must be taken into account for $U^{(2)}$, a temporal behavior resembling that which we have introduced in the first picture to define the S-matrix. In the limit $t_f - t_i \to \infty$, the equality (D.10) becomes

$$S_{ba}^{(1)} = S_{ba}^{(2)} \quad \text{(D.11)}$$

where

$$S_{ba}^{(2)} = \delta_{ba} - 2\,\pi\mathrm{i}\,\delta(E_b - E_a)\,\mathcal{T}_{ba}^{(2)} \quad \text{(D.12.a)}$$

$$\mathcal{T}_{ba}^{(2)} = \langle \varphi_b \,|\, H_I' \,|\, \varphi_a \rangle + \lim_{\varepsilon\to 0_+} \sum_c \frac{\langle \varphi_b \,|\, H_I' \,|\, \varphi_c \rangle \langle \varphi_c \,|\, H_I' \,|\, \varphi_a \rangle}{E_a - E_c + \mathrm{i}\varepsilon} + \cdots$$
$$\text{(D.12.b)}$$

The identity (D.11) between (D.3) and (D.12.a) involves finally the identity between $\mathcal{T}_{ba}^{(1)}$ and $\mathcal{T}_{ba}^{(2)}$ when $E_a = E_b$ (as a result of the presence of the delta function):

$$\mathcal{T}_{ba}^{(1)} = \mathcal{T}_{ba}^{(2)} \quad \text{when} \quad E_a = E_b\,. \quad \text{(D.13)}$$

The equality (D.13) can be established more rigorously by starting from the formal theory of scattering for all transformations F of the kind introduced in this chapter (see the references at the end of the chapter).

(*) It is possible to construct transformations F sufficiently singular so that $\dot{\lambda}(t)F$ can not be neglected (K. Haller, private communication). We assume that this is not the case for the transformations envisioned here.

The identity (D.13) between the transition matrices $\mathcal{T}^{(1)}$ and $\mathcal{T}^{(2)}$ on the energy shell is a remarkable result which simplifies most practical calculations considerably. The majority of the physically accessible experimental parameters (cross-sections, transition probabilities, etc.) are indeed expressible directly as functions of the transition-matrix elements, and it is quite convenient to have the option of making the calculation either with H_I or with H_I' in the *same state basis*, formed by the set of eigenstates of H_0, without bothering with the fact that H_0 is not invariant under T and that its eigenstates do not represent the same physical states in the two representations.

Remark

To stress the fact that the Hamiltonian is changed, but not the states, when one goes from the term on the left to the term on the right in (D.13), some authors call it a "hybrid transformation". Other authors call $H_I' - H_I$ a "pseudo-perturbation", since this difference has no effect on the transition matrix.

3. Comments on the Use of the Equivalence between the *S*-Matrices

To conclude this part, it is important to draw the reader's attention to the various dangers accompanying the hasty use of (D.13).

First of all, one must not forget that (D.13) is only applicable if the problem under study can be put in the form of a collision problem. The fact that one can use the same ket $|\varphi_a\rangle$ (or $|\varphi_b\rangle$) to describe the initial (or final) state in the two representations is not the result of a "miracle". It rather means that in a collisional process the interaction can be ignored in the remote past before the collision and in the remote future after it. It is indeed so as to stress this physical idea that we have preferred to follow here a qualitative approach based on the switching on and off of λ rather than giving a more rigorous demonstration based on the formal theory of scattering.

It is clear that if $|\varphi_b\rangle$ is not the final state of a scattering process, for example, if $|\varphi_b\rangle$ is an excited atomic state which can decay by spontaneous emission, the reasoning above is no longer valid, since one can no longer let t_f and t_f' tend to infinity while retaining the same state $|\varphi_b\rangle$. Thus it is important in the application of (D.13) that the initial and final physical states be stable or at least that their instabilities can be neglected in the problem being examined. If not, it is necessary to "embed" the process under study in a more complex process corresponding to a true scattering (*) or revert to the general method of §C.5 above.

(*) It is possible in this way to resolve completely the apparent contradictions in the results given by the two representations for the shape of the $2s \to 2p$ line of hydrogen, the Lamb transition. See Complement B_{IV}, §3.b.

Another error to avoid is applying (D.13) off the energy shell. $\mathcal{T}_{ba}^{(1)}$ has no reason to be equal to $\mathcal{T}_{ba}^{(2)}$ if E_b is different from E_a. On the other hand, the system can of course pass through intermediate states $|\varphi_c\rangle$ with unperturbed energy E_c different from $E_a = E_b$. This is what the higher-order terms of the Born expansions (D.4) and (D.12) address.

Note finally that the equality between $\mathcal{T}_{ba}^{(1)}$ and $\mathcal{T}_{ba}^{(2)}$ on the energy shell does not imply the equality of the contributions of a given intermediate path in the high-order terms of the expansions (D.4) and (D.12.b). In particular, even if one intermediate state plays a preponderant role as a result of a quasi-resonance, it is incorrect to systematically neglect all the other intermediate states. The transition-matrix elements calculated in the two representations retaining only the quasi-resonant intermediate state are in general different. It is quite possible that the approximation consisting of neglecting all the other intermediate states will be valid in one representation. It is rare for this to be the case in both simultaneously. (See Exercise 2.)

General References and Further Reading

For the Power–Zienau–Woolley transformation see the original articles: E. A. Power and S. Zienau, *Philos. Trans. Roy. Soc.*, **A251**, 427 (1959), and R. G. Woolley, *Proc. Roy. Soc. Lond.*, **A321**, 557 (1971). See also Power (Chapter 8) and Healy (Chapter 7). A simple discussion is given in E. A. Power and T. Thirunamachandran, *Am. J. Phys.*, **46**, 370 (1976).

For an introduction to the S-matrix and the formal theory of scattering see Goldberger and Watson (Chapter 5) and Schweber (Chapter 11).

For the study of the equivalence of the S-matrices through the formal theory of scattering (hybrid transformations, pseudo-perturbations, etc.) see K. Haller and S. B. Sohn, *Phys. Rev A*, **20**, 1541 (1979); Y. Aharonov and C. K. Au, *Phys. Rev. A*, **20**, 1553 (1979); E. Kazes, T. E. Feuchtwang, P. H. Cutler, and H. Grotch, *Ann. of Phys.*, **142**, 80 (1982).

COMPLEMENT A_{IV}

ELEMENTARY INTRODUCTION TO THE ELECTRIC DIPOLE HAMILTONIAN

We consider an ensemble of charged particles forming a localized system with spatial extension of the order of a. This is the case, for example, in atoms or molecules made up of electrons and nuclei in bound states whose spatial dimensions are of the order of a few Bohr radii. If one such system interacts with radiation with wavelength λ large with respect to a, it is legitimate to neglect the spatial variation of the electromagnetic field over the expanse of the system. This approximation, called the *long-wavelength approximation*, has been used in §B.3 of this chapter to get a simpler equivalent formulation of electrodynamics for a localized system of charges coupled to an external field. The corresponding transformation, the Göppert-Mayer transformation, has been presented as a change of Lagrangian or a gauge change. We again treat the same problem here (§A_{IV}.1) by directly studying the unitary transformation which permits one to get the electric dipole Hamiltonian $\mathbf{E} \cdot \mathbf{r}$ by starting with the usual Hamiltonian $\mathbf{A} \cdot \mathbf{p}$ (*). Next we generalize the Göppert-Mayer transformation to the case where the electromagnetic field is treated, not as an external field, but as a quantized system with its proper dynamics (§A_{IV}.2). Finally, we look at some possible extensions of the method used here (§A_{IV}.3).

1. The Electric Dipole Hamiltonian for a Localized System of Charges Coupled to an External Field

a) The Unitary Transformation Suggested by the Long-Wavelength Approximation

Let $\mathbf{A}_e(\mathbf{r}, t)$ be the potential vector describing the external radiation (the scalar potential $U_e(\mathbf{r}, t)$ is assumed to be zero). The long-wavelength approximation involves neglecting the spatial variation of $\mathbf{A}_e(\mathbf{r}, t)$ in the Hamiltonian. Thus one can replace $\mathbf{A}_e(\mathbf{r}_\alpha, t)$ by $\mathbf{A}_e(\mathbf{R}, t)$ in the kinetic energy term, where $\mathbf{R}$ is a point taken in the interior of the system of charges. In the following, we take $\mathbf{R}$ as a fixed point, which we choose as the origin of the coordinates $\mathbf{R} = \mathbf{0}$, which amounts to ignoring all the displacements of the atom or molecule as a whole (see, however, the Remark in §A_{IV}.1.*b* below). Under these conditions, the Hamiltonian is

(*) Since we are not using here the Lagrangian formalism, reading this complement does not require knowledge of the ideas introduced in Chapters II and IV.

written

$$H(t) = \sum_\alpha \frac{1}{2\,m_\alpha}[\mathbf{p}_\alpha - q_\alpha\,\mathbf{A}_e(\mathbf{0}, t)]^2 + V_{\text{Coul}} \tag{1}$$

where V_{Coul} is the Coulomb energy of the system.

The simple form which the kinetic energy term takes in the long-wavelength approximation suggests the application of a unitary transformation $T(t)$ which translates each operator $\mathbf{p}_\alpha$ by an amount $q_\alpha \mathbf{A}_e(\mathbf{0}, t)$:

$$T(t)\,\mathbf{p}_\alpha\,T^+(t) = \mathbf{p}_\alpha + q_\alpha\,\mathbf{A}_e(\mathbf{0}, t)\,. \tag{2}$$

The translation operator $T(t)$ which effects this transformation is given by

$$\begin{aligned} T(t) &= \exp\left[-\frac{\mathrm{i}}{\hbar}\sum_\alpha q_\alpha\,\mathbf{r}_\alpha \cdot \mathbf{A}_e(\mathbf{0}, t)\right] \\ &= \exp\left[-\frac{\mathrm{i}}{\hbar}\mathbf{d} \cdot \mathbf{A}_e(\mathbf{0}, t)\right] \end{aligned} \tag{3}$$

where

$$\mathbf{d} = \sum_\alpha q_\alpha\,\mathbf{r}_\alpha \tag{4}$$

is the electric dipole moment of the charge distribution with respect to the origin. Equation (3) coincides with that found in Part B above from the Lagrangian formalism [see Equation (B.34)].

Remark

We have not included in (1) the interaction terms of the particle spin magnetic moments with the magnetic field of the external radiation. These terms are actually smaller than the interaction terms in $\mathbf{A}_e \cdot \mathbf{p}$ by a factor of the order of $\hbar k/p$ [see Equation (D.15), Chapter III], that is to say, of the order of a/λ, since $\hbar/p \sim a$. They are thus of the same order of magnitude as the interaction terms which have been neglected by replacing $\mathbf{A}_e(\mathbf{r}_\alpha, t)$ with $\mathbf{A}_e(\mathbf{0}, t)$.

b) The Transformed Hamiltonian

In this new representation, the temporal evolution of the transformed state vector $|\psi'(t)\rangle = T(t)|\psi(t)\rangle$ is governed by the Hamiltonian

$$H'(t) = T(t)\,H(t)\,T^+(t) + \mathrm{i}\hbar\left[\frac{\mathrm{d}T(t)}{\mathrm{d}t}\right]T^+(t)\,. \tag{5}$$

By using (1) and (2), we find

$$T(t)\,H(t)\,T^+(t) = \sum_\alpha \frac{\mathbf{p}_\alpha^2}{2\,m_\alpha} + V_{\text{Coul}} \tag{6}$$

and by using (3)

$$i\hbar\left[\frac{dT(t)}{dt}\right]T^+(t) = \mathbf{d}\cdot\dot{\mathbf{A}}_e(\mathbf{0},t) = -\,\mathbf{d}\cdot\mathbf{E}_e(\mathbf{0},t) \tag{7}$$

which gives finally

$$H'(t) = \sum_\alpha \frac{\mathbf{p}_\alpha^2}{2\,m_\alpha} + V_{\text{Coul}} - \mathbf{d}\cdot\mathbf{E}_e(\mathbf{0},t)\,. \tag{8}$$

We have thus found that in the long-wavelength approximation, the interaction with the external field is simply described in this new representation by a coupling term between the dipole moment $\mathbf{d}$ of the atom and the external electric field evaluated at the position of the atom.

Remark

It is possible to take into account the global motion of the atom and to refer the positions of the charges q_α to a point $\mathbf{R}$ which, rather than being a fixed point, is taken at the center of mass of the atom (see Exercise 3). One then finds that if the total charge $Q = \Sigma_\alpha q_\alpha$ is zero, Equation (8) remains valid provided that one replaces $\mathbf{0}$ by $\mathbf{R}$ in $\mathbf{E}_e(\mathbf{0}, t)$. In contrast, if the system is an ion ($Q \neq 0$), new terms appear in the Hamiltonian $H'(t)$. They describe the coupling of the global motion of the ion to the external field $\mathbf{A}_e$. In all that follows in this complement, we only consider globally neutral systems:

$$Q = \sum_\alpha q_\alpha = 0\,. \tag{9}$$

c) The Velocity Operator in the New Representation

In the initial description, the velocity of the particle is represented by the operator

$$\mathbf{v}_\alpha = \frac{1}{m_\alpha}\left[\mathbf{p}_\alpha - q_\alpha\,\mathbf{A}_e(\mathbf{0},t)\right] \tag{10}$$

while in the new description it is represented by

$$\mathbf{v}'_\alpha = T(t)\,\mathbf{v}_\alpha\,T^+(t) \tag{11.a}$$

which, using (2) and (10), is equal to

$$\mathbf{v}'_\alpha = \frac{\mathbf{p}_\alpha}{m_\alpha}\,. \tag{11.b}$$

Thus, in the new description we find a much simpler relation between the momentum and the velocity.

2. The Electric Dipole Hamiltonian for a Localized System of Charges Coupled to Quantized Radiation

We now consider the radiation field as a quantized system with its own dynamics.

a) The Unitary Transformation

If the coupling between particles and radiation mainly involves the modes whose wavelength is large with respect to a, we can, in a first approximation, neglect the contribution of the other modes. The operator

$$\mathbf{A}(\mathbf{r}_\alpha) = \sum_j \mathscr{A}_{\omega_j}[a_j\, \boldsymbol{\varepsilon}_j\, e^{i\mathbf{k}_j \cdot \mathbf{r}_\alpha} + a_j^+\, \boldsymbol{\varepsilon}_j\, e^{-i\mathbf{k}_j \cdot \mathbf{r}_\alpha}] \tag{12}$$

can then be replaced by $\mathbf{A}(\mathbf{0})$, since for all the modes taken into account $|\mathbf{k}_j \cdot \mathbf{r}_\alpha| \ll 1$. In this approximation, the Hamiltonian in the Coulomb gauge is written

$$H = \sum_\alpha \frac{1}{2\,m_\alpha}[\mathbf{p}_\alpha - q_\alpha\,\mathbf{A}(\mathbf{0})]^2 + V_{\text{Coul}} + \sum_j \hbar\omega_j\left(a_j^+\,a_j + \frac{1}{2}\right). \tag{13}$$

An argument similar to that presented in §A_{IV}.1 suggests that one apply to (13) the unitary transformation

$$T = \exp\left[-\frac{i}{\hbar}\sum_\alpha q_\alpha\,\mathbf{r}_\alpha \cdot \mathbf{A}(\mathbf{0})\right] = \exp\left[-\frac{i}{\hbar}\mathbf{d}\cdot\mathbf{A}(\mathbf{0})\right] \tag{14}$$

which, while resembling (3), differs from it. In (14), $\mathbf{A}(\mathbf{0})$ is a time-independent field operator, while $\mathbf{A}_e(\mathbf{0}, t)$ in (3) is a classical function of time. It is indeed interesting in what follows to reexpress (14) with the aid of the operators a_j and a_j^+ using the expansion (12) for $\mathbf{A}(\mathbf{0})$. We then get

$$T = \exp\left\{\sum_j (\lambda_j^*\,a_j - \lambda_j\,a_j^+)\right\} \tag{15}$$

with

$$\lambda_j = \frac{i}{\sqrt{2\,\varepsilon_0\,\hbar\omega_j\,L^3}}\,\boldsymbol{\varepsilon}_j\cdot\mathbf{d}\,. \tag{16}$$

b) Transformation of the Physical Variables

Consider first the operators relating to the particles. Since $\mathbf{r}_\alpha$ commutes with T, this operator represents the position of the particle in both descriptions. As for the velocity, since the operator $\mathbf{A}(\mathbf{0})$ acts like a

c-number with respect to the particles, we get in a fashion identical to (10) and (11)

$$\mathbf{v}'_\alpha = T\mathbf{v}_\alpha T^+ = \frac{\mathbf{p}_\alpha}{m_\alpha}. \tag{17}$$

We now examine the field operators. Note first that the quantities λ_j introduced in (16) are purely atomic operators commuting between themselves and can then be considered as numbers with respect to the radiation operators a_j and a_j^+. The operator T in the form (15) thus appears as a translation operator for a_j and a_j^+ [see (C.61) and (C.66) of Chapter III]:

$$\begin{cases} Ta_j T^+ = a_j + \lambda_j & \text{(18.a)} \\ Ta_j^+ T^+ = a_j^+ + \lambda_j^*. & \text{(18.b)} \end{cases}$$

We are going to use these relations to calculate the operator $\mathbf{E}'_\perp(\mathbf{r})$ describing the transverse field in the new representation. Starting from the expression for the transverse electric field in the Coulomb gauge,

$$\mathbf{E}_\perp(\mathbf{r}) = \mathrm{i}\sum_j \mathscr{E}_{\omega_j}[a_j\, \boldsymbol{\varepsilon}_j\, \mathrm{e}^{\mathrm{i}\mathbf{k}_j\cdot\mathbf{r}} - a_j^+\, \boldsymbol{\varepsilon}_j\, \mathrm{e}^{-\mathrm{i}\mathbf{k}_j\cdot\mathbf{r}}] \tag{19}$$

we find

$$\begin{aligned} \mathbf{E}'_\perp(\mathbf{r}) &= T\mathbf{E}_\perp(\mathbf{r})\, T^+ \\ &= \mathrm{i}\sum_j \mathscr{E}_{\omega_j}[(a_j + \lambda_j)\, \boldsymbol{\varepsilon}_j\, \mathrm{e}^{\mathrm{i}\mathbf{k}_j\cdot\mathbf{r}} - (a_j^+ + \lambda_j^*)\, \boldsymbol{\varepsilon}_j\, \mathrm{e}^{-\mathrm{i}\mathbf{k}_j\cdot\mathbf{r}}] \\ &= \mathbf{E}_\perp(\mathbf{r}) - \sum_{\mathbf{k}_j\boldsymbol{\varepsilon}_j}\left[\frac{1}{2\,\varepsilon_0\, L^3}\, \boldsymbol{\varepsilon}_j(\boldsymbol{\varepsilon}_j\cdot\mathbf{d})\, \mathrm{e}^{\mathrm{i}\mathbf{k}_j\cdot\mathbf{r}} + c.c.\right] \end{aligned} \tag{20}$$

where $\mathbf{E}_\perp(\mathbf{r})$ is the mathematical operator given in (19). The last term of (20) will be interpreted later. Consider finally the magnetic field. Since $\mathbf{A}(\mathbf{r})$ and $\mathbf{A}(\mathbf{r}')$ commute for all $\mathbf{r}$ and $\mathbf{r}'$ [see (A.13) of Chapter III], $\mathbf{B}(\mathbf{r}) = \boldsymbol{\nabla} \times \mathbf{A}(\mathbf{r})$ commutes also with $\mathbf{A}(\mathbf{r}')$. It follows that $\mathbf{A}(\mathbf{r})$ and $\mathbf{B}(\mathbf{r})$ commute with T. In particular, the magnetic field operator retains the same form in both representations:

$$\mathbf{B}'(\mathbf{r}) = T\mathbf{B}(\mathbf{r})\, T^+ = \mathbf{B}(\mathbf{r}). \tag{21}$$

c) Polarization Density and Displacement

To interpret the last term of (20) physically and to identify the variable represented in the new representation by the mathematical operator $\mathbf{E}_\perp(\mathbf{r})$ given in (19), it is convenient to describe the localized system of charges by a polarization density **P(r)**, and then, starting from the electric field and the polarization density **P(r)**, to introduce the displacement **D(r)**. For a more complete discussion of this problem not limited to the lowest order in a/λ as here, the interested reader should refer to §C.1 of Chapter IV, which can be read independently of the rest of that chapter.

The charge density associated with the system of localized charges q_α is written in real space as

$$\rho(\mathbf{r}) = \sum_\alpha q_\alpha\, \delta(\mathbf{r} - \mathbf{r}_\alpha) \tag{22.a}$$

and in reciprocal space as

$$\rho(\mathbf{k}) = \left(\frac{1}{2\pi}\right)^{3/2} \sum_\alpha q_\alpha\, e^{-i\mathbf{k}.\mathbf{r}_\alpha}. \tag{22.b}$$

Since the charges are localized near the origin ($|\mathbf{r}_\alpha| \lesssim a$) and we are assuming that the coupling with the radiation takes place substantially with the long-wavelength modes ($ka \ll 1$), it is reasonable to expand the exponential of (22.b) and to take only the first nonzero term (a lowest-order calculation in a/λ). One then gets using (9) and (4)

$$\rho(\mathbf{k}) = -\left(\frac{1}{2\pi}\right)^{3/2} i\mathbf{k}\cdot\mathbf{d} \tag{23.a}$$

and by Fourier transformation

$$\rho(\mathbf{r}) = -\nabla\cdot[\mathbf{d}\,\delta(\mathbf{r})]. \tag{23.b}$$

Equation (23.b) suggests one introduce the *polarization density*

$$\mathbf{P}(\mathbf{r}) = \mathbf{d}\,\delta(\mathbf{r}) \tag{24.a}$$

corresponding to a dipole **d** localized at $\mathbf{r} = \mathbf{0}$, as well as its spatial Fourier transform

$$\mathcal{P}(\mathbf{k}) = \left(\frac{1}{2\pi}\right)^{3/2} \mathbf{d}. \tag{24.b}$$

Equation (23.b) is then written

$$\rho(\mathbf{r}) = -\nabla\cdot\mathbf{P}(\mathbf{r}). \tag{25}$$

The simple form of Equation (25) shows that if one introduces the displacement **D(r)** defined by

$$\mathbf{D}(\mathbf{r}) = \varepsilon_0\,\mathbf{E}(\mathbf{r}) + \mathbf{P}(\mathbf{r}), \tag{26}$$

where $\mathbf{E}(\mathbf{r})$ is the total electric field, then Maxwell's equation $\nabla \cdot \mathbf{E} = \rho/\varepsilon_0$ and Equation (25) imply

$$\nabla \cdot \mathbf{D}(\mathbf{r}) = 0 \tag{27.a}$$

which shows that $\mathbf{D}(\mathbf{r})$ is a transverse field. Equation (27.a) can be rewritten using (26):

$$\mathbf{D}(\mathbf{r}) = \mathbf{D}_\perp(\mathbf{r}) = \varepsilon_0\, \mathbf{E}_\perp(\mathbf{r}) + \mathbf{P}_\perp(\mathbf{r})\,. \tag{27.b}$$

The importance of $\mathbf{D}(\mathbf{r})$ rests with the fact that outside the origin $\mathbf{P}(\mathbf{r})$ is zero [see (24.a)], with the result that, from (26), $\mathbf{D}(\mathbf{r})$ coincides with $\varepsilon_0 \mathbf{E}(\mathbf{r})$. Thus, $\mathbf{D}(\mathbf{r})$ is a *transverse field which, outside the system of charges, coincides with the total electric field* to within a factor ε_0. Since the electric field is purely retarded, it follows that the displacement outside the atom is a retarded transverse field.

Return now to Equation (20). By transforming the discrete sum of the last line into an integral we get

$$-\frac{1}{\varepsilon_0}\int \mathrm{d}^3k \sum_{\boldsymbol{\varepsilon}\perp\mathbf{k}} \frac{1}{(2\,\pi)^3}\, \boldsymbol{\varepsilon}(\boldsymbol{\varepsilon}\cdot\mathbf{d})\, e^{i\mathbf{k}.\mathbf{r}}\,. \tag{28}$$

Comparison with (24.b) shows that (28) is just the Fourier transform of $-\mathscr{P}_\perp(\mathbf{k})/\varepsilon_0$, so that Equation (20) can be written

$$\mathbf{E}'_\perp(\mathbf{r}) = \mathbf{E}_\perp(\mathbf{r}) - \frac{1}{\varepsilon_0}\mathbf{P}_\perp(\mathbf{r})\,. \tag{29}$$

Finally we calculate the operator $\mathbf{D}'(\mathbf{r})$ which represents the displacement in the new description. Using (27.b), $\mathbf{D}'(\mathbf{r})$ is written

$$\mathbf{D}'(\mathbf{r}) = T\mathbf{D}(\mathbf{r})\,T^+ = \varepsilon_0\, T\mathbf{E}_\perp(\mathbf{r})\,T^+ + T\mathbf{P}_\perp(\mathbf{r})\,T^+\,. \tag{30}$$

Using (19), (20), (29), and the fact that $\mathbf{P}_\perp(\mathbf{r})$ commutes with T, one then gets

$$\frac{1}{\varepsilon_0}\mathbf{D}'(\mathbf{r}) = i\sum_j \mathscr{E}_{\omega_j}\left[a_j\,\boldsymbol{\varepsilon}_j\, e^{i\mathbf{k}_j.\mathbf{r}} - a_j^+\,\boldsymbol{\varepsilon}_j\, e^{-i\mathbf{k}_j.\mathbf{r}}\right]. \tag{31}$$

It appears then that the *same* mathematical operator, namely the linear combination of a_j and a_j^+ on the right in (31), describes two *different* physical variables, depending on the representation used: the transverse electric field in the initial representation, and the displacement (divided by ε_0) in the new one. The advantage of the latter representation is that the simple operator (31) describes a transverse field which outside the atom is purely retarded.

d) THE HAMILTONIAN IN THE NEW REPRESENTATION

The Hamiltonian H' in the new representation is given by

$$H' = THT^+ . \tag{32}$$

Here we are going to find the T-transformed expression for H given in (13). The physical interpretation of the results will then be obtained by attributing to the operators appearing in the expression for H' the physical meaning which they have in the new representation. The transform of the first term of (13) is simply

$$\sum_\alpha \frac{\mathbf{p}_\alpha^2}{2\,m_\alpha} \tag{33}$$

and according to (17) represents the kinetic energy of the particles. The second term of (13) is an atomic operator which depends only on the positions $\mathbf{r}_\alpha$ of the particles and is therefore unchanged in the transformation. It remains to find the transform of the third term of (13), that is, the operator called H_R in the initial representation and which describes in that representation the energy of the transverse field. Using (18), we get

$$\begin{aligned} H'_R &= TH_R\,T^+ = T\sum_j \hbar\omega_j\left(a_j^+\,a_j + \frac{1}{2}\right)T^+ \\ &= \sum_j \hbar\omega_j\left[(a_j^+ + \lambda_j^*)\,(a_j + \lambda_j) + \frac{1}{2}\right] \\ &= H_R + \sum_j \hbar\omega_j(\lambda_j\,a_j^+ + \lambda_j^*\,a_j) + \sum_j \hbar\omega_j\,\lambda_j^*\,\lambda_j\,. \end{aligned} \tag{34}$$

In addition to the operator H_R we get a linear operator in λ_j and λ_j^* and a bilinear operator in λ_j and λ_j^*. Consider first the linear term, which can be written according to (16) as

$$-\,\mathbf{d}\cdot\sum_j \mathrm{i}\sqrt{\frac{\hbar\omega_j}{2\,\varepsilon_0\,L^3}}\,(a_j\,\boldsymbol{\varepsilon}_j - a_j^+\,\boldsymbol{\varepsilon}_j)\,. \tag{35}$$

We get the scalar product of $\mathbf{d}$ (which is equal to $\mathbf{d}'$, since $\mathbf{r}'_\alpha = \mathbf{r}_\alpha$) with an operator coincident with $\mathbf{D}'(0)/\varepsilon_0$. The operator appearing in (35) must indeed be interpreted as the displacement (divided by ε_0), since we are studying the Hamiltonian in the new representation. Thus for the term of H'_R linear in a_j and a_j^+, we get

$$-\,\mathbf{d}\cdot\frac{\mathbf{D}'(0)}{\varepsilon_0}\,. \tag{36}$$

Consider finally the last term of (34). According to (16), it is equal to

$$\sum_j \hbar\omega_j \lambda_j^* \lambda_j = \sum_{\mathbf{k}_j \varepsilon_j} \frac{1}{2\,\varepsilon_0 L^3} (\varepsilon_j \cdot \mathbf{d})^2 . \tag{37}$$

This term depends only on the operator $\mathbf{d}$ (equal to $\mathbf{d}'$). Physically it represents a dipole self-energy of the system, denoted ε_{dip}. The expression (37) for ε_{dip} seems to diverge. It fact, one must limit the sum to the values of k_j for which the long-wavelength approximation is valid.

By regrouping the various terms, one gets the following expression for H':

$$H' = \sum_\alpha \frac{\mathbf{p}_\alpha^2}{2\,m_\alpha} + V_{\text{Coul}} + \varepsilon_{\text{dip}} + \sum_j \hbar\omega_j \left(a_j^+ a_j + \frac{1}{2} \right) - \mathbf{d} \cdot \frac{\mathbf{D}'(\mathbf{0})}{\varepsilon_0} . \tag{38}$$

The structure of the new Hamiltonian H' is very simple. One has first a purely atomic Hamiltonian representing the sum of the kinetic energy, the Coulomb energy, and the dipole self-energy. Then one has a proper radiation Hamiltonian H_R, and finally, an electric dipole interaction Hamiltonian coupling the dipole moment of the system of charges to the displacement at the origin $\mathbf{0}$. The fact that the new interaction Hamiltonian is purely linear in the field and no longer has quadratic terms (such as H_{I2} examined in §D.1 of Chapter III) is an important advantage of this new description.

Remark

In the new representation, the energy of the transverse field is given by H_R' and differs from H_R by the last two terms in (34). Since the operators $\mathbf{D}'(\mathbf{r})/\varepsilon_0$ (31) and $\mathbf{B}'(\mathbf{r})$ (21) have the same mathematical form in the new description as $\mathbf{E}(\mathbf{r})$ and $\mathbf{B}(\mathbf{r})$ in the earlier one, we have

$$H_R = \frac{1}{2} \int \mathrm{d}^3 r \left[\frac{\mathbf{D}'^2(\mathbf{r})}{\varepsilon_0} + \frac{\mathbf{B}'^2(\mathbf{r})}{\mu_0} \right] \tag{39}$$

(where $\varepsilon_0 c^2$ is replaced by $1/\mu_0$).

3. Extensions

We are now going to examine two possible extensions of the treatment above.

a) The Case of Two Separated Systems of Charges

Consider two systems of charges $\mathcal{S}_A$ and $\mathcal{S}_B$ localized about well-separated points $\mathbf{R}_A$ and $\mathbf{R}_B$, each of the systems being neutral. The transformation T which generalizes (14) is

$$T = \exp\left\{ -\frac{i}{\hbar}[\mathbf{d}_A \cdot \mathbf{A}(\mathbf{R}_A) + \mathbf{d}_B \cdot \mathbf{A}(\mathbf{R}_B)] \right\} \tag{40}$$

$\mathbf{d}_A$ and $\mathbf{d}_B$ being the respective dipole moments.

The Hamiltonian in the Coulomb gauge describing this system is in the long-wavelength approximation equal to

$$\begin{aligned} H = & \sum_\alpha \frac{1}{2\,m_\alpha}[\mathbf{p}_\alpha - q_\alpha\,\mathbf{A}(\mathbf{R}_A)]^2 + \sum_\beta \frac{1}{2\,m_\beta}[\mathbf{p}_\beta - q_\beta\,\mathbf{A}(\mathbf{R}_\beta)]^2 \\ & + V_{\text{Coul}}^{AA} + V_{\text{Coul}}^{BB} + V_{\text{dip dip}}^{AB} \\ & + \sum_i \hbar\omega_i\left(a_i^+\,a_i + \frac{1}{2}\right) \end{aligned} \tag{41}$$

where V_{Coul}^{AA} (V_{Coul}^{BB}) is the Coulomb energy of the system of charges $\mathscr{S}_A$ ($\mathscr{S}_B$), and $V_{\text{dip dip}}^{AB}$ is the electrostatic interaction energy between the dipoles $\mathbf{d}_A$ and $\mathbf{d}_B$ of the two systems of charges.

The Hamiltonian $H' = TH'T^+$ in the new representation is gotten with the aid of the transformation (40), and one finds it equal to

$$\begin{aligned} H' = & \sum_\alpha \frac{\mathbf{p}_\alpha^2}{2\,m_\alpha} + V_{\text{Coul}}^{AA} + \varepsilon_{\text{dip}}^A + \sum_\beta \frac{\mathbf{p}_\beta^2}{2\,m_\beta} + V_{\text{Coul}}^{BB} + \varepsilon_{\text{dip}}^B \\ & + H_R - \mathbf{d}'_A \cdot \frac{\mathbf{D}'(\mathbf{R}_A)}{\varepsilon_0} - \mathbf{d}'_B \cdot \frac{\mathbf{D}'(\mathbf{R}_B)}{\varepsilon_0}. \end{aligned} \tag{42}$$

The structure of the new Hamiltonian H' is very simple. We have first of all two purely atomic Hamiltonians for $\mathscr{S}_A$ and $\mathscr{S}_B$, representing in each case the sum of the kinetic energy, the Coulomb energy (inside $\mathscr{S}_A$ or $\mathscr{S}_B$), and the dipole self-energy. Then, we have a radiation Hamiltonian H_R, and finally, an electric dipole interaction Hamiltonian coupling the dipoles of $\mathscr{S}_A$ and $\mathscr{S}_B$ to the displacement at $\mathbf{R}_A$ and $\mathbf{R}_B$. The important point is that there are no longer instantaneous electrostatic interaction terms between the systems in (42). The term $V_{\text{dip dip}}^{AB}$ has in fact been compensated for by the dipole terms (37), which contain, besides the self-energy terms $\varepsilon_{\text{dip}}^A$ and $\varepsilon_{\text{dip}}^B$, cross terms simultaneously involving $\mathbf{d}_A$ and $\mathbf{d}_B$. The corresponding calculations are not detailed here, since they are treated again in the general case in Complement C_{IV}. One can however understand this point by noting that the new Hamiltonian contains the coupling between the electric dipole moment $\mathbf{d}'_A$ of $\mathscr{S}_A$ and the total displacement at $\mathbf{R}_A$, $\mathbf{D}'(\mathbf{R}_A)$, which includes in particular the displacement generated at $\mathbf{R}_A$ by $\mathscr{S}_B$. Now the displacement generated by

$\mathscr{S}_B$ outside $\mathscr{S}_B$ coincides with the total electric field generated by $\mathscr{S}_B$ outside $\mathscr{S}_B$. It follows that the coupling term $-\mathbf{d}'_A \cdot \mathbf{D}'(\mathbf{R}_A)/\varepsilon_0$ of (42) contains the interaction of $\mathscr{S}_A$ with the transverse electric field as well as the longitudinal field created by $\mathscr{S}_B$ at $\mathbf{R}_A$.

b) The Case of a Quantized Field Coupled to Classical Sources

In §A_{IV}.1 above we have examined the coupling of a system of quantized particles with a classical field. The opposite problem, which we now treat, is that of a quantized field interacting with classical currents. The Hamiltonian in Coulomb gauge for such a system has been given in Complement B_{III} [Equation (7)]:

$$H = H_R - \int \mathrm{d}^3r\, \mathbf{j}_{\mathrm{cl}}(\mathbf{r}, t) \cdot \mathbf{A}(\mathbf{r}) . \tag{43}$$

The current $\mathbf{j}_{\mathrm{cl}}(\mathbf{r}, t)$ is produced by classical particles with charge q_α whose positions and velocities are described by the classical functions $\mathbf{r}_\alpha(t)$ and $\dot{\mathbf{r}}_\alpha(t)$, with a given time dependence

$$\mathbf{j}_{\mathrm{cl}}(\mathbf{r}, t) = \sum_\alpha q_\alpha\, \dot{\mathbf{r}}_\alpha(t)\, \delta(\mathbf{r} - \mathbf{r}_\alpha(t)) . \tag{44}$$

If the charges creating the current are localized near the origin in a volume of dimension a and if one considers only the dynamics of the modes with wavelength long with respect to a, it is possible to simplify the interaction Hamiltonian

$$H_I = - \int \mathrm{d}^3r\, \mathbf{j}_{\mathrm{cl}}(\mathbf{r}, t) \cdot \mathbf{A}(\mathbf{r}) = - \sum_\alpha q_\alpha\, \dot{\mathbf{r}}_\alpha(t) \cdot \mathbf{A}(\mathbf{r}_\alpha) \tag{45}$$

by replacing $\mathbf{A}(\mathbf{r}_\alpha)$ with its value at the origin:

$$H_I \simeq - \sum_\alpha q_\alpha \dot{\mathbf{r}}_\alpha(t) \cdot \mathbf{A}(\mathbf{0}) = - \dot{\mathbf{d}}(t) \cdot \mathbf{A}(\mathbf{0}) \tag{46}$$

$\mathbf{d}(t)$ being the electric dipole moment of the charge distribution, equal to $\sum_\alpha q_\alpha \mathbf{r}_\alpha(t)$. We then apply to the Hamiltonian

$$H = H_R - \dot{\mathbf{d}}(t) \cdot \mathbf{A}(\mathbf{0}) \tag{47}$$

a unitary transformation similar to the transformations in §§A_{IV}.1 and A_{IV}.2 and aimed at giving rise to an electric dipole interaction term. Consider the transformation

$$T = \exp\left\{ -\frac{\mathrm{i}}{\hbar}\, \mathbf{d}(t) \cdot \mathbf{A}(\mathbf{0}) \right\} \tag{48}$$

analogous to (3) and (14), but where $\mathbf{d}(t)$ is now a classical function depending on time and $\mathbf{A}(\mathbf{0})$ an operator acting on the radiation variables. The transformed Hamiltonian is given by Equation (5). The calculation of TH_RT^+ is identical to that presented in (34) and gives

$$TH_R T^+ = H_R - \mathbf{d}(t) \cdot \frac{\mathbf{D}'(\mathbf{0})}{\varepsilon_0} + \varepsilon_{\text{dip}} \tag{49}$$

where $\mathbf{D}'(\mathbf{0})/\varepsilon_0$ corresponds to the operator of (31) evaluated at $\mathbf{r} = \mathbf{0}$ and where ε_{dip} is now a classical function of time gotten by replacing the operator $\mathbf{d}$ of (37) with the function $\mathbf{d}(t)$. Here TH_IT^+ is unchanged, since H_I commutes with T. Finally the term in $i\hbar(\partial T/\partial t)T^+$ is equal to $\dot{\mathbf{d}}(t) \cdot \mathbf{A}(0)$ and thus offsets H_I [see (46)]. The new Hamiltonian is then

$$H = H_R - \mathbf{d}(t) \cdot \frac{\mathbf{D}'(\mathbf{0})}{\varepsilon_0} + \varepsilon_{\text{dip}} \, . \tag{50}$$

The last term here is a classical function of time and can eventually be dropped.

Note that, just as in §A_{IV}.2, $\mathbf{D}'(\mathbf{r})$ represents the displacement. The interaction between the quantized field and the classical sources is thus, in the new representation, proportional to the product of the classical electric dipole with the displacement at the point about which the dipole is localized.

COMPLEMENT B_{IV}

ONE-PHOTON AND TWO-PHOTON PROCESSES: THE EQUIVALENCE BETWEEN THE INTERACTION HAMILTONIANS A · p AND E · r

We show in this complement that the transition amplitudes calculated using the interaction Hamiltonians $\mathbf{A} \cdot \mathbf{p}$ and $\mathbf{E} \cdot \mathbf{r}$ are identical for a resonant one- or two-photon absorption process. The equality of the two amplitudes is demonstrated by direct calculation of the evolution-operator matrix elements. We examine various possible extensions of this demonstration and show, in particular, how a one-photon nonresonant process can often be considered as part of a two-photon resonant process. This allows the resolution of several paradoxes involving the equivalence of the two Hamiltonians for calculating line shapes.

1. Notation. Principles of Calculation

Consider an atom at $\mathbf{R} = \mathbf{0}$ which interacts with an incident wave packet with central frequency ω. The external field associated with this wave packet is described by a vector potential whose value at $\mathbf{R} = \mathbf{0}$ is of the form

$$\mathbf{A}(\mathbf{0}, t) = A(t)\, \boldsymbol{\varepsilon}_i = \tilde{A}(t) \cos \omega t\, \boldsymbol{\varepsilon}_i \tag{1}$$

where $\boldsymbol{\varepsilon}_i$ is the polarization vector of the wave and $\tilde{A}(t)$ the envelope of $A(t)$ (see Figure 1). The transit time of the wave packet is characterized by a time T which is of the order of the half-maximum width of $\tilde{A}(t)$. We assume that the field oscillates many times during T, that is, that

$$\omega T \gg 1 . \tag{2}$$

We propose to examine one- and two-photon resonant absorption processes induced by such a field between two atomic levels a and b such that

$$E_b - E_a = \hbar\omega \tag{3.a}$$

and

$$E_b - E_a = 2\,\hbar\omega . \tag{3.b}$$

respectively. To this end, consider the matrix element of the evolution operator between a and b, $\langle b|U(t_2, t_1)|a\rangle$, t_1 and t_2 being the initial and final times for the transit of the wave packet (see Figure 1). It is clear that $\langle b|U(t_2, t_1)|a\rangle$ must present a resonant character for $\omega_{ba} = \omega$ or $\omega_{ba} = 2\omega$

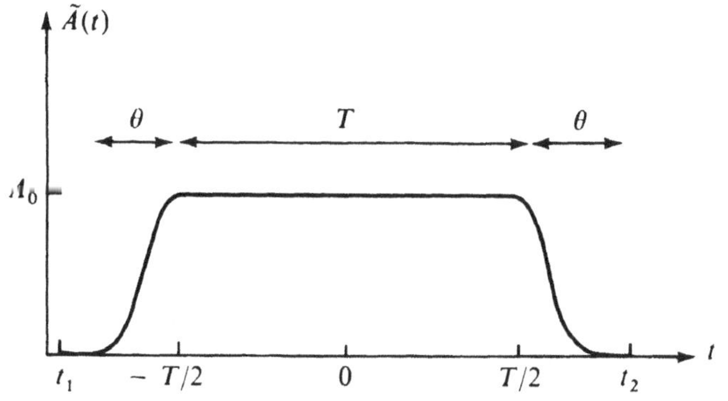

Figure 1. The shape of the envelope $\tilde{A}(t)$ of $A(t)$. The duration θ of the switching on and off of the incident field is assumed small compared to the interaction time T. Between $-T/2$ and $+T/2$, $\tilde{A}(t)$ is constant and equal to A_0. One assumes that T and θ are very large with respect to $1/\omega$.

(we have taken $E_b - E_a = \hbar\omega_{ba}$). In fact, it is more appropriate here to calculate the S-matrix, whose element S_{ba} is

$$S_{ba} = \lim_{\substack{t_2 \to +\infty \\ t_1 \to -\infty}} \langle b \,|\, e^{iH_0 t_2/\hbar}\, U(t_2, t_1)\, e^{-iH_0 t_1/\hbar} \,|\, a \rangle \tag{4}$$

where H_0 is the unperturbed Hamiltonian (the atomic Hamiltonian in the absence of an incident field).

Such a calculation can be made starting either with the Hamiltonian in the Coulomb gauge or with the Göppert-Mayer Hamiltonian [see (B.33) of Chapter IV or (8) of complement A_{IV}]. Let $U'(t_2, t_1)$ be the evolution operator in the new representation, leading to the new matrix S'. If we assume that $\tilde{A}(t)$ is zero for $t \le t_1$ and $t \ge t_2$ (Figure 1), the Hamiltonians and state vectors are the same in either representation for $t \le t_1$ or $t \ge t_2$. It follows then that the general treatment in Chapter IV (§B.3.*d*.i) leads to the identity

$$S_{ba} = S'_{ba}\,. \tag{5}$$

Rather than using this general proof here, we are going to calculate S_{ba} and S'_{ba} and show *directly*, using the resulting expressions, the equality of the two transition amplitudes.

2. Calculation of the Transition Amplitudes in the Two Representations

a) The Interaction Hamiltonian $\mathbf{A} \cdot \mathbf{p}$

Consider first the case of a one-photon resonant transition, $\omega_{ba} \simeq \omega$. To first order in q and with the long-wavelength approximation, the matrix element S_{ba} becomes

$$S_{ba} = \frac{1}{i\hbar}\left(-\frac{q}{m}\right)\int_{t_1}^{t_2} d\tau\, e^{i\omega_{ba}\tau} \langle b \mid \mathbf{p} \cdot \mathbf{A}(\mathbf{0}, \tau) \mid a \rangle \tag{6}$$

The resonant behavior of S_{ba} near $\omega = \omega_{ba}$ comes from the term $e^{-i\omega\tau}$ of $\cos \omega t$ appearing in Equation (1) for $\mathbf{A}(\mathbf{0}, \tau)$. If one neglects the contribution of the term $e^{i\omega\tau}$—which, in the neighborhood of $\omega = \omega_{ba}$, is much smaller and nonresonant—this gives

$$S_{ba} = \frac{1}{i\hbar} M_{ba} \int_{-\infty}^{+\infty} d\tau \frac{\tilde{A}(\tau)}{2} e^{i(\omega_{ba} - \omega)\tau} \tag{7.a}$$

where

$$M_{ba} = -\frac{q}{m} \langle b \mid \boldsymbol{\varepsilon}_i \cdot \mathbf{p} \mid a \rangle = -\frac{q}{m} (p_i)_{ba}\,. \tag{7.b}$$

We now calculate the value of S_{ba} when $\tilde{A}(t)$ corresponds to the curve given in Figure 1, $\tilde{A}(t)$ increasing from 0 to A_0 over the time interval θ, remaining constant between $-T/2$ and $T/2$, then decreasing from A_0 to 0 over the same interval θ. When the interaction time T, assumed much larger than θ, becomes very large, the integral over τ in (7.a) tends toward a delta function of width $1/T$, denoted $\delta^{(T)}$, and (7.a) can be written

$$S_{ba} = \frac{2\pi}{i\hbar} M_{ba} \frac{A_0}{2} \delta^{(T)}(\omega_{ba} - \omega)\,. \tag{7.c}$$

It is clear from (7.c) that S_{ba} is only important at resonance.

Assume now that the levels a and b cannot be connected by a one-photon transition. For example, a and b might have the same parity, so that the matrix element of the odd operator $\boldsymbol{\varepsilon}_i \cdot \mathbf{p}$ between $|a\rangle$ and $|b\rangle$ is zero. It is then necessary to calculate S_{ba} to higher orders in q. To order 2, the contribution of the term $q^2\mathbf{A}^2/2m$ is, in the dipole approximation, proportional to $\langle b|a\rangle$ and thus zero. The second term of the perturbation series for $-q\mathbf{A} \cdot \mathbf{p}/m$ gives

$$S_{ba} = \left(\frac{1}{i\hbar}\right)^2 \left(-\frac{q}{m}\right)^2 \int_{t_1}^{t_2} d\tau \int_{t_1}^{t_2} d\tau'\, \theta(\tau - \tau') \times$$

$$\times\, e^{i\omega_b\tau} \sum_r \left[\langle b \mid \mathbf{p} \cdot \mathbf{A}(\mathbf{0}, \tau) \mid r \rangle e^{-i\omega_r(\tau - \tau')} \langle r \mid \mathbf{p} \cdot \mathbf{A}(\mathbf{0}, \tau') \mid a \rangle\right] e^{-i\omega_a\tau'} \tag{8}$$

The sum Σ_r is taken on all the intermediate levels which are coupled to $|a\rangle$ and $|b\rangle$ in the electric dipole approximation. $\theta(\tau - \tau')$ is the Heaviside function, which is zero when $\tau < \tau'$ and 1 when $\tau > \tau'$. We have taken $\omega_i = E_i/\hbar$ $(i = a, b, r)$. By retaining only the terms in $e^{-i\omega\tau}$ and

$e^{-i\omega\tau'}$ for $\cos \omega\tau$ and $\cos \omega\tau'$ which give rise to a resonance for $\omega_{ba} = 2\omega$, we get

$$S_{ba} = \left(\frac{1}{i\hbar}\right)^2 \left(-\frac{q}{m}\right)^2 \int_{-\infty}^{+\infty} d\tau \int_{-\infty}^{+\infty} d\tau' \, \theta(\tau - \tau') \, e^{i(\omega_{ba} - 2\omega)\tau} \times$$

$$\times \sum_r (p_i)_{br} (p_i)_{ra} \frac{\tilde{A}(\tau)}{2} \frac{\tilde{A}(\tau')}{2} e^{-i(\omega_{ra} - \omega)(\tau - \tau')}. \quad (9)$$

To enable the integration in (9) to be carried out independently on τ and τ', it is useful to introduce the following integral representation:

$$e^{-i(\omega_{ra} - \omega)(\tau - \tau')} \theta(\tau - \tau') = -\frac{1}{2\pi i} \int_{-\infty}^{+\infty} d\Omega \frac{e^{-i\Omega(\tau - \tau')}}{(\Omega - \omega_{ra} + \omega) + i\varepsilon} \quad (10)$$

where ε is a vanishingly small positive number. Note that (10) can be demonstrated quite simply by integrating the second term by residues. Taking account of (10), Equation (9) becomes

$$S_{ba} = \left(\frac{q}{\hbar m}\right)^2 \frac{1}{2\pi i} \sum_r (p_i)_{br} (p_i)_{ra} \times$$

$$\iiint d\Omega \, d\tau \, d\tau' \frac{\tilde{A}(\tau) \tilde{A}(\tau')}{4} \times \frac{e^{i(\omega_{ba} - 2\omega - \Omega)\tau} \, e^{i\Omega\tau'}}{(\Omega - \omega_{ra} + \omega) + i\varepsilon}. \quad (11)$$

Finally, we transform (11) by introducing the Fourier transform $\tilde{\mathscr{A}}(\Omega)$ of $\tilde{A}(t)$ defined by

$$\tilde{\mathscr{A}}(\Omega) = \frac{1}{\sqrt{2\pi}} \int_{-\infty}^{+\infty} \tilde{A}(t) \, e^{i\Omega t} \, dt. \quad (12)$$

Equation (11) becomes

$$S_{ba} = \left(\frac{q}{\hbar m}\right)^2 \frac{1}{i} \sum_r (p_i)_{br} (p_i)_{ra} \int_{-\infty}^{+\infty} d\Omega \frac{\tilde{\mathscr{A}}(\Omega) \tilde{\mathscr{A}}(\omega_{ba} - 2\omega - \Omega)}{4[(\Omega - \omega_{ra} + \omega) + i\varepsilon]}. \quad (13)$$

The function $\tilde{\mathscr{A}}(\Omega)$ is centered about $\Omega = 0$. If the one-photon transitions are not resonant ($\omega - \omega_{ra} \neq 0$) and if the width of $\tilde{\mathscr{A}}(\Omega)$ is sufficiently narrow, one can replace Ω by 0 in the denominator of (13) and drop the factor $i\varepsilon$. If $\tilde{A}(t)$ has the form represented in Figure 1, $\tilde{\mathscr{A}}(\Omega)$ appears like a peak of width $1/T$ superimposed on a background of width $1/\theta$. For the following calculations to be valid we assume $|\omega - \omega_{ra}| \gg 1/\theta$. By using this approximation we find then

$$S_{ba} = \frac{1}{i\hbar} \left(\frac{q}{m}\right)^2 \left[\sum_r \frac{(p_i)_{br} (p_i)_{ra}}{\hbar(\omega - \omega_{ra})}\right] \int_{-\infty}^{+\infty} d\Omega \frac{\tilde{\mathscr{A}}(\Omega) \tilde{\mathscr{A}}(\omega_{ba} - 2\omega - \Omega)}{4}. \quad (14)$$

It remains only to use the Fourier transform of a convolution product to get

$$S_{ba} = \frac{1}{i\hbar} Q_{ba} \int_{-\infty}^{+\infty} d\tau \left[\frac{\tilde{A}(\tau)}{2}\right]^2 e^{i(\omega_{ba} - 2\omega)\tau} \tag{15.a}$$

with

$$Q_{ba} = \left(\frac{q}{m}\right)^2 \sum_r \frac{(p_i)_{br} (p_i)_{ra}}{\hbar(\omega - \omega_{ra})}. \tag{15.b}$$

Comparison of the two transition amplitudes (7.a) and (15.a) shows that things occur as if one had a perturbation $[\tilde{A}(t)/2]^2 e^{-2i\omega t}$ coupling the levels a and b (via all the intermediate levels r) with an effective matrix element Q_{ba} given in (15.b). By taking for $\tilde{A}(t)$ the function represented by the curve in Figure 1, we find

$$S_{ba} = \frac{2\pi}{i\hbar} Q_{ba} \left(\frac{A_0}{2}\right)^2 \delta^{(T)}(\omega_{ba} - 2\omega). \tag{15.c}$$

It is clear here that S_{ba} is only important at resonance ($\omega_{ba} = 2\omega$). S_{ba} is then a resonant two-photon transition amplitude.

Remark

Equation (15.a) shows that, in the case where one does not have an intermediate level r which can be resonantly excited by a one-photon process, the transition amplitude involves the square of the vector potential rather than the product of the vector potentials taken at two different times. This result arises from the fact that under these conditions the two photons are almost simultaneously absorbed.

b) The Interaction Hamiltonian $\mathbf{E} \cdot \mathbf{r}$

To compare the results from the two interaction Hamiltonians $\mathbf{A} \cdot \mathbf{p}$ and $\mathbf{E} \cdot \mathbf{r}$, it is necessary that $\mathbf{A}(t)$ and $\mathbf{E}(t)$ represent the same field. From the form (1) postulated for $\mathbf{A}(\mathbf{0}, t)$, we get

$$\mathbf{E}(\mathbf{0}, t) = [\tilde{A}(t)\, \omega \sin \omega t - \dot{\tilde{A}}(t) \cos \omega t]\, \mathbf{e}_i\,. \tag{16}$$

Remark

It would be more satisfactory a priori to start with the time dependence of the electric field, which is the quantity with a clear physical meaning, and then to derive the time dependence of the vector potential. Such a procedure would most certainly lead to equivalent results, but can also raise a problem. The fact that the electric field is zero for $t_1' \leq t_1$ and $t_2' \geq t_2$ does not assure that the vector potential will also vanish on the same time intervals. Since $\mathbf{A}(t)$ is gotten by integrating $\mathbf{E}(t)$, $\mathbf{A}(t_2') - \mathbf{A}(t_1')$ can be equal to a nonzero constant. It is not possible to nullify the interaction Hamilton $\mathbf{A} \cdot \mathbf{p}$ simultaneously for $t \leq t_1$ and $t \geq t_2$. In the general case, this is only a formal difficulty. The electric field oscillates a huge number of times between t_1 and t_2, and the constant depends, in fact, only on the contribution of a fraction of period whose relative effect becomes negligible when $T \to \infty$.

The calculation of the amplitude S'_{ba} is analogous to that in the preceding subsection, the interaction Hamiltonian being henceforth equal to $-q\mathbf{r} \cdot \mathbf{E}$, where $\mathbf{E}$ is given by (16). Note that $\mathbf{E}$ is the sum of two terms and the second term of $\mathbf{E}$ is, in order of magnitude, equal to $[\tilde{A}(t)/\theta]\cos \omega t$. It is then on average smaller than the first by a factor $\omega\theta$, so that its contribution in calculating the two-photon excitation probability will be negligible. In the limit where $T \gg \theta \gg 1/\omega$, we can then follow the proof in the subsection above and retain only the first term of (16). We then get for the one-photon resonant transitions

$$S'_{ba} = \frac{2\pi}{\mathrm{i}\hbar} M'_{ba} \frac{A_0}{2} \delta^{(T)}(\omega_{ba} - \omega) \tag{17.a}$$

$$M'_{ba} = -\mathrm{i}\omega q \langle b \mid \boldsymbol{\varepsilon}_i \cdot \mathbf{r} \mid a \rangle = -\mathrm{i}\omega q (r_i)_{ba} \tag{17.b}$$

which correspond to (7.c) and (7.b), and for the two-photon resonant transitions

$$S'_{ba} = \frac{2\pi}{\mathrm{i}\hbar} Q'_{ba}\left(\frac{A_0}{2}\right)^2 \delta^{(T)}(\omega_{ba} - 2\omega) \tag{18.a}$$

$$Q'_{ba} = -\omega^2 q^2 \sum_r \frac{(r_i)_{br}(r_i)_{ra}}{\hbar(\omega - \omega_{ra})} \tag{18.b}$$

which correspond to (15.c) and (15.b).

c) Direct Verification of the Identity of the Two Amplitudes

We will start with

$$(p_i)_{ba} = i\omega_{ba} m (r_i)_{ba} \tag{19}$$

relating the matrix elements of the operators $\mathbf{p}$ and $\mathbf{r}$ [see (B.43)]. Comparison of (7.b) and (17.b) gives, using (19),

$$M_{ba} = \frac{\omega_{ba}}{\omega} M'_{ba} \tag{20.a}$$

that is,

$$M_{ba} = M'_{ba} \quad \text{if} \quad \omega_{ba} = \omega . \tag{20.b}$$

The two amplitudes (7.c) and (17.a) are then identical: at resonance the equality results from (20.b), and off resonance, the delta functions being 0, the transition amplitude is zero in the two representations.

We will now show that the two-photon amplitudes (15.c) and (18.a) are identical. For this it suffices to show that

$$\mathcal{Q}_{ba} = \mathcal{Q}'_{ba} \quad \text{if} \quad \omega_{ba} = 2\,\omega\,. \tag{21}$$

For this we use the identity (19) to transform (15.b). $\mathcal{Q}_{ba}$ is equal to $\mathcal{Q}'_{ba}$ only if

$$\sum_r \frac{\omega_{br}\,\omega_{ra}(r_i)_{br}\,(r_i)_{ra}}{\omega - \omega_{ra}} = \omega^2 \sum_r \frac{(r_i)_{br}\,(r_i)_{ra}}{\omega - \omega_{ra}}\,. \tag{22}$$

To prove (22) algebraically, we introduce the difference D between the two terms of (22):

$$D = \sum_r \frac{\omega_{br}\,\omega_{ra} - \omega^2}{\omega - \omega_{ra}} (r_i)_{br}\,(r_i)_{ra} \tag{23}$$

and show that, when $\omega_{ba} = 2\omega$, D is zero. The resonance condition implies

$$\omega_{br} = 2\,\omega - \omega_{ra} \tag{24}$$

and thus

$$\omega_{br}\,\omega_{ra} - \omega^2 = -\,(\omega - \omega_{ra})^2 \tag{25}$$

D, defined in (23), can then be transformed into

$$D = -\sum_r (\omega - \omega_{ra})\,(r_i)_{br}\,(r_i)_{ra}\,. \tag{26}$$

Consider now the commutator $[r_i, p_i] = i\hbar$, and calculate $\langle b|[r_i, p_i]|a\rangle$. We find

$$\sum_r [(r_i)_{br}\,(p_i)_{ra} - (p_i)_{br}\,(r_i)_{ra}] = i\hbar \langle\, b \,|\, a \,\rangle = 0 \tag{27}$$

since $|b\rangle$ and $|a\rangle$ correspond to distinct eigenstates of H_0. Using (19), this yields

$$\sum_r (\omega_{br} - \omega_{ra})\,(r_i)_{br}\,(r_i)_{ra} = 0\,. \tag{28}$$

Using the two-photon resonance condition (24), we get finally

$$\sum_r (\omega - \omega_{ra})\,(r_i)_{br}\,(r_i)_{ra} = 0\,. \tag{29}$$

Comparison of (26) and (29) shows that $D = 0$, which implies that $Q_{ba} = Q'_{ba}$ if $\omega_{ba} = 2\omega$, and thus the equality of the amplitudes S_{ba} given in (15.c) and S'_{ba} in (18.a).

Remarks

(i) In everything above we have only calculated amplitudes. To get the transition probability from a to b it is necessary to take the square of the modulus of (7.c) or (17.a) for a one-photon excitation and of (15.c) or (18.a) for a two-photon excitation. One then gets the square of a function $\delta^{(T)}(x)$. To within terms in θ/T that are negligible because we have assumed $T \gg \theta$, $\delta^{(T)}(x)$ is

$$\delta^{(T)}(x) = \frac{1}{2\pi}\int_{-T/2}^{+T/2} e^{ixt}\,dt = \frac{1}{\pi}\frac{\sin xT/2}{x}. \tag{30}$$

We conclude that $[\delta^{(T)}(x)]^2$ is a function equal to $T^2/4\pi^2$ at its maximum (at $x = 0$), and whose width is of the order of $2\pi/T$. The area under this curve being proportional to T, we can see that for large T, $[\delta^{(T)}(x)]^2$ is proportional to $T\delta^{(T)}(x)$. A more precise calculation starting from (30) gives

$$[\delta^{(T)}(x)]^2 \simeq \frac{T}{2\pi}\,\delta^{(T)}(x). \tag{31}$$

It then appears that near resonance, i.e. for $\omega_{ba} = \omega$ or $\omega_{ba} = 2\omega$, the transition probability from a to b, $|S_{ba}|^2 = |S'_{ba}|^2$, is proportional to the duration of excitation, T, which allows us to define the transition probability per unit time,

$$w_{ba} = \frac{|S_{ba}|^2}{T} = \frac{|S'_{ba}|^2}{T}. \tag{32}$$

For resonant one-photon transitions, using (7.c), (17.a), and (31), one gets

$$\begin{aligned} w_{ba} &= \frac{2\pi}{\hbar}\left(\frac{A_0}{2}\right)^2 |M_{ba}|^2\,\delta^{(T)}(\hbar\omega_{ba} - \hbar\omega) \\ &= \frac{2\pi}{\hbar}\left(\frac{A_0}{2}\right)^2 |M'_{ba}|^2\,\delta^{(T)}(\hbar\omega_{ba} - \hbar\omega) \end{aligned} \tag{33}$$

where $\delta^{(T)}(\omega_{ba} - \omega)$ has been replaced by $\hbar\delta^{(T)}(\hbar\omega_{ba} - \hbar\omega)$, and for the two-photon resonant transitions, using (15.c), (18.a), and (31),

$$\begin{aligned} w_{ba} &= \frac{2\pi}{\hbar}\left(\frac{A_0}{2}\right)^4 |Q_{ba}|^2\,\delta^{(T)}(\hbar\omega_{ba} - 2\hbar\omega) \\ &= \frac{2\pi}{\hbar}\left(\frac{A_0}{2}\right)^4 |Q'_{ba}|^2\,\delta^{(T)}(\hbar\omega_{ba} - 2\hbar\omega). \end{aligned} \tag{34}$$

The results (33) and (34) have a form analogous to that of Fermi's "golden rule"

(ii) In practice it is often impossible to perform exactly the summation on all the intermediate levels in the series (15.b) and (18.b). It is thus useful to know which equation gives the best approximation when the sum is limited to a few intermediate levels. In the case of a two-photon transition between two bound levels, the equation derived from the interaction Hamiltonian $-q\mathbf{r}\cdot\mathbf{E}$ will most often give the best approximation if one restricts the sum to the essential intermediate levels. This holds because the ratio of the two terms corresponding to a given intermediate level and derived respectively from the Hamiltonians $\mathbf{A}\cdot\mathbf{p}$ and $\mathbf{E}\cdot\mathbf{r}$ is $(\omega_{br}\omega_{ra})/\omega^2$. All the levels situated above the levels $|a\rangle$ and $|b\rangle$ (see Figure 2) have a more important contribution in the series correspond-

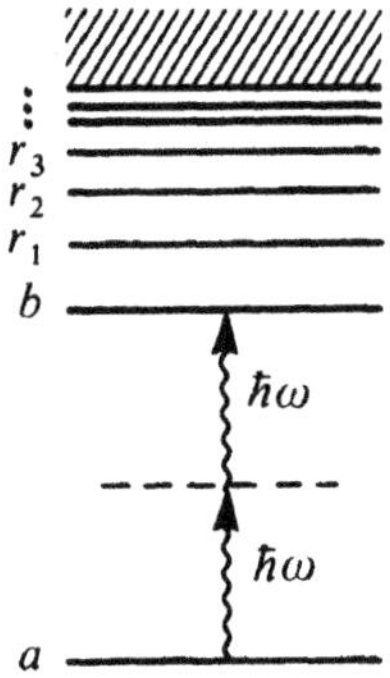

Figure 2. Two-photon resonant transitions between the levels a and b. The nonresonant levels $r_1, r_2, \ldots$ are the intermediate levels.

ing to the Hamiltonian $\mathbf{A}\cdot\mathbf{p}$ than that in the series corresponding to $\mathbf{E}\cdot\mathbf{r}$. Since the two sums are equal, it follows that the convergence of the first series is generally slower than that of the second. A classic example of this is the $1s$–$2s$ hydrogen transition (*). If one limits oneself to the $2p$ intermediate level in the sum, one gets zero with $\mathbf{A}\cdot\mathbf{p}$, and the value gotten with $\mathbf{E}\cdot\mathbf{r}$ has a 50% error from the exact result. If the sum is taken on the first three p-levels $(2p, 3p, 4p)$, one gets with $\mathbf{E}\cdot\mathbf{r}$ a result with a 20% error, while with $\mathbf{A}\cdot\mathbf{p}$ this sum is three times smaller than the true value. (See Exercise 2.)

(iii) The equality between the transition amplitudes at resonance is only true if exact wave functions are used. With approximate wave functions for the states $|b\rangle$ and $|a\rangle$, different results can be obtained from the two points of view. One of the two interaction Hamiltonians may give more accurate results than the other one for a particular type of approximate wave functions.

(*) The contribution of each intermediate level in the $\mathbf{A}\cdot\mathbf{p}$ and $\mathbf{E}\cdot\mathbf{r}$ approaches has been studied by F. Bassani, J. J. Forney, and A. Quattropani, *Phys. Rev. Lett.*, **39**, 107 (1977).

3. Generalizations

a) Extension to Other Processes

The discussion above can be extended to other processes, for example two-photon resonant absorption, where the atom interacts with two electromagnetic fields with *different* frequencies ω_1 and ω_2, the resonant condition becoming $\omega_{ba} = \omega_1 + \omega_2$. In this case, two types of amplitudes appear, corresponding to different time orders for the two interactions, either first with ω_1 and then ω_2 or in the reverse order. It is only when the two time orders are taken into account that the predictions with $\mathbf{A} \cdot \mathbf{p}$ and $\mathbf{E} \cdot \mathbf{r}$ agree.

Other important examples of two-photon processes are scattering processes: an incident photon vanishes, and a new photon ω' appears while the atom goes from state a to a'. Resonance is then given by $E_a + \hbar\omega = E_{a'} + \hbar\omega'$. Exercise 7 proposes a direct proof of the equality between the two transition amplitudes by using the general expression for these amplitudes given by (D.3), (D.4), and (D.12) of Chapter IV.

Other processes involving more than two photons can also be treated.

b) Nonresonant Processes

The equality of the transition matrix elements M_{ba} and M'_{ba} given by (7.b) and (17.b) for the one-photon transitions and Q_{ba} and Q'_{ba} given by (15.b) and (18.b) for the two-photon transitions and derived from the Hamiltonians $\mathbf{A} \cdot \mathbf{p}$ and $\mathbf{E} \cdot \mathbf{r}$ is only valid at *resonance*. The direct proof in §$B_{IV}.2$ above rests on the use of Equation (19) and on the resonance conditions (3.a) or (3.b), which express the conservation of energy. Recall also that the general reasoning of Part D concluded with the identity of the transition matrices on the energy shell.

What happens when one goes away from resonance? One can easily imagine, for example, that an atom is excited from a state a by a photon $\hbar\omega$ which does not have the requisite energy $\hbar\omega_{ba}$ to transfer the atom into an excited state b. If the states a and b are stable, the transition probability per unit time (33) is zero when $\omega \neq \omega_{ba}$ because of the delta function. If level b is unstable and has a width Γ, it can be tempting to generalize (33) by replacing the delta function $\delta^{(T)}(\hbar\omega_{ba} - \hbar\omega)$ with a Lorentzian with width $\hbar\Gamma$. However, such a step immediately leads to difficulty, since the matrix elements M_{ba} and M'_{ba} are not equal when $\omega \neq \omega_{ba}$ [see (20.a)]. The two lines of (33) are then not equal. Can one imagine, for an off-resonance excitation, nonidentical transition probabilities in the two approaches?

The answer is clearly no. The use of the S-matrix to find the transition rate assumes implicitly that the duration of the interaction with the electromagnetic field has been sufficiently long so that the uncertainty in the definition of its frequency will be smaller than Γ. This implies that the duration of the process has been much longer than the lifetime of the level b. It is therefore incorrect in this case to calculate the transition probability into b, since the latter is disintegrating to other states, for example, by spontaneous emission. The real process then is not a one-photon transition to an unstable state b, but rather a two-photon Rayleigh (Figure 3α) or Raman (Figure 3β) scattering, the atom ending in a final stable state a' (*).

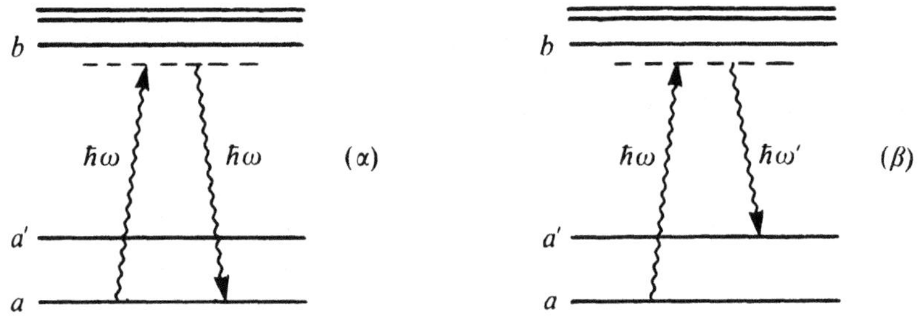

Figure 3. Nonresonant excitation. In the complete process, the incident photon $\hbar\omega$ disappears and a new photon appears (α) at the same frequency (Rayleigh scattering) or (β) at a different frequency (Raman scattering).

In seeking to define in a precise experimental fashion the absorption of a nonresonant photon, we are then led quite naturally to substitute for a nonresonant one-photon process a two-photon process conserving energy, for which we know that the transition probabilities are identical in the $\mathbf{A} \cdot \mathbf{p}$ and $\mathbf{E} \cdot \mathbf{r}$ approaches. It would be incorrect in this case to calculate the nonresonant one-photon amplitudes, since the "final" state in these amplitudes, that is, an atom in state b, is not a true final state, but rather an intermediate state. It should be also noted that the amplitudes associated with the processes in Figure 3α and β involve a summation on all the intermediate states b. Even if a particular excited level is very near resonance, neglecting the other states would lead to erroneous predictions and different results for each representation. The nonresonant process being only one step in a higher-order process, it is not possible to state with certainty through which intermediate level the atom passes, and only the sum on all the possible "paths" has physical meaning.

(*) An analogous problem is analyzed by G. Grynberg and E. Giacobino, *J. Phys.*, **B12**, L93 (1979), and by Y. Aharanov and C. K. Au, *Phys. Rev.* A **20**, 1553 (1979).

Another important, often studied example is the line shape of the $2s_{1/2}$–$2p_{1/2}$ hydrogen transition (*). We review the experimental situation briefly. Hydrogen atoms initially in the metastable $2s_{1/2}$ state are subjected to a microwave with frequency ω near that of the $2s_{1/2}$–$2p_{1/2}$ transition, ω_0. The microwave resonance is detected by the Lyman-α photons spontaneously emitted in the $2p_{1/2}$–$1s_{1/2}$ transition. As stated, the experiment seems to involve only the states $2s_{1/2}$ and $2p_{1/2}$. A calculation limited to these two states does not, however, give the same results in the two representations $\mathbf{A} \cdot \mathbf{p}$ and $\mathbf{E} \cdot \mathbf{r}$ for the line shape, that is, for the variations with $\omega - \omega_0$ of the $2s_{1/2}$–$2p_{1/2}$ transition probability. This anomaly is due to the fact that the $2p_{1/2}$ state is not a true final state, but an intermediate state among others in a two-photon resonant process, which consists of an induced emission of one photon ω and a spontaneous emission of one photon ω' with frequency near that of Lyman-α, the resonance condition being $\hbar\omega + \hbar\omega' = E(2s_{1/2}) - E(1s_{1/2})$, as shown in Figure 4. Even if the $2p_{1/2}$ level appears to have a major importance in the two-photon emission process, it is wrong to neglect the contribution of the other intermediate *np* levels (**).

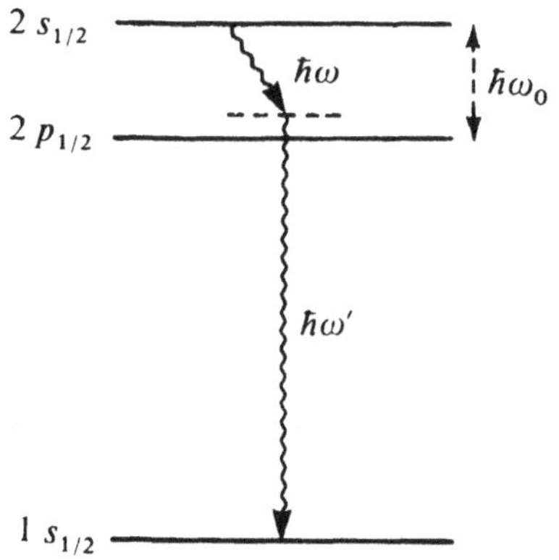

Figure 4. The nonresonant transition between the $2s_{1/2}$ and $2p_{1/2}$ states, the Lamb transition. The stimulated emission of the microwave photon $\hbar\omega$ is accompanied by the spontaneous emission of the ultraviolet photon $\hbar\omega'$.

(*) This problem was raised by W. E. Lamb, *Phys. Rev.*, **85**, 259 (1952).
(**) See Z. Fried, *Phys. Rev.* A, **8**, 2835 (1973).

COMPLEMENT C_{IV}

INTERACTION OF TWO LOCALIZED SYSTEMS OF CHARGES FROM THE POWER–ZIENAU–WOOLLEY POINT OF VIEW

The Power–Zienau–Woolley transformation has been presented in Chapter IV for the case where the positions of all the charges are measured from a single point of reference, taken as the origin. The generalization to the case of two subsystems $\mathscr{S}_A$ and $\mathscr{S}_B$, spatially separated and localized about two different points $\mathbf{R}_A$ and $\mathbf{R}_B$ respectively, presents no difficulties. We give in this complement the Power–Zienau–Woolley Hamiltonian for such a system with the assumption that $\mathscr{S}_A$ and $\mathscr{S}_B$ are both globally neutral. The essential motivation of this complement is to show that in this new Hamiltonian, all the terms describing instantaneous interactions of the particles of $\mathscr{S}_A$ with the particles of $\mathscr{S}_B$ have vanished. The two systems interact solely via retarded fields: the displacement and the magnetic field.

1. Notation

The charges of the system $\mathscr{S}_A$, denoted by index α, are grouped about point $\mathbf{R}_A$; their positions with respect to this point are denoted $\mathbf{s}_\alpha$:

$$\mathbf{s}_\alpha = \mathbf{r}_\alpha - \mathbf{R}_A . \tag{1}$$

Similarly, the charges of the system $\mathscr{S}_B$, denoted by index β, are confined about $\mathbf{R}_B$, and their positions with respect to this point are denoted $\mathbf{s}_\beta$:

$$\mathbf{s}_\beta = \mathbf{r}_\beta - \mathbf{R}_B . \tag{2}$$

The charge distribution of $\mathscr{S}_A$ can be described by its total charge in $\mathbf{R}_A$, which we have assumed zero, and by its polarization density

$$\mathbf{P}_A(\mathbf{r}) = \sum_\alpha \int_0^1 \mathrm{d}u\, q_\alpha\, \mathbf{s}_\alpha\, \delta(\mathbf{r} - \mathbf{R}_A - u\,\mathbf{s}_\alpha) \tag{3}$$

which generalizes (C.2). The longitudinal electric field created by $\mathscr{S}_A$ is given by (C.11):

$$\mathbf{E}_{A\parallel}(\mathbf{r}) = -\frac{1}{\varepsilon_0}\mathbf{P}_{A\parallel}(\mathbf{r}) . \tag{4}$$

Analogous expressions can be written for the polarization $\mathbf{P}_B(\mathbf{r})$ of the system $\mathscr{S}_B$ and for the longitudinal electric field created by $\mathscr{S}_B$.

The total polarization density $\mathbf{P}(\mathbf{r})$ is the sum of the polarizations relative to $\mathscr{S}_A$ and $\mathscr{S}_B$:

$$\mathbf{P}(\mathbf{r}) = \mathbf{P}_A(\mathbf{r}) + \mathbf{P}_B(\mathbf{r}) . \tag{5}$$

Likewise, the total magnetization density is the sum of the contributions of $\mathscr{S}_A$ and $\mathscr{S}_B$:

$$\mathbf{M}(\mathbf{r}) = \mathbf{M}_A(\mathbf{r}) + \mathbf{M}_B(\mathbf{r}) . \tag{6}$$

2. The Hamiltonian

In Part C of this chapter, we have derived the Power–Zienau–Woolley Lagrangian (C.26) by adding $\mathrm{d}F/\mathrm{d}t$ to the Lagrangian in the Coulomb gauge, F being given by (C.20):

$$F = - \int \mathrm{d}^3 r \, \mathbf{P}(\mathbf{r}) \cdot \mathbf{A}(\mathbf{r}) . \tag{7}$$

In this case, with two localized systems of charges, it is necessary to use the same transformation, but $\mathbf{P}(\mathbf{r})$ is now the total polarization given in (5). All the results gotten there remain valid under the condition that we replace $\mathbf{P}(\mathbf{r})$ and $\mathbf{M}(\mathbf{r})$ by their expressions (5) and (6). We get, in particular, for the Hamiltonian the following generalization of (C.37):

$$\begin{aligned} H_{L'} = & \sum_\alpha \frac{1}{2 m_\alpha} \left[\mathbf{p}_{\alpha L'} - \int_0^1 u \, \mathrm{d}u \, q_\alpha \, \mathbf{B}(\mathbf{R}_A + u \mathbf{s}_\alpha) \times \mathbf{s}_\alpha \right]^2 + \\ & + \sum_\beta \frac{1}{2 m_\beta} \left[\mathbf{p}_{\beta L'} - \int_0^1 u \, \mathrm{d}u \, q_\beta \, \mathbf{B}(\mathbf{R}_B + u \mathbf{s}_\beta) \times \mathbf{s}_\beta \right]^2 + \\ & + \int \mathrm{d}^3 k \left[\frac{| \pi_{L'} + \mathscr{P}_{\perp A} + \mathscr{P}_{\perp B} |^2}{\varepsilon_0} + \varepsilon_0 \, c^2 k^2 \, | \mathscr{A} |^2 \right] + \sum_\alpha \varepsilon_{\text{Coul}}^\alpha + \sum_\beta \varepsilon_{\text{Coul}}^\beta + \\ & \sum_{\alpha < \alpha'} \frac{q_\alpha q'_\alpha}{4 \pi \varepsilon_0 | \mathbf{r}_\alpha - \mathbf{r}_{\alpha'} |} + \sum_{\beta < \beta'} \frac{q_\beta q_{\beta'}}{4 \pi \varepsilon_0 | \mathbf{r}_\beta - \mathbf{r}_{\beta'} |} + \sum_{\alpha, \beta} \frac{q_\alpha q_\beta}{4 \pi \varepsilon_0 | \mathbf{r}_\alpha - \mathbf{r}_\beta |} . \end{aligned} \tag{8}$$

The different terms of H' can be regrouped as follows:

$$H_{L'} = H_{RL'} + H_{IL'}^A + H_{IL'}^B + H_{PL'}^A + H_{PL'}^B + V^{AB} \tag{9}$$

$H_{RL'}$ is identical to the Hamiltonian (C.39). $H_{IL'}^A$ and $H_{IL'}^B$ are the copies of the interaction Hamiltonian (C.40) for the systems $\mathscr{S}_A$ and $\mathscr{S}_B$ respectively. Likewise, $H_{PL'}^A$ and $H_{PL'}^B$ are the respective Hamiltonians of the particles in $\mathscr{S}_A$ and $\mathscr{S}_B$, given by expressions analogous to (C.38). The only new term is

$$V^{AB} = \frac{1}{2 \varepsilon_0} \int \mathrm{d}^3 k [\mathscr{P}_{\perp A} \cdot \mathscr{P}_{\perp B}^* + \mathscr{P}_{\perp A}^* \cdot \mathscr{P}_{\perp B}] + \sum_{\alpha, \beta} \frac{q_\alpha q_\beta}{4 \pi \varepsilon_0 | \mathbf{r}_\alpha - \mathbf{r}_\beta |} . \tag{10}$$

This is the sum of a self-energy term for the transverse polarization and the Coulomb interaction energy V_{Coul}^{AB} between the charges of $\mathscr{S}_A$ and those of $\mathscr{S}_B$.

We are now going to show that V^{AB} is identically zero. To this end, note that the final term of (10) can be written

$$\begin{aligned} V_{\text{Coul}}^{AB} &= V_{\text{Coul}}^{\text{Tot}} - V_{\text{Coul}}^{AA} - V_{\text{Coul}}^{BB} \\ &= \frac{\varepsilon_0}{2} \int \mathrm{d}^3 r [(\mathbf{E}_{\parallel A} + \mathbf{E}_{\parallel B})^2 - \mathbf{E}_{\parallel A}^2 - \mathbf{E}_{\parallel B}^2] \\ &= \varepsilon_0 \int \mathrm{d}^3 r \, \mathbf{E}_{\parallel A} \cdot \mathbf{E}_{\parallel B} \, . \end{aligned} \tag{11}$$

Using (4) and the analogous expression for $\mathbf{E}_{\parallel B}$, we get finally

$$V_{\text{Coul}}^{AB} = \frac{1}{\varepsilon_0} \int \mathrm{d}^3 r \, \mathbf{P}_{\parallel A} \cdot \mathbf{P}_{\parallel B} \tag{12}$$

with the result that (10) can be written in real space as

$$\begin{aligned} V^{AB} &= \frac{1}{\varepsilon_0} \int \mathrm{d}^3 r [\mathbf{P}_{\perp A}(\mathbf{r}) \cdot \mathbf{P}_{\perp B}(\mathbf{r}) + \mathbf{P}_{\parallel A}(\mathbf{r}) \cdot \mathbf{P}_{\parallel B}(\mathbf{r})] \\ &= \frac{1}{\varepsilon_0} \int \mathrm{d}^3 r \, \mathbf{P}_A(\mathbf{r}) \cdot \mathbf{P}_B(\mathbf{r}) \, . \end{aligned} \tag{13}$$

Since $\mathbf{P}_A(\mathbf{r})$ is rigorously zero outside $\mathscr{S}_A$, and $\mathbf{P}_B(\mathbf{r})$ outside $\mathscr{S}_B$, we see that V^{AB} reduces to a contact term which is zero when the distance $|\mathbf{R}_A - \mathbf{R}_B|$ is greater than the sum of the radii of the two systems of charges. Thus, in the new approach, there is no longer an instantaneous Coulomb interaction between two separate, globally neutral systems of charges. The interaction between these systems is only via the fields $\mathbf{B}(\mathbf{r})$ and $\mathbf{\Pi}_{L'}(\mathbf{r})$, which propagate between the systems with finite velocity c. (Recall that outside a system of charges, $\mathbf{\Pi}_{L'}$ corresponds to the total electric field to within a factor $-\varepsilon_0$).

Remark

The results above can be generalized to the case where the systems of charges $\mathscr{S}_A$ and $\mathscr{S}_B$ are not globally neutral. It is then necessary to add to V^{AB} the energy of the polarization density of $\mathscr{S}_A$ in the static field produced by the reference distribution ρ_{0B} of $\mathscr{S}_B$ and the symmetric term, as well as the Coulomb interaction energy between the reference distributions ρ_{0A} and ρ_{0B}.

COMPLEMENT D_{IV}

THE POWER–ZIENAU–WOOLLEY TRANSFORMATION AND THE POINCARÉ GAUGE

We are going to show in this complement that when the polarization of the system of charges is defined with respect to a single reference point, the Power–Zienau–Woolley transformation reduces to a gauge change from the Coulomb gauge ($\boldsymbol{\nabla} \cdot \mathbf{A} = 0$) to the *Poincaré gauge*, in which $\mathbf{A}'(\mathbf{r})$ is everywhere orthogonal to $\mathbf{r}$. Furthermore, in this new gauge, $\mathbf{A}'$ and U' can be simply expressed as functions of the $\mathbf{B}$ and $\mathbf{E}$ fields. The expressions for $\mathbf{A}'$ and U' generalize the expressions $\mathbf{A} = \mathbf{B}_0 \times \mathbf{r}/2$ and $U = -\mathbf{r} \cdot \mathbf{E}_0$, valid for uniform static fields $\mathbf{B}_0$ and $\mathbf{E}_0$.

1. The Power–Zienau–Woolley Transformation Considered as a Gauge Change (*)

In a gauge transformation defined by a function $\chi(r, t)$, the potentials become

$$\mathbf{A}'(\mathbf{r}, t) = \mathbf{A}(\mathbf{r}, t) + \boldsymbol{\nabla}\chi(\mathbf{r}, t) \tag{1.a}$$

$$U'(\mathbf{r}, t) = U(\mathbf{r}, t) - \frac{\partial}{\partial t}\chi(\mathbf{r}, t). \tag{1.b}$$

The Lagrangian is also transformed and becomes [see (B.11)]

$$L' = L + \frac{\mathrm{d}}{\mathrm{d}t}\left[\sum_\alpha q_\alpha \chi(\mathbf{r}_\alpha, t)\right]. \tag{2}$$

We know, in addition, that the Power–Zienau–Woolley Lagrangian can be derived from the standard Lagrangian L in the Coulomb gauge by the transformation (**) [see (C.20) and (C.22)]

$$L' = L - \frac{\mathrm{d}}{\mathrm{d}t}\int \mathrm{d}^3r\, \mathbf{P}(\mathbf{r}) \cdot \mathbf{A}_\perp(\mathbf{r}). \tag{3}$$

By using (C.2) for $\mathbf{P}(\mathbf{r})$ we get

$$L' = L - \frac{\mathrm{d}}{\mathrm{d}t}\sum_\alpha q_\alpha \int_0^1 \mathrm{d}u\, \mathbf{r}_\alpha \cdot \mathbf{A}_\perp(\mathbf{r}_\alpha u). \tag{4}$$

(*) R. G. Woolley, *J. Phys.*, **B7**, 488 (1974); M. Babiker and R. Loudon, *Proc. Roy. Soc.*, **A385**, 439 (1983).

(**) We reintroduce the notation $\mathbf{A}_\perp$, since in the Poincaré gauge $\mathbf{A}'$ is no longer transverse.

It is clear that (2) and (4) can be made identical if one takes for $\chi(\mathbf{r}, t)$ the following function:

$$\chi(\mathbf{r}, t) = -\int_0^1 du\, \mathbf{r} \cdot \mathbf{A}_\perp(\mathbf{r}u) . \tag{5}$$

[The dependence on t of the function $\chi(\mathbf{r}, t)$ is associated with the temporal variations of the dynamical variables $\mathbf{A}_\perp(\mathbf{r})$.] Note that we have introduced the gauge function (5) by using the definition of the polarization (C.2) valid for a system of charges referred to a single reference point (the origin). In the case where the system of charges involves several subsystems localized about several different points $\mathbf{R}_A, \mathbf{R}_B, \ldots$ (as for example in Complement C_{IV}), the proof above cannot be generalized. The expression which generalizes (C.2) is then

$$\begin{aligned}\mathbf{P}(\mathbf{r}) = \sum_\alpha \int_0^1 du\, q_\alpha(\mathbf{r}_\alpha - \mathbf{R}_A)\, \delta[\mathbf{r} - \mathbf{R}_A - u(\mathbf{r}_\alpha - \mathbf{R}_A)] + \\ + \sum_\beta \int_0^1 du\, q_\beta(\mathbf{r}_\beta - \mathbf{R}_B)\, \delta[\mathbf{r} - \mathbf{R}_B - u(\mathbf{r}_\beta - \mathbf{R}_B)] + \cdots\end{aligned} \tag{6}$$

where the index α designates the particles localized about $\mathbf{R}_A$ and the index β those localized about $\mathbf{R}_B$. When one substitutes (6) in (3), one discovers, because of the multiplicity of reference points, that the matching of (2) and (3) is no longer possible. The Power–Zienau–Woolley transformation reduces to a gauge change only if all the charges are localized about a single reference point. In this complement we treat only that situation.

2. Properties of the Vector Potential in the New Gauge

In the new gauge, the vector potential is no longer transverse, since we find using (1.a) and (5) the following expression for $\mathbf{A}'_\parallel$:

$$\mathbf{A}'_\parallel(\mathbf{r}) = -\nabla \int_0^1 du\, \mathbf{r} \cdot \mathbf{A}_\perp(\mathbf{r}u) . \tag{7.a}$$

The transverse part $\mathbf{A}_\perp$ of $\mathbf{A}'$ remains unchanged:

$$\mathbf{A}'_\perp(\mathbf{r}) = \mathbf{A}_\perp(\mathbf{r}) . \tag{7.b}$$

We take now the scalar product of $\mathbf{r}$ and $\mathbf{A}'$,

$$\begin{aligned}\mathbf{r} \cdot \mathbf{A}'(\mathbf{r}) &= \mathbf{r} \cdot \mathbf{A}_\perp(\mathbf{r}) + \mathbf{r} \cdot \mathbf{A}'_\parallel(\mathbf{r}) \\ &= \mathbf{r} \cdot \mathbf{A}_\perp(\mathbf{r}) - \int_0^1 du(\mathbf{r} \cdot \nabla)\,(\mathbf{r} \cdot \mathbf{A}_\perp(\mathbf{r}u)) .\end{aligned} \tag{8}$$

To transform (8) we use the identity

$$u \frac{\partial}{\partial u} f(\mathbf{r}u) = (\mathbf{r} \cdot \nabla) f(\mathbf{r}u) \tag{9}$$

which allows us to rewrite (8) in the form

$$\begin{aligned} \mathbf{r} \cdot \mathbf{A}'(\mathbf{r}) &= \mathbf{r} \cdot \mathbf{A}_\perp(\mathbf{r}) - \int_0^1 \mathrm{d}u \frac{1}{u} (\mathbf{r} \cdot \nabla) (\mathbf{r}u \cdot \mathbf{A}_\perp(\mathbf{r}u)) \\ &= \mathbf{r} \cdot \mathbf{A}_\perp(\mathbf{r}) - \int_0^1 \mathrm{d}u \frac{\partial}{\partial u} [\mathbf{r}u \cdot \mathbf{A}_\perp(\mathbf{r}u)] . \end{aligned} \tag{10}$$

The integral in Equation (10) is found immediately; its value $[-\mathbf{r} \cdot \mathbf{A}_\perp(\mathbf{r})]$ exactly cancels the first term of (10), so that

$$\mathbf{r} \cdot \mathbf{A}'(\mathbf{r}) = 0 . \tag{11}$$

It thus appears that the new vector potential is everywhere perpendicular to the vector **r**. The new gauge, called the Poincaré gauge (*), appears in some ways as symmetric with the Coulomb gauge: in the former, $\mathscr{A}(\mathbf{k})$ is normal to **k** at every point **k**, whereas in the latter, $\mathbf{A}'(\mathbf{r})$ is normal to **r** at every **r**.

3. The Potentials in the Poincaré Gauge

A major interest in the Poincaré gauge is that the potentials are easily expressed as a function of the magnetic field **B** and the electric field **E**. We are going to show that

$$\mathbf{A}'(\mathbf{r}) = - \int_0^1 u \, \mathrm{d}u \, \mathbf{r} \times \mathbf{B}(\mathbf{r}u) \tag{12.a}$$

$$U'(\mathbf{r}) = - \int_0^1 \mathrm{d}u \, \mathbf{r} \cdot \mathbf{E}(\mathbf{r}u) . \tag{12.b}$$

It is already interesting to note that these potentials generalize the well-known potentials for a uniform magnetic field $\mathbf{A}_0 = -\mathbf{r} \times \mathbf{B}_0/2$ and a uniform electric field $U_0 = -\mathbf{r} \cdot \mathbf{E}_0$. It is also clear that (12.a) satisfies the orthogonality relation (11) between $\mathbf{A}'$ and **r**.

We now detail the process for getting (12) from (1) and (5). First of all, consider the scalar potential. In the Coulomb gauge,

$$\mathbf{E}_\parallel(\mathbf{r}) = - \nabla U(\mathbf{r}) \tag{13}$$

(*) B. S. Skagerstam, *Am. J. Phys.*, **51**, 1148 (1983).

$U(\mathbf{r})$ can thus be gotten to within a global constant, as the line integral of $-\mathbf{E}_{\parallel}$ along the line joining the origin to $\mathbf{r}$:

$$U(\mathbf{r}) = -\int_0^1 \mathrm{d}(\mathbf{r}u) \cdot \mathbf{E}_{\parallel}(\mathbf{r}u). \tag{14}$$

Using (1.b), (5), (14), and the fact that $\mathbf{E}_{\perp} = -\dot{\mathbf{A}}_{\perp}$, we get

$$\begin{aligned} U'(\mathbf{r}) &= -\int_0^1 du\, \mathbf{r} \cdot \mathbf{E}_{\parallel}(\mathbf{r}u) - \int_0^1 du\, \mathbf{r} \cdot \mathbf{E}_{\perp}(\mathbf{r}u) \\ &= -\int_0^1 du\, \mathbf{r} \cdot \mathbf{E}(\mathbf{r}u). \end{aligned} \tag{15}$$

We now examine the vector potential and find first the gradient of χ which appears in (1.a) for A':

$$\begin{aligned} \nabla\chi(\mathbf{r}) &= -\nabla \int_0^1 du\, \mathbf{r} \cdot \mathbf{A}_{\perp}(\mathbf{r}u) \\ &= -\int_0^1 du\, \mathbf{A}_{\perp}(\mathbf{r}u) - \sum_i \int_0^1 du\, r_i\, \nabla A_{\perp i}(\mathbf{r}u). \end{aligned} \tag{16}$$

Integration of the first term of (16) by parts gives

$$\begin{aligned} -\int_0^1 du\, \mathbf{A}_{\perp}(\mathbf{r}u) &= -u\mathbf{A}_{\perp}(\mathbf{r}u)\Big|_0^1 + \int_0^1 du\, u \frac{\partial}{\partial u} \mathbf{A}_{\perp}(\mathbf{r}u) \\ &= -\mathbf{A}_{\perp}(\mathbf{r}) + \int_0^1 du (\mathbf{r} \cdot \nabla)\, \mathbf{A}_{\perp}(\mathbf{r}u) \end{aligned} \tag{17}$$

where the second term has been gotten using (9). We now use (1.a), (16), and (17) to get $\mathbf{A}'$. Since the first term of (17) cancels the first term of (1.a), we obtain

$$\begin{aligned} \mathbf{A}'(\mathbf{r}) &= \int_0^1 du \left[(\mathbf{r} \cdot \nabla)\, \mathbf{A}_{\perp}(\mathbf{r}u) - \sum_i r_i\, \nabla A_{\perp i}(\mathbf{r}u) \right] \\ &= -\int_0^1 du\, \mathbf{r} \times (\nabla \times \mathbf{A}_{\perp}(\mathbf{r}u)). \end{aligned} \tag{18}$$

It suffices then to note that

$$\nabla_{\mathbf{r}} \times \mathbf{A}_{\perp}(\mathbf{r}u) = u\, \nabla_{\mathbf{r}u} \times \mathbf{A}_{\perp}(\mathbf{r}u) = u\mathbf{B}(\mathbf{r}u) \tag{19}$$

to get finally

$$\mathbf{A}'(\mathbf{r}) = -\int_0^1 u\,\mathrm{d}u\;\mathbf{r} \times \mathbf{B}(\mathbf{r}u) \tag{20}$$

which is the expression sought (12.a).

Remark

It is possible to show directly from Equations (12) that

$$\nabla \times \mathbf{A}'(\mathbf{r}) = \mathbf{B}(\mathbf{r}) \tag{21.a}$$

$$-\nabla U'(\mathbf{r}) - \dot{\mathbf{A}}'(\mathbf{r}) = \mathbf{E}(\mathbf{r})\,. \tag{21.b}$$

The vector and scalar fields $\mathbf{A}'$ and U' then define a gauge. In addition, it is clear from (12.a) that $\mathbf{A}'(\mathbf{r})$ is everywhere orthogonal to $\mathbf{r}$. We will show that this latter condition uniquely defines the Poincaré gauge to within a constant. Assume that another gauge $\{\mathbf{A}'', U''\}$, related to the first $\{\mathbf{A}', U'\}$ by the function χ', has the same property. The relation $\mathbf{A}'' = \mathbf{A}' + \nabla\chi'$, along with $\mathbf{A}'' \cdot \mathbf{r} = 0$ and $\mathbf{A}' \cdot \mathbf{r} = 0$, shows that

$$\mathbf{r} \cdot \nabla\chi' = 0 \tag{22}$$

which gives in spherical coordinates

$$\frac{\partial \chi'}{\partial r} = 0\,. \tag{23}$$

The function χ' at any point in space is then equal to its value at the origin and thereby reduces to a constant.

COMPLEMENT E_{IV}

EXERCISES

Exercise 1. An example of the effect produced by sudden variations of the vector potential.

Exercise 2. Two-photon excitation of the hydrogen atom. Approximate results obtained with the Hamiltonians $\mathbf{A}\cdot\mathbf{p}$ and $\mathbf{E}\cdot\mathbf{r}$.

Exercise 3. The electric dipole Hamiltonian for an ion coupled to an external field.

Exercise 4. Scattering of a particle by a potential in the presence of laser radiation.

Exercise 5. The equivalence between the interaction Hamiltonians $\mathbf{A}\cdot\mathbf{p}$ and $\mathbf{Z}\cdot(\nabla V)$ for the calculation of transition amplitudes.

Exercise 6. Linear response and susceptibility. Application to the calculation of the radiation from a dipole.

Exercise 7. Nonresonant scattering. Direct verification of the equality of the transition amplitudes calculated from the Hamiltonians $\mathbf{A}\cdot\mathbf{p}$ and $\mathbf{E}\cdot\mathbf{r}$.

1. An example of the effect produced by sudden variations of the vector potential

Consider an atom located at $\mathbf{R}=\mathbf{0}$ and whose states $|a\rangle, |b\rangle, \ldots$ have energies $E_a, E_b, \ldots$.

a) This atom interacts with a field whose vector potential at $\mathbf{R}=\mathbf{0}$ is

$$\mathbf{A}_e(\mathbf{0},t) = \tilde{A}(t)\,\mathbf{e}_z \cos\omega t \tag{1}$$

with $\tilde{A}(t)=0$ for $t<-T/2$ and $t>T/2$ and $\tilde{A}(t)=A_0$ for $-T/2<t<T/2$. Find the transition amplitude from $|a\rangle$ to $|b\rangle$, $U_{ba}(T'/2,-T'/2)$, to first order in the field, using the Hamiltonian $-(q/m)\mathbf{p}\cdot\mathbf{A}_e(\mathbf{0},t)$. One takes $T'>T$.

b) Assume now that the atom interacts with the electric field

$$\mathbf{E}'_e(\mathbf{0},t) = \omega\tilde{A}(t)\,\mathbf{e}_z \sin\omega t\,. \tag{2}$$

Find the transition amplitude $U'_{ba}(T'/2, -T'/2)$, to first order in the field using the Hamiltonian $-q\mathbf{r} \cdot \mathbf{E}'_e(0, t)$.

c) Show that at time $t = -T'/2$, the ket $|a\rangle$ represents the same physical state from both points of view. Show that this same property is also true for the ket $|b\rangle$ at $T'/2$. Compare the transition amplitudes calculated in *a*) and *b*). Why are these different when $\omega \neq \omega_{ba}$?

Solution

a) Applying first-order, time-dependent perturbation theory, we find in the interaction representation

$$U_{ba}\left(\frac{T'}{2}, -\frac{T'}{2}\right) = \frac{1}{i\hbar}\left(-\frac{q}{m}\right)\langle b \mid p_z \mid a\rangle A_0 \int_{-T/2}^{T/2} dt\, e^{i\omega_{ba}t} \cos \omega t$$

$$= \frac{1}{i\hbar}\left(-\frac{q}{m}\right)\langle b \mid p_z \mid a\rangle A_0\left(\frac{\sin(\omega_{ba} - \omega)T/2}{(\omega_{ba} - \omega)} + \frac{\sin(\omega_{ba} + \omega)T/2}{(\omega_{ba} + \omega)}\right). \tag{3}$$

b) Proceeding in the same fashion with the Hamiltonian $-q\mathbf{r} \cdot \mathbf{E}'_e(0, t)$, we find

$$U'_{ba}\left(\frac{T'}{2}, -\frac{T'}{2}\right) = \frac{1}{i\hbar}(-q)\langle b \mid z \mid a\rangle \omega A_0 \int_{-T/2}^{T/2} dt\, e^{i\omega_{ba}t} \sin \omega t$$

$$= -\frac{q}{\hbar}\langle b \mid z \mid a\rangle \omega A_0\left(\frac{\sin(\omega_{ba} - \omega)T/2}{(\omega_{ba} - \omega)} - \frac{\sin(\omega_{ba} + \omega)T/2}{(\omega_{ba} + \omega)}\right). \tag{4}$$

c) At times $-T'/2$ and $T'/2$, the vector potential $\mathbf{A}_e(\mathbf{0}, t)$ is zero. The operator $\mathbf{p}/m$ then represents the velocity variable, and the particle Hamiltonian is associated with the same physical variable, the sum of the kinetic and potential energies, in both $\mathbf{A} \cdot \mathbf{p}$ and $\mathbf{E} \cdot \mathbf{r}$ descriptions. It then follows that $|a\rangle$ and $|b\rangle$ represent the same physical states in both approaches at $-T'/2$ and $T'/2$. If $\mathbf{A}_e(\mathbf{0}, t)$ and $\mathbf{E}'_e(\mathbf{0}, t)$ correspond to the same field, the transition amplitudes $U_{ba}(T'/2, -T'/2)$ and $U'_{ba}(T'/2, -T'/2)$ should be identical from (A.47), whatever ω is.

Now, if we transform (3) using (B.43) relating the matrix elements of p_z to those of z, we find

$$U_{ba}\left(\frac{T'}{2}, -\frac{T'}{2}\right) = -\frac{q}{\hbar}\langle b \mid z \mid a\rangle \omega_{ba} A_0\left(\frac{\sin(\omega_{ba} - \omega)T/2}{(\omega_{ba} - \omega)} + \frac{\sin(\omega_{ba} + \omega)T/2}{(\omega_{ba} + \omega)}\right) \tag{5}$$

which obviously does not coincide with (4) when $\omega_{ba} \neq \omega$. This difference is a result of the fact that $\mathbf{A}_e(\mathbf{0}, t)$ and $\mathbf{E}'_e(\mathbf{0}, t)$ do not represent the same field. More precisely

$$\mathbf{E}'_e(\mathbf{0}, t) \neq -\dot{\mathbf{A}}_e(\mathbf{0}, t). \tag{6}$$

The difference between $-\dot{\mathbf{A}}_e$ and $\mathbf{E}'_e$ arises from the discontinuities of $\mathbf{A}_e(\mathbf{0}, t)$ at $-T/2$ and $T/2$, which introduce into $\dot{\mathbf{A}}_e$ the functions $\delta(t + T/2)$ and $\delta(t - T/2)$. The field $\mathbf{E}_e(\mathbf{0}, t)$ associated with $\mathbf{A}_e(\mathbf{0}, t)$ is written

$$\mathbf{E}_e(\mathbf{0}, t) = \mathbf{E}'_e(\mathbf{0}, t) + A_0\, \mathbf{e}_z\,(\delta(t - T/2) - \delta(t + T/2))\cos \omega T/2. \tag{7}$$

The second term of (7) gives a contribution $U''_{ba}(T'/2, -T'/2)$ to the transition amplitude, equal in first order to

$$U''_{ba}\left(\frac{T'}{2}, -\frac{T'}{2}\right) = \frac{1}{i\hbar}(-q)\langle b \mid z \mid a\rangle A_0\left(2\,i \sin \omega_{ba}\frac{T}{2}\right)\cos\frac{\omega T}{2}$$

$$= -\frac{q}{\hbar}\langle b \mid z \mid a\rangle A_0[\sin(\omega_{ba} - \omega)T/2 + \sin(\omega_{ba} + \omega)T/2] \tag{8}$$

which one can then write

$$U''_{ba}\left(\frac{T'}{2}, -\frac{T'}{2}\right) = -\frac{q}{\hbar}\langle b \mid z \mid a \rangle A_0\left[\frac{\omega_{ba} - \omega}{\omega_{ba} - \omega}\sin(\omega_{ba} - \omega)T/2 + \right.$$

$$\left. + \frac{\omega_{ba} + \omega}{\omega_{ba} + \omega}\sin(\omega_{ba} + \omega)T/2\right]. \quad (9)$$

By adding (4) and (9), one finds a result identical to (5). This shows that the transition amplitude is the same in the two approaches at resonance and off resonance provided that $\mathbf{A}_e$ and $\mathbf{E}_e$ correspond to the same physical field.

2. Two-photon excitation of the hydrogen atom. Approximate results obtained with the Hamiltonians A · p and E · r

The purpose of this exercise is to compare various approximations used in the calculation of the two-photon excitation probability of the metastable $2s$ level of the hydrogen atom from the ground state $1s$.

One recalls that when an atom interacts with an incident wave of frequency ω, it can be excited from a state $|a\rangle$ to a state $|b\rangle$ when $E_b - E_a = 2\hbar\omega$. The transition amplitude is proportional to an effective matrix element whose form in the case of a wave polarized along the z-axis is

$$Q_{ba} = -q^2\sum_r \omega_{br}\,\omega_{ra}\frac{\langle b \mid z \mid r\rangle\langle r \mid z \mid a\rangle}{\hbar(\omega - \omega_{ra})} \quad (1.a)$$

in the $\mathbf{A}\cdot\mathbf{p}$ picture, and

$$Q'_{ba} = -q^2\omega^2\sum_r \frac{\langle b \mid z \mid r\rangle\langle r \mid z \mid a\rangle}{\hbar(\omega - \omega_{ra})} \quad (1.b)$$

in the $\mathbf{E}\cdot\mathbf{r}$ picture (see §B_{IV}.2). The values of Q_{ba} and Q'_{ba} are equal when one sums over all the intermediate levels $|r\rangle$, but can be different if the sum is limited to a finite number of intermediate levels. It is then important to know which result is closest to the exact result.

a) Show that the only discrete states which contribute to the sums (1.a) and (1.b) in the case of the $1s$–$2s$ transition are the np, $m = 0$ states (one ignores the electron spin here).

b) Evaluate the contribution of the discrete state np, $m = 0$ to the sums (1.a) and (1.b). One writes this contribution in the form

$$-q^2\frac{\omega^2}{3E_I}a_0^2 J_n \quad (2)$$

for the sum (2.a), and in analogous fashion where J_n is replaced by J'_n for the sum (2.b) (a_0 is the Bohr radius, and E_I the ionization energy of the hydrogen atom). Express the dimensionless quantities J'_n and J_n as functions of n and of the radial integrals

$$\int \mathrm{d}r\, r^3\, R_{n'l'}(r)\, R_{nl}(r) = a_0\, R_{nl}^{n'l'} \tag{3}$$

involving the hydrogen radial wave functions (see the end of the statement of this exercise).

c) Compare the approximate values obtained for Q_{ba} and Q'_{ba}, taking into account only one intermediate state, $2p$, or the two intermediate states $2p$ and $3p$. Compare these approximate values with the exact value taking into account all of the intermediate states in the discrete and continuous spectrum (*),

$$Q_{ba} = Q'_{ba} = -\, q^2 \frac{\omega^2}{3\,E_I}\, a_0^2 \times (11{,}8)\,. \tag{4}$$

d) A frequently used approximation to find the sums appearing in (1.a) and (1.b) involves replacing all the energy denominators $\hbar(\omega - \omega_{ra})$ by a single one, which will be denoted by xE_I, where x is a dimensionless quantity. Show that Q'_{ba} and Q_{ba} are then respectively proportional to $\langle 2s|r^2|1s\rangle$ and $\langle 2s|p^2|1s\rangle$.

Take $x = -0.5$. Is this a reasonable choice? Now find Q_{ba} and Q'_{ba} and compare the results with the exact result.

Data for the hydrogen atom:

— The energy of a state nlm: $E_n = -E_I/n^2$.

— A few radial wave functions:

$$R_{10}(r) = 2(a_0)^{-3/2}\, \mathrm{e}^{-r/a_0} \tag{5.a}$$

$$R_{20}(r) = 2(2\,a_0)^{-3/2}\,(1 - (r/2\,a_0))\, \mathrm{e}^{-r/2a_0}\,. \tag{5.b}$$

— A few radial integrals (**):

$$R_{10}^{n1} = 2^4\, n^{7/2} \frac{(n-1)^{n-(5/2)}}{(n+1)^{n+(5/2)}} \tag{6.a}$$

$$R_{20}^{n1} = 2^8\, n^{7/2} \sqrt{2(n^2-1)}\, \frac{(n-2)^{n-3}}{(n+2)^{n+3}} \quad \text{pour} \quad n \neq 2 \tag{6.b}$$

$$R_{20}^{21} = -\,3\sqrt{3}\,. \tag{6.c}$$

(*) F. Bassani, J. J. Forney, and A. Quattropani, *Phys. Rev. Lett.*, **39**, 1070 (1977).

(**) H. A. Bethe and E. E. Salpeter, *Quantum Mechanics of One- and Two-Electron Atoms*, (§63), Springer-Verlag, New York, 1957.

Solution

a) The vector operator **r** can only couple an *s*-level ($l = 0$) to a *p*-level ($l = 1$). The levels r of the sums (1.a) and (1.b) are then necessarily the *p*-levels. The calculation of the matrix element $\langle r|z|a\rangle$ is broken up into a product of a radial integral and an angular integral which is nonzero only for $m = 0$. More precisely, in the case of a discrete intermediate level,

$$\langle r \mid z \mid a \rangle = \int r^2 \,\mathrm{d}r\, \mathrm{d}\Omega\, R_{n1}(r)\, Y_1^{m*}(\theta, \varphi)\, r\cos\theta\, R_{10}(r)\, Y_0^0(\theta, \varphi)$$

$$= \int \mathrm{d}r\, r^3\, R_{n1}(r)\, R_{10}(r) \int \mathrm{d}\Omega\, Y_1^{m*}(\theta, \varphi) \cos\theta\, Y_0^0(\theta, \varphi) = \frac{R_{10}^{n1}\, a_0}{\sqrt{3}}\, \delta_{m,0}\,. \tag{7}$$

In an identical fashion, we find

$$\langle b \mid z \mid r \rangle = \frac{R_{20}^{n1}\, a_0}{\sqrt{3}}\, \delta_{m,0}\,. \tag{8}$$

b) The contribution of an intermediate level to the sum (1.b) is

$$Q'_{ba}(r) = -\, q^2\, \omega^2 \frac{\langle b \mid z \mid r \rangle \langle r \mid z \mid a \rangle}{\hbar(\omega - \omega_{ra})}. \tag{9}$$

The product of the matrix elements is calculated from (7) and (8). To find the energy denominator recall the two-photon resonance condition

$$2\,\hbar\omega = -\frac{E_I}{4} + E_I = \frac{3}{4} E_I \tag{10}$$

Hence $\hbar\omega = 3E_I/8$. With $\hbar\omega_{ra} = -(E_I/n^2) + E_I$, we can write (9) in the form

$$Q'_{ba}(r) = -\, q^2\, \omega^2 \frac{a_0^2}{3\,E_I} \frac{R_{10}^{n1}\, R_{20}^{n1}}{\left(-\frac{5}{8} + \frac{1}{n^2}\right)} \tag{11}$$

Comparing this expression with (2), we get

$$J'_n = \frac{R_{10}^{n1}\, R_{20}^{n1}}{-\frac{5}{8} + \frac{1}{n^2}}\,. \tag{12}$$

Between an element of the sum (1.b) and the corresponding element of the sum (1.a), there is a multiplying factor $(\omega_{br}\omega_{ra}/\omega^2)$. The relationships $E_n = -E_I/n^2$ and $\hbar\omega = 3E_I/8$ permit the calculation of the factor

$$\frac{\omega_{br}\,\omega_{ra}}{\omega^2} = \frac{E_I^2}{\hbar^2\,\omega^2}\left(\frac{1}{4} - \frac{1}{n^2}\right)\left(\frac{1}{n^2} - 1\right) = -\left(\frac{8}{3}\right)^2\left(\frac{1}{4} - \frac{1}{n^2}\right)\left(1 - \frac{1}{n^2}\right) \tag{13}$$

and lead to

$$J_n = -\, J'_n\left(\frac{8}{3}\right)^2\left(\frac{1}{4} - \frac{1}{n^2}\right)\left(1 - \frac{1}{n^2}\right). \tag{14}$$

c) To calculate J_n and J'_n (n being 2 or 3), we calculate the necessary radial integrals using (6.a) and (6.b). We find $R_{10}^{21} = 1.29$, $R_{10}^{31} = 0.52$, and $R_{20}^{31} = 3.06$, R_{20}^{21} being given by (6.c). Substituting these values in (12) and (14), we find $J'_2 = 17.9$, $J'_3 = -3.1$, $J_2 = 0$, and $J_3 = 2.7$.

These results are given in the following table, which includes also the contribution of the other intermediate *np* levels in order to clarify the discussion.

The contribution of the intermediate levels to the two-photon transition amplitude between the 1*s and* 2*s levels*

	$\mathbf{A}\cdot\mathbf{p}$ Hamiltonian	$\mathbf{E}\cdot\mathbf{r}$ Hamiltonian
Exact result (including continuum)	11.8	11.8
1 intermediate level, $2p$	0	17.9
2 intermediate levels, $2p, 3p$	2.7	14.8
3 intermediate levels, $2p, 3p, 4p$	3.6	14.1
10 intermediate discrete levels, $2p, 3p, \ldots, 11p$	4.5	13.5

It is clear from this table that, if a partial sum is used, the most satisfactory results are gotten from the $\mathbf{E}\cdot\mathbf{r}$ Hamiltonian. For example, $J_2' + J_3'$ differs from the exact result by only about 25%. Note that in the $\mathbf{E}\cdot\mathbf{r}$ description the contribution of the $2p$ level has a sign opposite to that of the more excited states. In this description, the sum on all the discrete levels is equal to 13.4, the contribution of the continuum being equal to -1.6.

With the $\mathbf{A}\cdot\mathbf{p}$ Hamiltonian, the sum on two intermediate levels leads to a very poor result, differing from the exact result by about one order of magnitude. Even if one takes into account all the discrete levels, the result (4.7) is still far from the exact result. This comes from the contribution of the continuum (7.1), which is predominant in this representation. It is related to the factor $\omega_{br}\omega_{ra}/\omega^2$, which favors the highly excited states and causes a much slower convergence in the summation over the intermediate levels.

d) When all the energy denominators are replaced by xE_I, Q'_{ba} becomes

$$\begin{aligned} Q'_{ba} &\simeq -\frac{q^2\omega^2}{xE_I}\sum_r \langle b\,|\,z\,|\,r\rangle\langle r\,|\,z\,|\,a\rangle \\ &= -\frac{q^2\omega^2}{xE_I}\langle b\,|\,z^2\,|\,a\rangle = -\left(\frac{q^2\omega^2}{3\,E_I}\right)\frac{\langle b\,|\,r^2\,|\,a\rangle}{x} \end{aligned} \tag{15}$$

since the levels $|b\rangle$ and $|a\rangle$ are s-states and thus rotationally invariant. To use the closure relation in an analogous way for the case of Q_{ba}, it suffices to express the matrix element $\langle r|z|a\rangle$ as a function of $\langle r|p_z|a\rangle$ using (B.43). We find then

$$\begin{aligned} Q_{ba} &= \frac{q^2}{m^2}\sum_r \frac{\langle b\,|\,p_z\,|\,r\rangle\langle r\,|\,p_z\,|\,a\rangle}{\hbar(\omega-\omega_{ra})} \\ &\simeq \frac{q^2}{m^2\,xE_I}\sum_r \langle b\,|\,p_z\,|\,r\rangle\langle r\,|\,p_z\,|\,a\rangle = \frac{q^2}{3\,m^2\,xE_I}\langle b\,|\,\mathbf{p}^2\,|\,a\rangle \end{aligned} \tag{16}$$

or finally

$$Q_{ba} \simeq \left(\frac{q^2}{3\,E_I}\right)\frac{2\,q^2}{4\,\pi\varepsilon_0\,mx}\left\langle b\left|\frac{1}{r}\right|a\right\rangle = \left(\frac{q^2}{3\,E_I}\right)\frac{2\,\alpha^2\,c^2\,a_0}{x}\left\langle b\left|\frac{1}{r}\right|a\right\rangle \tag{17}$$

since the matrix element of $H_0 = (\mathbf{p}^2/2m) - (q^2/4\pi\varepsilon_0 r)$ is zero between the orthogonal 1s and 2s states.

For the discrete intermediate levels, the energy denominators vary between $-\frac{3}{8}E_I$ for $n = 2$ and $-\frac{5}{8}E_I$ for $n = \infty$, the value $-0.5E_I$ being a reasonable compromise. Note however that this approximation overestimates the contribution of the continuum intermediate levels, for which the denominators have a modulus greater than $\frac{5}{8}E_I$ (and a fortiori than $0.5E_I$).

By taking $x = -0.5$ and using the radial wave functions (5.a) and (5.b) we find from (15) and (17)

$$Q'_{ba} \simeq -q^2 \frac{\omega^2}{3E_I} a_0^2 \times (6{,}0) \tag{18.a}$$

$$Q_{ba} \simeq -q^2 \frac{\omega^2}{3E_I} a_0^2 \times (23{,}8)\,. \tag{18.b}$$

If one compares these approximate results with the exact result (4), one sees that the amplitude gotten in the $\mathbf{E}\cdot\mathbf{r}$ description is two times too small, while in the $\mathbf{A}\cdot\mathbf{p}$ description it is two times too large. In both cases, the use of a unique energy denominator increases the contribution of the most excited intermediate levels and most particularly that of the continuum. In the $\mathbf{E}\cdot\mathbf{r}$ representation, the contribution of these levels has a sign opposite to that of the predominant $2p$ level. It is then reasonable that the result in this approximation will be smaller than the exact result. On the other hand, in the $\mathbf{A}\cdot\mathbf{p}$ representation, the contributions being of the same sign, one overestimates the result by taking a single value $0.5E_I$ for the energy denominator. It is of course possible, knowing the exact result, to improve the approximation by changing the denominator, but the choice, a priori, of a denominator twice as large or twice as small is rather arbitrary.

In conclusion, for the precise problem studied in this exercise, the best approximation consists of making the calculation with the $\mathbf{E}\cdot\mathbf{r}$ Hamiltonian and retaining the first discrete intermediate levels as seen in the table.

3. The Electric Dipole Hamiltonian for an Ion Coupled to an External Field

The purpose of this exercise is to show that if a localized system of charges coupled to an external field is not globally neutral, as in the case of an ion, then new terms appear in the electric dipole Hamiltonian which can be used to describe the system. These terms describe the coupling of the global motion of the ion's charge with the external field.

The various particles α with charge q_α, and mass m_α for an ion with total charge

$$Q = \sum_\alpha q_\alpha \tag{1}$$

are localized about the center of mass

$$\mathbf{R} = \frac{1}{M}\sum_\alpha m_\alpha \mathbf{r}_\alpha \quad (2.\text{a}); \qquad M = \sum_\alpha m_\alpha \quad (2.\text{b})$$

of the ion in a volume of dimension a. The ion interacts with an external field described by the vector potential $\mathbf{A}_e(\mathbf{r}_\alpha, t)$. The long-wavelength approximation ($\lambda \gg a$) involves replacing $\mathbf{A}_e(\mathbf{r}_\alpha, t)$ by $\mathbf{A}_e(\mathbf{R}, t)$ in the ion's Hamiltonian, which is then written

$$H(t) = \sum_\alpha \frac{1}{2\,m_\alpha}\,[\mathbf{p}_\alpha - q_\alpha\,\mathbf{A}_e(\mathbf{R},t)]^2 + V_{\text{Coul}}\,. \tag{3}$$

a) Consider the unitary transformation

$$T(t) = \exp\left\{ -\frac{\mathrm{i}}{\hbar}\sum_\alpha q_\alpha(\mathbf{r}_\alpha - \mathbf{R})\cdot\mathbf{A}_e(\mathbf{R},t)\right\} = \exp\left[-\frac{\mathrm{i}}{\hbar}\,\mathbf{d}\cdot\mathbf{A}_e(\mathbf{R},t)\right] \tag{4}$$

where

$$\mathbf{d} = \sum_\alpha q_\alpha(\mathbf{r}_\alpha - \mathbf{R}) \tag{5}$$

is the electric dipole moment of the ion with respect to the center of mass. Find $T(t)\mathbf{p}_\alpha T^+(t)$. One should take into account the dependence of $\mathbf{R}$ on $\mathbf{r}_\alpha$ in $T(t)$, but neglect the spatial derivatives of $\mathbf{A}_e(\mathbf{R},t)$, which introduce higher-order corrections in a/λ.

b) Find the Hamiltonian $H'(t)$ which describes the temporal evolution in the new representation. What is the physical interpretation of the new terms which appear, beyond those for the globally neutral system ($Q = 0$)?

Solution

a) Equations (4) and (5) generalize Equations (3) and (4) of Complement A_{IV}. Rather than being referred to a fixed point taken as the origin, the various charges q_α are now referred to the center of mass $\mathbf{R}$. Since $\mathbf{R}$ appears in (4) for $T(t)$, and since $\mathbf{R}$ depends through (2.a) on $\mathbf{r}_\alpha$, which does not commute with $\mathbf{p}_\alpha$, Equation (2) of Complement A_{IV} is not applicable and must be replaced by

$$\begin{aligned} T(t)\,\mathbf{p}_\alpha\,T^+(t) &= \mathbf{p}_\alpha + T(t)\,\frac{\hbar}{\mathrm{i}}\,\nabla_{\mathbf{r}_\alpha}\,T^+(t) \\ &= \mathbf{p}_\alpha + q_\alpha\,\mathbf{A}_e(\mathbf{R},t) - \sum_{\alpha'} q_{\alpha'}\frac{m_\alpha}{M}\,\mathbf{A}_e(\mathbf{R},t) \\ &= \mathbf{p}_\alpha + q_\alpha\,\mathbf{A}_e(\mathbf{R},t) - \frac{Q}{M}\,m_\alpha\,\mathbf{A}_e(\mathbf{R},t)\,. \end{aligned} \tag{6}$$

We have used (1) and neglected the terms arising from $\nabla_{\mathbf{r}_\alpha}\mathbf{A}_e(\mathbf{R},t)$, which are of higher order in a/λ.

b) The Hamiltonian which describes the temporal evolution in the new representation is equal to

$$H'(t) = T(t)\,H(t)\,T^+(t) + \mathrm{i}\hbar\left[\frac{\mathrm{d}T(t)}{\mathrm{d}t}\right]T^+(t)\,. \tag{7}$$

Equations (3) and (6) give

$$\begin{aligned} T(t)\,H(t)\,T^+(t) &= \sum_\alpha \frac{1}{2\,m_\alpha}\left[\mathbf{p}_\alpha - \frac{Q}{M}\,m_\alpha\,\mathbf{A}_e(\mathbf{R},t)\right]^2 + V_{\text{Coul}} \\ &= \sum_\alpha \frac{\mathbf{p}_\alpha^2}{2\,m_\alpha} + V_{\text{Coul}} - \frac{Q}{M}\,\mathbf{P}\cdot\mathbf{A}_e(\mathbf{R},t) + \frac{Q^2}{2\,M}\,\mathbf{A}_e^2(\mathbf{R},t) \end{aligned} \tag{8}$$

where

$$\mathbf{P} = \sum_\alpha \mathbf{p}_\alpha \tag{9}$$

is the total momentum of the system of charges. For the last term of (7), a calculation analogous to that of Complement A_{IV} gives, using (5),

$$i\hbar\left[\frac{dT(t)}{dt}\right]T^+(t) = \sum_\alpha q_\alpha(\mathbf{r}_\alpha - \mathbf{R}) \cdot \dot{\mathbf{A}}_e(\mathbf{R}, t) = -\,\mathbf{d} \cdot \mathbf{E}_e(\mathbf{R}, t). \tag{10}$$

Finally by adding (8) and (10), one gets

$$H'(t) = \sum_\alpha \frac{\mathbf{p}_\alpha^2}{2\,m_\alpha} + V_{\text{Coul}} - \mathbf{d} \cdot \mathbf{E}_e(\mathbf{R}, t) - \frac{Q}{M}\mathbf{P} \cdot \mathbf{A}_e(\mathbf{R}, t) + \frac{Q^2}{2\,M}\mathbf{A}_e^2(\mathbf{R}, t). \tag{11}$$

The last two terms of (11) are in addition to those of Equation (8) of Complement A_{IV}. They exist only for an ion and go to zero if $Q = 0$ (neutral atom). They depend only on the variables $\mathbf{R}$ and $\mathbf{P}$ of the center of mass. They describe the interaction of a fictitious particle of mass M, charge Q, position $\mathbf{R}$, and momentum $\mathbf{P}$ with the external field $\mathbf{A}_e$.

4. Scattering of a particle by a potential in the presence of laser radiation

A charged particle, scattered by a static potential $V(\mathbf{r})$ and simultaneously interacting with laser radiation, can absorb (or emit in a stimulated fashion) many laser photons in the course of the scattering process. The purpose of this exercise is to show how the Henneberger transformation allows one to calculate such processes simply in the lowest order in $V(\mathbf{r})$ (*).

A particle with mass m and charge q interacts, on one hand, with a static potential $V(\mathbf{r})$ localized about the origin, and on the other hand with incident monochromatic radiation taken as an external field. This latter is described by the vector potential $\mathbf{A}_e(\mathbf{r}, t)$, the scalar potential $U_e(\mathbf{r}, t)$ being zero. One neglects the interaction of the particle with the vacuum field and considers only the interaction with the incident radiation, whose wavelength is assumed long compared to the range of the static potential $V(\mathbf{r})$. The particle Hamiltonian is then written, within the long-wavelength approximation,

$$H = \frac{1}{2\,m}[\mathbf{p} - q\mathbf{A}_e(\mathbf{0}, t)]^2 + V(\mathbf{r}) \tag{1}$$

with

$$\mathbf{A}_e(\mathbf{0}, t) = A_0 \cos \omega t \; \mathbf{e}_z \tag{2}$$

the incident radiation having amplitude A_0, frequency ω, and polarization $\mathbf{e}_z$.

(*) See Y. Gontier and N. K. Rahman, *Lett. Nuovo Cim.*, **9**, 537 (1974), and N. M. Kroll and K. M. Watson, *Phys. Rev. A*, **8**, 804 (1973).

a) Carry out the unitary transformation

$$T(t) = \exp[\mathrm{i}F(t)/\hbar] \tag{3.a}$$

with

$$F(t) = \mathbf{p} \cdot \boldsymbol{\xi}(t) + \frac{q^2}{2m} \int^t \mathbf{A}_e^2(\mathbf{0}, t')\, \mathrm{d}t' \tag{3.b}$$

$$\boldsymbol{\xi}(t) = -\frac{qA_0}{m\omega} \sin \omega t \, \mathbf{e}_z \,. \tag{3.c}$$

Let $|\psi^{(1)}(t)\rangle$ and $|\psi^{(2)}(t)\rangle = T(t)|\psi^{(1)}(t)\rangle$ be the state vectors describing respectively the state of the system in the original representation and in the new one. Give the Hamiltonian $H^{(2)}$ describing the evolution of $|\psi^{(2)}(t)\rangle$. Show that $H^{(2)}$ can be written

$$H^{(2)}(t) = \frac{\mathbf{p}^2}{2m} + \tilde{V}(\mathbf{r}, t) \tag{4}$$

and give the expression for $\tilde{V}(\mathbf{r}, t)$.

b) Interpret $\boldsymbol{\xi}(t)$ physically. Derive a "geometric" interpretation for the transformation (3). What is the physical significance of the operator $\mathbf{r}$ (multiplication by x, y, z) in the new representation?

c) Expand the new potential $\tilde{V}(\mathbf{r}, t)$ in a Fourier time series of the form

$$\tilde{V}(\mathbf{r}, t) = \sum_{n=-\infty}^{+\infty} \mathrm{e}^{-\mathrm{i}n\omega t}\, \tilde{V}_n(\mathbf{r}) \,. \tag{5}$$

Give $\tilde{V}_n(\mathbf{r})$ as an integral involving the spatial Fourier transform $\mathcal{V}(\mathbf{k})$ of $V(\mathbf{r})$ and the Bessel function of order n, J_n. Recall that

$$\mathrm{e}^{-\mathrm{i}\alpha \sin\varphi} = \sum_{n=-\infty}^{+\infty} J_n(\alpha)\, \mathrm{e}^{-\mathrm{i}n\varphi} \tag{6}$$

d) We are interested here in the static part $\tilde{V}_0(\mathbf{r})$ of the expansion (5) for $\tilde{V}(\mathbf{r}, t)$. In addition, we limit ourselves to the Coulomb potential $V(\mathbf{r}) = -q^2/4\pi\varepsilon_0 r$.

Using the Poisson equation, find the charge distribution which one can associate with the potential $\tilde{V}_0(\mathbf{r})$. One can use the relationship

$$\int_{-\infty}^{+\infty} J_0(au)\, \mathrm{e}^{\mathrm{i}bu}\, \mathrm{d}u = \begin{cases} \dfrac{2}{\sqrt{a^2 - b^2}} & \text{if } |b| < |a| \\ 0 & \text{if } |a| < |b| . \end{cases} \tag{7}$$

What is the physical interpretation of the result?

e) Consider the problem of the scattering of an electron by the potential $V(\mathbf{r})$ in the presence of the external field $\mathbf{A}_e(\mathbf{r}, t)$. The electron goes from an initial state with energy E_i and momentum $\mathbf{p}_i$ to a final state with E_f and $\mathbf{p}_f$.

By applying the results of first-order time-dependent perturbation theory, show, starting with the expansion (5), that the possible final energies E_f associated with a given initial energy E_i form a series of discrete values labeled by an integer n. Interpret the result physically.

f) Calculate, to first order in $V(\mathbf{r})$, the scattering amplitude associated with a given transfer of energy and momentum. What happens to the elastic scattering amplitudes $(E_f = E_i)$ and the inelastic amplitudes $(E_f \neq E_i)$ when $A_0 \to 0$?

g) Starting with the preceding results, can you get, *without calculation*, an expression for the differential scattering cross-section $(\mathrm{d}\sigma/\mathrm{d}\Omega)_{i \to f}$ for an inelastic process associated with one of the energies E_f in the series defined in part *e*)? Relate this cross-section to the effective differential elastic cross-section $(\mathrm{d}\sigma/\mathrm{d}\Omega)_0$ associated with an electron scattered by the potential $V(\mathbf{r})$ in the absence of any external field. Do not try to calculate $(\mathrm{d}\sigma/\mathrm{d}\Omega)_0$.

Solution

a) Since $F(t)$ depends on t, the new Hamiltonian $H^{(2)}$ is written

$$H^{(2)} = T(t)\, HT^{+}(t) + i\hbar\left(\frac{\mathrm{d}}{\mathrm{d}t}\, T(t)\right) T^{+}(t)\,. \tag{8}$$

Since $\mathrm{d}F/\mathrm{d}t$ commutes with F, the last term of (8) is equal to

$$-\frac{\mathrm{d}F}{\mathrm{d}t} = -\,\mathbf{p}\cdot\dot{\boldsymbol{\xi}}(t) - \frac{q^2}{2m}\,\mathbf{A}_e^2(\mathbf{0}, t) \tag{9}$$

—that is, finally, from (3.c) and (2), to

$$-\frac{\mathrm{d}F}{\mathrm{d}t} = \frac{q}{m}\,\mathbf{p}\cdot\mathbf{A}_e(\mathbf{0}, t) - \frac{q^2}{2m}\,\mathbf{A}_e^2(\mathbf{0}, t)\,. \tag{10}$$

In addition, $T(t)$ is a translation operator for $\mathbf{r}$:

$$T(t)\,\mathbf{r}T^{+}(t) = \mathbf{r} + \boldsymbol{\xi}(t) \tag{11}$$

with the result that the first term of (8) is written, following (1), as

$$T(t)\, HT^{+}(t) = \frac{1}{2m}\left[\mathbf{p} - q\mathbf{A}_e(\mathbf{0}, t)\right]^2 + V(\mathbf{r} + \boldsymbol{\xi}(t))\,. \tag{12}$$

Finally, the sum of (10) and (12) gives

$$H^{(2)}(t) = \frac{\mathbf{p}^2}{2m} + V(\mathbf{r} + \boldsymbol{\xi}(t))\,. \tag{13}$$

The transformation $T(t)$ is nothing but the Henneberger transformation studied in §B.4, the last integral of (3.b) causing the term $q^2\mathbf{A}_e^2(\mathbf{0}, t)/2m$ to vanish in $H^{(2)}$. This term is actually a c-number.

b) As we have seen in §B.4, $\boldsymbol{\xi}(t)$ represents the vibrational motion of the particle in the incident wave. In the new representation, the operator **r** represents the mean position of the particle, about which it effects the vibrational motion described by $\boldsymbol{\xi}(t)$. Under the influence of the incident wave, the particle averages the static potential about the point **r** over an interval characterized by $\boldsymbol{\xi}(t)$. It is this effect which is described by the last term of (13).

c) Let $\mathscr{V}(\mathbf{k})$ be the spatial Fourier transform of $V(\mathbf{r})$:

$$V(\mathbf{r}) = \frac{1}{(2\pi)^{3/2}} \int \mathrm{d}^3k \, \mathscr{V}(\mathbf{k}) \, \mathrm{e}^{\mathrm{i}\mathbf{k}.\mathbf{r}} \tag{14}$$

Replace **r** by $\mathbf{r} + \boldsymbol{\xi}(t)$ in (14), and use (3.c) and (6). This gives

$$V(\mathbf{r} + \boldsymbol{\xi}(t)) = \frac{1}{(2\pi)^{3/2}} \int \mathrm{d}^3k \, \mathscr{V}(\mathbf{k}) \, \mathrm{e}^{\mathrm{i}\mathbf{k}.\mathbf{r}} \exp\left(-\mathrm{i}\frac{qA_0}{m\omega} \mathbf{k} \cdot \mathbf{e}_z \sin \omega t \right)$$

$$= \sum_n \mathrm{e}^{-\mathrm{i}n\omega t} \, \tilde{V}_n(\mathbf{r}) \tag{15}$$

with

$$\tilde{V}_n(\mathbf{r}) = \frac{1}{(2\pi)^{3/2}} \int \mathrm{d}^3k \, \mathscr{V}(\mathbf{k}) \, J_n(\mathbf{k} \cdot \boldsymbol{\xi}_0) \, \mathrm{e}^{\mathrm{i}\mathbf{k}.\mathbf{r}} \tag{16}$$

where we have set

$$\boldsymbol{\xi}_0 = \frac{qA_0}{m\omega} \mathbf{e}_z \, . \tag{17}$$

As a result of its vibration at the frequency ω in the potential $V(\mathbf{r})$, the particle "sees" a potential periodic in t whose Fourier series expansion is given by (15).

d) If $V(\mathbf{r}) = -q^2/4\pi\varepsilon_0 r$, then $\mathscr{V}(\mathbf{k}) = -q^2/(2\pi)^{3/2}\varepsilon_0 k^2$, so that from (16)

$$\tilde{V}_0(\mathbf{r}) = -\frac{q^2}{\varepsilon_0} \frac{1}{(2\pi)^3} \int \mathrm{d}^3k \frac{1}{k^2} J_0(\mathbf{k} \cdot \boldsymbol{\xi}_0) \, \mathrm{e}^{\mathrm{i}\mathbf{k}.\mathbf{r}} \, . \tag{18}$$

Let $\rho_0(\mathbf{r})$ be the effective charge density which creates, for the particle with charge q, the potential energy $\tilde{V}_0(\mathbf{r})$, and let $\rho_0(\mathbf{k})$ be the Fourier transform of $\rho_0(\mathbf{r})$. Poisson's equation $\Delta(\tilde{V}_0/q) = -\rho_0/\varepsilon_0$ is written in reciprocal space as $\rho_0 = \varepsilon_0 k^2 \tilde{\mathscr{V}}_0/q$, where $\tilde{\mathscr{V}}_0(\mathbf{k})$ is the Fourier transform of $\tilde{V}_0(\mathbf{r})$. Since $\tilde{\mathscr{V}}_0$ is given by (18), one concludes that

$$\rho_0(\mathbf{k}) = -\frac{q}{(2\pi)^{3/2}} J_0(\mathbf{k} \cdot \boldsymbol{\xi}_0) \tag{19.a}$$

and consequently

$$\rho_0(\mathbf{r}) = -q \frac{1}{(2\pi)^3} \int \mathrm{d}^3k \, \mathrm{e}^{\mathrm{i}\mathbf{k}.\mathbf{r}} J_0(k_z \xi_0) \tag{19.b}$$

where $\xi_0 = qA_0/m\omega$ is the maximum amplitude of the vibration. It is sufficient then to use (7) to get (*)

$$\rho_0(\mathbf{r}) = -q\,\delta(x)\,\delta(y) \times \begin{cases} \dfrac{1}{\pi\sqrt{\xi_0^2 - z^2}} & \text{if} \quad 0 < |z| < \xi_0 \\ 0 & \text{if} \quad 0 < \xi_0 < |z| \, . \end{cases} \tag{20}$$

(*) W. C. Henneberger, *Phys. Rev. Lett.*, **21**, 838 (1968).

To interpret (20), we note that in its rest frame, the particle sees the charge $-q$ creating the potential $V(\mathbf{r})$ oscillating at frequency ω along the z-axis with amplitude ξ_0. Equation (20) gives the apparent static charge distribution associated with this oscillating charge $-q$. It is clearly localized on the z-axis. In addition, the function of z given in (20) is just the classical probability density for finding the charge $-q$ at a point z on the z-axis, which clearly diverges at the turning points $-\xi_0$ and ξ_0, since the velocity of the charge $-q$ is zero at these points.

e) From first-order time-dependent perturbation theory, the transition amplitude for going from an initial state $|\varphi_i\rangle$, with energy E_i to a final state $|\varphi_f\rangle$ with energy E_f is proportional to the Fourier transform of $\langle\varphi_f|\tilde{V}(\mathbf{r}, t)|\varphi_i\rangle$ evaluated at the frequency $(E_f - E_i)/\hbar$. Since $\tilde{V}(\mathbf{r}, t)$ is a sum of exponentials $e^{-in\omega t}$, the transition amplitude will only be important for

$$E_f = E_i + n\hbar\omega \tag{21}$$

where n is an integer, positive, negative or zero. The case $n > 0$ corresponds to the absorption of n incident photons by the particle in the course of the scattering process, and the case $n < 0$ to the stimulated emission of n photons (*).

f) The scattering amplitude $\{\mathbf{p}_i = \hbar\mathbf{k}_i, E_i\} \rightarrow \{\mathbf{p}_f = \hbar\mathbf{k}_f, E_f\}$ with the absorption of n incident photons is given by

$$\frac{1}{i\hbar}\langle\varphi_f|\tilde{V}_n(\mathbf{r})|\varphi_i\rangle\int_{-T/2}^{+T/2} e^{i(E_f - E_i - n\hbar\omega)t/\hbar}\, dt \tag{22}$$

where T is the duration of the collision. The integral over t in (22) gives $2\pi\hbar\delta^{(T)}(E_f - E_i - n\hbar\omega)$. As for the matrix element of $\tilde{V}_n(\mathbf{r})$, it is proportional to

$$\mathscr{V}(\mathbf{k}_f - \mathbf{k}_i)\, J_n[(\mathbf{k}_f - \mathbf{k}_i)\cdot\xi_0]\,. \tag{23}$$

As a result of the conservation of energy, $\mathbf{k}_f$ and $\mathbf{k}_i$ are connected by the relation

$$\frac{\hbar^2\,\mathbf{k}_f^2}{2m} = \frac{\hbar^2\,\mathbf{k}_i^2}{2m} + n\,\hbar\omega\,. \tag{24}$$

The elastic scattering amplitude ($E_f = E_i$) corresponds to $n = 0$. Since $J_0(0) = 1$, one gets, in the limit $\xi_0 \rightarrow 0$, the Born amplitude $\mathscr{V}(\mathbf{k}_f - \mathbf{k}_i)$ with $|\mathbf{k}_f| = |\mathbf{k}_i|$. For $n \neq 0$, $J_n(0) = 0$ and the inelastic scattering amplitudes tend to zero with A_0 as A_0^n. The presence of terms of arbitrary order in A_0 shows the nonperturbative character of such a calculation.

g) Let $(d\sigma/d\Omega)_0$ be the effective differential cross-section in the absence of the external field. $(d\sigma/d\Omega)_0$ is proportional to $|\mathscr{V}(\mathbf{k}_f - \mathbf{k}_i)|^2$ and to the density of final states, $\rho(E_f = E_i)$. In the presence of an external field and for the process where n photons are absorbed in the course of the scattering, it is necessary to replace $\mathscr{V}(\mathbf{k}_f - \mathbf{k}_i)$ by (23). On the other hand, since $E_f \neq E_i$ as a result of (24), the density of final states is different. Since $\rho(E)$ is proportional to $\sqrt{E}$ for a particle of mass m, one has then

$$\left(\frac{d\sigma}{d\Omega}\right)_{i\rightarrow f} = \left(\frac{d\sigma}{d\Omega}(\mathbf{k}_f - \mathbf{k}_i)\right)_0 J_n^2[(\mathbf{k}_f - \mathbf{k}_i)\cdot\xi_0]\sqrt{\frac{E_f}{E_i}}\,. \tag{25}$$

(*) An experimental observation of this process is presented in A. Weingartshofer et al., *Phys. Rev. Lett.*, **39**, 269 (1977).

5. The equivalence between the interaction Hamiltonians $\mathbf{A} \cdot \mathbf{p}$ and $\mathbf{Z} \cdot (\nabla V)$ for the calculation of transition amplitudes

Consider a particle bound near the origin by a static potential $V(\mathbf{r})$. This particle interacts also with an external electromagnetic field described by a vector potential whose value at the origin is

$$\mathbf{A}(\mathbf{0}, t) = \tilde{A}(t)\,\mathbf{e}_z \cos \omega t \tag{1}$$

where $\tilde{A}(t)$ is a slowly varying function of time.

The Hamiltonian of this particle from the Henneberger point of view (see §B.4) is, in the dipole approximation, equal to

$$H^{(2)} = \frac{\mathbf{p}^2}{2m} + V\left(\mathbf{r} + \frac{q}{m}\mathbf{Z}(\mathbf{0}, t)\right) + \frac{q^2}{2m}\mathbf{A}^2(\mathbf{0}, t)\,. \tag{2}$$

Since $\tilde{A}(t)$ is assumed slowly varying, one can take

$$\mathbf{Z}(\mathbf{0}, t) \simeq -\frac{\tilde{A}(t)}{\omega}\,\mathbf{e}_z \sin \omega t\,. \tag{3}$$

In what follows, we will split $H^{(2)}$ into a particle Hamiltonian

$$H_0 = \frac{\mathbf{p}^2}{2m} + V(\mathbf{r}) \tag{4}$$

and a Hamiltonian of interaction with the electromagnetic field: $H'_I = H^{(2)} - H_0$. The purpose of this exercise is to show that the transition matrix element for one- or two-photon processes is, at resonance, the same in this representation as in the usual one where the interaction Hamiltonian is $\mathbf{A} \cdot \mathbf{p}$ (*).

a) Find the expansion of H'_I to second order in q.

b) Using the commutator $[p_z, H_0]$, establish the equation

$$\langle b \mid \partial V/\partial z \mid a \rangle = \mathrm{i}\omega_{ab} \langle b \mid p_z \mid a \rangle \tag{5}$$

relating the matrix elements of $\partial V/\partial z$ to those of p_z. The states $|a\rangle$ and $|b\rangle$ are two eigenstates of H_0 with eigenvalues E_a and E_b, and $\hbar\omega_{ba} = E_b - E_a$. Show that in the case of a resonant one-photon transition, the transition matrix elements are identical in the two representations.

c) In the case of a two-photon transition, the transition amplitude is proportional to the matrix element Q_{ba} of an operator connecting the initial level to the final level (see §B_{IV}.2.*a*). In the $\mathbf{A} \cdot \mathbf{p}$ representation,

(*) W. C. Henneberger, *Phys. Rev. Lett.*, **21**, 838 (1968).

this matrix element is written

$$Q_{ba} = \left(\frac{q}{m}\right)^2 \sum_r \frac{\langle b \mid p_z \mid r \rangle \langle r \mid p_z \mid a \rangle}{\hbar(\omega - \omega_{ra})}. \tag{6}$$

Explain why this expression should be replaced by

$$Q_{ba}^H = -\left(\frac{q}{m\omega}\right)^2 \left[\sum_r \frac{\langle b | \partial V/\partial z | r \rangle \langle r | \partial V/\partial z | a \rangle}{\hbar(\omega - \omega_{ra})} + \frac{1}{2}\left\langle b \left| \frac{\partial^2 V}{\partial z^2} \right| a \right\rangle\right] \tag{7}$$

if the calculations are done in the Henneberger representation.

d) By using Equation (5) and the commutator $[p_z, \partial V/\partial z]$, show directly that $Q_{ba} = Q_{ba}^H$ in the case of a two-photon resonant excitation (as a hint, look to the proof in §B_{IV}.2.*c*).

Solution

a) The expansion of $V(x, y, z + (q/m)Z_z(0, t))$ to second order in q,

$$V\left(x, y, z + \frac{q}{m} Z_z(0, t)\right) = V(\mathbf{r}) + \frac{q}{m} Z_z(0, t) \frac{\partial V(\mathbf{r})}{\partial z} + \frac{q^2}{2\,m^2} Z_z^2(0, t) \frac{\partial^2 V(\mathbf{r})}{\partial z^2} + \cdots \tag{8}$$

gives for H_I'

$$H_I' = \frac{q}{m} Z_z(0, t) \frac{\partial V(\mathbf{r})}{\partial z} + \frac{q^2}{2\,m^2} Z_z^2(0, t) \frac{\partial^2 V(\mathbf{r})}{\partial z^2} + \frac{q^2}{2\,m} \mathbf{A}^2(0, t). \tag{9}$$

b) One starts with the equality

$$-\,\mathrm{i}\hbar \frac{\partial V}{\partial z} = [p_z, H_0] \tag{10}$$

and calculates the matrix elements of both sides of (10) between two eigenstates $|a\rangle$ and $|b\rangle$ of H_0:

$$-\,\mathrm{i}\hbar \langle b \mid \partial V/\partial z \mid a \rangle = \langle b \mid [p_z, H_0] \mid a \rangle = (E_a - E_b) \langle b \mid p_z \mid a \rangle. \tag{11}$$

By dividing both sides of this equality by $-\mathrm{i}\hbar$, one finds the required relationship (5).

In the usual description, the transition amplitude S_{ba} is proportional to $M_{ba} = -q(p_z)_{ba}/m$ [see (7.a) and (7.b), Complement B_{IV}]. In the Henneberger representation, one gets for the transition amplitude S_{ba} an expression identical to (7.a) of Complement B_{IV}, but the matrix element involved is, using (3) and (9),

$$M_{ba}^H = -\frac{\mathrm{i}}{\omega}\frac{q}{m}\left\langle b \left| \frac{\partial V}{\partial z} \right| a \right\rangle. \tag{12}$$

Actually, of the three terms of (9), only the term linear in q having a component oscillating in $\mathrm{e}^{-\mathrm{i}\omega t}$ can induce a one-photon transition. Equation (5) yields

$$M_{ba}^H = -\frac{\omega_{ba}}{\omega}\frac{q}{m} \langle b \mid p_z \mid a \rangle \tag{13}$$

which is clearly equal to M_{ba} at resonance ($\omega_{ba} = \omega$).

c) In the new description, there are two contributions to the two-photon transition amplitude. The first involves the first term of (9) (one-photon operator) in second order and corresponds to the successive absorption of two photons through the intermediate states r of the atom. This contribution may be gotten by replacing $-(q/m)p_z[\tilde{A}(t)e^{-i\omega t}]/2$ by $(q/m)(\partial V/\partial z)[\tilde{A}(t)e^{-i\omega t}]/2i\omega$ in Equation (9) of Complement B_{IV}. Using (15) of the same complement thus leads to

$$[Q_{ba}^H]' = -\left(\frac{q}{m\omega}\right)^2 \sum_r \frac{\langle b | \partial V/\partial z | r \rangle \langle r | \partial V/\partial z | a \rangle}{\hbar(\omega - \omega_{ra})}. \tag{14}$$

There is a second contribution to the transition amplitude, which is provided by the second term of (9) (two-photon operator in first order) and which corresponds to a simultaneous absorption of two photons. The contribution S''_{ba} of this term is gotten through first-order perturbation theory by retaining only the oscillatory component in $e^{-2i\omega t}$ in $Z_z^2(\mathbf{0}, t)$ and leads to

$$S''_{ba} = \frac{1}{i\hbar} \frac{q^2}{2m^2} \left\langle b \left| \frac{\partial^2 V}{\partial z^2} \right| a \right\rangle \int dt \left(\frac{\tilde{A}(t)\, e^{-i\omega t}}{2\, i\omega} \right)^2 e^{i\omega_{ba} t} \tag{15}$$

from which we get

$$[Q_{ba}^H]'' = -\frac{q^2}{2\, m^2\, \omega^2} \left\langle b \left| \frac{\partial^2 V}{\partial z^2} \right| a \right\rangle. \tag{16}$$

The last term of (9) does not contribute to the two-photon transition, since it is a c-number. The sum of (14) and (16) is the same as Equation (7) and describes the two-photon transition in the new representation.

d) Denote by D' the difference

$$D' = \left[-\left(\frac{1}{\omega}\right)^2 \sum_r \frac{\langle b | \partial V/\partial z | r \rangle \langle r | \partial V/\partial z | a \rangle}{\hbar(\omega - \omega_{ra})} \right] - \sum_r \frac{\langle b | p_z | r \rangle \langle r | p_z | a \rangle}{\hbar(\omega - \omega_{ra})}. \tag{17.a}$$

The identity (5) allows one to write D' in the form

$$D' = \sum_r \left(\frac{\omega_{br}\, \omega_{ra}}{\omega^2} - 1 \right) \frac{\langle b | p_z | r \rangle \langle r | p_z | a \rangle}{\hbar(\omega - \omega_{ra})}. \tag{17.b}$$

Now, in the case of a resonant two-photon transition [see Equation (25) of Complement B_{IV}],

$$\omega_{br}\, \omega_{ra} - \omega^2 = -(\omega - \omega_{ra})^2. \tag{18}$$

Substituting (18) in (17.b), one finds

$$D' = -\frac{1}{\hbar\omega^2} \sum_r (\omega - \omega_{ra}) \langle b | p_z | r \rangle \langle r | p_z | a \rangle. \tag{19}$$

Consider now the commutator

$$\left[p_z, \frac{\partial V}{\partial z} \right] = -i\hbar \frac{\partial^2 V}{\partial z^2} \tag{20}$$

and take the matrix elements of both sides of (20) between $|b\rangle$ and $|a\rangle$. We get

$$\sum_r (\langle b | p_z | r \rangle \langle r | \partial V/\partial z | a \rangle - \langle b | \partial V/\partial z | r \rangle \langle r | p_z | a \rangle) = -i\hbar \langle b | \partial^2 V/\partial z^2 | a \rangle. \tag{21}$$

Equation (5) allows one to transform (21) into

$$\sum_r (\omega_{ra} - \omega_{br}) \langle b | p_z | r \rangle \langle r | p_z | a \rangle = \hbar \langle b | \partial^2 V/\partial z^2 | a \rangle \tag{22}$$

which can again be written, since $\omega_{br} = 2\omega - \omega_{ra}$,

$$2\sum_r (\omega_{ra} - \omega) \langle b \mid p_z \mid r \rangle \langle r \mid p_z \mid a \rangle = \hbar \langle b \mid \partial^2 V/\partial z^2 \mid a \rangle . \tag{23}$$

Comparison of (19) and (23) then gives

$$D' = \frac{1}{2\,\omega^2} \left\langle b \left| \frac{\partial^2 V}{\partial z^2} \right| a \right\rangle . \tag{24}$$

Consider finally $Q^H_{ba} - Q_{ba}$. Taking (6) and (7) into account and the definition (17.a) for D', we have

$$Q^H_{ba} - Q_{ba} = \left(\frac{q}{m}\right)^2 \left[D' - \frac{1}{2\,\omega^2} \left\langle b \left| \frac{\partial^2 V}{\partial z^2} \right| a \right\rangle \right] \tag{25}$$

which from (24) is zero.

Thus, it is theoretically equivalent to calculate a two-photon transition with Equation (15.b) or (18.b) of Complement B_{IV} in the $\mathbf{A} \cdot \mathbf{p}$ and $\mathbf{E} \cdot \mathbf{r}$ descriptions or Equation (7) in the Henneberger description, as long as one carries out the summations over *all* the intermediate levels. It should be noted however that, as a result of the ratio $-\omega_{br}\omega_{ra}/\omega^2$ between the contributions of each intermediate level r in the first term of (7) and (6), the summation on the intermediate levels in the new description converges much more slowly that in the $\mathbf{A} \cdot \mathbf{p}$ description, which itself, for two-photon absorption processes between two bound states, is less well suited than the $\mathbf{E} \cdot \mathbf{r}$ description (see Exercise 2).

6. Linear response and susceptibility. Application to the calculation of the radiation from a dipole

Consider a system described by the Hamiltonian H_0 and whose equilibrium state is described by the density operator ρ_{eq} (which commutes with H_0). This system is subjected to a perturbation.

$$V(t) = -\lambda(t)\, M \tag{1}$$

where $\lambda(t)$ is a time-dependent parameter and M an observable of the system. The purpose of this exercise is to find the linear response (*) of the system to this perturbation as evidenced in the evolution of the mean values of other observables, and to apply the results to the electromagnetic field.

Let $\rho(t)$ be the density operator describing the perturbed system at time t.

a) Find the equation of motion of the density operator in the interaction representation

$$\rho_I(t) = \exp(\mathrm{i}H_0 t/\hbar)\, \rho \exp(-\mathrm{i}H_0 t/\hbar). \tag{2}$$

as a function of the operator M in the interaction representation

$$M_I(t) = \exp(\mathrm{i}H_0 t/\hbar)\, M(t) \exp(-\mathrm{i}H_0 t/\hbar). \tag{3}$$

(*) See for example, P. Martin, in *Many-Body Physics*, Les Houches 1967, C. de Witt and B. Balian, eds., Gordon and Breach, New York, 1968.

b) Integrate the equation of motion for ρ_I to first order in V by assuming that the system is in the state ρ_{eq} at $t = -\infty$. Show that the mean value of an observable N at time t in the perturbed state, to first order in V, can be written

$$\langle N \rangle_t = \langle N \rangle_{eq} + \int_{-\infty}^{+\infty} dt' \, \chi_{NM}(t - t') \, \lambda(t') \tag{4}$$

where $\langle N \rangle_{eq}$ is the equilibrium mean value of N in the unperturbed state ρ_{eq} and where χ_{NM} can be written as the mean value of an operator in ρ_{eq}. Give the expression for $\chi_{NM}(t - t')$.

c) The Hamiltonian of a quantized field coupled to a classical source formed by a classical electric dipole $\mathbf{d}(t)$ localized near the origin in a region of extent a can be written (see Complement A_{IV}, §3.b)

$$H = H_R - \frac{1}{\varepsilon_0} \mathbf{d}(t) \cdot \mathbf{D}'(\mathbf{0}) \tag{5}$$

where $H_R = \sum_i \hbar\omega_i (a_i^+ a_i + \frac{1}{2})$ is the radiation Hamiltonian and where $\mathbf{D}'(0)$ is the displacement at the origin. Recall that

$$\frac{\mathbf{D}'(\mathbf{r})}{\varepsilon_0} = \sum_i i\mathscr{E}_{\omega_i} (\boldsymbol{\varepsilon}_i \, a_i \, e^{i\mathbf{k}_i \cdot \mathbf{r}} - \boldsymbol{\varepsilon}_i \, a_i^+ \, e^{-i\mathbf{k}_i \cdot \mathbf{r}}) \tag{6}$$

coincides with the total electric field outside the system of charges.

By using the results of the foregoing question and the commutators from Complement C_{III}, find the mean value of the total electric field radiated by the dipole at a point $\mathbf{r}$ such that $r \gg a$. Give the field as a function of the dipole components, their velocity, and their acceleration. Show that the field radiated in the far zone is proportional to the acceleration of the dipole.

d) Apply these results to the case where the dipole is oscillating,

$$\mathbf{d}(t) = q\mathbf{a}_0 \cos \omega_0 t \,. \tag{7}$$

and compare the result with that of Exercise 6, Chapter I.

Solution

a) Take the derivative of Equation (2) with respect to time:

$$\frac{d}{dt}\rho_I(t) = \frac{i}{\hbar} H_0 \, \rho_I(t) - \frac{i}{\hbar} \rho_I(t) \, H_0 + \exp\left(i \frac{H_0}{\hbar} t\right)\left[\frac{d}{dt}\rho(t)\right]\exp\left(-i\frac{H_0}{\hbar} t\right). \tag{8}$$

The rate of variation of $\rho(t)$ is given by the equation of evolution of the density matrix under the action of the perturbed Hamiltonian $H_0 + V(t)$,

$$i\hbar \frac{d}{dt}\rho(t) = [H_0, \rho(t)] + [V(t), \rho(t)] \,. \tag{9}$$

This gives for the last term of (8)

$$\exp\left(\mathrm{i}\frac{H_0}{\hbar}t\right)\left[\frac{\mathrm{d}}{\mathrm{d}t}\rho(t)\right]\exp\left(-\mathrm{i}\frac{H_0}{\hbar}t\right) = -\frac{\mathrm{i}}{\hbar}\left[H_0, \exp\left(\mathrm{i}\frac{H_0}{\hbar}t\right)\rho(t)\exp\left(-\mathrm{i}\frac{H_0}{\hbar}t\right)\right] +$$

$$+\left(-\frac{\mathrm{i}}{\hbar}\right)\exp\left(\mathrm{i}\frac{H_0}{\hbar}t\right)[V(t), \rho(t)]\exp\left(-\mathrm{i}\frac{H_0}{\hbar}t\right)$$

$$= -\frac{\mathrm{i}}{\hbar}[H_0, \rho_I(t)] - \frac{\mathrm{i}}{\hbar}[-\lambda(t)M_I(t), \rho_I(t)] \tag{10}$$

and consequently

$$\frac{\mathrm{d}}{\mathrm{d}t}\rho_I(t) = \frac{\mathrm{i}}{\hbar}\lambda(t)[M_I(t), \rho_I(t)]. \tag{11}$$

The initial condition is given by $\rho(t) \to \rho_{eq}$ for $t \to -\infty$. Since ρ_{eq} commutes with H_0, one gets

$$\rho_I(-\infty) = \rho_{eq}. \tag{12}$$

b) If $\lambda(t)$ remains identically zero, $\mathrm{d}\rho_I/\mathrm{d}t$ is also zero, and

$$\rho_I(t) = \rho_I(-\infty) = \rho_{eq}. \tag{13}$$

To first order in λ, one can replace $\rho_I(t)$ by ρ_{eq} in Equation (11), which is then written

$$\frac{\mathrm{d}}{\mathrm{d}t}\rho_I(t) = \frac{\mathrm{i}\lambda(t)}{\hbar}[M_I(t), \rho_{eq}] \tag{14}$$

and integrated to give

$$\rho_I(t) - \rho_{eq} = \frac{\mathrm{i}}{\hbar}\int_{-\infty}^{t}\mathrm{d}t'\,\lambda(t')[M_I(t'), \rho_{eq}]. \tag{15}$$

We return to $\rho(t)$ by using (2) and the fact that ρ_{eq} commutes with H_0:

$$\rho(t) = \rho_{eq} + \frac{\mathrm{i}}{\hbar}\int_{-\infty}^{t}\mathrm{d}t'\,\lambda(t')\left[\exp\left\{\mathrm{i}\frac{H_0}{\hbar}(t'-t)\right\}M\exp\left\{-\mathrm{i}\frac{H_0}{\hbar}(t'-t)\right\}, \rho_{eq}\right]. \tag{16}$$

One can extend the integral on t' to $+\infty$ by multiplying the term to be integrated by the Heaviside function $\theta(t-t')$, which is zero for $t' > t$. The mean of the observable N at time t is then found:

$$\langle N\rangle_t = \mathrm{Tr}(\rho(t)N) = \mathrm{Tr}(\rho_{eq}N) + \frac{\mathrm{i}}{\hbar}\int_{-\infty}^{\infty}\mathrm{d}t'\,\theta(t-t')\lambda(t')\,\mathrm{Tr}\{N[M_I(t'-t), \rho_{eq}]\}$$

$$= \langle N\rangle_{eq} + \int_{-\infty}^{\infty}\mathrm{d}t'\,\lambda(t')\chi_{NM}(t-t') \tag{17}$$

where

$$\chi_{NM}(\tau) = \frac{\mathrm{i}}{\hbar}\mathrm{Tr}\{NM_I(-\tau)\rho_{eq} - N\rho_{eq}M_I(-\tau)\}\,\theta(\tau). \tag{18}$$

Using trace invariance under circular permutation of the operators in the second term, we get

$$\chi_{NM}(\tau) = \frac{\mathrm{i}}{\hbar}\mathrm{Tr}\{NM_I(-\tau)\rho_{eq} - M_I(-\tau)N\rho_{eq}\}\,\theta(\tau)$$

$$= \frac{\mathrm{i}}{\hbar}\langle[N, M_I(-\tau)]\rangle_{eq}\,\theta(\tau). \tag{19}$$

We note that $M_I(-\tau)$ is nothing more than the observable M having evolved in the Heisenberg picture from 0 to $-\tau$ under the action of H_0 (free evolution). Thus, χ_{NM} is the mean in state ρ_{eq} of a commutator involving unperturbed operators taken at two different times (N at time 0, M at time $-\tau$). Equation (17) shows that χ_{NM} defines the linear response of the system in state ρ_{eq} to a weak perturbation. It is thus a linear susceptibility.

c) We now look for the linear response of the field to the perturbation caused by its coupling to the classical dipole $\mathbf{d}(t)$. This is measured through the components of the total electric field $\mathbf{E}'(\mathbf{r})$ at point $\mathbf{r}$ situated well apart from the region of dimension a about the origin where the charges forming the dipole are located. It has been shown in Complement A_{IV} that one then has

$$\mathbf{E}'(\mathbf{r}) = \frac{\mathbf{D}'(\mathbf{r})}{\varepsilon_0} \tag{20}$$

where $\mathbf{D}'$ is the displacement given by (6). One can then apply the theory of linear response in part b) by taking

$$N = E'_n(\mathbf{r}) = \frac{D'_n(\mathbf{r})}{\varepsilon_0} \quad (n = x, y, z) \tag{21.a}$$

$$\lambda(t)\, M = \mathbf{d}(t) \cdot \frac{\mathbf{D}'(\mathbf{0})}{\varepsilon_0} = \sum_m \mathrm{d}_m(t) \frac{D'_m(\mathbf{0})}{\varepsilon_0} \quad (m = x, y, z) \tag{21.b}$$

$$\rho_{eq} = |\,0\,\rangle\langle\,0\,|\,. \tag{21.c}$$

Note first of all that, according to (6), the mean of $\mathbf{D}'(\mathbf{r})$ in the vacuum is zero. The second term of (17) then remains, which is given here by

$$\langle\, E'_n(\mathbf{r})\,\rangle_t = \int_{-\infty}^{+\infty} \mathrm{d}t' \sum_m \chi_{nm}(t - t')\, \mathrm{d}_m(t') = \int_{-\infty}^{\infty} \mathrm{d}\tau \sum_m \chi_{nm}(\tau)\, \mathrm{d}_m(t - \tau) \tag{22}$$

where $\tau = t - t'$ and

$$\chi_{nm}(\tau) = \frac{\mathrm{i}}{\hbar} \left\langle 0 \left| \left[\frac{D'_n(\mathbf{r}, 0)}{\varepsilon_0}, \frac{D'_m(\mathbf{0}, -\tau)}{\varepsilon_0} \right] \right| 0 \right\rangle \theta(\tau)\,. \tag{23}$$

Equation (6) shows that the mathematical expression for $\mathbf{D}'/\varepsilon_0$ is, in the new representation, identical to that of $\mathbf{E}_\perp$ in the Coulomb gauge. The radiation Hamiltonian, H_R, is also the same in both cases. As a result, the value of the commutator in (23) is identical to that for the commutator of the components of $\mathbf{E}_\perp$ found in Complement C_{III} (Equation 22), where one puts $\mathbf{r}_1 = \mathbf{r}$, $t_1 = 0$, $\mathbf{r}_2 = \mathbf{0}$, and $t_2 = -\tau$. We get in this way

$$\chi_{nm}(\tau) = \frac{\mathrm{i}}{\hbar}\,\theta(\tau)\,\frac{\mathrm{i}\hbar c}{4\pi\varepsilon_0} \left\{ \left(\frac{3\, r_n r_m}{r^2} - \delta_{nm} \right) \left(\frac{\delta'(r - c\tau) - \delta'(r + c\tau)}{r^2} - \frac{\delta(r - c\tau) - \delta(r + c\tau)}{r^3} \right) - \left(\frac{r_n r_m}{r^2} - \delta_{nm} \right) \left(\frac{\delta''(r - c\tau) - \delta''(r + c\tau)}{r} \right) \right\}. \tag{24}$$

The functions having $r + c\tau$ as an argument do not contribute, since $\theta(\tau)$ gives 0 for $\tau = -r/c$.

Substituting (24) in (22) and using the properties of the δ-function and its derivatives,

$$\int c\, \mathrm{d}\tau\, \delta(r - c\tau)\, \mathrm{d}_m(t - \tau) = \mathrm{d}_m\!\left(t - \frac{r}{c}\right) \tag{25.a}$$

$$\int c\, \mathrm{d}\tau\, \delta'(r - c\tau)\, \mathrm{d}_m(t - \tau) = -\frac{1}{c}\,\dot{\mathrm{d}}_m\!\left(t - \frac{r}{c}\right) \tag{25.b}$$

$$\int c\, \mathrm{d}\tau\, \delta''(r - c\tau)\, \mathrm{d}_m(t - \tau) = \frac{1}{c^2}\,\ddot{\mathrm{d}}_m\!\left(t - \frac{r}{c}\right). \tag{25.c}$$

we obtain

$$\langle E'_n(\mathbf{r}) \rangle_t = \frac{1}{4\pi\varepsilon_0}\left\{\left(\frac{3\,r_n r_m}{r^2} - \delta_{nm}\right)\left(\frac{d_m\left(t - \frac{r}{c}\right)}{r^3} + \frac{\dot{d}_m\left(t - \frac{r}{c}\right)}{cr^2}\right) + \right.$$

$$\left. + \left(\frac{r_n r_m}{r^2} - \delta_{nm}\right)\frac{\ddot{d}_m\left(t - \frac{r}{c}\right)}{c^2 r}\right\}. \quad (26)$$

At large distance only the term in $1/r$ remains, which describes the radiation field. It is proportional to the dipole's acceleration or rather to the part of it that is orthogonal to the radius vector $\mathbf{r}$.

d) Applying (26) to the case where the dipole is given by (7), one gets, setting $k_0 = \omega_0/c$,

$$\langle \mathbf{E}'(\mathbf{r}) \rangle_t = \frac{1}{4\pi\varepsilon_0}\left[\frac{3\,\mathbf{r}(\mathbf{a}_0 \cdot \mathbf{r})}{r^2} - \mathbf{a}_0\right]\left[\frac{\cos(\omega_0 t - k_0 r)}{r^3} - \frac{k_0 \sin(\omega_0 t - k_0 r)}{r^2}\right] -$$

$$- \frac{q}{4\pi\varepsilon_0}\left[\frac{\mathbf{r}(\mathbf{a}_0 \cdot \mathbf{r})}{r^2} - \mathbf{a}_0\right] k_0^2 \frac{\cos(\omega_0 t - k_0 r)}{r}.$$

One then again gets the results from Exercise 6 of Chapter I [Equation (14)]: the mean of the quantized field is identical to the classical field.

7. Nonresonant scattering. Direct verification of the equality of the transition amplitudes calculated from the Hamiltonians A · p and E · r

Consider an electron bound near the origin by a static potential $V(\mathbf{r})$. This electron can pass from a state $|a\rangle$ with energy E_a to a state $|a'\rangle$ with energy $E_{a'}$ in a scattering process in the course of which an incident photon with wave vector $\mathbf{k}$ and polarization $\boldsymbol{\varepsilon}$ is absorbed and a photon $\mathbf{k}', \boldsymbol{\varepsilon}'$ is emitted in a different mode ($\mathbf{k} \neq \mathbf{k}'$). Denote by $|\varphi_i\rangle = |a, \mathbf{k}\boldsymbol{\varepsilon}\rangle$ and $|\varphi_f\rangle = |a', \mathbf{k}'\boldsymbol{\varepsilon}'\rangle$ the initial and final states of the global system with energies $E_i = E_a + \hbar\omega$ and $E_f = E_{a'} + \hbar\omega'$; there are no photons in the other modes, and $|\mathbf{k}\boldsymbol{\varepsilon}\rangle = a_{\varepsilon}^+(\mathbf{k})|0\rangle$, $|0\rangle$ being the vacuum state. Assume that the scattering process is nonresonant, that is to say, there is no discrete atomic level $|b\rangle$ whose energy E_b is equal to E_i. The purpose of this exercise is to show that the transition amplitude to second order in the electric charge q,

$$\mathcal{T}_{fi} = \langle \varphi_f | H'_I | \varphi_i \rangle + \lim_{\varepsilon \to 0_+} \sum_j \frac{\langle \varphi_f | H_I | \varphi_j \rangle \langle \varphi_j | H_I | \varphi_i \rangle}{E_i - E_j + i\varepsilon} \quad (1)$$

has the same value in the $\mathbf{A} \cdot \mathbf{p}$ and $\mathbf{E} \cdot \mathbf{r}$ representations, on the energy shell, that is, when $E_i = E_f$ (*). The energies E_i, E_f, and E_j are those of

(*) See also Dirac (§64).

the global system atom + radiation. In the long-wavelength approximation, the interaction Hamiltonians are respectively

$$H_I = -\frac{q}{m}\mathbf{p}\cdot\mathbf{A}(\mathbf{0}) + \frac{q^2}{2m}\mathbf{A}^2(\mathbf{0}) \tag{2.a}$$

$$H_I' = -q\mathbf{r}\cdot\mathbf{E}_\perp(\mathbf{0}) \tag{2.b}$$

the quantized fields at $\mathbf{R} = \mathbf{0}$ being

$$\mathbf{A}(\mathbf{0}) = \int \mathrm{d}^3k \sum_\varepsilon \frac{\mathscr{E}_\omega}{\omega}[a_\varepsilon(\mathbf{k}) + a_\varepsilon^+(\mathbf{k})]\,\boldsymbol{\varepsilon} \tag{2.c}$$

$$\mathbf{E}_\perp(\mathbf{0}) = \int \mathrm{d}^3k \sum_\varepsilon \mathrm{i}\mathscr{E}_\omega[a_\varepsilon(\mathbf{k}) - a_\varepsilon^+(\mathbf{k})]\,\boldsymbol{\varepsilon} \tag{2.d}$$

where $\mathscr{E}_\omega = [\hbar\omega/2\varepsilon_0(2\pi)^3]^{1/2}$.

a) Find $\mathcal{T}_{fi}'$ in the $\mathbf{E}\cdot\mathbf{r}$ representation to second order in the electric charge q. Show that the result is of the form

$$\mathcal{T}_{fi}' = C\sum_b\left[\frac{(\mathbf{r}\cdot\boldsymbol{\varepsilon}')_{a'b}(\mathbf{r}\cdot\boldsymbol{\varepsilon})_{ba}}{\hbar(\omega-\omega_{ba})} - \frac{(\mathbf{r}\cdot\boldsymbol{\varepsilon})_{a'b}(\mathbf{r}\cdot\boldsymbol{\varepsilon}')_{ba}}{\hbar(\omega'+\omega_{ba})}\right] \tag{3}$$

where C is a constant to be found, and where the sum is over the electronic levels $|b\rangle$.

b) Find $\mathcal{T}_{fi}$ to the same order in the electric charge, using the interaction Hamiltonian H_I given in (2.a). Denote by $\mathcal{T}_{fi}(2)$ the contribution of the quadratic term of H_I, and by $\mathcal{T}_{fi}(1)$ the contribution of the linear term to second order.

c) Let D be

$$D = \sum_b(\omega_{a'b}\,\omega_{ba} + \omega\omega')\left(\frac{(\mathbf{r}\cdot\boldsymbol{\varepsilon}')_{a'b}(\mathbf{r}\cdot\boldsymbol{\varepsilon})_{ba}}{\omega-\omega_{ba}} - \frac{(\mathbf{r}\cdot\boldsymbol{\varepsilon})_{a'b}(\mathbf{r}\cdot\boldsymbol{\varepsilon}')_{ba}}{\omega'+\omega_{ba}}\right). \tag{4}$$

Show that when $E_{a'} + \hbar\omega' = E_a + \hbar\omega$, D is equal to

$$D = \frac{\hbar}{m}\boldsymbol{\varepsilon}\cdot\boldsymbol{\varepsilon}'\,\delta_{a'a} \tag{5}$$

One can, to begin with, prove that $\omega_{a'b}\omega_{ba} + \omega\omega' = (\omega - \omega_{ba})(\omega' + \omega_{ba})$.

d) Show that the values of $\mathcal{T}_{fi}$ gotten in *a*) and *b*) are equal on the energy shell.

Solution

a) From the $\mathbf{E}\cdot\mathbf{r}$ point of view, the interaction Hamiltonian (2.b) is linear in a and a^+ and cannot directly couple $|\varphi_i\rangle$ and $|\varphi_f\rangle$. The transition between these states is exclusively via an intermediate state [second term of (1)]. This state $|\varphi_j\rangle$ can have no photon ($|b,0\rangle$) or two photons ($|b,\mathbf{k}\boldsymbol{\varepsilon},\mathbf{k}'\boldsymbol{\varepsilon}'\rangle$). In the first case, one has absorption of the photon $\mathbf{k}\boldsymbol{\varepsilon}$ followed by emission of the photon $\mathbf{k}'\boldsymbol{\varepsilon}'$, the energy E_j of the intermediate state being E_b. In the second case, the emission of the photon $\mathbf{k}'\boldsymbol{\varepsilon}'$ precedes the absorption of the photon $\mathbf{k}\boldsymbol{\varepsilon}$, and the energy E_j of the intermediate state is $E_b + \hbar\omega + \hbar\omega'$.

We now calculate $\langle b,0|H_I'|a,\mathbf{k}\boldsymbol{\varepsilon}\rangle$. By using (2.b) and (2.d), we obtain

$$\langle b,0\,|\,H_I'\,|\,a,\mathbf{k}\boldsymbol{\varepsilon}\rangle = -\,q\int \mathrm{d}^3k'' \sum_{\boldsymbol{\varepsilon}''} \mathrm{i}\mathscr{E}_{\omega''}\,\langle 0\,|\,a_{\boldsymbol{\varepsilon}''}(\mathbf{k}'')\,|\,\mathbf{k}\boldsymbol{\varepsilon}\rangle\,(\mathbf{r}\cdot\boldsymbol{\varepsilon}'')_{ba} \tag{6}$$

Since $|\mathbf{k}\boldsymbol{\varepsilon}\rangle = a_{\boldsymbol{\varepsilon}}^+(\mathbf{k})|0\rangle$, by using the commutator

$$[a_{\boldsymbol{\varepsilon}''}(\mathbf{k}''), a_{\boldsymbol{\varepsilon}}^+(\mathbf{k})] = \delta_{\boldsymbol{\varepsilon}\boldsymbol{\varepsilon}''}\,\delta(\mathbf{k}-\mathbf{k}'') \tag{7}$$

we get

$$\langle b,0\,|\,H_I'\,|\,a,\mathbf{k}\boldsymbol{\varepsilon}\rangle = -\,q(\mathbf{r}\cdot\boldsymbol{\varepsilon})_{ba}\,\mathrm{i}\mathscr{E}_\omega\,. \tag{8}$$

A similar calculation gives

$$\begin{aligned}\langle a',\mathbf{k}'\boldsymbol{\varepsilon}'\,|\,H_I'\,|\,b,0\rangle &= -\,q\int \mathrm{d}^3k'' \sum_{\boldsymbol{\varepsilon}''}(-\,\mathrm{i}\mathscr{E}_{\omega''})\,\langle \mathbf{k}'\boldsymbol{\varepsilon}'\,|\,a_{\boldsymbol{\varepsilon}''}^+(\mathbf{k}'')\,|\,0\rangle\,(\mathbf{r}\cdot\boldsymbol{\varepsilon}'')_{a'b}\\ &= q(\mathbf{r}\cdot\boldsymbol{\varepsilon}')_{a'b}\,\mathrm{i}\mathscr{E}_{\omega'}\,.\end{aligned} \tag{9}$$

The matrix elements

$$\langle b,\mathbf{k}\boldsymbol{\varepsilon},\mathbf{k}'\boldsymbol{\varepsilon}'|H_I'|a,\mathbf{k}\boldsymbol{\varepsilon}\rangle \quad \text{and} \quad \langle a',\mathbf{k}'\boldsymbol{\varepsilon}'|H_I'|b,\mathbf{k}\boldsymbol{\varepsilon},\mathbf{k}'\boldsymbol{\varepsilon}'\rangle$$

are found in an identical fashion, which gives finally for $\mathcal{T}'_{fi}$

$$\mathcal{T}'_{fi} = q^2\,\mathscr{E}_\omega\,\mathscr{E}_{\omega'}\sum_b\left[\frac{(\mathbf{r}\cdot\boldsymbol{\varepsilon}')_{a'b}(\mathbf{r}\cdot\boldsymbol{\varepsilon})_{ba}}{\hbar(\omega-\omega_{ba})} - \frac{(\mathbf{r}\cdot\boldsymbol{\varepsilon})_{a'b}(\mathbf{r}\cdot\boldsymbol{\varepsilon}')_{ba}}{\hbar(\omega'+\omega_{ba})}\right]. \tag{10}$$

This equation is the same as Equation (3) if one takes

$$C = q^2\,\mathscr{E}_\omega\,\mathscr{E}_{\omega'} = \frac{q^2\,\hbar}{2\,\varepsilon_0(2\,\pi)^3}\sqrt{\omega\omega'}\,. \tag{11}$$

We have assumed that no denominator in (10) can vanish, which allows us to take $\varepsilon = 0$ in Equation (1).

b) In contrast to the above, it is now possible to directly couple $|\varphi_i\rangle$ and $|\varphi_f\rangle$ using the quadratic term in q in H_I. One first finds this initial contribution to $\mathcal{T}_{fi}$, called $\mathcal{T}_{fi}(2)$. Since $q^2\mathbf{A}^2(\mathbf{0})/2m$ does not act on the particles, it is clear that $\mathcal{T}_{fi}(2)$ is proportional to $\delta_{a,a'}$. To find the matrix element, it is sufficient to retain in $\mathbf{A}^2(\mathbf{0})$, expanded with the aid of (2.c), the products $a_{\boldsymbol{\varepsilon}_1''}^+(\mathbf{k}_1'')a_{\boldsymbol{\varepsilon}_2''}(\mathbf{k}_2'')$ and $a_{\boldsymbol{\varepsilon}_1''}(\mathbf{k}_1'')a_{\boldsymbol{\varepsilon}_2''}^+(\mathbf{k}_2'')$. After an integration over $\mathbf{k}_1''$ and $\mathbf{k}_2''$, we get using (7)

$$\mathcal{T}_{fi}(2) = \frac{q^2}{m}\,\frac{\mathscr{E}_\omega\,\mathscr{E}_{\omega'}}{\omega\omega'}\,\boldsymbol{\varepsilon}\cdot\boldsymbol{\varepsilon}'\,\delta_{a',a}\,. \tag{12}$$

The second term $\mathcal{T}''_{fi}(1)$ requires a sum on the intermediate states $|\varphi_j\rangle$. This second term is found like $\mathcal{T}'_{fi}$ in *a*) by replacing $-q\mathbf{r}\cdot\mathbf{E}$ with $-q\mathbf{p}\cdot\mathbf{A}/m$. Using (2.c) for $\mathbf{A}(\mathbf{0})$, one finds

$$\mathcal{T}_{fi}(1) = \frac{q^2}{m^2}\,\frac{\mathscr{E}_\omega\,\mathscr{E}_{\omega'}}{\omega\omega'}\sum_b\left[\frac{(\mathbf{p}\cdot\boldsymbol{\varepsilon}')_{a'b}(\mathbf{p}\cdot\boldsymbol{\varepsilon})_{ba}}{\hbar(\omega-\omega_{ba})} - \frac{(\mathbf{p}\cdot\boldsymbol{\varepsilon})_{a'b}(\mathbf{p}\cdot\boldsymbol{\varepsilon}')_{ba}}{\hbar(\omega'+\omega_{ba})}\right]. \tag{13}$$

Equation (B.43), relating the matrix elements of $\mathbf{p}\cdot\boldsymbol{\varepsilon}$ to those of $\mathbf{r}\cdot\boldsymbol{\varepsilon}$, allows one to transform (13) into

$$\mathcal{T}_{fi}(1) = q^2 \mathcal{E}_\omega \mathcal{E}_{\omega'} \sum_b \left(-\frac{\omega_{a'b}\,\omega_{ba}}{\omega\omega'}\right)\left[\frac{(\mathbf{r}\cdot\boldsymbol{\varepsilon}')_{a'b}(\mathbf{r}\cdot\boldsymbol{\varepsilon})_{ba}}{\hbar(\omega-\omega_{ba})} - \frac{(\mathbf{r}\cdot\boldsymbol{\varepsilon})_{a'b}(\mathbf{r}\cdot\boldsymbol{\varepsilon}')_{ba}}{\hbar(\omega'+\omega_{ba})}\right]. \tag{14}$$

c) To show

$$\omega_{a'b}\,\omega_{ba} + \omega\omega' = (\omega - \omega_{ba})(\omega' + \omega_{ba}) \tag{15}$$

note that when $E_{a'} = E_a + \hbar(\omega - \omega')$, one can replace $\omega_{a'b}$ by

$$\omega_{a'b} = \omega - \omega' - \omega_{ba} \tag{16}$$

in the first term of (15) and equate the two sides of (15). Equation (15) then allows us to rewrite (4) in the form

$$D = \sum_b \left[(\omega' + \omega_{ba})(\mathbf{r}\cdot\boldsymbol{\varepsilon}')_{a'b}(\mathbf{r}\cdot\boldsymbol{\varepsilon})_{ba} - (\omega - \omega_{ba})(\mathbf{r}\cdot\boldsymbol{\varepsilon})_{a'b}(\mathbf{r}\cdot\boldsymbol{\varepsilon}')_{ba}\right]. \tag{17}$$

Now, on the energy shell, $\omega - \omega_{ba} = \omega' + \omega_{a'b}$. By transforming the second term in the brackets with this equality, we find

$$D = \omega' \sum_b \left[(\mathbf{r}\cdot\boldsymbol{\varepsilon}')_{a'b}(\mathbf{r}\cdot\boldsymbol{\varepsilon})_{ba} - (\mathbf{r}\cdot\boldsymbol{\varepsilon})_{a'b}(\mathbf{r}\cdot\boldsymbol{\varepsilon}')_{ba}\right] -$$
$$- \frac{i}{m}\sum_b \left[(\mathbf{r}\cdot\boldsymbol{\varepsilon}')_{a'b}(\mathbf{p}\cdot\boldsymbol{\varepsilon})_{ba} - (\mathbf{p}\cdot\boldsymbol{\varepsilon})_{a'b}(\mathbf{r}\cdot\boldsymbol{\varepsilon}')_{ba}\right] \tag{18}$$

after having again used (B.43) to write the second sum on $|b\rangle$. Using the closure relation this then becomes

$$D = \omega' \langle a' \,|\, [(\mathbf{r}\cdot\boldsymbol{\varepsilon}'), (\mathbf{r}\cdot\boldsymbol{\varepsilon})] \,|\, a \rangle - \frac{i}{m}\langle a' \,|\, [(\mathbf{r}\cdot\boldsymbol{\varepsilon}'), (\mathbf{p}\cdot\boldsymbol{\varepsilon})] \,|\, a \rangle. \tag{19}$$

The first commutator is zero, and the second is a *c*-number equal to $i\hbar\boldsymbol{\varepsilon}\cdot\boldsymbol{\varepsilon}'$, which proves Equation (5).

d) Using (4), (10), (12), and (14), one can write

$$\mathcal{T}'_{fi} - \mathcal{T}_{fi} = \frac{q^2 \mathcal{E}_\omega \mathcal{E}_{\omega'}}{\hbar\omega\omega'} D - \frac{q^2 \mathcal{E}_\omega \mathcal{E}_{\omega'}}{m\omega\omega'} \boldsymbol{\varepsilon}\cdot\boldsymbol{\varepsilon}'\,\delta_{a'a}. \tag{20}$$

The equality (5) then gives

$$\mathcal{T}'_{fi} = \mathcal{T}_{fi}. \tag{21}$$

The transition matrix on the energy shell is thus the same in the $\mathbf{A}\cdot\mathbf{p}$ and $\mathbf{E}\cdot\mathbf{r}$ representations. Note that the proof of this equality assumes nothing about the values of ω and ω', with the exception of the long-wavelength approximation and the absence of resonant atomic levels. The preceding result can then be applied to various problems such as Rayleigh scattering, Thomson scattering, or Raman scattering.

CHAPTER V

Introduction to the Covariant Formulation of Quantum Electrodynamics

The purpose of this last chapter is to give the reader a first view of relativistic quantum electrodynamics and to introduce the necessary background for the study of more advanced books in this field.

Thus far, by choosing the Coulomb gauge, we have deliberately renounced the use of manifestly covariant equations for the electromagnetic field. We have also treated the particles nonrelativistically. Such an approach is sufficient for low-energy physics and simplifies the theoretical formalism as much as possible.

We now return to these two limitations and try to give some idea about other more elaborate approaches which are essential at high energy (when the particles have a high velocity or the incident photons have a frequency large compared to $m_\alpha c^2/h$) or for the study of radiative corrections (virtual emissions and reabsorptions of high-frequency photons, covariant introduction of renormalized charges and masses). This chapter essentially treats the *covariant formulation* of classical and quantum electrodynamics in the Lorentz gauge. For simplification, our treatment will be limited either to the *free field* in the absence of sources, or to the field interacting with *external sources* whose motions are given a priori. The relativistic description of the particles and of their coupling to the field is a much broader and more complex problem. The particles must then be considered as the elementary excitations of a *relativistic quantized matter field* coupled to the photon field. We will give in Complement A_V an elementary introduction to the theory of coupled Dirac and Maxwell fields in the Lorentz gauge. We will also show in Complement B_V how, starting with a completely relativistic theory, it is possible to justify the nonrelativistic, Coulomb-gauge Hamiltonians used in the rest of this book.

Recall first of all the procedures followed in Chapter II to quantize electrodynamics. Starting from the standard Lagrangian which leads to the Maxwell–Lorentz equations, we eliminated all the redundant degrees of freedom of the potentials to transform this Lagrangian into an equivalent one where only the essential dynamical variables, each having a conjugate momentum, appear. The canonical quantization of the theory is then straightforward.

If it has the advantage of simplicity, such a procedure also has the disadvantage of not retaining the manifest covariance of the field. Although it does not alter the fundamental relativistic nature of the field, the elimination of the scalar potential has broken the symmetry between the four components of the potential four-vector. In the same way, the dynamical variables of the vector potential in the Coulomb gauge, that is, the transverse components $\mathscr{A}_\varepsilon(\mathbf{k})$ and $\mathscr{A}_{\varepsilon'}(\mathbf{k})$ in reciprocal space, do not transform simply under Lorentz transformations. Now there are problems in quantum electrodynamics, such as renormalization and the elimination of divergent quantities, for which it is essential to deal only with manifestly covariant equations.

Here we are going to treat the vector potential $\mathbf{A}$ and the scalar potential U symmetrically. The simple approach in Chapter II must thus be abandoned. Indeed, the symmetry which we desire to preserve between $\mathbf{A}$ and U prohibits us from eliminating the redundant degrees of freedom of the potentials in the Lagrangian itself, as in Chapter II. We are now obliged to consider $\mathbf{A}$ and U as *independent* dynamical variables in the Lagrangian, each with a conjugate momentum. As a result, after canonical quantization, there are four kinds of photons associated with the four components of the potential four-vector. The problem of the redundancy of the potentials, which we have ignored in the Lagrangian, necessarily arises later. We will indeed see that the solutions of the equations of motion no longer coincide in general with the solutions of the Maxwell equations. It is then necessary to introduce, in addition to the Lagrangian, a *subsidiary condition* allowing one to select from all the possible solutions those that have a physical meaning. One of the essential purposes of this chapter is precisely to discuss the problems brought up by the realization of such a project in quantum theory and to give an idea of the way in which these problems can be resolved by constructing a state space for the radiation with four kinds of independent photons and then characterizing a subspace of physical states by means of the subsidiary condition.

We begin (Part A) by introducing a new, manifestly covariant Lagrangian for the classical fields, differing from the standard Lagrangian in the sense that it contains $\dot{U}$, so that U has a conjugate momentum. This Lagrangian leads to equations of motion which agree with Maxwell's equations only if a subsidiary condition—that is, the Lorentz condition—is imposed on the potentials. Starting with this Lagrangian, we then calculate the momenta conjugate with the potentials, the Hamiltonian, and the classical normal variables of the field, that is, the variables evolving independently of one another in the absence of sources. The canonical quantization of this theory is done in Part B, the normal variables becoming the creation and annihilation operators for transverse, longitudinal, and scalar photons. We establish manifestly covariant expressions for the commutators of the free potentials, and we analyze the difficulties arising in quantum theory and due to the Lorentz condition. We then study (Part C) a possible solution to these difficulties, involving the introduction of an indefinite metric in Hilbert radiation space. The preceding ideas are finally illustrated by a simple example, the effect on the field of its interaction with two fixed charges. This leads to a new derivation of the Coulomb interaction and its interpretation as resulting from an exchange of photons between the two charges (Part D).

A—CLASSICAL ELECTRODYNAMICS IN THE LORENTZ GAUGE

1. Lagrangian Formalism

a) Covariant Notation. Ordinary Notation

Before beginning we give the notation to be used for the potentials.

The *covariant notation* A^μ for the potential four-vector uses a Greek index μ which can take on four values: 1, 2, 3 for the spatial components and 0 for the time component. It is desirable then to distinguish the *contravariant components* A^μ with superscript index from the *covariant components* A_μ with subscript index, related by

$$A_\mu = \sum_\nu g_{\mu\nu} A^\nu \tag{A.1}$$

where $g_{\mu\nu}$ is the diagonal metric tensor ($g_{00} = +1$, $g_{11} = g_{22} = g_{33} = -1$).

The ordinary notation for the vector potential, A_j, uses a Latin index which can take on the values x, y, z. For the scalar potential U we also use the notation

$$A_s = \frac{U}{c} \tag{A.2}$$

so as to have a potential with the same dimension as the vector potential. The upper or lower position of the indices j and s has no importance in A_j and A_s.

Finally we give the equations relating the different components which have just been introduced:

$$\begin{aligned} A^0 &= A_0 = A_s = U/c \\ A^1 &= -A_1 = A_x \\ A^2 &= -A_2 = A_y \\ A^3 &= -A_3 = A_z . \end{aligned} \tag{A.3}$$

b) Selection of a New Lagrangian for the Field

In the standard Lagrangian [(B.5) of Chapter II], the scalar potential U does not have a conjugate momentum, since $\partial\mathscr{L}/\partial\dot{U}$ is identically zero. If we want to treat $U(\mathbf{r})$ and $A_j(\mathbf{r})$ symmetrically, it is necessary to modify the standard Lagrangian so that $U(\mathbf{r})$ has a conjugate momentum. For this property to be true a priori for the free field, the radiation Lagrangian density $\mathcal{L}_R$ has to be modified. We must impose two conditions on $\mathcal{L}_R$:

(i) $\mathcal{L}_R$ must be manifestly covariant

(ii) $\mathcal{L}_R$ must contain $\dot{U}$.

We consider then the following Lagrangian density:

$$\mathcal{L}_R = -\frac{\varepsilon_0 c^2}{2} \sum_{\mu\nu} (\partial_\mu A^\nu)(\partial^\mu A_\nu) . \tag{A.4}$$

which is also written in the usual notation as

$$\mathcal{L}_R = \frac{\varepsilon_0}{2}\left[\dot{\mathbf{A}}^2 - \left(\frac{\dot{U}}{c^2}\right)^2 - c^2 \sum_{ij} (\partial_i A_j)^2 + (\nabla U)^2\right]. \tag{A.5}$$

Equation (A.4) is clearly covariant, and U has a conjugate momentum, since $\dot{U}$ appears explicitly in (A.5).

We analyze now several properties of $\mathcal{L}_R$. Note first of all that $\mathcal{L}_R$ involves only quadratic functions of A_μ. This insures that the Lagrange equations derived from (A.4) or (A.5) are linear with respect to the potentials. In addition, $\mathcal{L}_R$ involves only the first-order derivatives of the potentials, which is not surprising, since only the time derivatives of first order are permitted in the Lagrangian formalism and the covariance imposes the same condition for spatial derivatives. Note finally, and this is an important point, that the new Lagrangian is *not equivalent* to the standard Lagrangian, whose free-field density $\mathcal{L}_R^{st}$ is written [see Equation (B.26), Chapter II]

$$\mathcal{L}_R^{st} = -\frac{\varepsilon_0 c^2}{4} \sum_{\mu,\nu} F_{\mu\nu} F^{\mu\nu} \tag{A.6}$$

where

$$F_{\mu\nu} = \partial_\mu A_\nu - \partial_\nu A_\mu \tag{A.7}$$

is the electromagnetic field tensor. We show in Part C below that the Lagrange equations associated with (A.4) differ in general from those derived from (A.6), which, as shown in Chapter II, are the Maxwell equations. One thus gets the Maxwell equations again only if a *subsidiary condition* is imposed on the potentials.

Heretofore, we have only considered the free field. To describe the interaction between the field and the particles, we retain the same density $\mathcal{L}_I$ as that in the standard Lagrangian [see (B.4.e) of Chapter II],

$$\mathcal{L}_I = \mathbf{j} \cdot \mathbf{A} - \rho U = -\sum_\mu j_\mu A^\mu \tag{A.8}$$

where j_μ is the current four-vector $(c\rho, \mathbf{j})$. In what follows in this chapter, we will be interested principally in the fields (the equation of motion of the fields, the commutation relations of the fields, etc.), and we will use only the Lagrangian densities $\mathscr{L}_R$ and $\mathscr{L}_I$.

Remarks

(i) In relativistic quantum electrodynamics, the particles themselves are described by a relativistic field—the Dirac field, if the particles are electrons and positrons. The Lagrangian of the particles, L_P, is the Lagrangian of the free Dirac field, and the current $j_\mu(x)$ is equal to $qc\bar{\psi}(x)\gamma_\mu\psi(x)$, where the γ_μ are the Dirac matrices, and ψ and $\bar{\psi}$ the Dirac field and its relativistic adjoint. After quantization, the Dirac field becomes a quantized field whose elementary excitations describe the electrons and positrons (see Complement A_V).

(ii) In the covariant formulations of quantum electrodynamics, one frequently uses another Lagrangian density for the field, called the *Fermi Lagrangian*,

$$\mathscr{L}_R^F = -\varepsilon_0 c^2\left[\sum_{\mu\nu}\frac{1}{4}F_{\mu\nu}F^{\mu\nu} + \frac{1}{2}\left(\sum_\mu \partial_\mu A^\mu\right)^2\right] \tag{A.9}$$

which differs from $\mathscr{L}_R$, but is equivalent to it. An elementary calculation shows that (A.4) and (A.9) differ by a four-divergence

$$\mathscr{L}_R - \mathscr{L}_R^F = \frac{\varepsilon_0 c^2}{2}\sum_{\mu\nu}\partial_\mu[A^\mu\partial_\nu A^\nu - A^\nu\partial_\nu A^\mu]\,. \tag{A.10}$$

The first term of (A.9) is the standard Lagrangian density $\mathscr{L}_R^{\text{st}}$. Thus, to go from $\mathscr{L}_R^{\text{st}}$ to $\mathscr{L}_R^F$, it is necessary to add a term proportional to $(\Sigma_\mu\partial_\mu A^\mu)^2$, which is not a four-divergence. The densities $\mathscr{L}_R^{\text{st}}$ and $\mathscr{L}_R^F$, and as a result $\mathscr{L}_R^{\text{st}}$ and $\mathscr{L}_R$, are thus not equivalent.

c) Lagrange Equations for the Field

Since the new Lagrangian density $\mathscr{L} = \mathscr{L}_R + \mathscr{L}_I$ is not equivalent to $\mathscr{L}_R^{\text{st}} + \mathscr{L}_I$, there is no reason that the new Lagrange equations for the field should coincide with the Maxwell equations [(A.11.a, b) of Chapter I]. The latter can be put in the form

$$\Box\mathbf{A} = \frac{1}{\varepsilon_0 c^2}\mathbf{j} - \nabla\Lambda \tag{A.11.a}$$

$$\Box U = \frac{1}{\varepsilon_0}\rho + \frac{\partial}{\partial t}\Lambda \tag{A.11.b}$$

where $\square$ is the d'Alembertian and where

$$\Lambda = \frac{1}{c^2}\frac{\partial U}{\partial t} + \nabla \cdot \mathbf{A}. \qquad \text{(A.11.c)}$$

The form (A.11) for the Maxwell equations makes it easier to compare them with the new Lagrange equations.

The new Lagrange density is the sum of (A.5) and (A.8),

$$\mathscr{L} = \frac{\varepsilon_0}{2}\left[\dot{\mathbf{A}}^2 - \left(\frac{\dot{U}}{c}\right)^2 - c^2 \sum_{ij} (\partial_i A_j)^2 + (\nabla U)^2\right] - \rho U + \mathbf{j} \cdot \mathbf{A}. \qquad \text{(A.12)}$$

Consider first the Lagrange equation relative to the components $A_i(\mathbf{r})$ of the vector potential. $\dot{A}_i$ and the space derivatives of A_i arise only in the Lagrangian density of the field (A.5), whereas A_i arises only in the interaction term. It follows that

$$\partial\mathscr{L}/\partial\dot{A}_i = \varepsilon_0\, \dot{A}_i \qquad \text{(A.13.a)}$$

$$\partial\mathscr{L}/\partial(\partial_j A_i) = -\, \varepsilon_0\, c^2\, \partial_j A_i \qquad \text{(A.13.b)}$$

$$\partial\mathscr{L}/\partial A_i = j_i\,. \qquad \text{(A.13.c)}$$

By applying the Lagrange equations relative to a system having a continuous infinity of degrees of freedom [Equation (A.39) of Chapter II], we then find

$$\varepsilon_0\left[\ddot{A}_i - c^2 \sum_j \frac{\partial^2 A_i}{\partial x_j^2}\right] = j_i \qquad \text{(A.14)}$$

that is,

$$\square \mathbf{A} = \frac{1}{\varepsilon_0 c^2}\mathbf{j}\,. \qquad \text{(A.15.a)}$$

An analogous calculation gives for the Lagrange equation relative to the scalar potential

$$\square U = \frac{1}{\varepsilon_0}\rho\,. \qquad \text{(A.15.b)}$$

The equations (A.15.a, b) differ from the Maxwell equations (A.11.a, b).

Remark

We have considered here the Lagrange equations for the fields. If one takes for the particle Lagrangian the Lagrangian of the Dirac field $\psi(x)$, and if one takes as the current $j_\mu = qc\bar{\psi}\gamma_\mu\psi$ (see Remark i of §A.1.*b*), the Lagrange equation relative to ψ is just the Dirac equation in the presence of the potential A_μ (see exercises 5 and 6).

d) The Subsidiary Condition

There is actually a choice of potentials for which Equations (A.11) and (A.15) agree. It suffices to set

$$\Lambda = \nabla \cdot \mathbf{A} + \frac{1}{c^2}\dot{U} = 0 \tag{A.16.a}$$

that is, to use the Lorentz gauge. The condition (A.16.a) is also written in covariant notation as

$$\sum_\mu \partial_\mu A^\mu = 0\,. \tag{A.16.b}$$

The two approaches (standard Lagrangian, new Lagrangian) then lead to the same result if the Lorentz condition (A.16) is imposed as a subsidiary condition.

We will now verify that (A.16) is compatible with the equations of motion (A.15). A combination of (A.15.a) and (A.15.b) leads to the following equation of evolution for Λ:

$$\Box\Lambda = \frac{1}{\varepsilon_0 c^2}\left(\nabla \cdot \mathbf{j} + \frac{\partial \rho}{\partial t}\right) = 0 \tag{A.17}$$

as a result of conservation of charge. Thus, if initially $\Lambda = \dot{\Lambda} = 0$, then Λ remains identically zero at all times. Now it is always possible at the initial instant to impose the condition $\Lambda = 0$ between the generalized coordinates $\mathbf{A}$, U and the generalized velocities $\dot{\mathbf{A}}$, $\dot{U}$. As for $\dot{\Lambda}$, one can find its value at the initial time by using the equation of motion (A.15.b) to reexpress $\ddot{U}$ as a function of ΔU and ρ. One gets

$$\dot{\Lambda} = \nabla \cdot (\dot{\mathbf{A}} + \nabla U) + \frac{\rho}{\varepsilon_0} = -\nabla \cdot \mathbf{E} + \frac{\rho}{\varepsilon_0} \tag{A.18}$$

which is also zero if the initial conditions are such that $\mathbf{E}$ obeys the equation $\nabla \cdot \mathbf{E} = \rho/\varepsilon_0$. One can then take $\Lambda = \dot{\Lambda} = 0$ at the initial time and have subsequently, from (A.17), $\Lambda = 0$ at all times.

Note finally that the condition (A.16) still allows certain gauge changes:

$$A_\mu \to A'_\mu = A_\mu - \partial_\mu f\,. \tag{A.19}$$

It suffices that

$$\partial_\mu \partial^\mu f = \Box f = 0 \tag{A.20}$$

in which case A'_μ is also a potential satisfying the Lorentz condition.

e) The Lagrangian Density in Reciprocal Space

To conclude this part, we will write the Lagrangian density $\bar{\mathscr{L}}_R$ of the field when the dynamical variables are expanded in reciprocal space:

$$L_R = \int \mathrm{d}^3r \, \mathscr{L}_R = \int \mathrm{d}^3k \, \overline{\mathscr{L}}_R . \tag{A.21}$$

Recall that the **k** integral in (A.21) is only over a half space, as in §B.1.*b* of Chapter II. By using the Parseval–Plancherel relation to transform the **r** integral of (A.5) and by using A_s and its Fourier transform $\mathscr{A}_s$ in place of U and $\mathscr{U}$ [see (A.2)], one gets

$$\overline{\mathscr{L}}_R = \varepsilon_0[\dot{\mathscr{A}}^* \cdot \dot{\mathscr{A}} - \omega^2 \mathscr{A}^* \cdot \mathscr{A} - \dot{\mathscr{A}}_s^* \dot{\mathscr{A}}_s + \omega^2 \mathscr{A}_s^* \mathscr{A}_s] \tag{A.22}$$

with $\omega = ck$. The Lagrangian density (A.22) describes four independent harmonic oscillators associated respectively with the three spatial components and the time component of the four-potential. For $\mathscr{A}_s$ the sign differs from the usual sign, which, as we shall see, has important consequences for the quantization.

2. Hamiltonian Formalism

a) Conjugate Momenta of the Potentials

Let π_j and π_s be the conjugate momenta of $\mathscr{A}_j$ and $\mathscr{A}_s$. By using the definition (A.54) of Chapter II for the conjugate momenta and (A.22), one gets (with $j = x, y, z$)

$$\pi_j = \partial\overline{\mathscr{L}}_R/\partial\dot{\mathscr{A}}_j^* = \varepsilon_0 \dot{\mathscr{A}}_j \tag{A.23.a}$$

$$\pi_s = \partial\overline{\mathscr{L}}_R/\partial\dot{\mathscr{A}}_s^* = -\varepsilon_0 \dot{\mathscr{A}}_s \tag{A.23.b}$$

which gives in real space

$$\Pi_j = \varepsilon_0 \dot{A}_j \tag{A.24.a}$$

$$\Pi_s = -\varepsilon_0 \dot{A}_s . \tag{A.24.b}$$

Equations (A.23) and (A.24) remain valid in the presence of interaction with particles, since the interaction Lagrangian does not involve $\dot{\mathscr{A}}_j$ and $\dot{\mathscr{A}}_s$ and therefore does not contribute to π_j and π_s.

It is useful for what follows to rewrite the Lorentz subsidiary condition

(A.16.a) in reciprocal space by using (A.23.b) to reexpress $\dot{\mathscr{U}} = \dot{\mathscr{A}}_s/c$ as a function of π_s:

$$\mathrm{i}\mathbf{k} \cdot \mathscr{A} = \frac{1}{\varepsilon_0\, c}\, \pi_s\,. \tag{A.25}$$

b) THE HAMILTONIAN OF THE FIELD

The Hamiltonian H_R of the field alone can be gotten from $\mathscr{L}_R$. Then H_R is given by the integral over a reciprocal half space of the Hamiltonian density $\overline{\mathscr{H}}_R$ equal to [see Equation (A.59), Chapter II]

$$\overline{\mathscr{H}}_R = [\boldsymbol{\pi} \cdot \dot{\boldsymbol{\mathscr{A}}}^* + \boldsymbol{\pi}^* \cdot \dot{\boldsymbol{\mathscr{A}}} + \pi_s\, \dot{\mathscr{A}}_s^* + \pi_s^*\, \dot{\mathscr{A}}_s] - \overline{\mathscr{L}}_R \tag{A.26}$$

that is, also from (A.23),

$$\overline{\mathscr{H}}_R = \varepsilon_0 \left[\frac{1}{\varepsilon_0^2}\, \boldsymbol{\pi}^* \cdot \boldsymbol{\pi} + \omega^2\, \boldsymbol{\mathscr{A}}^* \cdot \boldsymbol{\mathscr{A}} - \frac{1}{\varepsilon_0^2}\, \pi_s^*\, \pi_s - \omega^2\, \mathscr{A}_s^*\, \mathscr{A}_s \right]. \tag{A.27}$$

We identify in (A.27) four harmonic-oscillator Hamiltonians, three with the + sign for the three components of the vector potential ($\mathscr{A}_j$, π_j) and one with the − sign for the scalar potential ($\mathscr{A}_s$, π_s). This negative sign seems to pose a problem, since the energy of the radiation can apparently become negative. We will see later that the subsidiary condition (A.25) prevents that from happening.

Remarks

(i) The Lagrangian of the field, L_R, is invariant under spatial translation. A treatment analogous to that of Complement $\mathrm{B_{II}}$ (§ 4) allows one to find the constant of the motion associated with this invariance of L_R, which is just the global momentum $\mathbf{P}_R$ of the radiation (see also Exercise 5 of Chapter II). $\mathbf{P}_R$ is given by

$$\mathbf{P}_R = -\,\mathrm{i} \int \mathrm{d}^3k\; \mathbf{k}(\boldsymbol{\pi}^* \cdot \boldsymbol{\mathscr{A}} + \pi_s^*\, \mathscr{A}_s - \boldsymbol{\pi} \cdot \boldsymbol{\mathscr{A}}^* - \pi_s\, \mathscr{A}_s^*)\,. \tag{A.28}$$

(ii) Since we have not explicitly given the expression for the Lagrangian L_P of the particles, it is impossible here to find the total Hamiltonian H. If, in contrast, the sources have an externally imposed motion, L_P no longer appears in the Lagrangian, which reduces to $L_R + L_I$. One then finds for the Hamiltonian H of the field coupled to such external sources

$$H = H_R - L_I = H_R + \int \mathrm{d}^3k\,(j_{e\mu}^*\, \mathscr{A}^\mu + j_{e\mu}\, \mathscr{A}^{\mu *}) \tag{A.29}$$

where the currents $j_{e\mu}$ are given functions of $\mathbf{k}$ and t.

c) Hamilton–Jacobi Equations for the Free Field

Application of Equations (A.60) of Chapter II to the Hamiltonian density (A.27) gives the equations of evolution of $\mathscr{A}_j$ and π_j,

$$\dot{\mathscr{A}}_j = \partial \mathscr{H}_R / \partial \pi_j^* = \frac{1}{\varepsilon_0} \pi_j \tag{A.30.a}$$

$$\dot{\pi}_j = - \partial \overline{\mathscr{H}}_R / \partial \mathscr{A}_j^* = - \varepsilon_0 \omega^2 \mathscr{A}_j \tag{A.30.b}$$

and those of $\mathscr{A}_s$ and π_s,

$$\dot{\mathscr{A}}_s = \partial \overline{\mathscr{H}}_R / \partial \pi_s^* = - \frac{1}{\varepsilon_0} \pi_s \tag{A.31.a}$$

$$\dot{\pi}_s = - \partial \overline{\mathscr{H}}_R / \partial \mathscr{A}_s^* = \varepsilon_0 \omega^2 \mathscr{A}_s . \tag{A.31.b}$$

Equations (A.30.a) and (A.31.a) have already been seen in (A.23.a) and (A.23.b). It is important to note the sign difference between the right-hand sides of (A.30) and (A.31), due to the sign difference between the first and last terms of (A.27).

Remark

In the presence of external sources, it is necessary to use (A.29). Equations (A.30.a) and (A.31.a) remain unchanged. It is necessary to add j_{ej} to the right side of (A.30.b), and $-j_{e0} = -c\rho_e$ to the right side of (A.31.b), which gives

$$\dot{\pi}_j = - \varepsilon_0 \omega^2 \mathscr{A}_j + j_{ej} \tag{A.30.b'}$$

$$\dot{\pi}_s = \varepsilon_0 \omega^2 \mathscr{A}_s - c\rho_e . \tag{A.31.b'}$$

3. Normal Variables of the Classical Field

a) Definition

The normal variables of the classical field are linear combinations of the dynamical field variables and their conjugate momenta which have the property of evolving independently of one another in the absence of sources, i.e., for the free field. Thus if one takes

$$\alpha_j = \sqrt{\frac{\varepsilon_0}{2 \hbar \omega}} \left[\omega \mathscr{A}_j + \frac{i}{\varepsilon_0} \pi_j \right] \tag{A.32}$$

it follows from (A.30.a) and (A.30.b) that

$$\dot{\alpha}_j + i\omega \alpha_j = 0 . \tag{A.33}$$

In the absence of sources, the normal variable $\alpha_j(\mathbf{k}, t)$ is only coupled with itself and evolves as $e^{-i\omega t}$. Note that the definition (A.32) agrees with that used in Chapters I and II [see (C.46) of Chapter II].

On the other hand, because of the sign difference between the right-hand sides of (A.30) and (A.31), it is necessary to take

$$\alpha_s = \sqrt{\frac{\varepsilon_0}{2\hbar\omega}}\left[\omega\mathscr{A}_s - \frac{i}{\varepsilon_0}\pi_s\right] \quad \text{(A.34)}$$

to have for α_s an equation analogous to (A.33),

$$\dot{\alpha}_s + i\omega\alpha_s = 0 . \quad \text{(A.35)}$$

If one took the + sign on the right in (A.34) as in (A.32), α_s would evolve as $e^{+i\omega t}$ and not as $e^{-i\omega t}$.

The solutions of (A.33) and (A.35) are written

$$\alpha_j(\mathbf{k}, t) = \alpha_j(\mathbf{k})\, e^{-i\omega t} \quad \text{(A.36.a)}$$

$$\alpha_s(\mathbf{k}, t) = \alpha_s(\mathbf{k})\, e^{-i\omega t} . \quad \text{(A.36.b)}$$

Remark

Equations (A.32) and (A.34) continue to define α_j and α_s in the presence of interaction. But the equations of motion (A.33) and (A.35) then contain source terms. For example, with external sources, (A.33) and (A.35) must be replaced by [see (A.30.b′) and (A.31.b′)]

$$\dot{\alpha}_j + i\omega\alpha_j = \frac{i}{\sqrt{2\varepsilon_0\hbar\omega}}\, j_{ej} \quad \text{(A.33′)}$$

$$\dot{\alpha}_s + i\omega\alpha_s = \frac{i}{\sqrt{2\varepsilon_0\hbar\omega}}\, c\rho_e . \quad \text{(A.35′)}$$

b) Expansion of the Potential in Normal Variables

Recall first that the potentials A_j and A_s are real, which implies

$$\mathscr{A}_j(\mathbf{k}) = \mathscr{A}_j^*(-\mathbf{k}) \quad \text{(A.37.a)}$$

$$\mathscr{A}_s(\mathbf{k}) = \mathscr{A}_s^*(-\mathbf{k}) \quad \text{(A.37.b)}$$

and analogous equations for π_j and π_s.

We then replace $\mathbf{k}$ by $-\mathbf{k}$ in (A.32) and (A.34) and take the complex conjugate. Using (A.37) and taking $\alpha_{j-}^* = \alpha_j^*(-\mathbf{k}, t)$, $\alpha_{s-}^* = \alpha_s^*(-\mathbf{k}, t)$, we get

$$\alpha^*_{j-} = \sqrt{\frac{\varepsilon_0}{2\,\hbar\omega}}\left[\omega\mathscr{A}_j - \frac{\mathrm{i}}{\varepsilon_0}\,\pi_j\right] \tag{A.38.a}$$

$$\alpha^*_{s-} = \sqrt{\frac{\varepsilon_0}{2\,\hbar\omega}}\left[\omega\mathscr{A}_s + \frac{\mathrm{i}}{\varepsilon_0}\,\pi_s\right]. \tag{A.38.b}$$

Starting with (A.32), (A.34), (A.38.a), and (A.38.b), we then derive the Fourier components of the potentials as functions of the normal variables:

$$\mathscr{A}_j(\mathbf{k}, t) = \sqrt{\frac{\hbar}{2\,\varepsilon_0\,\omega}}\,[\alpha_j(\mathbf{k}, t) + \alpha^*_j(-\,\mathbf{k}, t)] \tag{A.39.a}$$

$$\mathscr{A}_s(\mathbf{k}, t) = \sqrt{\frac{\hbar}{2\,\varepsilon_0\,\omega}}\,[\alpha_s(\mathbf{k}, t) + \alpha^*_s(-\,\mathbf{k}, t)]\,. \tag{A.39.b}$$

Finally, we substitute (A.39) in the Fourier integrals defining $A_j(\mathbf{r}, t)$ and $A_s(\mathbf{r}, t)$ and change $\mathbf{k}$ to $-\mathbf{k}$ in the integrals of the second term in (A.39). We get the equations

$$\left\{\begin{aligned} A_j(\mathbf{r}, t) &= \int \mathrm{d}^3k\,\sqrt{\frac{\hbar}{2\,\varepsilon_0\,\omega(2\,\pi)^3}}\,[\alpha_j(\mathbf{k}, t)\,\mathrm{e}^{\mathrm{i}\mathbf{k}.\mathbf{r}} + \alpha^*_j(\mathbf{k}, t)\,\mathrm{e}^{-\,\mathrm{i}\mathbf{k}.\mathbf{r}}] && \text{(A.40.a)}\\ A_s(\mathbf{r}, t) &= \int \mathrm{d}^3k\,\sqrt{\frac{\hbar}{2\,\varepsilon_0\,\omega(2\,\pi)^3}}\,[\alpha_s(\mathbf{k}, t)\,\mathrm{e}^{\mathrm{i}\mathbf{k}.\mathbf{r}} + \alpha^*_s(\mathbf{k}, t)\,\mathrm{e}^{-\,\mathrm{i}\mathbf{k}.\mathbf{r}}] && \text{(A.40.b)}\end{aligned}\right.$$

which give the expansion of the potentials in normal variables. In the special case of the free field, it is possible to use Equations (A.36) and to get

$$\left\{\begin{aligned} A_j(\mathbf{r}, t) &= \int \mathrm{d}^3k\,\sqrt{\frac{\hbar}{2\,\varepsilon_0\,\omega(2\,\pi)^3}}\,[\alpha_j(\mathbf{k})\,\mathrm{e}^{\mathrm{i}(\mathbf{k}.\mathbf{r}-\omega t)} + \alpha^*_j(\mathbf{k})\,\mathrm{e}^{-\,\mathrm{i}(\mathbf{k}.\mathbf{r}-\omega t)}] && \text{(A.40.c)}\\ A_s(\mathbf{r}, t) &= \int \mathrm{d}^3k\,\sqrt{\frac{\hbar}{2\,\varepsilon_0\,\omega(2\,\pi)^3}}\,[\alpha_s(\mathbf{k})\,\mathrm{e}^{\mathrm{i}(\mathbf{k}.\mathbf{r}-\omega t)} + \alpha^*_s(\mathbf{k})\,\mathrm{e}^{-\,\mathrm{i}(\mathbf{k}.\mathbf{r}-\omega t)}]\,. && \text{(A.40.d)}\end{aligned}\right.$$

Equations (A.40.c) and (A.40.d) give the expansions of the free vector and scalar potentials in traveling plane waves.

Since the vector and scalar potentials are considered as independent dynamical variables in the new Lagrangian, there are at each point **k** of reciprocal space four independent degrees of freedom, described by the three Cartesian components $\alpha_j(\mathbf{k})$ of $\boldsymbol{\alpha}(\mathbf{k})$ and by $\alpha_s(\mathbf{k})$. Rather than project $\boldsymbol{\alpha}(\mathbf{k})$ on a fixed Cartesian basis in reciprocal space, one can project $\boldsymbol{\alpha}(\mathbf{k})$ on the two unit transverse vectors $\boldsymbol{\varepsilon}$ and $\boldsymbol{\varepsilon}'$, perpendicular to **k** and to one another (see Figure 1 of Chapter 1), and on the unit vector $\boldsymbol{\kappa} = \mathbf{k}/k$ along **k**. The transverse normal variables $\alpha_\varepsilon(\mathbf{k}) = \boldsymbol{\varepsilon} \cdot \boldsymbol{\alpha}(\mathbf{k})$ and $\alpha_{\varepsilon'}(\mathbf{k}) = \boldsymbol{\varepsilon}' \cdot \boldsymbol{\alpha}(\mathbf{k})$ are the same as those used in Chapters I and II. In addition to these transverse variables one now has a longitudinal normal variable

$$\alpha_l(\mathbf{k}) = \boldsymbol{\kappa} \cdot \boldsymbol{\alpha}(\mathbf{k}) = \frac{1}{k}(k_x\,\alpha_x + k_y\,\alpha_y + k_z\,\alpha_z) \tag{A.41}$$

and also the normal variable $\alpha_s(\mathbf{k})$ associated with the scalar potential. For each value of **k**, there are then four normal modes of vibration of the free potential—two transverse, one longitudinal, and one scalar—described by the set

$$\{\, \alpha_\varepsilon(\mathbf{k}), \alpha_{\varepsilon'}(\mathbf{k}), \alpha_l(\mathbf{k}), \alpha_s(\mathbf{k}) \,\}\,. \tag{A.42}$$

The elementary excitations of these four types of modes give rise after quantization to four kinds of photons for each value of **k**.

c) Form of the Subsidiary Condition for the Free Classical Field. Gauge Arbitrariness

Until now we have considered the vector potential and the scalar potential as independent dynamical variables. We will now introduce the subsidiary condition (A.16.a), which is written for the free field using (A.40.c) and (A.40.d):

$$\boldsymbol{\nabla} \cdot \mathbf{A} + \frac{\dot{A}_s}{c} = \int \mathrm{d}^3k \sqrt{\frac{\hbar}{2\,\varepsilon_0\,\omega(2\,\pi)^3}}\; \mathrm{i}(\mathbf{k}.\boldsymbol{\alpha} - k\alpha_s)\, \mathrm{e}^{\mathrm{i}(\mathbf{k}.\mathbf{r} - \omega t)} + c.c. \tag{A.43}$$

If one wishes the subsidiary condition to be satisfied for all **r** and all t, it is necessary that the coefficient of each exponential in (A.43) be zero, that is, using the definition (A.41) for α_l (which implies $\mathbf{k} \cdot \boldsymbol{\alpha} = k\alpha_l$),

$$\alpha_l(\mathbf{k}) - \alpha_s(\mathbf{k}) = 0 \quad \forall \mathbf{k}\,. \tag{A.44}$$

The subsidiary condition then takes a very simple form for the free field in terms of the normal variables. Among all the solutions (A.40.c) and

(A.40.d) of the equations of motion, only those for which at every point **k** the normal longitudinal variable $\alpha_l(\mathbf{k})$ is equal to the normal scalar variable $\alpha_s(\mathbf{k})$ have a physical meaning.

The very simple form for the subsidiary condition (A.44) suggests introducing for each value of **k** two orthogonal linear combinations of α_l and α_s, one of them being precisely $\alpha_l - \alpha_s$. We thus take

$$\alpha_d = \frac{\mathrm{i}}{\sqrt{2}}(\alpha_l - \alpha_s) \tag{A.45.a}$$

$$\alpha_g = \frac{1}{\sqrt{2}}(\alpha_l + \alpha_s) . \tag{A.45.b}$$

We have introduced two new types of normal variables, α_d and α_g. With this new notation, the condition (A.44) is written

$$\alpha_d = 0 . \tag{A.46}$$

The condition (A.46) restricts the number of degrees of freedom for the physical field: the normal variable α_d is zero for a free physical field. We will now examine how the gauge arbitrariness associated with the gauge transformations (A.19) satisfying (A.20) is evidenced. A given physical field can be described by many sets of normal variables (A.42) satisfying (A.44) and derived from one another by a gauge transformation. Since all real functions f satisfying (A.20) can be written

$$f = \int \mathrm{d}^3k \sqrt{\frac{\hbar}{2\,\varepsilon_0\,\omega(2\,\pi)^3}}\,\mathscr{F}(\mathbf{k})\,\mathrm{e}^{\mathrm{i}(\mathbf{k}.\mathbf{r}-\omega t)} + c.c. \tag{A.47}$$

the gauge transformation associated with f $[\mathbf{A}' = \mathbf{A} + \nabla f,\ A'_s = A_s - (\partial f/c\,\partial t)]$ involves, for the normal variables (A.42), the following transformation:

$$\begin{cases} \alpha'_\varepsilon = \alpha_\varepsilon \quad \alpha'_{\varepsilon'} = \alpha_{\varepsilon'} \\ \alpha'_l = \alpha_l + \mathrm{i}k\mathscr{F} \\ \alpha'_s = \alpha_s + \mathrm{i}k\mathscr{F} \end{cases} \tag{A.48}$$

Expressed as a function of the variables α_d and α_g defined by (A.45), the equations (A.48) become

$$\begin{aligned} \alpha'_\varepsilon &= \alpha_\varepsilon \quad \alpha'_{\varepsilon'} = \alpha_{\varepsilon'} \\ \alpha'_d &= \alpha_d \\ \alpha'_g &= \alpha_g + \mathrm{i}\sqrt{2}\,k\mathscr{F} \end{aligned} \tag{A.49}$$

A gauge transformation does not modify α_d, which is zero for a physical field, and transforms only the normal variable α_g. Finally, the gauge arbitrariness appears only in the value of α_g.

Remark

Since $\alpha_d = 0$, one sees that by taking $\mathscr{F} = -\alpha_g/ik\sqrt{2}$, one can always in a given Lorentz frame cancel α'_g and thereby α'_l and α'_s. This shows, as in Chapters I and II, that the relevant degrees of freedom of the physical field are described at each point **k** by the two transverse normal variables α_ε and $\alpha_{\varepsilon'}$. It is however impossible to cancel α_l and α_s in all the Lorentz frames, since if A_s is zero in one frame, it no longer is in another. This explains why the construction of a manifestly covariant theory necessitates the retention of four types of normal variables combined with the condition (A.46) and the arbitrariness of gauge (A.49).

It will be useful in what follows to reexpress the free potentials as a function of the variables α_d and α_g. For this we use (A.40.c) and (A.40.d), which can also be written in their covariant form (*)

$$A_\mu(x^\nu) = \int d^3k \sqrt{\frac{\hbar}{2\,\varepsilon_0\,\omega(2\,\pi)^3}}\,[\alpha_\mu(\mathbf{k})\,e^{-ik_\nu x^\nu} + \alpha_\mu^*(\mathbf{k})\,e^{ik_\nu x^\nu}] \quad (A.50)$$

where k^μ is the four-vector $(\omega/c, \mathbf{k})$ with $\omega = c|\mathbf{k}|$ satisfying

$$\sum_\nu k^\nu k_\nu = 0\,. \quad (A.51)$$

To give α_μ and α_μ^* as a function of the variables α_ε, $\alpha_{\varepsilon'}$, α_d, and α_g, we start from

$$\alpha_\mu(\mathbf{k}) = \alpha_\varepsilon(\mathbf{k})\,\varepsilon_\mu + \alpha_{\varepsilon'}(\mathbf{k})\,\varepsilon'_\mu + \alpha_l(\mathbf{k})\,\kappa_\mu + \alpha_s(\mathbf{k})\,\eta_\mu \quad (A.52)$$

ε_μ and ε'_μ are two four-vectors having only spatial components and constructed from the transverse vectors $\boldsymbol{\varepsilon}$ and $\boldsymbol{\varepsilon}'$:

$$\begin{aligned} \varepsilon^\mu &= (0, \boldsymbol{\varepsilon}) \\ \varepsilon'^\mu &= (0, \boldsymbol{\varepsilon}') \end{aligned} \quad (A.53)$$

κ_μ is a four-vector having only spatial components and constructed from the longitudinal vector $\boldsymbol{\kappa}$:

$$\kappa^\mu = (0, \boldsymbol{\kappa}) \quad (A.54)$$

(*) To simplify the notation, we use in the exponentials the convention of summation on repeated indices ($k_\nu x^\nu = \sum_\nu k_\nu x^\nu$).

Finally, η_μ is a four-vector having only a time component:

$$\eta^\mu = (1, \mathbf{0}) . \qquad (A.55)$$

We express then the last two terms of (A.52) as a function of α_d and α_g thanks to (A.45). This gives

$$\alpha_l(\mathbf{k})\, \kappa_\mu + \alpha_s(\mathbf{k})\, \eta_\mu = \frac{1}{\sqrt{2}} \alpha_g(\mathbf{k})\,(\kappa_\mu + \eta_\mu) + \frac{\mathrm{i}}{\sqrt{2}} \alpha_d(\mathbf{k})\,(\eta_\mu - \kappa_\mu) . \qquad (A.56)$$

We finally get

$$A_\mu = A_\mu^T + A_\mu^G + A_\mu^D \qquad (A.57)$$

where A_μ^T, A_μ^G, and A_μ^D are respectively the contributions to A_μ of the terms in α_ε and $\alpha_{\varepsilon'}$, in α_g, and in α_d from α_μ. A_μ^T is the transverse vector potential, having only spatial transverse components. It agrees with that studied in Chapter I,

$$A_\mu^T(x^\nu) = \int \mathrm{d}^3k \sqrt{\frac{\hbar}{2\,\varepsilon_0\,\omega(2\,\pi)^3}} \{ [\varepsilon_\mu\, \alpha_\varepsilon(\mathbf{k}) + \varepsilon'_\mu\, \alpha_{\varepsilon'}(\mathbf{k})]\, \mathrm{e}^{-\mathrm{i}k_\nu x^\nu} + [\varepsilon_\mu\, \alpha_\varepsilon^*(\mathbf{k}) + \varepsilon'_\mu\, \alpha_{\varepsilon'}^*(\mathbf{k})]\, \mathrm{e}^{\mathrm{i}k_\nu x^\nu} \} . \qquad (A.58)$$

Before giving A_μ^G, we note that the components of the four-vector $\kappa^\mu + \eta^\mu$ which multiplies α_g in (A.56) are, from (A.54) and (A.55), equal to $(1, \boldsymbol{\kappa}) = k^{-1}(k, \mathbf{k})$, which are just the components of the four-vector k^μ/k. Since on the other hand $k_\mu \exp(-\mathrm{i}k_\nu x^\nu) = \mathrm{i}\,\partial_\mu \exp(-\mathrm{i}k_\nu x^\nu)$, one sees that

$$A_\mu^G(x^\nu) = -\,\partial_\mu f(x^\nu) \qquad (A.59)$$

where

$$f(x^\nu) = -\int \mathrm{d}^3k \sqrt{\frac{\hbar}{2\,\varepsilon_0\,\omega(2\,\pi)^3}}\, \frac{\mathrm{i}}{k\sqrt{2}} [\alpha_g(\mathbf{k})\,\mathrm{e}^{-\mathrm{i}k_\nu x^\nu} - \alpha_g^*(\mathbf{k})\, \mathrm{e}^{\mathrm{i}k_\nu x^\nu}] \qquad (A.60)$$

The function f satisfies

$$\Box f = \sum_\nu \partial_\nu\, \partial^\nu f = 0 \qquad (A.61)$$

since $\Sigma_\nu k_\nu k^\nu = 0$. Thus A_μ^G has the structure of a gauge term.

Finally, using (A.56), A_μ^D is written

$$A_\mu^D(x^\nu) = \frac{\mathrm{i}}{\sqrt{2}} \int \mathrm{d}^3k \sqrt{\frac{\hbar}{2\,\varepsilon_0\,\omega(2\,\pi)^3}} \times (\eta_\mu - \kappa_\mu)\, [\alpha_d(\mathbf{k})\, \mathrm{e}^{-\mathrm{i}k_\nu x^\nu} - \alpha_d^*(\mathbf{k})\, \mathrm{e}^{\mathrm{i}k_\nu x^\nu}] . \qquad (A.62)$$

From (A.46), a physical state of the field is characterized by $A_\mu^D = 0$.

We consider finally the electric and magnetic fields, that is, the tensor $F_{\mu\nu} = \partial_\mu A_\nu - \partial_\nu A_\mu$. Now A_μ^G does not contribute to $F_{\mu\nu}$, by virtue of (A.59), as expected for a gauge term. We find finally for a physical state $(\alpha_d = 0)$

$$F_{\mu\nu} = \partial_\mu A_\nu^T - \partial_\nu A_\mu^T = F_{\mu\nu}^T . \qquad \text{(A.63)}$$

The free electric and magnetic fields are purely transverse, and their expressions as functions of the variables α_ε, $\alpha_{\varepsilon'}$, α_ε^*, and $\alpha_{\varepsilon'}^*$ agree with those of Chapter I, since this is the case for the transverse potential.

d) Expression of the Field Hamiltonian

To get the expression for H_R in terms of the normal variables, it suffices to substitute into (A.27) the expressions (A.39) for $\mathscr{A}_j$ and $\mathscr{A}_s$ and the analogous expressions for π_j and π_s derived from (A.32), (A.38.a), (A.34), and (A.38.b). As in §C.4 of Chapter I, one retains the order between α and α^* as it arises in the calculation, so as to get a result immediately generalizable to quantum theory. One then gets, after a process analogous to that in Chapter I,

$$H_R = \int \mathrm{d}^3k \, \frac{\hbar\omega}{2} [(\alpha_\varepsilon^* \, \alpha_\varepsilon + \alpha_\varepsilon \, \alpha_\varepsilon^*) + (\alpha_{\varepsilon'}^* \, \alpha_{\varepsilon'} + \alpha_{\varepsilon'} \, \alpha_{\varepsilon'}^*) + \\ + (\alpha_l^* \, \alpha_l + \alpha_l \, \alpha_l^*) - (\alpha_s^* \, \alpha_s + \alpha_s \, \alpha_s^*)] . \qquad \text{(A.64)}$$

One then sees clearly how the difficulty of the negative sign in (A.64) is resolved by the subsidiary condition. Although the energy (A.64) can be negative for some values of the normal variables, the subsidiary condition (A.44) implies that, for a physical state, the energy associated with the longitudinal variables compensates exactly that associated with the scalar variables, the only nonzero contribution being provided, as it must be, by the transverse variables.

Remarks

(i) By expressing α_l and α_s as functions of α_d and α_g by means of (A.45), we can write H_R in the form

$$H_R = \int \mathrm{d}^3k \, \frac{\hbar\omega}{2} [(\alpha_\varepsilon^* \, \alpha_\varepsilon + \alpha_\varepsilon \, \alpha_\varepsilon^*) + (\alpha_{\varepsilon'}^* \, \alpha_{\varepsilon'} + \alpha_{\varepsilon'} \, \alpha_{\varepsilon'}^*) + \\ + \mathrm{i}(\alpha_d^* \, \alpha_g - \alpha_g^* \, \alpha_d) + \mathrm{i}(\alpha_g \, \alpha_d^* - \alpha_d \, \alpha_g^*)] . \qquad \text{(A.65)}$$

It thus appears clear that for a physical state $(\alpha_d = 0)$, α_g does not contribute to the energy of the free field, which is then exclusively due to the transverse field and so is positive.

(ii) An analogous calculation for the momentum of the field, $\mathbf{P}_R$ given in (A.28), yields, after symmetrization of the products,

$$\mathbf{P}_R = \int d^3k \frac{\hbar \mathbf{k}}{2} [(\alpha_\varepsilon^* \alpha_\varepsilon + \alpha_\varepsilon \alpha_\varepsilon^*) + (\alpha_{\varepsilon'}^* \alpha_{\varepsilon'} + \alpha_{\varepsilon'} \alpha_{\varepsilon'}^*) + \\ + (\alpha_l^* \alpha_l + \alpha_l \alpha_l^*) - (\alpha_s^* \alpha_s + \alpha_s \alpha_s^*)] . \quad (A.66)$$

As for the energy H_R, only the transverse normal variables contribute to $\mathbf{P}_R$ in a physical state. By using covariant notation, P^μ for the momentum–energy of the field (ordinary components: H_R/c, $\mathbf{P}_R$) and k^μ for the four-wave-vector $(\omega/c, \mathbf{k})$, the two expressions (A.64) and (A.66) can be regrouped in the form

$$P^\mu = \int d^3k \frac{\hbar k^\mu}{2} [(\alpha_\varepsilon^* \alpha_\varepsilon + \alpha_\varepsilon \alpha_\varepsilon^*) + (\alpha_{\varepsilon'}^* \alpha_{\varepsilon'} + \alpha_{\varepsilon'} \alpha_{\varepsilon'}^*) + \\ + (\alpha_l^* \alpha_l + \alpha_l \alpha_l^*) - (\alpha_s^* \alpha_s + \alpha_s \alpha_s^*)] . \quad (A.67)$$

(iii) Before quantizing the theory, we return to the definition (A.34) of α_s and imagine that we take the same plus sign as in (A.32), then replacing (A.34) by

$$\alpha_s' = \sqrt{\frac{\varepsilon_0}{2\hbar\omega}} \left[\omega \mathscr{A}_s + \frac{i}{\varepsilon_0} \pi_s \right] . \quad (A.34')$$

What will be the modification of the results found in this subsection? First of all, as we have already indicated above, α_s' evolves as $e^{i\omega t}$ and not as $e^{-i\omega t}$. The cancellation of the coefficient of $\exp[i(\mathbf{k} \cdot \mathbf{r} - \omega t)]$ in the subsidiary condition gives then in place of (A.44)

$$\alpha_l(\mathbf{k}) - \alpha_s'^*(-\mathbf{k}) = 0 \quad (A.44')$$

which is less satisfactory than (A.44). Finally, one can easily prove that Equation (A.64) for H_R remains unchanged, although Equation (A.66) for $\mathbf{P}_R$ will be modified, all the signs becoming positive, which then prohibits regrouping the two expressions into one as in (A.67). All these reasons indicate that the definition (A.34) for α_s should be preferred to (A.34') in establishing a covariant theory.

B—DIFFICULTIES RAISED BY THE QUANTIZATION OF THE FREE FIELD

1. Canonical Quantization

In this section, we will proceed with the canonical quantization of the preceding theory without wondering at this stage about physics, that is to say, by considering all the degrees of freedom as independent. Then, in the next section, we will discuss the difficulties which arise in quantum theory when one tries to introduce the subsidiary condition and to construct physical states with an arbitrary number of photons.

a) Canonical Commutation Relations

As in §A.2.*e* of Chapter II, we associate operators with the dynamical variables and their conjugate momenta. The reality conditions in classical theory become for the quantum operators

$$A_j(\mathbf{r}) = A_j^+(\mathbf{r}) \tag{B.1.a}$$

$$\mathscr{A}_j(\mathbf{k}) = \mathscr{A}_j^+(-\mathbf{k}) \tag{B.1.b}$$

and the analogous relations for A_s and $\mathscr{A}_s$, Π_j and π_j, and Π_s and π_s. When the range of variation of $\mathbf{k}$ is limited to a reciprocal half space, all the foregoing dynamical variables can be thought of as independent. Quantization then is accomplished by means of the canonical commutation relation (A.61) of Chapter II:

$$[\mathscr{A}_i(\mathbf{k}), \pi_j^+(\mathbf{k}')] = i\hbar\, \delta_{ij}\, \delta(\mathbf{k} - \mathbf{k}') \tag{B.2.a}$$

$$[\mathscr{A}_s(\mathbf{k}), \pi_s^+(\mathbf{k}')] = i\hbar\, \delta(\mathbf{k} - \mathbf{k}') \tag{B.2.b}$$

all other commutators being zero. The extension of equations (B.2) when $\mathbf{k}$ and $\mathbf{k}'$ vary over all space is done as in §C.4.*a* of Chapter II, using (B.1.b) and analogous expressions for π_j, $\mathscr{A}_s$, and π_s.

The commutators (B.2) are equal-time commutators (Schrödinger approach). Canonical quantization thus favors time, which, at first sight, seems poorly adapted to a covariant formalism. We will see below, however, that it leads in the Heisenberg picture to manifestly covariant expressions for the commutators of free potentials at any two points in space–time, $\mathbf{r}$, t and $\mathbf{r}'$, t'.

b) Annihilation and Creation Operators

The classical normal variables α_j and α_s become after canonical quantization the *annihilation operators* a_j and a_s, which are related to the operators $\mathscr{A}_j$, π_j, $\mathscr{A}_s$, and π_s by expressions identical to (A.32) and (A.34):

$$a_j(\mathbf{k}) = \sqrt{\frac{\varepsilon_0}{2\hbar\omega}}\left[\omega\mathscr{A}_j(\mathbf{k}) + \frac{\mathrm{i}}{\varepsilon_0}\pi_j(\mathbf{k})\right] \tag{B.3.a}$$

$$a_s(\mathbf{k}) = \sqrt{\frac{\varepsilon_0}{2\hbar\omega}}\left[\omega\mathscr{A}_s(\mathbf{k}) - \frac{\mathrm{i}}{\varepsilon_0}\pi_s(\mathbf{k})\right]. \tag{B.3.b}$$

The adjoint operators of a_j and a_s, a_j^+ and a_s^+, are the creation operators.

The sign difference between the last terms on the right in (B.3.a) and (B.3.b) introduces important changes in the commutators between the creation and annihilation operators. Indeed, the canonical commutation relations (B.2) for the operators defined in (B.3) and their adjoints (**k** and **k**′ now vary over all space) imply

$$[a_i(\mathbf{k}), a_j^+(\mathbf{k}')] = \delta_{ij}\,\delta(\mathbf{k} - \mathbf{k}') \tag{B.4.a}$$

$$[a_s(\mathbf{k}), a_s^+(\mathbf{k}')] = -\,\delta(\mathbf{k} - \mathbf{k}') \tag{B.4.b}$$

all other commutators being zero. For the three spatial degrees of freedom ($i, j = x, y, z$), one gets the usual commutation relations for a quantum harmonic oscillator. On the other hand, for the scalar degree of freedom, (B.4.b) differs from the usual relation by a $-$ sign. We will see below the difficulties which arise from this $-$ sign in the construction of state space. Note finally that Equations (B.4.a) and (B.4.b) can be regrouped into a single expression with covariant notation

$$[a_\mu(\mathbf{k}), a_\nu^+(\mathbf{k}')] = -\,g_{\mu\nu}\,\delta(\mathbf{k} - \mathbf{k}'). \tag{B.5}$$

Remarks

(i) a_μ is the operator associated with the normal variable α_μ. We should point out that, in spite of the notation used, the α_μ are not the components of a four-vector.

(ii) The operators a_x, a_y, and a_z can always be replaced by a_ε, $a_{\varepsilon'}$, and a_l. Since the corresponding transformation is orthogonal, Equation (B.4.a) becomes

$$[a_\varepsilon(\mathbf{k}), a_{\varepsilon'}^+(\mathbf{k}')] = \delta_{\varepsilon\varepsilon'}\,\delta(\mathbf{k} - \mathbf{k}') \tag{B.6.a}$$

$$[a_l(\mathbf{k}), a_l^+(\mathbf{k}')] = \delta(\mathbf{k} - \mathbf{k}') \tag{B.6.b}$$

all other commutators being zero.

It is also clear that the important physical variables, such as the field energy H_R, the momentum $\mathbf{P}_R$ and the fields, can be given as functions of the creation and annihilation operators. Thus, the energy H_R of the field is gotten by replacing the normal variables α and α^* with the operators a and a^+ in (A.64), which gives

$$H_R = \int \mathrm{d}^3k \frac{\hbar\omega}{2} [(a_\varepsilon^+ a_\varepsilon + a_\varepsilon a_\varepsilon^+) + (a_{\varepsilon'}^+ a_{\varepsilon'} + a_{\varepsilon'} a_{\varepsilon'}^+) + + (a_l^+ a_l + a_l a_l^+) - (a_s^+ a_s + a_s a_s^+)] \quad \text{(B.7)}$$

and more generally, starting from (A.67),

$$P_\mu = - \int \mathrm{d}^3k \frac{1}{2} \hbar k_\mu \sum_{\nu\rho} g^{\nu\rho}(a_\nu^+ a_\rho + a_\rho a_\nu^+) . \quad \text{(B.8)}$$

c) Covariant Commutation Relations between the Free Potentials in the Heisenberg Picture

The free potentials in the Heisenberg picture are gotten by replacing α_μ and α_μ^* with a_μ and a_μ^+ in the classical expression (A.50), which gives, using covariant notation,

$$A_\mu(x^\nu) = \int \mathrm{d}^3k \sqrt{\frac{\hbar}{2\,\varepsilon_0\,\omega(2\,\pi)^3}} [a_\mu(\mathbf{k})\,\mathrm{e}^{-ik_\nu x^\nu} + a_\mu^+(\mathbf{k})\,\mathrm{e}^{ik_\nu x^\nu}] . \quad \text{(B.9)}$$

The commutation relations (B.5) then show that

$$[A_\mu(\mathbf{r}, t), A_\nu(\mathbf{r}', t')] = = - g_{\mu\nu} \int \mathrm{d}^3k \frac{\hbar}{2\,\varepsilon_0\,\omega(2\,\pi)^3} [\mathrm{e}^{i\mathbf{k}.(\mathbf{r}-\mathbf{r}') - i\omega(t-t')} - c.c.] \quad \text{(B.10)}$$

The triple integral of (B.10) can be transformed into a quadruple integral $\int \mathrm{d}^4k$, the constraint (A.51) being introduced through a delta function

$$\delta(k^\mu k_\mu) = \delta(k_0^2 - \mathbf{k}^2) = \frac{\delta(k_0 - |\mathbf{k}|)}{2|\mathbf{k}|} + \frac{\delta(k_0 + |\mathbf{k}|)}{2|\mathbf{k}|} , \quad \text{(B.11)}$$

which has the additional advantage of absorbing the factor $1/\omega = 1/c|\mathbf{k}|$ in (B.10). By introducing the sign function of k_0, $\eta(k_0)$, one can finally rewrite (B.10) in the manifestly covariant form

$$[A_\mu(\mathbf{r}, t), A_\nu(\mathbf{r}', t')] = \frac{i\hbar}{\varepsilon_0\, c} g_{\mu\nu}\, D(\mathbf{r} - \mathbf{r}', t - t') \quad \text{(B.12)}$$

with

$$D(\mathbf{r}, t) = \frac{i}{(2\pi)^3} \int d^4k \, e^{-ik_\nu x^\nu} \, \delta(k_\mu k^\mu) \, \eta(k^0) . \tag{B.13}$$

Note finally that it is possible to calculate the triple integral of (B.10) directly and thus to establish

$$D(\mathbf{r}, t) = \frac{1}{4\pi r} [\delta(r - ct) - \delta(r + ct)] \tag{B.14}$$

which shows that D is zero throughout except on the light cone.

Remarks

(i) Note that the function D has been already introduced in Complement C_{III}. It is possible to get the commutators between the components of the electromagnetic field by using (B.12). One then gets expressions identical to those found in Complement C_{III} [Equations (20)].

(ii) Starting from (B.8) for P_μ and the commutation relations (B.5), it is possible to show that

$$[P_\mu, a_\nu(\mathbf{k})] = -\hbar k_\mu a_\nu(\mathbf{k}) \tag{B.15}$$

and also, using (B.9),

$$[P_\mu, A_\nu(\mathbf{r}, t)] = -i\hbar \, \partial_\mu A_\nu(\mathbf{r}, t) . \tag{B.16}$$

The P_μ appear then as the generators of space–time translations in state space. We have derived (B.16) from (B.5) here, that is, from the canonical commutation relations. A completely covariant quantization would follow the inverse path. It would start from the study of the symmetries of the Lagrangian, establishing the expressions for the physical variables associated with the generators of the Lorentz group; one would then postulate (B.16) and the analogous equations for the other group generators, to finally derive (B.5) from (B.16).

2. Problems of Physical Interpretation Raised by Covariant Quantization

The approach followed in the preceding section permits us to derive manifestly covariant commutation relations like (B.12) or (B.16) and then to construct a theoretical framework better adapted to relativity than those of Chapter II, where the symmetry between **A** and U was broken. We are now going to examine the problems of physical interpretation posed by covariant quantization. Some difficulties appear at this level, which are much more serious than in the classical theory examined in the section above, where it was sufficient to impose the subsidiary condition (A.44) to get good physical states.

a) THE FORM OF THE SUBSIDIARY CONDITION IN QUANTUM THEORY

A treatment identical to that of §A.3.*c* above for the free field leads to a condition analogous to (A.44), where α_l and α_s are replaced by a_l and a_s respectively. It is clear however that an identity like (A.44) cannot be satisfied by operators like a_l and a_s, which act in different subspaces of the state space—the subspace of longitudinal photons and the subspace of scalar photons. *In the covariant quantum theory, Maxwell's equations can no longer hold between operators*, since it is not possible to impose the subsidiary condition as an operator identity.

One can then try to use the subsidiary condition to select the physical states $|\psi\rangle$ by requiring these states to be eigenvectors of $\Sigma_\mu \partial_\mu A^\mu$ with the eigenvalue zero:

$$\sum_\mu \partial_\mu A^\mu \,|\, \psi \rangle = 0 . \qquad \text{(B.17)}$$

This relation must be true for all $\mathbf{r}$ and for all t, so that an expansion analogous to (A.43) leads to the following two conditions (valid for all $\mathbf{k}$):

$$\begin{cases} [a_l(\mathbf{k}) - a_s(\mathbf{k})] \,|\, \psi \rangle = 0 & \text{(B.18.a)} \\ [a_l^+(\mathbf{k}) - a_s^+(\mathbf{k})] \,|\, \psi \rangle = 0 . & \text{(B.18.b)} \end{cases}$$

In fact, such conditions are too strong, and it is possible to show that the equations (B.17) or (B.18) do not have a physical solution. A condition less strong than (B.17) or (B.18) is that for all $\mathbf{r}$ and all t one has

$$\langle \psi \,|\sum_\mu \partial_\mu A^\mu(\mathbf{r}, t) \,|\, \psi \rangle = 0 \quad \forall \mathbf{r}, t . \qquad \text{(B.19)}$$

The physical states $|\psi\rangle$ are then such that the subsidiary condition is satisfied for the mean value in these states. The concern with (B.19) is that its solutions do not necessarily form a vector subspace of the state space. For this reason one prefers to use the subsidiary condition in the form

$$\sum_\mu \partial_\mu A^{(+)\mu} \,|\, \psi \rangle = 0 \qquad \text{(B.20)}$$

where $A^{(+)\mu}$ is the positive-frequency component of A^μ, containing only the terms in $e^{-i\omega t}$. For the free field, (B.9) shows that this condition is satisfied for all $\mathbf{r}$ and all t if

$$[a_l(\mathbf{k}) - a_s(\mathbf{k})] \,|\, \psi \rangle = 0 \quad \forall \mathbf{k} . \qquad \text{(B.21)}$$

The solutions of (B.20) or (B.21) form a vector subspace of state space and are all solutions of (B.19).

It is in the form (B.21) that we hereafter use the subsidiary condition to characterize the physical states of the free field (*).

b) Problems Raised by the Construction of State Space

It seems logical first to postulate the existence of a state $|0\rangle$ with no photons, on which the action of every annihilation operator $a_\mu(\mathbf{k})$ gives zero, since it is impossible to remove a photon from the vacuum:

$$a_\mu(\mathbf{k})\,|\,0\,\rangle = 0 \qquad \forall \mu, \mathbf{k}\,. \tag{B.22}$$

Note in particular that (B.22) implies that the vacuum obeys (B.21) and is therefore a physical state, a satisfying result.

The next stage then involves trying to construct states having any number of photons by the repeated action of the creation operator a_μ^+ on the vacuum. For the transverse and longitudinal photons this poses no problem, since the commutation relations (B.4.a) or (B.6) have the usual form. For example, a state having $n_\varepsilon(\mathbf{k})$ photons $\mathbf{k}$, ε, $n_{\varepsilon'}(\mathbf{k}')$ photons $\mathbf{k}'$, ε', and $n_l(\mathbf{k}'')$ photons $\mathbf{k}''$, $\boldsymbol{\kappa}''$ is written with simplified notation as

$$|\,n_\varepsilon, n_{\varepsilon'}, n_l\,\rangle = \frac{(a_\varepsilon^+)^{n_\varepsilon}\,(a_{\varepsilon'}^+)^{n_{\varepsilon'}}\,(a_l^+)^{n_l}}{\sqrt{n_\varepsilon\,!\,n_{\varepsilon'}\,!\,n_l\,!}}\,|\,0\,\rangle\,. \tag{B.23}$$

On the other hand, for scalar photons, serious difficulties appear. Calculate for example the norm of the state having one scalar photon,

$$|\,\psi\,\rangle = \int \mathrm{d}^3k\; g(\mathbf{k})\; a_s^+(\mathbf{k})\,|\,0\,\rangle\,. \tag{B.24}$$

By using the commutation relation (B.4.b) and (B.22), one gets

$$\begin{aligned}\langle\,\psi\,|\,\psi\,\rangle &= \iint \mathrm{d}^3k\,\mathrm{d}^3k'\; g^*(\mathbf{k}')\, g(\mathbf{k})\,\langle\,0\,|\,a_s(\mathbf{k}')\,a_s^+(\mathbf{k})\,|\,0\,\rangle \\ &= -\,\langle\,0\,|\,0\,\rangle \int \mathrm{d}^3k\,|\,g(\mathbf{k})\,|^2\,.\end{aligned} \tag{B.25}$$

(*) In §D.3.*a* below and in §$\mathrm{A_V}$.3.*c*, we indicate how one can generalize (B.21) for a quantized radiation field coupled to external sources or to the Dirac field.

It appears then that the vacuum $|0\rangle$ and the state $|\psi\rangle$ having one scalar photon have norms with opposite signs, which is wholly unacceptable in a Hilbert space, where all the norms must be positive to permit a probabilistic interpretation.

It seems necessary then to generalize the notion of norm in the state space so as to include in this space states with negative norms, besides the physical states which are distinguished by the subsidiary condition and which should have finite and positive norms.

Remark

One may be tempted to think that the interpretation of a_s (a_s^+) as an annihilation (creation) operator is erroneous and that one should exchange their roles. We take then

$$\begin{cases} b_s(\mathbf{k}) = a_s^+(\mathbf{k}) & \text{(B.26.a)} \\ b_s^+(\mathbf{k}) = a_s(\mathbf{k}) & \text{(B.26.b)} \end{cases}$$

and treat b_s as an annihilation operator and b_s^+ as a creation operator satisfying a commutation relation identical to (B.4.a):

$$[b_s(\mathbf{k}), b_s^+(\mathbf{k}')] = \delta(\mathbf{k} - \mathbf{k}') \tag{B.27}$$

and derived from (B.4.b) and (B.26). The scalar photon vacuum then satisfies

$$b_s(\mathbf{k})\,|\,0\,\rangle = a_s^+(\mathbf{k})\,|\,0\,\rangle = 0\,. \tag{B.28}$$

On the other hand, the subsidiary condition (B.21) gotten by cancelling the coefficient of $\exp[i(\mathbf{k}\cdot\mathbf{r} - \omega t)]$ in $\partial_\mu A^{(+)\mu}$ remains unchanged and is written, with the notation (B.26),

$$[a_l(\mathbf{k}) - b_s^+(\mathbf{k})]\,|\,\psi\,\rangle = 0 \quad \forall\mathbf{k}\,. \tag{B.29}$$

But new difficulties then arise. First, the new vacuum defined in (B.28) no longer satisfies (B.29) and can no longer be thought of as a physical state. Additionally, it is possible to show (see Exercise 2) that the solutions of (B.29) with this new interpretation of $a_s = b_s^+$ as a creation operator are not normalizable. Thus the $-$ sign in (B.4.b) leads us either to a one-scalar-photon state with negative norm (if we take a_s as an annihilation operator and a_s^+ as a creation operator), or to physical states with infinite norm (if we reverse the interpretations of a_s and a_s^+). The second eventuality is in fact much worse than the first, since it involves physical states, which is not the case for the one-scalar-photon states (B.24). It is for this reason that we revert to the original interpretation of a_s and a_s^+.

C—COVARIANT QUANTIZATION WITH AN INDEFINITE METRIC

In this third part we present a method which solves the difficulties mentioned in the discussion of §B.2. First, we show (§C.1) that it is possible to introduce in a Hilbert space a second scalar product leading to a second norm not necessarily positive definite (indefinite metric). This new scalar product allows us to define a new adjoint for each operator and a new mean value. Canonical quantization is then done (§C.2) by replacing the Hermitian conjugate operators with the new adjoints throughout the preceding theory, the new metric being chosen so as to resolve the difficulties associated with the scalar potential. We then construct (§C.3) the physical kets obeying the subsidiary condition (B.21), and show finally (§C.4) that for these kets all of the predictions about physical variables only involve the transverse degrees of freedom and conform to the usual quantum interpretation.

1. Indefinite Metric in Hilbert Space

Consider a Hilbert space with the usual Dirac notation (ket $|\psi\rangle$, bra $\langle\phi|$), with the usual scalar product $\langle\phi|\psi\rangle = \langle\psi|\phi\rangle^*$, linear with respect to $|\psi\rangle$ and antilinear with respect to $\langle\phi|$, and with a norm $\langle\psi|\psi\rangle$ which is positive definite, that is, strictly positive, and zero if and only if $|\psi\rangle = 0$.

Starting with a Hermitian, unitary linear operator M in this space, that is, such that

$$M = M^+ = M^{-1} \tag{C.1}$$

we introduce a second scalar product defined by

$$\subset \phi | \psi \supset \; = \langle \phi | M | \psi \rangle . \tag{C.2}$$

We will use the "round" Dirac notation $| \supset$ and $\subset |$ for this new scalar product. It is equivalent to say that one associates with the old ket $|\psi\rangle$ and bra $\langle\phi|$ the new ket $|\psi\supset$ and bra $\subset\phi|$ defined by

$$\begin{cases} |\psi\supset \; = |\psi\rangle & \text{(C.3.a)} \\ \subset\phi| = \langle\phi| M . & \text{(C.3.b)} \end{cases}$$

From (C.1) and (C.2) it follows that

$$\subset \phi | \psi \supset \; = \; \subset \psi | \phi \supset^* \tag{C.4}$$

and that $\subset\phi|\psi\supset$ is linear with respect to $|\psi\supset$ and antilinear with respect to $\subset\phi|$. The new scalar product then has some of the usual properties of an ordinary scalar product. In contrast, the new norm $\subset\psi|\psi\supset$ is not necessarily positive definite. To see this, consider the eigenvectors $|m_i\rangle$ of M with eigenvalues m_i. As a result of (C.1), m_i is real and necessarily equal to $+1$ or -1. Replacing M by $\Sigma_i m_i|m_i\rangle\langle m_i|$ in (C.2), one gets

$$\subset\psi|\psi\supset = \langle\psi|M|\psi\rangle = \sum_i m_i |\langle m_i|\psi\rangle|^2 . \tag{C.5}$$

Since some m_i can be equal to -1, it appears clear from (C.5) that $\subset\psi|\psi\supset$ although real can take zero or negative values. The new metric associated with M is called *indefinite* in this case.

Starting from the new scalar product, one can introduce new matrix elements for a linear operator A,

$$\subset\phi|A|\psi\supset = \langle\phi|MA|\psi\rangle \tag{C.6}$$

and a new adjoint of A, which we denote $\bar{A}$ to distinguish it from the old A^+ and which is defined by

$$\subset\phi|\bar{A}|\psi\supset = \subset\psi|A|\phi\supset^* \tag{C.7}$$

for all ψ and ϕ. Equations (C.4) and (C.7) also imply that

$$|\psi'\supset = A|\psi\supset \quad\Leftrightarrow\quad \subset\psi'| = \subset\psi|\bar{A}. \tag{C.8}$$

What is the relationship between the new adjoint $\bar{A}$ and the old A^+? To get this relationship, it is sufficient to express the two terms of (C.7) in terms of the usual kets:

$$\subset\phi|\bar{A}|\psi\supset = \langle\phi|M\bar{A}|\psi\rangle \tag{C.9.a}$$

$$\subset\psi|A|\phi\supset^* = \langle\psi|MA|\phi\rangle^* = \langle\phi|A^+M^+|\psi\rangle . \tag{C.9.b}$$

Comparison of (C.9.a) and (C.9.b) then gives $M\bar{A} = A^+M^+$—that is, finally, since $M^2 = 1$ and $M = M^+$,

$$\bar{A} = MA^+M \tag{C.10}$$

A is Hermitian in the new metric if

$$A = \bar{A} \tag{C.11}$$

and then has, by (C.7), real diagonal elements. By definition, the *new mean value* in state ψ of an operator A is the quantity

$$\subset A \supset_{\psi} = \frac{\subset \psi \mid A \mid \psi \supset}{\subset \psi \mid \psi \supset} \tag{C.12}$$

which is real if $A = \bar{A}$ and which generalizes in the new metric the well-known usual mean value. We suppose of course that $\subset \psi | \psi \supset$ is different from zero in (C.12).

Remarks

(i) The notion of eigenvector and eigenvalue is independent of any metric. The eigenvalues λ_i of a linear operator A are therefore the same whether one uses the old or the new metric. One can also say from (C.3.a) that the equation

$$A \mid \varphi_i \rangle = \lambda_i \mid \varphi_i \rangle \tag{C.13.a}$$

implies

$$A \mid \varphi_i \supset = \lambda_i \mid \varphi_i \supset . \tag{C.13.b}$$

Note incidentally that the eigenvalue λ_i of A can be thought of as the old mean value of A as well as the new in the eigenstate φ_i. Actually, by projecting (C.13.a) on $\langle \varphi_i |$ and (C.13.b) on $\subset \varphi_i |$, one gets, if $\subset \phi_i | \phi_i \supset$ is nonzero (recall that $\langle \varphi_i | \varphi_i \rangle$ is always nonzero except if $| \phi_i \rangle$ is zero),

$$\lambda_i = \frac{\langle \varphi_i \mid A \mid \varphi_i \rangle}{\langle \varphi_i \mid \varphi_i \rangle} \tag{C.14.a}$$

$$\lambda_i = \frac{\subset \varphi_i \mid A \mid \varphi_i \supset}{\subset \varphi_i \mid \varphi_i \supset} . \tag{C.14.b}$$

If $\subset \varphi_i | \phi_i \supset$ is zero, the second equality (C.14) is no longer valid (indeterminate form).

(ii) Assume that $A = \bar{A}$ but $A \neq A^+$ (for example $A = -A^+$, the operator A being antihermitian in the usual sense). Since

$$\langle \varphi_i | A | \varphi_i \rangle = \langle \varphi_i | A^+ | \varphi_i \rangle^* = -\langle \varphi_i | A | \varphi_i \rangle^*$$

is then purely imaginary, the first equality (C.14) implies that λ_i is purely imaginary. In contrast, since $A = \bar{A}$, $\subset \varphi_i | A | \phi_i \supset$ is real and the second equality (C.14) seems to indicate that λ_i is real. This contradiction is only apparent, insomuch as one can show (see Exercise 3) that if $A = \bar{A}$ and if λ_i is not real, then $\subset \phi_i | \phi_i \supset$ is necessarily zero, with the result that (C.14.b) is no longer valid.

(iii) Let $\{ | u_n \rangle \}$ be a basis of the state space, orthonormal in the usual sense and satisfying the closure relationship

$$\sum_n | u_n \rangle \langle u_n | = \mathbb{1} . \tag{C.15.a}$$

Using $\langle u_n| = \subset u_n|M$, which follows from (C.3.b) and $M^2 = 1$, this relation can then be written

$$\sum_n |u_n \supset \subset u_n| M = \mathbb{1}. \tag{C.15.b}$$

Equations (C.15.a) and (C.15.b) are useful in relating the two types of components $\langle u_n|\psi\rangle$ and $\subset u_n|\psi\supset$ of a vector $|\psi\rangle$ in the state space (Exercise 3).

We must finally emphasize that the new mean value does not have the same physical content as the old. Even if $A = \bar{A}$, it is not possible in general to give a probabilistic interpretation to (C.12) like that given to usual mean values of a Hermitian operator in the old sense, $A = A^+$. Recall this interpretation: $\langle A\rangle_\psi = \langle\psi|A|\psi\rangle/\langle\psi|\psi\rangle$ is, when $A = A^+$, the average of the eigenvalues λ_i of A (which are real, since $A = A^+$) weighted by the probabilities $\pi_i = |\langle\varphi_i|\psi\rangle|^2/\langle\psi|\psi\rangle$ of finding λ_i for the system in state $|\psi\rangle$. When $A = \bar{A}$ and $A \neq A^+$, it can happen that certain eigenvalues of A are not real, so that it is out of the question to interpret them as results of a measurement. One can then question the interest of introducing (C.12). The new mean value is in fact interesting for variables like the potentials, which, although real in classical theory, are not truly physical variables in the sense that their precise value varies according to the gauge. The quantum measurement postulates do not apply in fact to potentials, and it is not absurd to associate them with operators $A \neq A^+$. (This indeed can be very useful, as we shall see below.) However, it is important that the potentials satisfy $A = \bar{A}$, so that their new mean value (C.12) is real. Such a reality condition is actually essential if one wants the mean values of the quantum equations of motion to coincide with the classical equations of motion, where the potentials are real functions of $\mathbf{r}$ and t. If we abandon the Hermiticity of the potentials in the sense of the old metric and replace it by Hermiticity in the new sense, we will have to check afterwards that the old and new mean values of measurable physical variables like the electric and magnetic fields agree when they are taken in physical states.

Finally, the generalization given here offers the possibility of admitting new states with negative norm and of considering operators that are non-Hermitian in the old sense ($A \neq A^+$) but Hermitian in the new ($A = \bar{A}$), which allows one to associate with these operators new real mean values. It is this flexibility in the formalism which will allow us to resolve the difficulties associated with the scalar potential in what follows.

2. Choice of the New Metric for Covariant Quantization

To resolve the difficulties mentioned at the end of Part B, we are now going to introduce a new metric in the radiation state space. Since all the mean values will ultimately be calculated in this new metric and these new mean values must be real for the potentials, we must require that the potential operators be Hermitian in the new sense

$$A_\mu(\mathbf{r}) = \overline{A_\mu}(\mathbf{r}) \tag{C.16.a}$$

$$\mathscr{A}_\mu(\mathbf{k}) = \overline{\mathscr{A}_\mu}(-\mathbf{k}) \tag{C.16.b}$$

and analogous conditions for Π_μ and π_μ. All the calculations in Part B above remain valid under the condition that all the old adjoints are replaced by the new throughout, in particular a_μ^+ by $\bar{a}_\mu$. For example, starting from Equations (A.40.a), (A.40.b), (B.4), (B.7), one gets

$$A_\mu(\mathbf{r}) = \int \mathrm{d}^3k \sqrt{\frac{\hbar}{2\,\varepsilon_0\,\omega(2\,\pi)^3}}\,[a_\mu(\mathbf{k})\,\mathrm{e}^{\mathrm{i}\mathbf{k}.\mathbf{r}} + \bar{a}_\mu(\mathbf{k})\,\mathrm{e}^{-\mathrm{i}\mathbf{k}.\mathbf{r}}] \tag{C.17}$$

$$[a_i(\mathbf{k}), \bar{a}_j(\mathbf{k}')] = \delta_{ij}\,\delta(\mathbf{k} - \mathbf{k}') \tag{C.18.a}$$

$$[a_s(\mathbf{k}), \bar{a}_s(\mathbf{k}')] = -\,\delta(\mathbf{k} - \mathbf{k}') \tag{C.18.b}$$

$$H_R = \int \mathrm{d}^3k\,\frac{\hbar\omega}{2}\,[(\bar{a}_\varepsilon\,a_\varepsilon + a_\varepsilon\,\bar{a}_\varepsilon) + (\bar{a}_{\varepsilon'}\,a_{\varepsilon'} + a_{\varepsilon'}\,\bar{a}_{\varepsilon'}) + \\ + (\bar{a}_l\,a_l + a_l\,\bar{a}_l) - (\bar{a}_s\,a_s + a_s\,\bar{a}_s)] \tag{C.19}$$

How can one choose the metric M? Since all the difficulties come from the $-$ sign in (C.18.b), the simplest idea is to use the possible difference between a_s^+ and $\bar{a}_s$ to correct this sign. Assume, for example, that one has succeeded in finding M such that

$$\bar{a}_j = M a_j^+\,M = a_j^+ \tag{C.20.a}$$

$$\bar{a}_s = M a_s^+\,M = -\,a_s^+\,. \tag{C.20.b}$$

In other words, M commutes with a_j but anticommutes with a_s. The commutation relations (C.18.a) remain valid for the operators a_i and a_j^+ relative to the spatial degrees of freedom, whereas (C.18.b) becomes

$$[a_s(\mathbf{k}), a_s^+(\mathbf{k}')] = \delta(\mathbf{k} - \mathbf{k}') \tag{C.21}$$

a_s and a_s^+ then become completely "normal" annihilation and creation operators and allow one to construct the state space of scalar photons without difficulty. In addition, since (C.20.a) implies $\bar{a}_\varepsilon = a_\varepsilon^+$, $\bar{a}_{\varepsilon'} = a_{\varepsilon'}^+$, and $\bar{a}_l = a_l^+$, the Hamiltonian (C.19) written as a function of the old adjoints becomes, using (C.20.b),

$$H_R = \int d^3k \frac{\hbar\omega}{2} [(a_\varepsilon^+ a_\varepsilon + a_\varepsilon a_\varepsilon^+) + (a_{\varepsilon'}^+ a_{\varepsilon'} + a_{\varepsilon'} a_{\varepsilon'}^+) + $$
$$+ (a_l^+ a_l + a_l a_l^+) + (a_s^+ a_s + a_s a_s^+)] \quad \text{(C.22)}$$

and has only + signs. The energy then becomes positive definite (recall that the eigenvalues of an operator are independent of the metric; see Remark i of §C.1 above).

Before going farther, note that $\bar{a}_s = -a_s^+$ implies that the scalar potential A_s is now anti-Hermitian in the old sense ($A_s^+ = -A_s$), since one requires it to be Hermitian in the new [Equations (C.17)]. The reestablishment of the + sign in (C.21) is then achieved at the price of abandoning the hermiticity of A_s in the usual sense. It is thus no longer possible to apply the quantum-mechanical postulates to A_s. This is not troublesome, however, since A_s is not a truly physical variable.

We now show how it is possible to satisfy (C.20). First, we introduce a basis of states for each scalar mode (*),

$$| n_s \rangle = \frac{(a_s^+)^{n_s}}{\sqrt{n_s\,!}} | 0_s \rangle \quad \text{(C.23)}$$

normalized in the usual sense:

$$\langle n_s | n_s' \rangle = \delta_{n_s n_s'} . \quad \text{(C.24)}$$

The action of a_s and a_s^+ on $|n_s\rangle$ is well known, since the commutation relation $[a_s, a_s^+] = 1$ is "normal":

$$a_s^+ | n_s \rangle = \sqrt{n_s + 1}\, | n_s + 1 \rangle \quad \text{(C.25.a)}$$
$$a_s | n_s \rangle = \sqrt{n_s}\, | n_s - 1 \rangle \quad \text{(C.25.b)}$$
$$a_s | 0_s \rangle = 0 \quad \text{(C.25.c)}$$

and that of $\bar{a}_s$ is gotten from (C.20.b):

$$\bar{a}_s | n_s \rangle = -\sqrt{n_s + 1}\, | n_s + 1 \rangle . \quad \text{(C.26)}$$

(*) To ease the notation, we omit the index **k** for the mode. We also assume that the field is quantized in a box so as to have discrete modes **k**.

Consider now the Hermitian unitary operator M defined by

$$M \mid n_s \rangle = (-1)^{n_s} \mid n_s \rangle . \tag{C.27}$$

From (C.25.a), (C.26), and (C.27) it follows that

$$Ma^+ \mid n_s \rangle = (-1)^{n_s+1} \sqrt{n_s + 1} \mid n_s + 1 \rangle = -a^+ M \mid n_s \rangle \tag{C.28}$$

which, since the set $\{|n_s\rangle\}$ forms a basis, implies that $Ma_s^+ = -a_s^+ M$ and thus proves (C.20.b). Equation (C.27) then permits the calculation of new scalar products

$$\subset n_s \mid n_s' \supset \; = \langle n_s \mid M \mid n_s' \rangle = (-1)^{n_s} \delta_{n_s n_s'} . \tag{C.29}$$

The vacuum has a new positive norm, but the one-photon scalar states have a new negative norm. The generalization presented in this section then allows us to include in the formalism situations like those mentioned above (§B.2).

Remark

Suppose that we change the phase factors of the basis vectors (C.23) by taking

$$| n_s \rangle = (-1)^{n_s} \frac{(a^+)^{n_s}}{\sqrt{n_s\, !}} \mid 0_s \rangle \tag{C.30}$$

so as to have a more satisfactory expression for $|n_s\rangle$ as a function of $\bar{a}_s$ and $|0_s\rangle$,

$$| n_s \rangle = \frac{(\bar{a}_s)^{n_s}}{\sqrt{n_s !}} \mid 0_s \rangle . \tag{C.31}$$

Equations (C.26), (C.25.b), and (C.25.a) then become

$$\bar{a}_s \mid n_s \rangle = \sqrt{n_s + 1} \mid n_s + 1 \rangle \tag{C.32.a}$$

$$a_s \mid n_s \rangle = -\sqrt{n_s} \mid n_s - 1 \rangle \tag{C.32.b}$$

$$a_s^+ \mid n_s \rangle = -\sqrt{n_s + 1} \mid n_s + 1 \rangle . \tag{C.32.c}$$

Equation (C.27), which does not depend on the phase factor of $|n_s\rangle$, remains unchanged, as does (C.29), which follows from it. Depending on whether we choose (C.23) or (C.30), one can then say that, with respect to the usual harmonic-oscillator theory, it is sufficient to attach a $-$ sign either to the matrix elements of $\bar{a}_s$ [if (C.23) is chosen] or to the matrix elements of a_s [if (C.30) is chosen].

3. Construction of the Physical Kets

The very simple form of the subsidiary condition (B.21) suggests the introduction for each value of **k** of two linear "orthogonal" combinations of a_l and a_s, thereby generalizing the normal variables (A.45) to the quantum case. We thus take

$$a_d = \frac{i}{\sqrt{2}}(a_l - a_s) \tag{C.33.a}$$

$$a_g = \frac{1}{\sqrt{2}}(a_l + a_s). \tag{C.33.b}$$

Using (C.18.a), (C.20.a), and (C.21), the new operators a_d and a_g thus introduced satisfy the equations

$$[a_d(\mathbf{k}), a_d^+(\mathbf{k}')] = \delta(\mathbf{k} - \mathbf{k}') \tag{C.34.a}$$

$$[a_g(\mathbf{k}), a_g^+(\mathbf{k}')] = \delta(\mathbf{k} - \mathbf{k}') \tag{C.34.b}$$

$$[a_d, a_g^+] = [a_g, a_d^+] = 0. \tag{C.34.c}$$

The relationships (C.33) then allow the introduction for each value of **k** of two new types of modes, *d* and *g*, with two types of photons, the "*d*-photons" and the "*g*-photons" respectively. Note incidentally that the vacuum of *d* and *g* photons coincides with the vacuum of *l* and *s* photons, since the expressions (C.33) imply

$$a_d \mid 0_l\, 0_s \rangle = 0 \tag{C.35.a}$$

$$a_g \mid 0_l\, 0_s \rangle = 0 \tag{C.35.b}$$

which shows that

$$\mid 0_d\, 0_g \rangle = \mid 0_l\, 0_s \rangle. \tag{C.36}$$

With this new notation the subsidiary condition (B.21) is written

$$a_d \mid \psi \rangle = 0 \tag{C.37}$$

which generalizes (A.46) and implies that the *physical kets have no d-photon*. The subsidiary condition specifies nothing on the other hand about the state of the *g*-mode, which can be anything (nor, obviously, about the state of the transverse modes ε and ε′).

We now construct a basis for the space of physical kets. For each value of **k** we have

$$\mid n_\varepsilon\, n_{\varepsilon'}\, 0_d\, n_g \rangle = \frac{(a_\varepsilon^+)^{n_\varepsilon}(a_{\varepsilon'}^+)^{n_{\varepsilon'}}(a_g^+)^{n_g}}{\sqrt{n_\varepsilon\,!\, n_{\varepsilon'}\,!\, n_g\,!}} \mid 0 \rangle. \tag{C.38}$$

The numbers of transverse photons, n_ε and $n_{\varepsilon'}$, can be anything, as can the number of *g*-photons. On the other hand, the number of *d*-photons is necessarily zero from (C.37). The set of vectors (C.38) forms an orthonormal basis in the usual sense in the subspace of physical kets:

$$\langle n_\varepsilon\, n_{\varepsilon'}\, 0_d\, n_g \mid n'_\varepsilon\, n'_{\varepsilon'}\, 0_d\, n'_g \rangle = \delta_{n_\varepsilon n'_\varepsilon}\, \delta_{n_{\varepsilon'} n'_{\varepsilon'}}\, \delta_{n_g n'_g}\,. \tag{C.39}$$

What can be said, on the other hand, about the new scalar products involving the basis vectors (C.38)? Since $\subset n_\varepsilon n_{\varepsilon'} | n'_\varepsilon n'_{\varepsilon'} \supset = \langle n_\varepsilon n_{\varepsilon'} | n'_\varepsilon n'_{\varepsilon'} \rangle = \delta_{n_\varepsilon n'_\varepsilon} \delta_{n'_{\varepsilon'} n_{\varepsilon'}}$, it suffices to find $\subset 0_d n_g | 0_d n'_g \supset$. According to (C.38), $|n_g \supset = |n_g\rangle$ is expressible as a function of a_g^+. In order to find $\subset 0_d n_g|$, we must first determine the new adjoint $\bar{a}_g^+$ of a_g^+. Now the equations (C.33) give

$$\bar{a}_d = -\frac{i}{\sqrt{2}}(\bar{a}_l - \bar{a}_s) = -\frac{i}{\sqrt{2}}(a_l^+ + a_s^+) = -\,i a_g^+ \tag{C.40.a}$$

$$\bar{a}_g = \frac{1}{\sqrt{2}}(\bar{a}_l + \bar{a}_s) = \frac{1}{\sqrt{2}}(a_l^+ - a_s^+) = +\,i a_d^+ \tag{C.40.b}$$

which shows that

$$\bar{a}_g^+ = -\,i a_d\,. \tag{C.41}$$

Also, since from (C.34.c) a_d commutes with a_g^+, and from (C.40.a) $a_g^+ = i\bar{a}_d$, one has

$$[a_d, \bar{a}_d] = 0\,. \tag{C.42}$$

One can then write, using (C.40.a), (C.41), and (C.42),

$$\begin{aligned}
\sqrt{n_g!\, n'_g!}\, \subset 0_d\, n_g \mid 0_d\, n'_g \supset &= \subset 0_d\, 0_g \mid (\bar{a}_g^+)^{n_g} (a_g^+)^{n'_g} \mid 0_d\, 0_g \supset \\
&= (i)^{(n'_g - n_g)} \subset 0_d\, 0_g \mid (a_d)^{n_g} (\bar{a}_d)^{n'_g} \mid 0_d\, 0_g \supset \\
&= (i)^{(n'_g - n_g)} \subset 0_d\, 0_g \mid (\bar{a}_d)^{n'_g} (a_d)^{n_g} \mid 0_d\, 0_g \supset = \delta_{n_g 0}\, \delta_{n'_g 0}\,.
\end{aligned} \tag{C.43}$$

Finally, all the basis vectors of the subspace of physical kets have a zero new scalar product and a zero new norm unless $n_g = 0$:

$$\subset n_\varepsilon\, n_{\varepsilon'}\, 0_d\, n_g \mid n'_\varepsilon\, n'_{\varepsilon'}\, 0_d\, n'_g \supset = \delta_{n_\varepsilon n'_\varepsilon}\, \delta_{n_{\varepsilon'} n'_{\varepsilon'}}\, \delta_{n_g 0}\, \delta_{n'_g 0}\,. \tag{C.44}$$

Consider now, for one value of **k**, a physical radiation ket

$$|\psi\rangle = |\psi_T\rangle \otimes |\psi_{LS}\rangle \tag{C.45}$$

describing a situation where the transverse degrees of freedom are in the state $|\psi_T\rangle$ and the longitudinal and scalar degrees of freedom are in the physical state

$$|\psi_{LS}\rangle = \sum_n c_n |0_d\, n_g\rangle \tag{C.46}$$

in which n_d is always zero. Using (C.44), the new norm of $|\psi\rangle$ becomes

$$\begin{aligned}\subset \psi | \psi \supset &= \subset \psi_T | \psi_T \supset \, \subset \psi_{LS} | \psi_{LS} \supset \\ &= \langle \psi_T | \psi_T \rangle |c_0|^2 .\end{aligned} \tag{C.47}$$

This relation can easily be generalized to a situation where $|\psi\rangle$ is expanded on all the transverse modes $\mathbf{k}$, and $|\psi_{LS}\rangle$ on all the modes $(\mathbf{k}, g)$, with n_d being zero always. Equation (C.47) remains valid, c_0 representing the component of $|\psi_{LS}\rangle$ on the vacuum of all the modes $(\mathbf{k}, d)$ and $(\mathbf{k}, g)$. It appears then that the new norm of a physical ket is proportional to the old norm of its transverse component, the coefficient of proportionality $|c_0|^2$ being the square of the modulus of the component of $|\psi_{LS}\rangle$ on the vacuum of modes d and g—or equivalently, from (C.36), on the vacuum of modes l and s.

4. Mean Values of the Physical Variables in a Physical Ket

Having characterized the physical kets by the condition (C.37), we now find the mean value of the various physical variables (potentials, fields, energy) in these kets.

a) Mean Values of the Potentials and the Fields

As in (A.57), we write the potentials A_μ in the form

$$A_\mu = A_\mu^T + A_\mu^G + A_\mu^D \tag{C.48}$$

where A_μ^T, A_μ^G, A_μ^D are given by expressions identical to (A.58), (A.59), and (A.62) except that the normal classical variables α_ε, $\alpha_{\varepsilon'}$, α_d, α_g, α_ε^*, $\alpha_{\varepsilon'}^*$, α_d^*, α_g^* are replaced by the operators a_ε, $a_{\varepsilon'}$, a_d, a_g, $\bar{a}_\varepsilon$, $\bar{a}_{\varepsilon'}$, $\bar{a}_d$, $\bar{a}_g$.

Since the operators appearing in A_μ^T act only on the transverse degrees of freedom, it follows that

$$\begin{aligned}\frac{\subset \psi | A_\mu^T | \psi \supset}{\subset \psi | \psi \supset} &= \frac{\subset \psi_T | A_\mu^T | \psi_T \supset}{\subset \psi_T | \psi_T \supset} \, \frac{\subset \psi_{LS} | \psi_{LS} \supset}{\subset \psi_{LS} | \psi_{LS} \supset} \\ &= \frac{\langle \psi_T | A_\mu^T | \psi_T \rangle}{\langle \psi_T | \psi_T \rangle} .\end{aligned} \tag{C.49}$$

The new mean value of the transverse potential agrees with the old. Such a result is satisfying, since A^T_μ, which is gauge invariant, can be considered as a truly physical variable.

From (A.62), A^D_μ is a linear superposition of the operators a_d and $\bar{a}_d$. Since the subsidiary condition (C.37) implies that $a_d|\psi \supset = 0$ and $\subset \psi|\bar{a}_d = 0$, the new mean value of A^D_μ in the physical ket $|\psi\rangle$ is zero:

$$\frac{\subset \psi \,|\, A^D_\mu \,|\, \psi \supset}{\subset \psi \,|\, \psi \supset} = 0\,. \tag{C.50}$$

Consider finally A^G_μ. Since A^G_μ does not act on the transverse degrees of freedom, its mean value involves only $|\psi_{LS}\rangle$ and using (A.59) is written

$$\frac{\subset \psi \,|\, A^G_\mu \,|\, \psi \supset}{\subset \psi \,|\, \psi \supset} = -\,\partial_\mu \subset f \supset \tag{C.51}$$

where $\subset f \supset$ is the new mean value in the state $|\psi\rangle$, of the operator f, gotten by replacing α_g and α^*_g with a_g and $\bar{a}_g$ in (A.60).

Finally, the new mean value of (C.48) is written using (C.49), (C.50), and (C.51) as

$$\frac{\subset \psi \,|\, A_\mu \,|\, \psi \supset}{\subset \psi \,|\, \psi \supset} = \frac{\langle \psi_T \,|\, A^T_\mu \,|\, \psi_T \rangle}{\langle \psi_T \,|\, \psi_T \rangle} - \partial_\mu \subset f \supset . \tag{C.52}$$

It differs from the old mean value of the transverse potential only by a gauge term.

The mean value of the electromagnetic field tensor $F_{\mu\nu} = \partial_\mu A_\nu - \partial_\nu A_\mu$ can be deduced from the previous results: $\subset F^D_{\mu\nu} \supset = 0$ because of (C.50), and $\subset F^G_{\mu\nu} \supset = 0$ because $\subset A^G_\mu \supset$ is a gauge term according to (C.51). It follows that

$$\frac{\subset \psi \,|\, F_{\mu\nu} \,|\, \psi \supset}{\subset \psi \,|\, \psi \supset} = \frac{\langle \psi_T \,|\, F^T_{\mu\nu} \,|\, \psi_T \rangle}{\langle \psi_T \,|\, \psi_T \rangle}\,. \tag{C.53}$$

The new mean value of the fields agrees with the old one found by taking account of only the transverse degrees of freedom.

b) Gauge Arbitrariness and Arbitrariness of the Kets Associated with a Physical State

The above results permit us to understand the role played by the excitation of the *g*-modes, about which the subsidiary condition does not provide any information.

To a given *physical state* of the transverse field there corresponds a set of *physical kets* of the form (C.45). For these kets, $|\psi_T\rangle$ is fixed but $|\psi_{LS}\rangle$

can describe an arbitrary excitation of the g-modes, the d-modes in contrast being always in the vacuum state. This arbitrariness corresponds to the remaining gauge arbitrariness on the potentials. Two different excitations of the g-modes correspond to two different gauges, and thus to two different mean values of the potentials but nevertheless to the same mean values for the truly physical variables $F_{\mu\nu}$ and A^T_μ.

Covariant quantization with an indefinite metric finally allows us to introduce in the physical kets themselves the gauge arbitrariness, while preserving for the measurable physical variables predictions identical to those of Chapters II and III.

c) Mean Value of the Hamiltonian

Starting from (C.33), it is possible to show that

$$a_l^+ a_l + a_s^+ a_s = a_d^+ a_d + a_g^+ a_g . \tag{C.54}$$

Then, using (C.22), one can write

$$H_R = H_R^T + H_R^{LS} \tag{C.55}$$

where H_R^T is the purely transverse Hamiltonian of Chapters II and III, and where H_R^{LS} is written for a given value of $\mathbf{k}$ (and by omitting the zero-point energy) as

$$\begin{aligned} H_R^{LS} &= \hbar\omega(a_g^+ a_g + a_d^+ a_d) \\ &= \mathrm{i}\,\hbar\omega(\bar{a}_d a_g - \bar{a}_g a_d) . \end{aligned} \tag{C.56}$$

where (C.40) has been used to replace a_g^+ by $\mathrm{i}\bar{a}_d$ and a_d^+ by $-\mathrm{i}\bar{a}_g$. This is a natural generalization of the classical expression (A.65). Since in a physical state $a_d|\psi\supset$ and $\subset\psi|\bar{a}_d$ are zero, it follows that

$$\frac{\subset \psi | H_R^{LS} | \psi \supset}{\subset \psi | \psi \supset} = 0 \tag{C.57}$$

and also

$$\frac{\subset \psi | H_R | \psi \supset}{\subset \psi | \psi \supset} = \frac{\langle \psi_T | H_R^T | \psi_T \rangle}{\langle \psi_T | \psi_T \rangle} . \tag{C.58}$$

The new mean value of the total Hamiltonian H_R in a physical ket then agrees with the old mean value of the energy of the transverse field.

Finally, by a generalization of the metric in the state space of the radiation, we have been able to solve all the problems posed by the subsidiary condition and by the sign of the commutation relation relative to the scalar potential. We have identified the different physical kets corresponding to a single given physical state, related the multiplicity of these kets to the gauge arbitrariness, and shown the equivalence of the theory thereby presented with that of Chapters II and III for all the predictions concerning physical variables and physical states.

D—A SIMPLE EXAMPLE OF INTERACTION: A QUANTIZED FIELD COUPLED TO TWO FIXED EXTERNAL CHARGES

In the preceding parts we were essentially dealing with the free field. The main purpose was to establish how it is possible to reconcile the presence in the covariant theory of four kinds of photons with the fact that the only physical photons are free transverse photons. However, since we had not given the form of the particle Lagrangian, it was impossible to study relativistic quantum electrodynamics in the presence of interactions.

We have however treated the simpler case of fields coupled to external sources, that is, sources with a given time dependence. It suffices then for the study of the evolution of the field to add to the Lagrangian or Hamiltonian of the free field an interaction term [see for example (A.29)].

In this final part, we consider a particularly simple example of this type of situations: a quantized field coupled to two external fixed charges. Our purpose is to illustrate here the role played in the interaction by the new types of photons (longitudinal and scalar) introduced by the covariant theory.

1. Hamiltonian for the Problem

Consider two fixed charges q_1 and q_2 situated at $\mathbf{r}_1$ and $\mathbf{r}_2$. The corresponding external charge density is given in real space by

$$\rho_e(\mathbf{r}) = q_1\,\delta(\mathbf{r}-\mathbf{r}_1) + q_2\,\delta(\mathbf{r}-\mathbf{r}_2) \tag{D.1}$$

and in reciprocal space by

$$\rho_e(\mathbf{k}) = \frac{1}{(2\pi)^{3/2}}\left[q_1\,e^{-i\mathbf{k}.\mathbf{r}_1} + q_2\,e^{-i\mathbf{k}.\mathbf{r}_2}\right]. \tag{D.2}$$

In the presence of a current four-vector $j_{e\mu}$, it is necessary to add to the Hamiltonian of the free field, H_R, the coupling term $V = \Sigma_\mu \int d^3r\, j_{e\mu} A^\mu$ [see Equation (A.29)]. The Hamiltonian H which describes the evolution of the field in the presence of the two charges is then written

$$H = H_R + V \tag{D.3}$$

where H_R is given in (C.19) and where, since the only nonzero component of $j_{e\mu}$ is $j_{e0} = c\rho_e$,

$$V = \int d^3r \, c\rho_e(\mathbf{r}) \, A_s(\mathbf{r})$$
$$= cq_1 \, A_s(\mathbf{r}_1) + cq_2 \, A_s(\mathbf{r}_2) \tag{D.4}$$

By using the expansion (C.17) of the potentials, one gets

$$V = \int d^3k \, c \sqrt{\frac{\hbar}{2\,\varepsilon_0\,\omega(2\,\pi)^3}} \left[a_s(\mathbf{k}) \left(q_1 \, e^{i\mathbf{k}.\mathbf{r}_1} + q_2 \, e^{i\mathbf{k}.\mathbf{r}_2} \right) + \right.$$
$$\left. + \, \bar{a}_s(\mathbf{k}) \left(q_1 \, e^{-i\mathbf{k}.\mathbf{r}_1} + q_2 \, e^{-i\mathbf{k}.\mathbf{r}_2} \right) \right] \tag{D.5}$$

that is, finally, using (D.2),

$$V = \int d^3k \, c \sqrt{\frac{\hbar}{2\,\varepsilon_0\,\omega}} \left[a_s(\mathbf{k}) \, \rho_e^*(\mathbf{k}) + \bar{a}_s(\mathbf{k}) \, \rho_e(\mathbf{k}) \right]. \tag{D.6}$$

2. Energy Shift of the Ground State of the Field. Reinterpretation of Coulomb's Law

In the absence of sources ($\rho_e = 0$), the ground state of H_R is the photon vacuum $|0\rangle$. When one introduces the two charges q_1 and q_2 at $\mathbf{r}_1$ and $\mathbf{r}_2$, the new ground state of the field, that is of (D.3), is going to be modified and shifted by an amount ΔE depending on q_1 and q_2. We first derive here a perturbative expression for ΔE by studying the shift of the ground state of H_R to second order in V. We show that one gets in this way the Coulomb energy of the system of two charges. Coulomb's law can thus be reinterpreted, in this approach, as being due to an exchange of scalar photons between the two charges. Finally, we shall see that the expression gotten for ΔE to second order in V is in fact valid to all orders.

a) Perturbative Calculation of the Energy Shift

We have already mentioned that the eigenvalues of an operator are independent of the metric (Remark i of §C.1). To determine the energy shift ΔE of the ground state $|0\rangle$ of H_R due to the potential V, we can thus apply the usual expression given by perturbation theory, with the matrix elements of V evaluated in the usual Hilbert-space metric. To second order in V, we then get, since $|0\rangle$ is not degenerate,

$$\Delta E = \langle\, 0 \,|\, V \,|\, 0 \,\rangle + \langle\, 0 \,|\, V \frac{Q}{E_0 - H_R} V \,|\, 0 \,\rangle \tag{D.7}$$

where E_0 is the unperturbed energy of $|0\rangle$ and where Q is the projection operator on the subspace orthogonal to $|0\rangle$. Since the diagonal elements of a_s and $\bar{a}_s$ which arise in the expression (D.6) for V are zero, the first term of (D.7) is zero. In addition, since V can only create or destroy a scalar photon, the only type of intermediate state which can arise in the second term of (D.7) is the state $|\mathbf{k}s\rangle$ of a scalar photon $\mathbf{k}$. The energy of such a state is greater than that of the vacuum by $\hbar\omega$, so that the energy denominator of (D.7) becomes $E_0 - (E_0 + \hbar\omega) = -\hbar\omega$. Finally, we get for ΔE

$$\Delta E = \int \mathrm{d}^3k \frac{\langle 0 | V | \mathbf{k}s \rangle \langle \mathbf{k}s | V | 0 \rangle}{-\hbar\omega}. \tag{D.8}$$

According to the selection rules for the operators a_s and $\bar{a}_s$,

$$\langle 0 | V | \mathbf{k}s \rangle = c \int \mathrm{d}^3k' \sqrt{\frac{\hbar}{2\,\varepsilon_0\,\omega'}} \langle 0 | a_s(\mathbf{k}') | \mathbf{k}s \rangle\, \rho_e^*(\mathbf{k}') \tag{D.9.a}$$

$$\langle \mathbf{k}s | V | 0 \rangle = c \int \mathrm{d}^3k'' \sqrt{\frac{\hbar}{2\,\varepsilon_0\,\omega''}} \langle \mathbf{k}s | \bar{a}_s(\mathbf{k}'') | 0 \rangle\, \rho_e(\mathbf{k}''). \tag{D.9.b}$$

Finally, we have seen above that the matrix elements of a_s and $\bar{a}_s$ always have opposite signs, whatever may be the convention chosen for the states $|n_s\rangle$ [see (C.25.b) and (C.26) or (C.32)], so that

$$\langle 0 | a_s(\mathbf{k}') | \mathbf{k}s \rangle \langle \mathbf{k}s | \bar{a}_s(\mathbf{k}'') | 0 \rangle = -\,\delta(\mathbf{k} - \mathbf{k}')\,\delta(\mathbf{k} - \mathbf{k}''). \tag{D.10}$$

Substituting (D.9) and (D.10) in (D.8) gives finally

$$\Delta E = \int \mathrm{d}^3k \frac{\rho_e^*(\mathbf{k})\,\rho_e(\mathbf{k})}{2\,\varepsilon_0\,k^2}. \tag{D.11}$$

This is the expression for the Coulomb energy of the charge distribution characterized by $\rho_e(\mathbf{k})$ [see (B.32) and (B.37) of Chapter I], which can be also written

$$\Delta E = V_{\text{Coul}} = \varepsilon^1_{\text{Coul}} + \varepsilon^2_{\text{Coul}} + \frac{q_1\,q_2}{4\,\pi\varepsilon_0\,|\mathbf{r}_1 - \mathbf{r}_2|}. \tag{D.12}$$

The first two terms of (D.12) represent the Coulomb self-energy of the two charges, q_1 and q_2, and the last term the Coulomb interaction energy between them.

Remark

One should note the importance of the $-$ sign appearing in (D.10), which implies that the product of the two matrix elements of V appearing in the numerator of (D.8) is negative, so that ΔE is ultimately positive for charges with the same sign (see also D.11). If the operator associated with the scalar potential were Hermitian in the usual sense, so would V be, and the numerator of (D.8) would be positive, leading to an energy shift $\Delta E < 0$. This result simply recalls the fact that the shift of the ground state due to a Hermitian potential is always negative to second order, the ground state being pushed downwards by the excited states. It is because the operator associated with the scalar potential is anti-Hermitian in the initial metric, but Hermitian in the new one, that we finally find a positive Coulomb energy.

b) Physical Discussion. Exchange of Scalar Photons between the Two Charges

The Coulomb energy appears then as associated with a second-order perturbation term (D.8). The structure of this expression then suggests the following physical interpretation. The field, initially in the vacuum state, makes under the effect of V a transition to the intermediate state $|\mathbf{k}s\rangle$ and then returns to its initial state. In other words, a scalar photon $\mathbf{k}$ is emitted virtually and then reabsorbed.

The term in $q_1 q_2$ of ΔE is obtained either by taking the term in q_1 of ρ in the matrix element $\langle \mathbf{k}s|V|0\rangle$ and the term in q_2 in $\langle 0|V|\mathbf{k}s\rangle$, in which case it is the charge q_1 which emits a virtual scalar photon reabsorbed subsequently by q_2, or the inverse process, in which case it is q_2 which emits a virtual scalar photon reabsorbed by q_1. Quantum electrodynamics in the Lorentz gauge thus lets us interpret the Coulomb interaction between two fixed charges as resulting from the exchange of scalar photons between them.

Note finally that the terms in q_1^2 (or q_2^2) of ΔE—that is, $\varepsilon^1_{\text{Coul}}$ (or $\varepsilon^2_{\text{Coul}}$)—are gotten by taking the same term in q_1 (or q_2) in the two matrix elements of V. The Coulomb self-energy of a particle appears then as due to the virtual emission and reabsorption of a scalar photon by this same particle.

c) Exact Calculation

Since the coupling Hamiltonian V is linear in a_s and $\bar{a}_s$, it is in fact possible to diagonalize exactly the total Hamiltonian $H = H_R + V$. Consider the part H_S of H relative to the scalar modes (V acts only on these modes). It is written

$$H_s = \int d^3k \, H_s(\mathbf{k}) \tag{D.13}$$

with

$$H_s(\mathbf{k}) = \hbar\omega\left[- \bar{a}_s(\mathbf{k})\, a_s(\mathbf{k}) + \lambda^*(\mathbf{k})\, a_s(\mathbf{k}) + \lambda(\mathbf{k})\, \bar{a}_s(\mathbf{k}) + \frac{1}{2}\right] \tag{D.14}$$

$\lambda(k)$ being given according to (D.6) by

$$\lambda(\mathbf{k}) = \frac{c}{\hbar\omega}\sqrt{\frac{\hbar}{2\,\varepsilon_0\,\omega}}\,\rho_e(\mathbf{k})\,. \tag{D.15}$$

The principle for the diagonalization of the Hamiltonian $H_s(\mathbf{k})$ for the scalar mode $\mathbf{k}$ involves imposing a translation on a_s and $\bar{a}_s$ so as to cause the linear terms in a_s and $\bar{a}_s$ in (D.14) to vanish and to get a harmonic-oscillator Hamiltonian.

For this we introduce the new operators b_s and $\bar{b}_s$:

$$b_s(\mathbf{k}) = a_s(\mathbf{k}) - \lambda(\mathbf{k}) \tag{D.16.a}$$

$$\bar{b}_s(\mathbf{k}) = \bar{a}_s(\mathbf{k}) - \lambda^*(\mathbf{k})\,. \tag{D.16.b}$$

The commutation relation (C.18.b) gives for b_s and $\bar{b}_s$

$$[b_s(\mathbf{k}), \bar{b}_s(\mathbf{k}')] = [a_s(\mathbf{k}), \bar{a}_s(\mathbf{k}')] = -\,\delta(\mathbf{k} - \mathbf{k}') \tag{D.17}$$

so that b_s and $\bar{b}_s$ can in fact be considered as annihilation and creation operators (*). In addition, since

$$\bar{b}_s\,b_s = \bar{a}_s\,a_s - \lambda^*\,a_s - \lambda\bar{a}_s + \lambda^*\,\lambda \tag{D.18}$$

it is possible to rewrite $H_s(\mathbf{k})$ in the form

$$H_s(\mathbf{k}) = \hbar\omega\left[-\bar{b}_s(\mathbf{k})\,b_s(\mathbf{k}) + \lambda^*(\mathbf{k})\,\lambda(\mathbf{k}) + \frac{1}{2}\right] \tag{D.19}$$

which is, to within a constant, a harmonic-oscillator Hamiltonian. Let $|\tilde{0}\rangle$ be the state defined by

$$b_s(\mathbf{k})\,|\,\tilde{0}\,\rangle = 0 \tag{D.20}$$

so that

$$a_s(\mathbf{k})\,|\,\tilde{0}\,\rangle = \lambda(\mathbf{k})\,|\,\tilde{0}\,\rangle\,. \tag{D.21}$$

The state $|\tilde{0}\rangle$ is then a coherent state of the mode $\mathbf{k}s$, that is, an eigenvector of $a_s(\mathbf{k})$ with eigenvalue $\lambda(\mathbf{k})$. Using (D.20), $|\tilde{0}\rangle$ is an eigenstate of $H_s(\mathbf{k})$:

$$H_s(\mathbf{k})\,|\,\tilde{0}\,\rangle = \hbar\omega\left[\lambda^*(\mathbf{k})\,\lambda(\mathbf{k}) + \frac{1}{2}\right]|\tilde{0}\,\rangle \tag{D.22}$$

(*) Since b_s and $\bar{b}_s$ obey the same commutation relations as a_s and $\bar{a}_s$, one can actually find a transformation T, unitary in the new sense (that is, such that $T\bar{T} = \bar{T}T = 1$), which transforms the operators a_s and $\bar{a}_s$ into b_s and $\bar{b}_s$ (see Exercise 4).

with an eigenvalue $\hbar\omega[\frac{1}{2} + \lambda^*(\mathbf{k})\lambda(\mathbf{k})]$. One concludes that the exact shift ΔE of the ground state of the field due to the presence of charges is the sum over all the scalar modes of $\hbar\omega\lambda^*(\mathbf{k})\lambda(\mathbf{k})$, so that, using (D.15),

$$\Delta E = \int \mathrm{d}^3k\ \hbar\omega\lambda^*(\mathbf{k})\ \lambda(\mathbf{k}) = \int \mathrm{d}^3k\ \frac{\rho_e^*(\mathbf{k})\ \rho_e(\mathbf{k})}{2\ \varepsilon_0\ k^2} \tag{D.23}$$

which coincides with (D.11). The result of the second-order perturbative calculation for ΔE agrees with the exact value of this shift.

Remark

It is easy to get the other states and eigenvalues of H_s. Starting with the commutation relation (D.17) and from (D.20), one can show that

$$H_s[b_s(\mathbf{k})]^p \mid \tilde{0}\,\rangle = \hbar\omega\left[p + \frac{1}{2} + \lambda^*(\mathbf{k})\ \lambda(\mathbf{k})\right][\bar{b}_s(\mathbf{k})]^p \mid \tilde{0}\,\rangle \tag{D.24}$$

The state $[\bar{b}_s(\mathbf{k})]^p|\tilde{0}\rangle$ is then a new eigenstate of H_s at a distance $p\hbar\omega$ above the new ground state $|\tilde{0}\rangle$. All the levels of the harmonic oscillator associated with the scalar mode **k** are then displaced together by the same amount ΔE.

3. Some Properties of the New Ground State of the Field

a) The Subsidiary Condition in the Presence of the Interaction. The Physical Character of the New Ground State

To see how it is possible to generalize (B.20) and (B.21), which characterize the physical states for the free field, one begins by calculating $\Sigma_\mu \partial_\mu A^\mu$ starting from the expansion (C.17) for A_μ in a and $\bar{a}$:

$$\sum_\mu \partial_\mu\ A^\mu = \int \mathrm{d}^3k\ \sqrt{\frac{\hbar}{2\ \varepsilon_0\ \omega(2\ \pi)^3}}\left(ika_l + \frac{\dot{a}_s}{c}\right)\mathrm{e}^{i\mathbf{k}.\mathbf{r}} + h.c. \tag{D.25}$$

Consider now at a given time t_0 the states $|\psi\rangle$ which are for all **k** solutions of the equation

$$\left[ika_l(\mathbf{k}) + \frac{\dot{a}_s(\mathbf{k})}{c}\right]\mid\psi\,\rangle = 0 \quad \forall\mathbf{k}\,. \tag{D.26}$$

It follows from (D.25) that the mean value of the operator $\Sigma_\mu \partial_\mu A^\mu$ in these states is zero at t_0. It is possible on the other hand to express the velocity $\dot{a}_s = [a_s, H]/i\hbar$ appearing in (D.26) by means of the equation of motion for a_s, which is the quantum generalization of the equation of motion (A.35′) for the classical normal variable α_s:

$$\dot{a}_s(\mathbf{k}) + \mathrm{i}\omega\, a_s(\mathbf{k}) = \frac{\mathrm{i}}{\sqrt{2\,\varepsilon_0\,\hbar\omega}}\, c\rho_e(\mathbf{k})\,. \tag{D.27}$$

Putting (D.27) into (D.26) then gives, using (D.15) for $\lambda(k)$,

$$[a_l(\mathbf{k}) - a_s(\mathbf{k}) + \lambda(\mathbf{k})]\,|\,\psi\,\rangle = 0 \quad \forall\mathbf{k} \tag{D.28}$$

which reduces to (B.21) in the absence of sources ($\lambda = 0$).

An important property of (D.28) is that its solutions are independent of the time t_0. In other words, the subspace of physical states selected by (D.28) is stable over time; a physical state $|\psi\rangle$ at t_0 remains a physical state at all subsequent times. We will prove this important point in §A_V.3.*c* by examining the temporal evolution of the operator $a_l - a_s + \lambda$ which appears in (D.28). This property assures that $\Sigma_\mu \partial_\mu A^\mu$ keeps a zero mean value for all $\mathbf{r}$ and all t, and allows us to consider (D.28) as the generalization of (B.21) in the presence of an interaction.

We return now to the new ground state of the field $|\tilde{0}\rangle$ defined in (D.21). It is easy to see that $|\tilde{0}\rangle$ satisfies (D.28) and is therefore a physical state. First, since the longitudinal modes are not excited by the fixed charges

$$a_l(\mathbf{k})\,|\,\tilde{0}\,\rangle = 0 \quad \forall\mathbf{k}\,. \tag{D.29}$$

It is sufficient then to use (D.29) and the definition (D.21) of $|\tilde{0}\rangle$ to see that $|\tilde{0}\rangle$ truly satisfies (D.28).

b) The Mean Value of the Scalar Potential in the New Ground State of the Field

Starting from the expansion of $A_s(\mathbf{r})$ in the scalar modes, one gets for the mean value (in the sense of the new metric) of $A_s(\mathbf{r})$ in the new state $|\tilde{0}\rangle$

$$\frac{\subset\tilde{0}\,|\,A_s(\mathbf{r})\,|\,\tilde{0}\supset}{\subset\tilde{0}\,|\,\tilde{0}\supset} = \frac{1}{\subset\tilde{0}\,|\,\tilde{0}\supset}\int \mathrm{d}^3k\,\sqrt{\frac{\hbar}{2\,\varepsilon_0\,\omega(2\,\pi)^3}}\,\mathrm{e}^{\mathrm{i}\mathbf{k}.\mathbf{r}} \subset\tilde{0}\,|\,a_s(\mathbf{k})\,|\,\tilde{0}\supset + c.c. \tag{D.30}$$

which is then written, using (D.21),

$$\frac{\langle \tilde{0} \,|\, A_s(\mathbf{r}) \,|\, \tilde{0} \rangle}{\langle \tilde{0} \,|\, \tilde{0} \rangle} = \int \mathrm{d}^3k \, \sqrt{\frac{\hbar}{2\,\varepsilon_0\,\omega(2\,\pi)^3}} \, \lambda(\mathbf{k}) \, \mathrm{e}^{\mathrm{i}\mathbf{k}.\mathbf{r}} + c.c. \,. \tag{D.31}$$

The requirement that $\rho_e(\mathbf{r})$ be real implies that $\lambda(\mathbf{k}) = \lambda^*(-\mathbf{k})$, which, by a change of $\mathbf{k}$ to $-\mathbf{k}$ in the integral (D.31), allows one to show that the complex conjugate term duplicates the first term. Using (D.15), one then gets for the mean value of $U(\mathbf{r}) = cA_s(\mathbf{r})$

$$\frac{\langle \tilde{0} \,|\, U(\mathbf{r}) \,|\, \tilde{0} \rangle}{\langle \tilde{0} \,|\, \tilde{0} \rangle} = \frac{1}{(2\,\pi)^{3/2}} \int \mathrm{d}^3k \, \frac{\rho_e(\mathbf{k})}{\varepsilon_0\, k^2} \, \mathrm{e}^{\mathrm{i}\mathbf{k}.\mathbf{r}} \tag{D.32}$$

which is just the Fourier transform of $\rho_e(\mathbf{k})/\varepsilon_0 k^2$, that is, the Coulomb potential associated with the charge distribution $\rho_e(\mathbf{r})$. To see this it suffices to note that on Fourier transformation the Poisson equation $\Delta U + \rho_e/\varepsilon_0 = 0$ gives $\mathscr{U} = \rho_e/\varepsilon_0 k^2$.

4. Conclusion and Generalization

We have shown that it is possible to transform the Hamiltonian of the quantized field coupled to fixed charges so as to make explicit the Coulomb interaction between charges. We have also proved that the (new) mean value of the scalar potential in the perturbed ground state of the field coincides with the Coulomb potential created by the charges.

Such a treatment can be extended to the case where the sources are particles forming a dynamical system. We will see in Complement $\mathrm{B_V}$ that it is possible to apply a unitary transformation (in the new sense) to the Hamiltonian of coupled Dirac and Maxwell fields which generalizes the transformation studied here and which makes explicit the Coulomb interaction between the particles. Along with the subsidiary condition selecting the physical states, such a transformation establishes a correspondence between the two possible formulations of quantum electrodynamics examined in this book, that in the Lorentz gauge and that in the Coulomb gauge.

General references and additional reading

Bogoliubov and Shirkov (§13), Heitler (§10), Itzykson and Zuber (§3-2), Jauch and Rohrlich (Chapter 6).

Quantization with an indefinite metric: S. N. Gupta, *Proc. Roy. Soc.*, **63**, 681 (1950), and K. Bleuler, *Helv. Phys. Acta.*, **23**, 567 (1950).

COMPLEMENT A_V

AN ELEMENTARY INTRODUCTION TO THE THEORY OF THE ELECTRON–POSITRON FIELD COUPLED TO THE PHOTON FIELD IN THE LORENTZ GAUGE

The description of particles used in the preceding chapters is valid only when the particles are moving at velocities small compared to the velocity of light. Furthermore, the theoretical framework which has been established does not allow a treatment of cases where the total number of particles varies through pair creation and particle–antiparticle annihilation. It also appears that matter and radiation are not treated in the same way: a relativistic quantum field describes radiation with an arbitrary number of elementary excitations, the photons, whereas we consider only a fixed number of charged particles represented by nonrelativistic wave functions.

The purpose of this complement is to show in a very elementary way how matter, especially electrons and positrons, can be described by a quantized relativistic field. We begin (§A_V.1) with a brief review of the Dirac equation treated as a one-electron relativistic wave equation (*). We then quantize the wave function of this equation, following the usual procedure for second quantization (§A_V.2) and thus get the quantized Dirac field, whose elementary excitations describe electrons (e^-) and positrons (e^+). Finally, we introduce (§A_V.3) the expression for the Hamiltonian describing the interaction of the quantized Dirac and Maxwell fields. This Hamiltonian is expressed in terms of the creation and annihilation operators of electrons, positrons, and photons. It forms the starting point for all calculations in quantum electrodynamics.

1. A Brief Review of the Dirac Equation

a) Dirac Matrices

A heuristic procedure to get a wave equation for a particle involves starting from the dispersion relation $E = f(\mathbf{p})$ between the energy E and the momentum $\mathbf{p}$ and making the substitution

$$E \to i\hbar \frac{\partial}{\partial t} \qquad \qquad \mathbf{p} \to - i\hbar \nabla . \tag{1.a}$$

(*) The reader will find more detailed discussions in the books referred to at the end of this complement.

In the presence of an electromagnetic field described by the potentials **A** and U, one uses the following substitutions:

$$i\hbar \frac{\partial}{\partial t} \rightarrow i\hbar \frac{\partial}{\partial t} - qU \qquad - i\hbar \nabla \rightarrow - i\hbar \nabla - q\mathbf{A} \qquad (1.\text{b})$$

where q is the particle's charge. This rule is related to the possibility of locally changing the phase of the wave function (see Exercise 5).

If one seeks a first-order differential equation, like the Schrödinger equation, but in which **r** and t play symmetric roles as in relativistic treatments, then one is naturally led to a linear relation in E and **p** of the form

$$E = \beta mc^2 + c\boldsymbol{\alpha} \cdot \mathbf{p} \qquad (2)$$

where β and $\boldsymbol{\alpha}$ are real and dimensionless. In addition, Equation (2) must be compatible with the well-known relativistic dispersion relation

$$E^2 = m^2 c^4 + p^2 c^2 . \qquad (3)$$

The square of (2) gives

$$E^2 = m^2 c^4 \beta^2 + mc^3 \sum_i (\alpha_i \beta + \beta \alpha_i) p_i + c^2 \sum_i \sum_j p_i p_j \alpha_i \alpha_j \qquad (4)$$

where $i, j = x, y, z$. Comparison with (3) leads to the following relations:

$$\begin{cases} \beta^2 = 1 & (5.\text{a}) \\ \alpha_i \beta + \beta \alpha_i = 0 & (5.\text{b}) \\ \alpha_i \alpha_j + \alpha_j \alpha_i = 2\, \delta_{ij} . & (5.\text{c}) \end{cases}$$

This shows clearly that β and $\boldsymbol{\alpha}$ cannot be numbers. On the other hand, one can find matrices of rank at least 4 which satisfy these equations. The wave function ψ is then necessarily a spinor, with at least four components, which implies the existence of internal degrees of freedom for the particle, described by $\boldsymbol{\alpha}$, β, in addition to its external degrees of freedom described by **r**, **p**.

The Dirac equation corresponds to the four-dimensional realization of Equations (5). One can check that the four *Dirac matrices*

$$\beta = \left(\begin{array}{c|c} \mathbb{1} & 0 \\ \hline 0 & -\mathbb{1} \end{array}\right) \qquad \alpha_i = \left(\begin{array}{c|c} 0 & \sigma_i \\ \hline \sigma_i & 0 \end{array}\right) \qquad (6)$$

in which $\mathbb{1}$ is the two-dimensional unit matrix and σ_i is one of the Pauli matrices

$$\sigma_x = \begin{pmatrix} 0 & 1 \\ 1 & 0 \end{pmatrix} \qquad \sigma_y = \begin{pmatrix} 0 & -i \\ i & 0 \end{pmatrix} \qquad \sigma_z = \begin{pmatrix} 1 & 0 \\ 0 & -1 \end{pmatrix} \tag{7}$$

satisfy (5).

Let's give finally a few relationships satisfied by the matrices (6) and which will be used later. From the well-known commutation relations for the Pauli matrices,

$$\sigma_i\,\sigma_j - \sigma_j\,\sigma_i = 2\,i \sum_k \varepsilon_{ijk}\,\sigma_k \tag{8}$$

(ε_{ijk} is the completely antisymmetric tensor), one gets the following relations for the matrices α_i:

$$\alpha_i\,\alpha_j - \alpha_j\,\alpha_i = 2\,i \sum_k \varepsilon_{ijk}\,\hat{\sigma}_k \tag{9.a}$$

where

$$\hat{\sigma}_k = \left(\begin{array}{c|c} \sigma_k & 0 \\ \hline 0 & \sigma_k \end{array}\right). \tag{9.b}$$

Joined with (5.c), this equation gives

$$\alpha_i\,\alpha_j = \delta_{ij} + \mathrm{i} \sum_k \varepsilon_{ijk}\,\hat{\sigma}_k\,. \tag{10}$$

If **A** and **B** are two vectors which do not act on the internal variables, the following expression generalizes (10):

$$(\boldsymbol{\alpha} \cdot \mathbf{A})\,(\boldsymbol{\alpha} \cdot \mathbf{B}) = \mathbf{A} \cdot \mathbf{B} + \mathrm{i}\hat{\boldsymbol{\sigma}} \cdot (\mathbf{A} \times \mathbf{B})\,. \tag{11}$$

b) The Dirac Hamiltonian. Charge and Current Density

The Dirac equation describes the temporal evolution of the spinor ψ with components ψ_λ ($\lambda = 1, 2, 3, 4$):

$$\mathrm{i}\hbar \frac{\partial}{\partial t} \psi = \mathscr{H}_D\,\psi\,. \tag{12}$$

Following (1) and (2), $\mathscr{H}_D$ is written

$$\mathscr{H}_D = \beta m c^2 + c\boldsymbol{\alpha} \cdot \mathbf{p} \tag{13}$$

for a free electron, and

$$\mathscr{H}_D = \beta m c^2 + c\boldsymbol{\alpha} \cdot [\mathbf{p} - q\mathbf{A}(\mathbf{r})] + qU(\mathbf{r}) \tag{14}$$

for an electron in the presence of an electromagnetic field. Let $\tilde{\psi}$ be the transpose of ψ (it is a row vector), and $\tilde{\psi}^*$ its adjoint. From Equation (12) and its adjoint one can show that the densities

$$\rho = q\tilde{\psi}^*(\mathbf{r})\,\psi(\mathbf{r}) = q\sum_{\lambda}\psi_{\lambda}^*(\mathbf{r})\,\psi_{\lambda}(\mathbf{r}) \qquad (15.a)$$

$$\mathbf{j} = qc\tilde{\psi}^*(\mathbf{r})\,\boldsymbol{\alpha}\psi(\mathbf{r}) = qc\sum_{\lambda\lambda'}\psi_{\lambda}^*(\mathbf{r})\,\boldsymbol{\alpha}_{\lambda\lambda'}\,\psi_{\lambda'}(\mathbf{r}) \qquad (15.b)$$

satisfy the continuity equation

$$\frac{\partial}{\partial t}\rho + \nabla\cdot\mathbf{j} = 0\,. \qquad (16)$$

One interprets ρ and $\mathbf{j}$ as the charge and current densities respectively. It is noticeable that Equations (15) and (16) retain the same form in the presence of an electromagnetic field.

c) Connection with the Covariant Notation

Instead of the matrices α_i and β introduced above, the matrices γ^μ are used frequently to get a more symmetric form of the Dirac equation:

$$\gamma^0 = \beta \qquad (17.a)$$

$$\gamma^i = \beta\alpha_i\,. \qquad (17.b)$$

By multiplying (12) by $\beta/\hbar c$, we can rewrite the equation so obtained in the form

$$\left[\sum_{\mu} \mathrm{i}\gamma^\mu\,\partial_\mu - \frac{mc}{\hbar}\right]\psi = 0 \qquad (18)$$

in the case of a free electron, and

$$\left[\sum_{\mu} \mathrm{i}\gamma^\mu\,D_\mu - \frac{mc}{\hbar}\right]\psi = 0 \qquad (19.a)$$

with

$$D_\mu = \partial_\mu + \mathrm{i}\,\frac{q}{\hbar}\,A_\mu \qquad (19.b)$$

in the case of an electron in a field defined by the four-potential A_μ.

Rather than the adjoint $\tilde{\psi}^*$ of ψ, one then uses a different quantity called the relativistic adjoint and defined by

$$\overline{\psi} = \tilde{\psi}^*\,\gamma^0\,. \qquad (20)$$

With this notation, Equations (15.a) and (15.b) can be reassembled in the form of a current four-vector

$$j^\mu = qc\overline{\psi}(\mathbf{r})\,\gamma^\mu\,\psi(\mathbf{r}) \qquad (21)$$

d) Energy Spectrum of the Free Particle

For a free electron, **p** and $\mathcal{H}_D$ commute, so that they have a common system of eigenvectors, the plane waves. For each eigenvalue **p**, $\mathcal{H}_D$ has two eigenvalues

$$E_p = \pm\sqrt{m^2 c^4 + p^2 c^2}\,. \tag{22}$$

The energy spectrum of $\mathcal{H}_D$ is therefore made up of two continua, one above mc^2 and the other below $-mc^2$.

The form of the eigenstates is particularly simple for $\mathbf{p} = \mathbf{0}$, since $\mathcal{H}_D$ then reduces to βmc^2. One finds two eigenvectors

$$u_{0+} = \begin{vmatrix} 1 \\ 0 \\ 0 \\ 0 \end{vmatrix} \qquad u_{0-} = \begin{vmatrix} 0 \\ 1 \\ 0 \\ 0 \end{vmatrix} \quad \text{with eigenvalue } +mc^2 \tag{23.a}$$

and two others

$$v_{0+} = \begin{vmatrix} 0 \\ 0 \\ 1 \\ 0 \end{vmatrix} \qquad v_{0-} = \begin{vmatrix} 0 \\ 0 \\ 0 \\ 1 \end{vmatrix} \quad \text{with eigenvalue } -mc^2 \tag{23.b}$$

Thus an electron with momentum zero and energy mc^2 can exist in two internally different states corresponding to the two states of a spin $\frac{1}{2}$. This result persists for states with a momentum $\mathbf{p} \neq \mathbf{0}$. The corresponding spinors have in general four nonzero components. Those relative to the eigenvalue $+\sqrt{p^2c^2 + m^2c^4}$ are derived from the spinors (23.a) by the transformation

$$T(\mathbf{p}) = \left[\cos\frac{\theta}{2} - \frac{\beta\boldsymbol{\alpha}\cdot\mathbf{p}}{p}\sin\frac{\theta}{2}\right]\exp\frac{i\mathbf{p}\cdot\mathbf{r}}{\hbar} \tag{24.a}$$

$$\theta = \arctan\frac{p}{mc}\,. \tag{24.b}$$

We denote these by $u_p(\mathbf{r})$. To simplify the notation, the index p designates collectively the three components of **p** and the + or − spin component. The spinors relative to the opposite eigenvalue are derived from (23.b) by $T(\mathbf{p})$ and denoted $v_p(\mathbf{r})$:

$$u_p(\mathbf{r}) = T(\mathbf{p})\,u_{0\pm} \qquad\qquad v_p(\mathbf{r}) = T(\mathbf{p})\,v_{0\pm}\,. \tag{25}$$

Remark

The spinors u_p and v_p defined in (25) are normalized as plane waves, since the transformation $T(\mathbf{p})$ is unitary. In the covariant formulation, one uses other spinors which are derived from $u_{0\pm}$ and $v_{0\pm}$ by a Lorentz transformation which is not unitary, so that they have a different normalization.

e) Negative-Energy States. Hole Theory

The existence of negative eigenvalues down to $-\infty$ for $\mathscr{H}_D$ poses a problem of physical interpretation.

One can first try to think of the corresponding states as mathematical solutions without physical significance and retain only the positive-energy states to describe the electron. However, the interaction with the radiation field couples the positive-energy states to the negative-energy states. An electron initially in a positive-energy state can, by photon emission, fall into a negative-energy state.

This difficulty led Dirac to imagine that all negative-energy states are occupied. Since electrons are fermions, the Pauli exclusion principle then prevents the electrons with positive energy from falling into the already occupied negative-energy states. The stability of the positive-energy states is then reestablished. In addition, this point of view suggests new ideas. The absence of an electron with negative energy E, charge q, momentum $\mathbf{p}$, and spin μ is equivalent to the presence of a particle with positive energy $-E$, charge $-q$, momentum $-\mathbf{p}$, and spin $-\mu$. Such a particle is nothing else than the positron, the electron antiparticle, which appears then as a hole in the "sea" of electrons occupying the continuum of negative-energy states. Other predictions flow directly. Through photon absorption, a negative-energy electron can be promoted into a positive-energy state, leaving a hole in the negative-energy continuum. Such a process corresponds to electron–positron (e^+–e^-) pair formation. The inverse process, pair annihilation, is interpreted as the recombination of an electron and a hole.

The preceding considerations show clearly that the Dirac equation cannot describe coherently a single relativistic particle. In relativity, mass is a form of energy and particles can be created or destroyed. The ad hoc approach of Dirac is actually one way of meeting this requirement with the introduction of electrons preexisting in the negative-energy states. To avoid the introduction of these not very physical electrons, it is preferable to quantize the Dirac wave function $\psi(\mathbf{r})$ according to the second-quantization procedure. The excitations of the quantized field $\Psi(r)$ then allow the description of an arbitrary number of particles and antiparticles.

2. Quantization of the Dirac Field

a) Second Quantization

We will follow the general procedure also used in nonrelativistic quantum mechanics to describe a set of identical fermions.

We begin by expanding the Dirac wave function $\psi(\mathbf{r})$ in an orthonormal basis. Such a basis can be, for example, the plane-wave basis (25) formed by the eigenstates of $\mathscr{H}_D$ and $\mathbf{p}$:

$$\psi(\mathbf{r}) = \sum_p \left[\gamma_p\, u_p(\mathbf{r}) + \gamma_{\bar{p}}\, v_{\bar{p}}(\mathbf{r})\right]. \tag{26}$$

The index $\bar{p}$ designates the set of quantum numbers $(-E, -\mathbf{p}, -\mu)$ opposite to those designated by p. As we have seen in the subsection above, a hole with wave function $v_{\bar{p}}$ describes, in the Dirac approach, a positron with quantum numbers p.

The wave function $\psi(\mathbf{r})$ is then quantized by replacing the coefficients γ_p and $\gamma_{\bar{p}}$ of (26) with the operators c_p and $c_{\bar{p}}$ annihilating an electron in the corresponding state:

$$\Psi(\mathbf{r}) = \sum_p \left[c_p\, u_p(\mathbf{r}) + c_{\bar{p}}\, v_{\bar{p}}(\mathbf{r})\right]. \tag{27}$$

Since electrons are fermions, anticommutation relations are imposed on these operators:

$$\begin{aligned} &[c_p, c_q]_+ = [c_{\bar{p}}, c_{\bar{q}}]_+ = [c_p, c_{\bar{q}}]_+ = [c_p, c_{\bar{q}}^+]_+ = 0 \\ &[c_p, c_q^+]_+ = \delta_{pq}; \qquad\qquad [c_{\bar{p}}, c_{\bar{q}}^+]_+ = \delta_{pq}. \end{aligned} \tag{28}$$

These anticommutation relations are necessary to preserve the positive character of the energy of the field (see the end of §A_V.2.*b* following).

Following the general line of §A_V.1.*d*, we now reinterpret the operators $c_{\bar{p}}$ and $c_{\bar{p}}^+$. Annihilating an electron $(-\mathbf{p}, -E_p, -\mu)$ is equivalent to creating a positron $(\mathbf{p}, E_p, \mu)$. We thus take

$$c_{\bar{p}} = b_p^+ \tag{29.a}$$

and inversely

$$c_{\bar{p}}^+ = b_p \tag{29.b}$$

where b_p and b_p^+ are the annihilation and creation operators of a positron with momentum $\mathbf{p}$, energy E_p, and spin μ. Finally, the quantized Dirac field $\Psi(\mathbf{r})$ and its Hermitian conjugate $\Psi^+(\mathbf{r})$ are given by

$$\Psi(\mathbf{r}) = \sum_p \left[c_p\, u_p(\mathbf{r}) + b_p^+\, v_{\bar{p}}(\mathbf{r})\right] \tag{30.a}$$

$$\Psi^+(\mathbf{r}) = \sum_p \left[c_p^+\, u_p^*(\mathbf{r}) + b_p\, v_{\bar{p}}^*(\mathbf{r})\right] \tag{30.b}$$

Ψ^+ is the spinor whose components Ψ_λ^+ are the Hermitian conjugates of those of Ψ. The anticommutation relations of c and b follow directly from (28) and (29):

$$[c_p, c_q]_+ = [b_p, b_q]_+ = [c_p, b_p]_+ = [c_p, b_p^+]_+ = 0$$
$$[c_p, c_q^+]_+ = \delta_{pq} \qquad [b_p, b_q^+]_+ = \delta_{pq} \,. \tag{31}$$

One can also extract from (30) and (31) the anticommutation relations of the field itself by utilizing the fact that the ensemble $(u_p, v_{\bar{p}})$ forms a complete basis of the wave-function space:

$$[\Psi_\lambda(\mathbf{r}), \Psi_{\lambda'}(\mathbf{r}')]_+ = [\Psi_\lambda^+(\mathbf{r}), \Psi_{\lambda'}^+(\mathbf{r}')]_+ = 0$$
$$[\Psi_\lambda(\mathbf{r}), \Psi_{\lambda'}^+(\mathbf{r}')]_+ = \delta_{\lambda\lambda'}\, \delta(\mathbf{r} - \mathbf{r}') \,. \tag{32}$$

b) The Hamiltonian of the Quantized Field. Energy Levels

The mean value of the Dirac Hamiltonian in the state described by the function $\psi(\mathbf{r})$ is

$$\langle \mathcal{H}_D \rangle = \int d^3r\, \tilde{\psi}^*(\mathbf{r})\, \mathcal{H}_D\, \psi(\mathbf{r})$$
$$= \sum_{\lambda\lambda'} \int d^3r\, \psi_\lambda^*(\mathbf{r}) \left[\beta_{\lambda\lambda'}\, mc^2 + \frac{\hbar c}{i} \boldsymbol{\alpha}_{\lambda\lambda'} \cdot \boldsymbol{\nabla} \right] \psi_{\lambda'}(\mathbf{r}) \,. \tag{33}$$

The second-quantized Hamiltonian H_D is gotten by replacing the wave function $\psi(\mathbf{r})$ with the field $\Psi(\mathbf{r})$:

$$H_D = \int d^3r\, \tilde{\Psi}^+(\mathbf{r})\, \mathcal{H}_D\, \Psi(\mathbf{r})$$
$$= \int d^3r\, \tilde{\Psi}^+(\mathbf{r}) \left[\beta mc^2 + \frac{\hbar c}{i} \boldsymbol{\alpha} \cdot \boldsymbol{\nabla} \right] \Psi(\mathbf{r}) \,. \tag{34}$$

Use of the expansions (30) and of the fact that u_p and $v_{\bar{p}}$ are the eigenfunctions of $\mathcal{H}_D$ with eigenvalues E_p and $-E_p$ leads to the following form for H_D:

$$H_D = \sum_p E_p\, c_p^+\, c_p + \sum_p (-E_p)\, b_p\, b_p^+ \,. \tag{35}$$

Now, from (26), $b_p b_p^+ = 1 - b_p^+ b_p$, so that

$$H_D = E_0 + \sum_p E_p\, c_p^+\, c_p + \sum_p E_p\, b_p^+\, b_p \,. \tag{36}$$

The physical interpretation of (36) is quite clear. The energy $E_0 =$

$\Sigma_p(-E_p)$ is the vacuum energy and is not directly observable. $c_p^+ c_p$ is the number of electrons of energy E_p, which contribute $E_p c_p^+ c_p$ to the total energy. The eigenvalues of $c_p^+ c_p$ can be only 0 and 1. This property, established in Exercise 8 of Chapter II, is a consequence of the anticommutation relations and justifies the Pauli exclusion principle. Likewise, $b_p^+ b_p$ is the number of positrons with energy E_p, which contribute in a similar fashion to the total energy with a positive sign. It is notable that if one had imposed commutation rather than anticommutation relations on b_p and b_p^+, the $-$ sign in (35) would have persisted in (36) and the total energy would have been able to become infinitely negative. The anticommutation relations are thus necessary to prevent such an unphysical situation.

Remarks

(i) The vacuum energy E_0 has no true physical significance. One could remove it by defining H_D in a more symmetric way between particles and antiparticles.

(ii) In the presence of an external field, it is necessary to replace, in the equation (34) for H_D, the operator $\mathscr{H}_D$ by the expression (14) rather than (13).

To classify the energy levels of the Dirac field, it is interesting to introduce the total charge

$$Q = q \sum_p c_p^+ c_p - q \sum_p b_p^+ b_p . \tag{37}$$

The importance of this quantity is due to the fact that it is conserved even in the presence of an interaction with the electromagnetic field.

The lowest energy levels of the field are illustrated in Figure 1. The

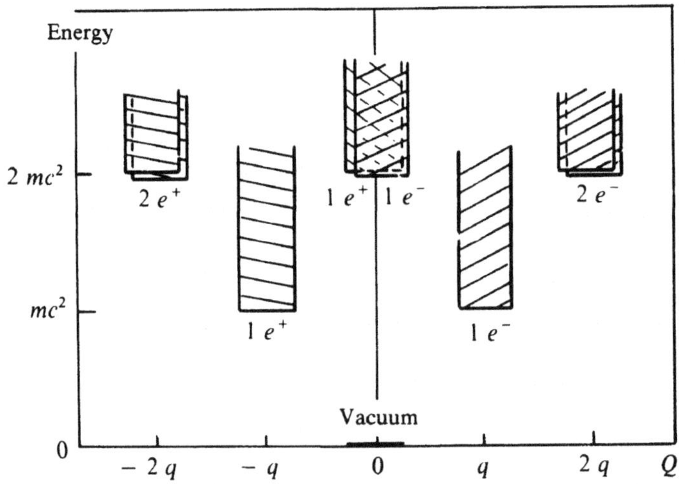

Figure 1. The lowest states of the quantized Dirac field.

vacuum has energy E_0, which we take as the origin, and the corresponding value of Q is 0. It is the ground state of the field. The lowest excited states, characterized by $Q = +q$ and $Q = -q$, form two continua beginning at $E = mc^2$ and stretching to $+\infty$. These are the one-electron and one-positron states respectively. Three double continua begin at $E = 2mc^2$ with $Q = 2q, 0, -2q$, describing respectively two electrons, an electron–positron pair, and two positrons. One gets next the three-particle states: some with charge $+q$ (above the one-electron states in Figure 1), others with charge $3q$, and so on.

The nonrelativistic formalism used in the foregoing chapters allows us to describe processes whose development occurs entirely inside one of the manifolds we have described. In the present relativistic theory, transitions between different manifolds with the same value of Q (vertical transitions in Figure 1) can take place, as we shall see in §$A_V.3$ by studying the coupling Hamiltonian with the radiation field.

c) Temporal and Spatial Translations

To conclude this section devoted to the quantization of the Dirac field, we will show that the Hamiltonian H_D and the momentum operator $\mathbf{P}$, whose expression we will give, are the generators of the temporal and spatial translations of the field, as for all quantized free systems.

For H_D, this property is evident, since the Heisenberg equation for the field operator is written

$$[H_D, \Psi(\mathbf{r}, t)] = \frac{\hbar}{\mathrm{i}} \frac{\partial}{\partial t} \Psi(\mathbf{r}, t) \tag{38}$$

which integrates to

$$\exp\left(\frac{\mathrm{i}}{\hbar} H_D \tau\right) \Psi(\mathbf{r}, t) \exp\left(-\frac{\mathrm{i}}{\hbar} H_D \tau\right) = \Psi(\mathbf{r}, t + \tau) . \tag{39}$$

The momentum operator of the field, $\mathbf{P}$, is gotten via a method identical with that which we used for H_D (see also Exercise 6):

$$\mathbf{P} = \int \mathrm{d}^3 r' \, \tilde{\Psi}^+(\mathbf{r}') \left[\frac{\hbar}{\mathrm{i}} \nabla \Psi(\mathbf{r}')\right] . \tag{40}$$

We take the commutator with $\Psi_\lambda(\mathbf{r})$,

$$[\mathbf{P}, \Psi_\lambda(\mathbf{r})] = \sum_{\lambda'} \int \mathrm{d}^3 r' \left\{ \Psi_{\lambda'}^+(\mathbf{r}') \left[\frac{\hbar}{\mathrm{i}} \nabla \Psi_{\lambda'}(\mathbf{r}')\right] \Psi_\lambda(\mathbf{r}) - \right.$$

$$\left. - \Psi_\lambda(\mathbf{r}) \, \Psi_{\lambda'}^+(\mathbf{r}') \left[\frac{\hbar}{\mathrm{i}} \nabla \Psi_{\lambda'}(\mathbf{r}')\right] \right\} \tag{41}$$

$\Psi_\lambda(\mathbf{r})$ anticommutes with $\Psi_{\lambda'}(\mathbf{r}')$ and therefore with $\nabla\Psi_{\lambda'}(\mathbf{r}')$. By transposing Ψ_λ to the left in the first term, one gets the anticommutator of $\Psi_{\lambda'}^+(\mathbf{r}')$ and $\Psi_\lambda(\mathbf{r})$, which is equal to $\delta(\mathbf{r}-\mathbf{r}')\delta_{\lambda\lambda'}$. Following integration over $\mathbf{r}'$, (41) becomes

$$[\mathbf{P}, \Psi_\lambda(\mathbf{r})] = -\frac{\hbar}{i}\nabla\Psi_\lambda(\mathbf{r}) \tag{42}$$

$\mathbf{P}$ is thus the generator of the spatial translations

$$\exp\left(-\frac{i}{\hbar}\mathbf{P}\cdot\mathbf{a}\right)\Psi(\mathbf{r})\exp\left(\frac{i}{\hbar}\mathbf{P}\cdot\mathbf{a}\right) = \Psi(\mathbf{r}+\mathbf{a}). \tag{43}$$

Note that (38) and (42) can be condensed into a single expression by taking $P^0 = H_D/c$:

$$[P^\mu, \Psi(x)] = \frac{\hbar}{i}\sum_\nu g^{\mu\nu}\,\partial_\nu\Psi(x), \quad x = (ct, \mathbf{r}). \tag{44}$$

In a more elaborate approach to quantization, this fundamental relationship, which we have proven here starting from the anticommutation relations for the field, is instead postulated, so that the translations are represented by a unitary transformation of the field operators generated by P^μ. The use of anticommutation relations for the fields is one of the ways to meet this requirement. The other involves using commutation relations, but it leads to an energy unbounded below and must therefore be rejected.

3. The Interacting Dirac and Maxwell Fields

a) The Hamiltonian of the Total System. The Interaction Hamiltonian

In this chapter we have studied the quantized Maxwell field in the Lorentz gauge in the presence of external sources $\mathbf{j}_e(\mathbf{r}, t)$ and $\rho_e(\mathbf{r}, t)$, whose dynamics are imposed. The free-field Hamiltonian H_R is given by (C.19) or (C.22). The Hamiltonian for the interaction with the sources is written [see (A.29) and (A.8)]

$$H_I = \int d^3r \sum_\mu j_{e\mu}(\mathbf{r}, t)\,A^\mu(\mathbf{r}) = \int d^3r[-\mathbf{j}_e(\mathbf{r}, t)\cdot\mathbf{A}(\mathbf{r}) + c\rho_e(\mathbf{r}, t)\,A_s(\mathbf{r})]. \tag{45}$$

On the other hand, the Hamiltonian of the Dirac field, in the presence of external electric and magnetic fields whose dynamics are imposed, is gotten by using in (34) the expression (14) for $\mathscr{H}_D$ rather than (13). This amounts to adding to the Hamiltonian H_D of the free field the following term:

$$
\begin{aligned}
H_I &= \int \mathrm{d}^3r\, \tilde{\Psi}^+(\mathbf{r})\,[-cq\boldsymbol{\alpha}\cdot\mathbf{A}_e(\mathbf{r},t) + qU_e(\mathbf{r},t)]\,\Psi(\mathbf{r},t) \\
&= \int \mathrm{d}^3r[-\,\mathbf{j}(\mathbf{r})\cdot\mathbf{A}_e(\mathbf{r},t) + \rho(\mathbf{r})\,U_e(\mathbf{r},t)]
\end{aligned}
\tag{46}
$$

where the density of charge and the density of current operators are the analogues of (15.a) and (15.b) in second quantization:

$$\rho(\mathbf{r}) = q\tilde{\Psi}^+(\mathbf{r})\,\Psi(\mathbf{r}) \tag{47.a}$$

$$\mathbf{j}(\mathbf{r}) = qc\tilde{\Psi}^+(\mathbf{r})\,\boldsymbol{\alpha}\Psi(\mathbf{r})\,. \tag{47.b}$$

When each of the two systems, radiation field and matter field, has its proper dynamics, it seems natural to generalize (45) or (46) by replacing the external variables $\mathbf{j}_e(\mathbf{r},t)$, $\rho_e(\mathbf{r},t)$, $\mathbf{A}_e(\mathbf{r},t)$, and $U_e(\mathbf{r},t)/c$ with the quantum operators $\mathbf{j}(\mathbf{r})$, $\rho(\mathbf{r})$, $\mathbf{A}(\mathbf{r})$, and $A_s(\mathbf{r})$. One then gets for the Hamiltonian of the two interacting quantized fields the following expression, which can be justified more precisely starting from the Lagrangian formalism (see Exercise 6):

$$H = H_D + H_R + H_I \tag{48}$$

In (48), H_D is the Hamiltonian (36) of the free particles (we omit the constant E_0),

$$H_D = \sum_p E_p\, c_p^+\, c_p + \sum_p E_p\, b_p^+\, b_p \tag{49}$$

H_R is the proper Hamiltonian of the field expanded into transverse, longitudinal, and temporal modes,

$$H_R = \int \mathrm{d}^3k \sum_{\lambda=\varepsilon,\varepsilon',l} \hbar\omega\left[\bar{a}_\lambda(\mathbf{k})\,a_\lambda(\mathbf{k}) + \frac{1}{2}\right] + \int \mathrm{d}^3k\,\hbar\omega\left[-\,\bar{a}_s(\mathbf{k})\,a_s(\mathbf{k}) + \frac{1}{2}\right] \tag{50}$$

and H_I is the interaction Hamiltonian

$$H_I = \int \mathrm{d}^3r[-\,\mathbf{j}(\mathbf{r})\cdot\mathbf{A}(\mathbf{r}) + c\rho(\mathbf{r})\,A_s(\mathbf{r})]\,. \tag{51}$$

This last operator is linear in a and $\bar{a}$ for the photons, and involves products such as $c_p^+c_{p'}$, $b_pb_{p'}^+$, $c_p^+b_{p'}^+$, and $b_pc_{p'}$ for the particles. The interaction thus changes the number of photons by 1, and the number of particles by 0 (terms in $c_p^+c_{p'}$ and $b_pb_{p'}^+$) or by 2 (the products $c_p^+b_{p'}^+$ and $b_pc_{p'}$ create and destroy electron–positron pairs respectively).

With slight technical changes, such as the use of products in the normal order, (48) is the basic Hamiltonian for calculating any process in quantum electrodynamics.

b) HEISENBERG EQUATIONS FOR THE FIELDS

As we have not derived (48) from first principles, it is interesting to study the Heisenberg equations for the quantized fields which can be derived from this Hamiltonian.

For the electromagnetic field, we determine the equation of motion of the operator $a_\mu(\mathbf{k})$:

$$\begin{aligned}\dot{a}_\mu(\mathbf{k}) &= \frac{1}{\mathrm{i}\hbar}[a_\mu(\mathbf{k}), H] \\ &= \frac{1}{\mathrm{i}\hbar}[a_\mu(\mathbf{k}), H_R] + \frac{1}{\mathrm{i}\hbar}[a_\mu(\mathbf{k}), H_I]. \end{aligned} \tag{52}$$

The first term gives simply $-\mathrm{i}\omega a_\mu$. In the second term, it is necessary to replace H_I by its expression (51), in which $\mathbf{A}(\mathbf{r})$ and $A_s(\mathbf{r})$ are expanded as functions of the operators $a_{\mu'}(\mathbf{k}')$ and $\bar{a}_{\mu'}(\mathbf{k}')$ [Equation (C.17) of this chapter]. Only the commutator $[a_\mu(\mathbf{k}), \bar{a}_{\mu'}(\mathbf{k}')]$ is nonzero, being $-g_{\mu\mu'}\delta(\mathbf{k}-\mathbf{k}')$. The integral over $\mathbf{r}$ in (51) for H_I projects the four-vector $j^\mu = (c\rho, \mathbf{j})$ on the mode $\mathbf{k}$:

$$\dot{a}_\mu(\mathbf{k}) = -\mathrm{i}\omega a_\mu(\mathbf{k}) + \frac{\mathrm{i}}{\sqrt{2\varepsilon_0\hbar\omega}} j_\mu(\mathbf{k}) \tag{53}$$

where

$$j_\mu(\mathbf{k}) = \frac{1}{\sqrt{(2\pi)^3}}\int \mathrm{d}^3r\, j_\mu(\mathbf{r})\,\mathrm{e}^{-\mathrm{i}\mathbf{k}.\mathbf{r}} \tag{54}$$

Equation (53) is similar to Equations (A.33′) and (A.35′), which are themselves equivalent to Maxwell's equations in the Lorentz gauge. Equation (53) is thus the correct equation of motion for the quantized fields in the subspace of states satisfying the subsidiary Lorentz condition (which will be made more precise in §A_V.3.*c* below). Note also that the source j_μ is here an operator in particle space and not a *c*-number.

For the Dirac field, the Heisenberg equation is

$$\begin{aligned}\mathrm{i}\hbar\dot{\Psi}(\mathbf{r}) &= [\Psi(\mathbf{r}), H] \\ &= [\Psi(\mathbf{r}), H_D + H_I] \\ &= \left[\Psi(\mathbf{r}), \int \mathrm{d}^3r'\,\Psi^+(\mathbf{r}')\{\mathcal{H}_D + qcA_s(\mathbf{r}') - qc\boldsymbol{\alpha}\cdot\mathbf{A}(\mathbf{r}')\}\Psi(\mathbf{r}')\right]. \end{aligned} \tag{55}$$

By using the anticommutation relations in a manner analogous to that for passing from (41) to (42), one gets

$$i\hbar\dot{\Psi}(\mathbf{r}) = \left\{ \beta mc^2 + qcA_s(\mathbf{r}) + c\boldsymbol{\alpha}\cdot\left(\frac{\hbar}{i}\nabla - q\mathbf{A}(\mathbf{r})\right)\right\}\Psi(\mathbf{r}). \tag{56}$$

This equation is like the Dirac equation in the presence of an electromagnetic field described by the potentials $A_s(\mathbf{r})$ and $\mathbf{A}(\mathbf{r})$. But these two quantities are now Maxwell field operators, Ψ itself being a Dirac field operator. Starting from (56), one can derive easily the equation of continuity satisfied by the densities ρ and $\mathbf{j}$ defined in (47.a) and (47.b):

$$\dot{\rho}(\mathbf{r}) + \nabla\cdot\mathbf{j}(\mathbf{r}) = 0 \tag{57.a}$$

or again in reciprocal space,

$$\dot{\rho}(\mathbf{k}) + i\mathbf{k}\cdot\mathbf{j}(\mathbf{k}) = 0. \tag{57.b}$$

c) The Form of the Subsidiary Condition in the Presence of Interaction

In the case of the free field, we have shown that there are physical state vectors $|\chi\rangle$ such that the Lorentz condition

$$\langle\chi|\sum_\mu \partial_\mu A^\mu(\mathbf{r},t)|\chi\rangle = 0 \tag{58}$$

is satisfied on the average for all $\mathbf{r}$ and t. For this, we have used the expansion of the free field in traveling plane waves (B.9) to identify the spatial and temporal Fourier components of $\sum_\mu \partial_\mu A^\mu$ and to cancel their mean value by the condition (B.21) expressed for all values of $\mathbf{k}$. We have thus explicitly used the simple temporal evolution of the operators a and a^+ for the free field.

In the presence of an interaction, the evolution of the fields in the Heisenberg representation is no longer simple, and the identification of the Fourier components of given frequency requires complete knowledge of the evolution of the interacting systems (matter field + electromagnetic field). Quite happily, even though the field A_μ has a complex evolution, the operator

$$\Lambda(\mathbf{r},t) = \sum_\mu \partial_\mu A^\mu(\mathbf{r},t) \tag{59}$$

which appears in (58) behaves simply. To see this, it suffices to refer to Equation (A.17) satisfied by the corresponding classical variable. This equation results on one hand from the equation of motion of A_μ, and on the other from the conservation of charge. These equations remain true for operators in quantum theory, and Λ then satisfies the equation

$$\Box\, \Lambda(\mathbf{r}, t) = 0\,. \tag{60}$$

In reciprocal space the spatial Fourier transform $\mathcal{A}(\mathbf{k})$ of $\Lambda(\mathbf{r})$ satisfies

$$\ddot{\mathcal{A}}(\mathbf{k}) = -\,\omega^2\, \mathcal{A}(\mathbf{k}) \qquad (\omega = ck) \tag{61}$$

so that $\mathcal{A}(\mathbf{k})$ has a simple evolution, characterized by the two exponentials $e^{\pm i\omega t}$.

We will identify more precisely the parts of $\mathcal{A}(\mathbf{k})$ associated with each of the frequencies $\pm\omega$. With the help of (C.17), $\mathcal{A}(\mathbf{k})$ is expressed as a function of the operators a and $\bar{a}$:

$$\mathcal{A}(\mathbf{k}) = \sqrt{\frac{\hbar}{2\,\varepsilon_0\,\omega}}\left[\mathrm{i}\mathbf{k}\cdot \boldsymbol{a}(\mathbf{k}) + \mathrm{i}\mathbf{k}\cdot \bar{\boldsymbol{a}}(-\,\mathbf{k}) + \frac{1}{c}\,\dot{a}_s(\mathbf{k}) + \frac{1}{c}\,\dot{\bar{a}}_s(-\,\mathbf{k})\right]. \tag{62}$$

The velocity $\dot{a}_s$ can be replaced by its value given by (53), and $\mathrm{i}\mathbf{k}\cdot\mathbf{a}$ by $\mathrm{i}ka_l$. We will show that the operator $\Lambda^{(+)}(\mathbf{k})$ defined by

$$\begin{aligned}\mathcal{A}^{(+)}(\mathbf{k}) &= \sqrt{\frac{\hbar}{2\,\varepsilon_0\,\omega}}\left[\mathrm{i}\mathbf{k}\cdot \boldsymbol{a}(\mathbf{k}) + \frac{1}{c}\,\dot{a}_s(\mathbf{k})\right] \\ &= \sqrt{\frac{\hbar}{2\,\varepsilon_0\,\omega}}\,\frac{\mathrm{i}\omega}{c}\left[a_l(\mathbf{k}) - a_s(\mathbf{k}) + \frac{c}{\omega\sqrt{2\,\varepsilon_0\,\hbar\omega}}\,\rho(\mathbf{k})\right]\end{aligned} \tag{63}$$

evolves over time under the action of the *total Hamiltonian H* like $e^{-i\omega t}$. For this, we will calculate its derivative $\dot{\Lambda}^{(+)}(\mathbf{k})$. It involves $\dot{a}_l$, $\dot{a}_s$, and $\dot{\rho}$, whose values are given in (53) and (57.b). One then gets

$$\begin{aligned}\dot{\mathcal{A}}^{(+)}(\mathbf{k}) = \sqrt{\frac{\hbar}{2\,\varepsilon_0\,\omega}}\,\frac{\mathrm{i}\omega}{c}\Bigg[&-\mathrm{i}\omega a_l(\mathbf{k}) + \frac{\mathrm{i}\boldsymbol{\kappa}\cdot \boldsymbol{j}(\mathbf{k})}{\sqrt{2\,\varepsilon_0\,\hbar\omega}} + \mathrm{i}\omega a_s(\mathbf{k}) - \\ &-\frac{\mathrm{i}c\rho(\mathbf{k})}{\sqrt{2\,\varepsilon_0\,\hbar\omega}} - \frac{\mathrm{i}c\mathbf{k}\cdot \boldsymbol{j}(\mathbf{k})}{\omega\sqrt{2\,\varepsilon_0\,\hbar\omega}}\Bigg] = -\,\mathrm{i}\omega\mathcal{A}^{(+)}(\mathbf{k})\end{aligned} \tag{64}$$

which demonstrates the desired property.

Likewise

$$\begin{aligned}\mathcal{A}^{(-)}(\mathbf{k}) &= \sqrt{\frac{\hbar}{2\,\varepsilon_0\,\omega}}\left[\mathrm{i}\mathbf{k}\cdot \bar{\boldsymbol{a}}(-\,\mathbf{k}) + \frac{1}{c}\,\dot{\bar{a}}_s(-\,\mathbf{k})\right] \\ &= \sqrt{\frac{\hbar}{2\,\varepsilon_0\,\omega}}\,\frac{-\,\mathrm{i}\omega}{c}\left[\bar{a}_l(-\,\mathbf{k}) - \bar{a}_s(-\,\mathbf{k}) + \frac{c}{\omega\sqrt{2\,\varepsilon_0\,\hbar\omega}}\,\rho^+(-\,\mathbf{k})\right]\end{aligned} \tag{65}$$

evolves as $e^{+i\omega t}$.

Thus, the condition (58) for all **r** and t reduces to two initial conditions to be satisfied for all **k**:

$$\langle \chi \,|\, a_l(\mathbf{k}) - a_s(\mathbf{k}) + \frac{c}{\omega\sqrt{2\,\varepsilon_0\,\hbar\omega}}\rho(\mathbf{k}) \,|\, \chi \rangle = 0 \qquad (66.\text{a})$$

$$\langle \chi \,|\, \bar{a}_l(\mathbf{k}) - \bar{a}_s(\mathbf{k}) + \frac{c}{\omega\sqrt{2\,\varepsilon_0\,\hbar\omega}}\rho^+(\mathbf{k}) \,|\, \chi \rangle = 0\,. \qquad (66.\text{b})$$

These conditions are realized for the states of the subspace defined by

$$[a_l(\mathbf{k}) - a_s(\mathbf{k}) + \lambda(\mathbf{k})] \,|\, \chi \rangle = 0 \quad \forall \mathbf{k} \qquad (67)$$

where

$$\lambda(\mathbf{k}) = \frac{c}{\omega\sqrt{2\,\varepsilon_0\,\hbar\omega}}\rho(\mathbf{k}) \qquad (68)$$

which makes up the subspace of physical states. This equation generalizes Equations (B.21) and (D.28), discussed in this chapter for the free field or the field in the presence of external sources.

General References and Additional Reading

For the Dirac equation see for example Berestetski, Lifshitz, and Pitayevski (Chapter 4), Bjorken and Drell (Chapters 1 to 7), Messiah (Chapter XX).

For the quantization of the Dirac field and relativistic quantum electrodynamics see for example Berestetski, Lifshitz, and Pitayevski; Bogoliubov and Shirkov; Feynman; Heitler; Itzykson and Zuber; Jauch and Rohrlich; Schweber.

COMPLEMENT B_V

JUSTIFICATION OF THE NONRELATIVISTIC THEORY IN THE COULOMB GAUGE STARTING FROM RELATIVISTIC QUANTUM ELECTRODYNAMICS

In Complement A_V above we have given the Hamiltonian describing the dynamics of the coupled Dirac and Maxwell fields. This Hamiltonian uses the *Lorentz gauge* for the electromagnetic field, with the result that the Coulomb interaction between particles does not appear explicitly in the Hamiltonian. In addition, *the number of positrons and electrons is not a constant of the motion* and can vary over time. Such a theoretical framework seems quite removed from that established in the earlier chapters. There we considered particles, fixed in number, described by Schrödinger wave functions (or by two-component Pauli spinors), and used for the radiation field the Coulomb gauge, or gauges derived from it, which *introduce the Coulomb interaction explicitly in the particle Hamiltonian*. The purpose of this complement is to tie these two treatments together and to show how the nonrelativistic theory of Chapters I to IV can be justified starting from relativistic quantum electrodynamics in the Lorentz gauge, which we have broadly sketched in Complement A_V.

We are going to do this in two stages. While retaining the quantized Dirac field description of the particles, we begin by applying to the Hamiltonian of Complement A_V a transformation which is unitary with respect to the new norm and which yields the Coulomb interaction between particles. If, in addition, one uses the subsidiary condition characterizing the subspace of physical states, such a transformation amounts to passing from the Lorentz gauge to the Coulomb gauge (§B_V.1) (*). In the nonrelativistic limit, the coupling between states with different numbers of particles (electrons + positrons) is weak and can be treated as a perturbation. We will show then (§B_V.2) that, for a given number of particles, the dynamics of the system particles + photons is equivalent to those described by the nonrelativistic Hamiltonians of the preceding chapters.

(*) A procedure of this type is followed in K. Haller and R. B. Sohn, *Phys. Rev. A*, **20**, 1541 (1979).

1. Transition from the Lorentz Gauge to the Coulomb Gauge in Relativistic Quantum Electrodynamics

a) Transformation on the Scalar Photons Yielding the Coulomb Interaction

We start with the Hamiltonian (48) of Complement A_V, which we rewrite in the form

$$H = H_D + H_R^T + H_R^L + H_R^S + H_I^T + H_I^L + H_I^S \tag{1}$$

separating, in the radiation Hamiltonian H_R and in the interaction Hamiltonian H_I, the contributions of the transverse, longitudinal, and scalar photons (labeled by the superscripts T, L, and S respectively).

Consider first the part of H involving the scalar photons. It reduces to $H_R^S + H_I^S$. By using (50) and (51) of Complement A_V, as well as the expansion of the scalar potential $A_s(\mathbf{r})$ in a_s and $\bar{a}_s$ [see (C.17) in this chapter], we get

$$H_R^S + H_I^S = \int d^3k\ \hbar\omega\left[-\bar{a}_s(\mathbf{k})\,a_s(\mathbf{k}) + \lambda^*(\mathbf{k})\,a_s(\mathbf{k}) + \lambda(\mathbf{k})\,\bar{a}_s(\mathbf{k}) + \frac{1}{2}\right] \tag{2}$$

where

$$\lambda(\mathbf{k}) = \frac{c}{\hbar\omega}\sqrt{\frac{\hbar}{2\,\varepsilon_0\,\omega}}\,\rho(\mathbf{k}) = \frac{c}{\hbar\omega}\sqrt{\frac{\hbar}{2\,\varepsilon_0\,\omega(2\,\pi)^3}}\int d^3r\,\rho(\mathbf{r})\,e^{-i\mathbf{k}.\mathbf{r}} \tag{3}$$

has already been introduced in Complement A_V [see (68)]. As in §D.2.*c* of Chapter V, we can then rewrite the bracket of (2) in the form

$$-\left[\bar{a}_s(\mathbf{k}) - \lambda^*(\mathbf{k})\right]\left[a_s(\mathbf{k}) - \lambda(\mathbf{k})\right] + \lambda^*(\mathbf{k})\,\lambda(\mathbf{k}) + \frac{1}{2}\,. \tag{4}$$

The particle operators $\lambda^*(\mathbf{k})$ and $\lambda(\mathbf{k})$ commute with the radiation operators $a_s(\mathbf{k})$ and $\bar{a}_s(\mathbf{k})$ and also among themselves, since they depend only on $\rho(\mathbf{r})$ according to (3), and one can show (see Exercise 7) that $\rho(\mathbf{r})$ commutes with $\rho(\mathbf{r}')$. It follows that the operators $\bar{a}_s - \lambda^*$ and $a_s - \lambda$ satisfy the same commutation relations as $\bar{a}_s$ and a_s. There must then be a "translation operator" T transforming $\bar{a}_s - \lambda^*$ into $\bar{a}_s$ and $a_s - \lambda$ into a_s,

$$T[\bar{a}_s(\mathbf{k}) - \lambda^*(\mathbf{k})]\,T^{-1} = \bar{a}_s(\mathbf{k}) \tag{5.a}$$

$$T[a_s(\mathbf{k}) - \lambda(\mathbf{k})]\,T^{-1} = a_s(\mathbf{k}) \tag{5.b}$$

and generalizing the translation operators introduced in §C.4.*d* of Chapter III for the operators a and a^+, which are adjoints of each other with respect to the usual scalar product. It is indeed possible to prove (see Exercise 4) that the operator

$$T = \exp\left\{ \int d^3k [\lambda(\mathbf{k})\,\bar{a}_s(\mathbf{k}) - \lambda^*(\mathbf{k})\,a_s(\mathbf{k})] \right\} \tag{6}$$

which is unitary with respect to the new scalar product

$$T\bar{T} = \bar{T}T = 1 \tag{7}$$

satisfies the relations (5). Using (2), (4), (5), and (3), one then gets

$$\begin{aligned} T(H_R^S + H_I^S)\,\bar{T} &= \int d^3k\,\hbar\omega\left[-\bar{a}_s(\mathbf{k})\,a_s(\mathbf{k}) + \frac{1}{2}\right] + \int d^3k\,\hbar\omega\lambda^*(\mathbf{k})\,\lambda(\mathbf{k}) \\ &= H_R^S + \int d^3k\,\frac{\rho^*(\mathbf{k})\,\rho(\mathbf{k})}{2\,\varepsilon_0\,k^2}. \end{aligned} \tag{8}$$

The last term of (8) is just the Coulomb energy of the system of charges [see Equations (B.32) and (B.33) of Chapter I],

$$\begin{aligned} V_{\text{Coul}} &= \int d^3k\,\frac{\rho^*(\mathbf{k})\,\rho(\mathbf{k})}{2\,\varepsilon_0\,k^2} = \iint d^3r\,d^3r'\,\frac{\rho(\mathbf{r})\,\rho(\mathbf{r}')}{8\,\pi\varepsilon_0\,|\mathbf{r}-\mathbf{r}'|} \\ &= \frac{q^2}{8\,\pi\varepsilon_0}\iint d^3r\,d^3r'\,\frac{\tilde{\Psi}^+(\mathbf{r})\,\Psi(\mathbf{r})\,\tilde{\Psi}^+(\mathbf{r}')\,\Psi(\mathbf{r}')}{|\mathbf{r}-\mathbf{r}'|}. \end{aligned} \tag{9}$$

Therefore the transformation T has allowed us to eliminate the interaction term H_I^S with the scalar photons and to replace it by the Coulomb interaction

$$T(H_R^S + H_I^S)\,\bar{T} = H_R^S + V_{\text{Coul}}. \tag{10}$$

Remark

In the new representation, the scalar potential $A_s(\mathbf{r})$ is represented by the operator

$$\begin{aligned} A_s'(\mathbf{r}) &= TA_s(\mathbf{r})\,\bar{T} = \\ &= A_s(\mathbf{r}) + \int d^3k\,\sqrt{\frac{\hbar}{2\,\varepsilon_0\,\omega(2\,\pi)^3}}\,[\lambda(\mathbf{k})\,e^{i\mathbf{k}\cdot\mathbf{r}} + \lambda^*(\mathbf{k})\,e^{-i\mathbf{k}\cdot\mathbf{r}}]. \end{aligned} \tag{11}$$

We have used the expansion of $A_s(\mathbf{r})$ in a_s and $\bar{a}_s$ [see Equation (C.17)] as well as Equations (5). The last term of (11) is easily found starting from (3) for $\lambda(\mathbf{k})$. It coincides with the scalar potential A_s^P created by the charge density $\rho(\mathbf{r})$ of the particles

$$A_s^P(\mathbf{r}) = \frac{1}{c}\int \mathrm{d}^3r' \frac{\rho(\mathbf{r}')}{4\pi\varepsilon_0 |\mathbf{r}-\mathbf{r}'|} \tag{12}$$

Substituting (12) in (11), we get

$$A_s'(\mathbf{r}) = A_s(\mathbf{r}) + A_s^P(\mathbf{r}). \tag{13}$$

It is clear from (13) that the mathematical operator $A_s(\mathbf{r})$ given by Equation (C.17), which describes in the initial representation the total scalar potential, now describes, in the new representation, the physical variable $A_s'(\mathbf{r}) - A_s^P(\mathbf{r}) = A_s'(\mathbf{r}) - [A_s^P(\mathbf{r})]'$, since A_s^P commutes with T and is thus equal to $(A_s^P)'$. In the new representation, the operator $A_s(\mathbf{r})$ thus describes the difference between the total scalar potential and the scalar potential created by the particles.

b) Effect of the Transformation on the Other Terms of the Hamiltonian in the Lorentz Gauge

To study how the other terms of (1) are transformed by T, we will rewrite T by substituting in (6) the expression (3) for $\lambda(\mathbf{k})$. This gives

$$T = \exp\left\{-\frac{\mathrm{i}c}{\hbar}\int \mathrm{d}^3r\,\rho(\mathbf{r})\,S(\mathbf{r})\right\} \tag{14}$$

where

$$S(\mathbf{r}) = \int \mathrm{d}^3k \sqrt{\frac{\hbar}{2\varepsilon_0\omega(2\pi)^3}}\left[\frac{a_s(\mathbf{k})}{\mathrm{i}\omega}\,\mathrm{e}^{\mathrm{i}\mathbf{k}.\mathbf{r}} - \frac{\bar{a}_s(\mathbf{k})}{\mathrm{i}\omega}\,\mathrm{e}^{-\mathrm{i}\mathbf{k}.\mathbf{r}}\right] \tag{15}$$

is a quantum field which acts only on the scalar photons and which is self-commuting:

$$[S(\mathbf{r}), S(\mathbf{r}')] = 0 \tag{16}$$

T acts on the particles, since the density $\rho(\mathbf{r})$ appears in (14). The simplicity of the commutation relations for $\rho(\mathbf{r}) = q\Psi^+(\mathbf{r})\Psi(\mathbf{r})$ with the Dirac operators $\Psi_\lambda(\mathbf{r})$ and $\Psi_\lambda^+(\mathbf{r})$ and the property (16) permit us to find simply the transforms of Ψ_λ and Ψ_λ^+ by T (see Exercise 7). Since $T^{-1} = \bar{T}$ from (7), Equation (6) of Exercise 7 can be written

$$\begin{cases} T\Psi_\lambda(\mathbf{r})\bar{T} = \exp\left[\mathrm{i}\frac{qc}{\hbar}S(\mathbf{r})\right]\Psi_\lambda(\mathbf{r}) & (17.\mathrm{a}) \\ T\Psi_\lambda^+(\mathbf{r})\bar{T} = \exp\left[-\mathrm{i}\frac{qc}{\hbar}S(\mathbf{r})\right]\Psi_\lambda^+(\mathbf{r}). & (17.\mathrm{b}) \end{cases}$$

We will return now to the Hamiltonian (1). Since H_R^T and H_R^L depend only on the transverse and longitudinal field variables on which T does not act,

$$T(H_R^T + H_R^L)\overline{T} = H_R^T + H_R^L . \tag{18}$$

One can rewrite $H_I^T + H_I^L$ in the form

$$H_I^T + H_I^L = -q\int \mathrm{d}^3r \sum_\lambda \sum_{\lambda'} \Psi_\lambda^+(\mathbf{r})\,[c(\boldsymbol{\alpha})_{\lambda\lambda'}\cdot \mathbf{A}(\mathbf{r})]\,\Psi_{\lambda'}(\mathbf{r}) \tag{19}$$

where $\mathbf{A}$ is the transverse and longitudinal vector potential, so that

$$T(H_I^T + H_I^L)\overline{T} = -q\int \mathrm{d}^3r \sum_\lambda \sum_{\lambda'}$$

$$\exp\left[-\mathrm{i}\frac{qc}{\hbar}S(\mathbf{r})\right]\Psi_\lambda^+(\mathbf{r})\,T[c(\boldsymbol{\alpha})_{\lambda\lambda'}\cdot \mathbf{A}(\mathbf{r})]\,\overline{T}\exp\left[\mathrm{i}\frac{qc}{\hbar}S(\mathbf{r})\right]\Psi_{\lambda'}(\mathbf{r}) . \tag{20}$$

Since T commutes with $\mathbf{A}$, and $S(\mathbf{r})$ commutes with $\mathbf{A}$ as well as with Ψ_λ^+ and Ψ_λ, the integral of (20) reduces to that of (19), so that

$$T(H_I^T + H_I^L)\overline{T} = H_I^T + H_I^L . \tag{21}$$

An analogous treatment can be applied to the term in βmc^2 of H_D [see (34) of Complement $\mathrm{A_V}$]:

$$T\left\{\int \mathrm{d}^3r\,\tilde{\Psi}^+(\mathbf{r})\,\beta mc^2\,\Psi(\mathbf{r})\right\}\overline{T} = \int \mathrm{d}^3r\,\tilde{\Psi}^+(\mathbf{r})\,\beta mc^2\,\Psi(\mathbf{r}) . \tag{22}$$

It only remains to study the term in $(\hbar c/\mathrm{i})\boldsymbol{\alpha}\cdot\boldsymbol{\nabla}$ of H_D:

$$T\left\{\frac{\hbar c}{\mathrm{i}}\int \mathrm{d}^3r\,\tilde{\Psi}^+(\mathbf{r})\,\boldsymbol{\alpha}\cdot[\boldsymbol{\nabla}\Psi(\mathbf{r})]\right\}\overline{T} =$$

$$= \frac{\hbar c}{\mathrm{i}}\int \mathrm{d}^3r \exp\left[-\frac{\mathrm{i}qc}{\hbar}S(\mathbf{r})\right]\tilde{\Psi}^+(\mathbf{r})\,\boldsymbol{\alpha}\cdot\left\{\boldsymbol{\nabla}\left[\exp\left[\mathrm{i}\frac{qc}{\hbar}S(\mathbf{r})\right]\Psi(\mathbf{r})\right]\right\}$$

$$= \frac{\hbar c}{\mathrm{i}}\int \mathrm{d}^3r\,\tilde{\Psi}^+(\mathbf{r})\,\boldsymbol{\alpha}\cdot\boldsymbol{\nabla}\Psi(\mathbf{r}) + qc^2\int \mathrm{d}^3r\,\tilde{\Psi}^+(\mathbf{r})\,\boldsymbol{\alpha}\cdot[\boldsymbol{\nabla}S(\mathbf{r})]\,\Psi(\mathbf{r}) . \tag{23}$$

Regrouping (22) and (23) gives

$$TH_D\overline{T} = H_D + qc^2\int \mathrm{d}^3r\,\tilde{\Psi}^+(\mathbf{r})\,\boldsymbol{\alpha}\cdot[\boldsymbol{\nabla}S(\mathbf{r})]\,\Psi(\mathbf{r}) . \tag{24}$$

Finally, using (10), (18), (21), and (24), we have shown that

$$T H \bar{T} = H_P + H_R^T + H_I^T + H_R^L + H_R^S + H_I^{LS} \tag{25}$$

where

$$H_P = H_D + V_{\text{Coul}} \tag{26}$$

acts only on the particle variables and where

$$\begin{aligned} H_I^{LS} &= H_I^L + qc^2 \int \mathrm{d}^3 r\, \tilde{\Psi}^+(\mathbf{r})\, \boldsymbol{\alpha} \cdot [\boldsymbol{\nabla} S(\mathbf{r})]\, \Psi(\mathbf{r}) = \\ &= -\,qc \int \mathrm{d}^3 r\, \tilde{\Psi}^+(\mathbf{r})\, \boldsymbol{\alpha} \cdot [\mathbf{A}_{\parallel}(\mathbf{r}) - c\boldsymbol{\nabla} S(\mathbf{r})]\, \Psi(\mathbf{r}) \end{aligned} \tag{27}$$

acts on the particle variables and on those of the longitudinal and scalar photons. Finally one can reexpress $\mathbf{A}_{\parallel}$ as a function of a_l and $\bar{a}_l$, and $\boldsymbol{\nabla} S$ as a function of a_s and $\bar{a}_s$, starting from (15). We get

$$\begin{aligned} &\mathbf{A}_{\parallel}(\mathbf{r}) - c\,\boldsymbol{\nabla} S(\mathbf{r}) = \\ &= \int \mathrm{d}^3 k \sqrt{\frac{\hbar}{2\,\varepsilon_0\,\omega(2\,\pi)^3}} \left\{ [a_l(\mathbf{k}) - a_s(\mathbf{k})]\, \boldsymbol{\kappa}\, \mathrm{e}^{\mathrm{i}\mathbf{k}.\mathbf{r}} + [\bar{a}_l(\mathbf{k}) - \bar{a}_s(\mathbf{k})]\, \boldsymbol{\kappa}\, \mathrm{e}^{-\mathrm{i}\mathbf{k}.\mathbf{r}} \right\} \end{aligned} \tag{28}$$

—that is, on introducing the operators $a_d(\mathbf{k})$ and $\bar{a}_d(\mathbf{k})$ defined in Chapter V [Equation (C.33.a)],

$$\mathbf{A}_{\parallel}(\mathbf{r}) - c\,\boldsymbol{\nabla} S(\mathbf{r}) = -\mathrm{i}\sqrt{2} \int \mathrm{d}^3 k \sqrt{\frac{\hbar}{2\,\varepsilon_0\,\omega(2\,\pi)^3}} \left[a_d(\mathbf{k})\, \boldsymbol{\kappa}\, \mathrm{e}^{\mathrm{i}\mathbf{k}.\mathbf{r}} - \bar{a}_d(\mathbf{k})\, \boldsymbol{\kappa}\, \mathrm{e}^{-\mathrm{i}\mathbf{k}.\mathbf{r}} \right]. \tag{29}$$

It appears then that in the new representation (that is, after the transformation T has been applied), the new Hamiltonian describing the interaction with the longitudinal and scalar photons depends only on the operators a_d and $\bar{a}_d$. We will see the importance of this below.

Remark

There is an analogy between the field $S(\mathbf{r})$ defined in (15) and the field $\mathbf{Z}(\mathbf{r})$ introduced in Chapter IV [Equation (B.63)]. For free fields in the Heisenberg approach, $S(\mathbf{r}, t)$ is, to within a sign, the time integral of the transverse scalar potential $A_s(\mathbf{r}, t)$, just as $\mathbf{Z}(\mathbf{r}, t)$ is the time integral of the transverse vector potential $\mathbf{A}_{\perp}(\mathbf{r}, t)$. Similarly, one can show that there is a certain analogy between the transformation studied here and the Pauli–Fierz–Kramers transformation mentioned in §B.4.d of Chapter IV. The Pauli–Fierz–Kramers transformation tries to remove from the total transverse vector potential the transverse vector potential "bound" to the particles (*) in the same way as T here removes the scalar potential "bound" to the particles (see the remark at the end of §B_V.1.*a* above).

c) Subsidiary Condition. Absence of Physical Effects of the Scalar and Longitudinal Photons

We have seen in Complement A_V that, in the initial representation in Lorentz gauge, the physical states $|\chi\rangle$ must satisfy the condition

$$[a_l(\mathbf{k}) - a_s(\mathbf{k}) + \lambda(\mathbf{k})]\,|\,\chi\,\rangle = 0 \quad \forall\mathbf{k} \tag{30}$$

where $\lambda(\mathbf{k})$ is given in (3). To find what becomes of this condition for the transformed states

$$|\,\chi'\,\rangle = T\,|\,\chi\,\rangle \tag{31}$$

we will multiply (30) on the left by T, insert $\bar{T}T = 1$ between the bracket and $|\chi\rangle$, and use (5.b). This gives

$$[a_l(\mathbf{k}) - a_s(\mathbf{k})]\,|\,\chi'\,\rangle = 0 \quad \forall\mathbf{k} \tag{32}$$

—that is, finally, in terms of the operators a_d and $\bar{a}_d$,

$$\begin{cases} a_d(\mathbf{k})\,|\,\chi'\,\rangle = 0 & (33.\text{a}) \\ \langle\,\chi'\,|\,\bar{a}_d(\mathbf{k}) = 0 & (33.\text{b}) \end{cases} \quad \forall\mathbf{k}.$$

In the new representation, the subsidiary condition then has the same form as for the free field [see Equation (C.37) of Chapter V and the adjoint equation].

Equations (33), joined with the equation

$$[a_d(\mathbf{k}), \bar{a}_d(\mathbf{k}')] = 0 \tag{34}$$

which we have derived in Chapter V [Equation (C.42)], entail that the interaction Hamiltonian H_I^{LS} introduced in the preceding subsection [see (27) and (29)] does not contribute to the transition amplitude between an initial physical state $|\chi_i'\rangle$ and a final physical state $|\chi_f'\rangle$, both obeying Equations (33). To see this, it suffices to note that such an amplitude is a sum of terms involving the matrix elements between $|\chi_i'\rangle$ and $\langle\chi_f'|$ of a product of interaction Hamiltonians in the interaction representation $\tilde{H}_I(t_1)$, $\tilde{H}_I(t_2)$, ... Now, in the expression (25) for the new Hamiltonian, the interaction Hamiltonian is $H_I^T + H_I^{LS}$. On the other hand, $\tilde{H}_I^T$ is given as a function of $a_\varepsilon\,e^{-i\omega t}$, $a_{\varepsilon'}\,e^{-i\omega t}$, $\bar{a}_\varepsilon\,e^{i\omega t}$, $\bar{a}_{\varepsilon'}\,e^{i\omega t}$, whereas, from (27) and (29), H_I^{LS} is expressed as a function of $a_d\,e^{-i\omega t}$ and $\bar{a}_d\,e^{i\omega t}$. Since a_d

(*) See for example, Cohen-Tannoudji, Dupont-Roc, and Grynberg, Complement B_{II}.

and $\bar{a}_d$ commute with each other according to (34), as well as with all the transverse photon operators a_ε, $a_{\varepsilon'}$, $\bar{a}_\varepsilon$, and $\bar{a}_{\varepsilon'}$, it is possible, as soon as H_I^{LS} appears in order 1 or higher, to transfer a_d to the extreme right or $\bar{a}_d$ to the extreme left in the product $\tilde{H}_I(t_1)\tilde{H}_I(t_2)\cdots$ and then get a zero transition amplitude using Equations (33). The only nonzero terms between physical states are then of order 0 in H_I^{LS}. This shows that for all physical calculations H_I^{LS} can be ignored.

d) Conclusion: The Relativistic Quantum Electrodynamics Hamiltonian in the Coulomb Gauge

We have shown that H_I^{LS} can be ignored in calculating the transition amplitude between two physical states $|\chi_i'\rangle$ and $|\chi_f'\rangle$. We will take then, as initial states $|\chi_i'\rangle$, states which have no longitudinal and no scalar photons. Such states satisfy Equation (33), since $a_d|0_l0_s\rangle = 0$. Since one can ignore H_I^{LS} in (25) and since none of the other terms of (25) can create longitudinal or scalar photons, n_l and n_s remain zero over time and H_R^L and H_R^S then reduce to constants in (25).

Finally, for all physical processes, it is possible to ignore completely all the terms relative to longitudinal and scalar photons in (25) and retain only the three terms related to the particles and the transverse field. We find in this way, by combining the transformation T applied to the Hamiltonian in Lorentz gauge and the subsidiary condition characterizing the physical states, that the real independent degrees of freedom of the field are the transverse degrees of freedom. The Hamiltonian

$$H = H_P + H_R^T + H_I^T \tag{35}$$

where

$$H_P = \int \mathrm{d}^3r\, \bar{\Psi}^+(\mathbf{r})\left[\beta mc^2 + \frac{\hbar c}{\mathrm{i}}\boldsymbol{\alpha}\cdot\nabla\right]\Psi(\mathbf{r}) + \int\!\!\int \mathrm{d}^3r\,\mathrm{d}^3r'\,\frac{\rho(\mathbf{r})\,\rho(\mathbf{r}')}{8\,\pi\varepsilon_0\,|\mathbf{r}-\mathbf{r}'|} \tag{36}$$

$$H_R^T = \int \mathrm{d}^3k \sum_{\varepsilon,\varepsilon'} \hbar\omega\left[a_\varepsilon^+(\mathbf{k})\,a_\varepsilon(\mathbf{k}) + \frac{1}{2}\right] \tag{37}$$

$$H_I^T = -\,qc\int \mathrm{d}^3r\,\bar{\Psi}^+(\mathbf{r})\,\boldsymbol{\alpha}\cdot\mathbf{A}_\perp(\mathbf{r})\,\Psi(\mathbf{r}) \tag{38}$$

with

$$\mathbf{A}_\perp(\mathbf{r}) = \int \mathrm{d}^3k \sum_{\varepsilon,\varepsilon'}\sqrt{\frac{\hbar}{2\,\varepsilon_0\,\omega(2\,\pi)^3}}\,[a_\varepsilon(\mathbf{k})\,\boldsymbol{\varepsilon}\,\mathrm{e}^{\mathrm{i}\mathbf{k}.\mathbf{r}} + a_\varepsilon^+(\mathbf{k})\,\boldsymbol{\varepsilon}\,\mathrm{e}^{-\mathrm{i}\mathbf{k}.\mathbf{r}}] \tag{39}$$

is the relativistic quantum electrodynamic Hamiltonian in Coulomb gauge.

2. The Nonrelativistic Limit in Coulomb Gauge: Justification of the Pauli Hamiltonian for the Particles (*)

a) The Dominant Term H_0 of the Hamiltonian in the Nonrelativistic Limit: Rest Mass Energy of the Particles

The Hamiltonian (35) is strictly equivalent to the relativistic Hamiltonian (48) of Complement A_V. The particles can have a kinetic energy large with respect to mc^2, and the photon frequencies are subject to no restriction.

We will turn our attention now to states describing nonrelativistic situations (slow particles, photon frequency small compared to mc^2). The principal term of the Hamiltonian is then the rest mass energy of the particles, which is very large compared to their kinetic energy or to the photon energy. In the Hamiltonian (35), it is described by the term

$$H_0 = \int d^3r \, \tilde{\Psi}^+(\mathbf{r}) \, \beta mc^2 \, \Psi(\mathbf{r}) \tag{40}$$

of (36). We are going to treat H_0 as a first approximation of H and examine its energy levels. The rest of the Hamiltonian, which we denote V, will then be treated as a perturbation. It involves the kinetic energy of the particles and their interaction with the transverse field,

$$H_1 = \int d^3r \, \tilde{\Psi}^+(\mathbf{r}) \, c\boldsymbol{\alpha} \cdot \left[\frac{\hbar}{i} \nabla - q\mathbf{A}_\perp(\mathbf{r}) \right] \Psi(\mathbf{r}) \tag{41}$$

($\mathbf{A}_\perp$ is given by 39), as well as the Coulomb energy of the particles and the proper Hamiltonian of the transverse field:

$$V = H_1 + V_{\text{Coul}} + H_R^T . \tag{42}$$

The total Hamiltonian is then written

$$H = H_0 + V . \tag{43}$$

In complement A_V, we have used an expansion of the quantized Dirac field in the basis of spinor functions adapted to the Hamiltonian $\mathscr{H}_D$ given by (13) of complement A_V. It is preferable here to make another

(*) This problem is also discussed by Cohen-Tannoudji (§6) and by I. Bialynicki-Birula in *Quantum Electrodynamics and Quantum Optics*, A. O. Barut, ed., Plenum Press, New York, 1984, p. 63.

expansion, better adapted to H_0. For this, consider the spinor basis $u^{(0)}_{\mathbf{p}\sigma}(\mathbf{r})$ and $v^{(0)}_{\mathbf{p}\sigma}(\mathbf{r})$, formed by the common eigenstates of the operators βmc^2, σ_z, and $(\hbar/\mathrm{i})\boldsymbol{\nabla}$:

$$u^{(0)}_{\mathbf{p}\sigma}(\mathbf{r}) = \frac{1}{\sqrt{L^3}}\, u^{(0)}_{\sigma} \exp\left(\mathrm{i}\frac{\mathbf{p}\cdot\mathbf{r}}{\hbar}\right) \tag{44.a}$$

$$v^{(0)}_{\mathbf{p}\sigma}(\mathbf{r}) = \frac{1}{\sqrt{L^3}}\, v^{(0)}_{\sigma} \exp\left(\mathrm{i}\frac{\mathbf{p}\cdot\mathbf{r}}{\hbar}\right) \tag{44.b}$$

with

$$u^{(0)}_{+} = \begin{bmatrix}1\\0\\0\\0\end{bmatrix}, \quad u^{(0)}_{-} = \begin{bmatrix}0\\1\\0\\0\end{bmatrix}, \quad v^{(0)}_{+} = \begin{bmatrix}0\\0\\1\\0\end{bmatrix}, \quad v^{(0)}_{-} = \begin{bmatrix}0\\0\\0\\1\end{bmatrix}. \tag{45}$$

The expansion of $\Psi(\mathbf{r})$ is then written (denoting $\overline{\mathbf{p}\sigma} = -\mathbf{p}, -\sigma$)

$$\Psi(\mathbf{r}) = \sum_{\mathbf{p}\sigma} C_{\mathbf{p}\sigma}\, u^{(0)}_{\mathbf{p}\sigma}(\mathbf{r}) + B^{+}_{\mathbf{p}\sigma}\, v^{(0)}_{\overline{\mathbf{p}\sigma}}(\mathbf{r}) \tag{46}$$

where $C_{\mathbf{p}\sigma}$ ($B^{+}_{\mathbf{p}\sigma}$ respectively) is the destruction operator of a particle (creation operator of an antiparticle) with momentum $\mathbf{p}$ and spin σ. The operators $C_{\mathbf{p}\sigma}$ and $B_{\mathbf{p}\sigma}$ anticommute. Using a path analogous to that of $A_V.2.b$, one can put H_0 in the form

$$H_0 = E'_0 + \sum_{\mathbf{p}\sigma} mc^2 [C^{+}_{\mathbf{p}\sigma} C_{\mathbf{p}\sigma} + B^{+}_{\mathbf{p}\sigma} B_{\mathbf{p}\sigma}]\,. \tag{47}$$

The operators C^+C and B^+B have 0 and 1 as eigenvalues, with the result that up to the nonobservable constant $E'_0 = \Sigma_{\mathbf{p}\sigma}(-mc^2)$, the spectrum of H_0 is a discrete spectrum made up of a set of integer multiples of mc^2. If one introduces the total charge

$$Q = \sum_{\mathbf{p}\sigma}\left[qC^{+}_{\mathbf{p}\sigma} C_{\mathbf{p}\sigma} + (-q)\, B^{+}_{\mathbf{p}\sigma} B_{\mathbf{p}\sigma}\right] \tag{48}$$

then the diagram of energy levels of H_0, classed according to the values of Q, has the shape given in Figure 1. In this figure, each of the levels has a large degeneracy, since it is necessary to specify the momentum and spin of the particles as well as the number of photons of the transverse field.

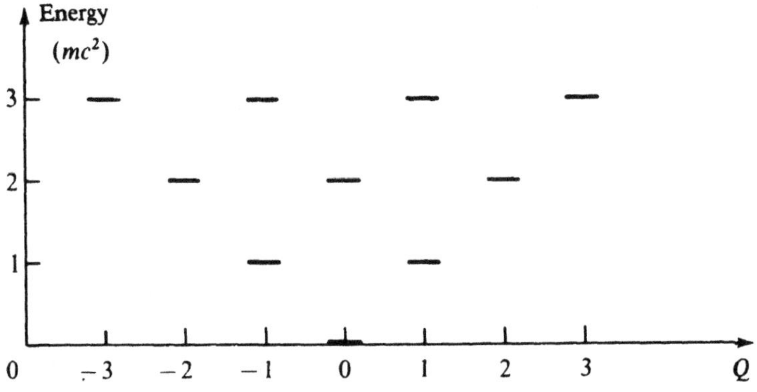

Figure 1. Energy levels of the Hamiltonian H_0 describing the rest mass energy of the particles of the quantized Dirac field.

It appears clearly in Equations (44) that the spinors $u_{\mathbf{p}\sigma}^{(0)}$ have only their first two components nonzero, as is the case for the last two components for $v_{\mathbf{p}\sigma}^{(0)}$ (this was not the case in Complement A_V, since the spinors u_p and v_p have their four components nonzero for $\mathbf{p} \neq \mathbf{0}$). In the expansion (46), the operators C and B are then associated with the first two and last two components of the field $\Psi(\mathbf{r})$ respectively. Taking account of the different roles which these two types of operators play, it is useful to decompose $\Psi(\mathbf{r})$ into two spinors with two components each, $\Phi(\mathbf{r})$ and $\Omega^+(\mathbf{r})$:

$$\Psi(\mathbf{r}) = \begin{bmatrix} \Phi(\mathbf{r}) \\ \Omega^+(\mathbf{r}) \end{bmatrix}. \tag{49}$$

Their components $\Phi_\sigma(\mathbf{r})$ and $\Omega_\sigma^+(\mathbf{r})$ ($\sigma = +1$ or -1) are given as functions of the operators C and B^+ by using the expression (44) for the spinors $u_{\mathbf{p}\sigma}^{(0)}$ and $v_{\mathbf{p}\sigma}^{(0)}$:

$$\Phi_\sigma(\mathbf{r}) = \sum_{\mathbf{p}} \frac{1}{\sqrt{L^3}} C_{\mathbf{p}\sigma} \exp\left(\mathrm{i}\frac{\mathbf{p}\cdot\mathbf{r}}{\hbar}\right) \tag{50.a}$$

$$\Omega_\sigma^+(\mathbf{r}) = \sum_{\mathbf{p}} \frac{1}{\sqrt{L^3}} B_{\overline{\mathbf{p}\sigma}}^+ \exp\left(\mathrm{i}\frac{\mathbf{p}\cdot\mathbf{r}}{\hbar}\right) \tag{50.b}$$

One should note the presence of $\overline{\mathbf{p}\sigma} = -\mathbf{p}, -\sigma$ in the expansion for $\Omega_\sigma^+(\mathbf{r})$. The charge density

$$\rho(\mathbf{r}) = q \sum_\lambda \Psi_\lambda^+(\mathbf{r})\, \Psi_\lambda(\mathbf{r}) \tag{51}$$

can be put in the form

$$\rho(\mathbf{r}) = q n_e(\mathbf{r}) + (-q)\, n_p(\mathbf{r}) \tag{52}$$

where

$$n_e(\mathbf{r}) = \sum_\sigma \Phi_\sigma^+(\mathbf{r})\, \Phi_\sigma(\mathbf{r}) \tag{53.a}$$

$$\begin{aligned} n_p(\mathbf{r}) &= \sum_\sigma (-)\, \Omega_\sigma(\mathbf{r})\, \Omega_\sigma^+(\mathbf{r}) \\ &= \sum_\sigma \Omega_\sigma^+(\mathbf{r})\, \Omega_\sigma(\mathbf{r}) - n_0 \,. \end{aligned} \tag{53.b}$$

It decomposes into two densities of opposite sign associated with the new fields Φ and Ω. The constant n_0, which appears when one reestablishes the normal order for the operators Ω, represents the charge density of the vacuum. It is not a physical quantity and vanishes when one defines ρ symmetrically with respect to the particles and antiparticles or defines the observables directly in terms of products in the normal order.

b) The Effective Hamiltonian inside a Manifold

We will now return to the total Hamiltonian (43). The supplementary term V, describing the kinetic energy of the particles, their coupling to the transverse field, and the Hamiltonian of the transverse field modifies the energy diagram of Figure 1. Insofar as the manifolds are well separated and we consider situations characterized by small values of V, it is possible to evaluate the effect of V with perturbation theory. To this end, we must find its matrix elements in the basis of eigenstates of H_0.

We start from (42) for V and consider H_R^T first. It is clear that H_R^T commutes with H_0 and Q, and thus has matrix elements only within each of the manifolds in Figure 1.

The operator V_{Coul} can be rewritten as a function of the densities n_e and n_p,

$$V_{\text{Coul}} = \frac{q^2}{8\pi\varepsilon_0} \iint \mathrm{d}^3r\, \mathrm{d}^3r' \frac{[n_e(\mathbf{r}) - n_p(\mathbf{r})]\,[n_e(\mathbf{r}') - n_p(\mathbf{r}')]}{|\mathbf{r} - \mathbf{r}'|}. \tag{54}$$

The densities $n_e(\mathbf{r})$ and $n_p(\mathbf{r})$ commute with $n_e(\mathbf{r}')$ and $n_p(\mathbf{r}')$ [the proof is the same as that given in Exercise 7 for $\rho(\mathbf{r})$]. They also commute with the particle numbers N_e and N_p, which are the integrals over all space of the preceding densities. Since H_0 and Q are functions of N_e and N_p, they commute with V_{Coul}, which thus has no matrix elements between different manifolds.

As for H_1, its expression as a function of Φ and Ω is written using (6) of Complement A_V for the matrices $\boldsymbol{\alpha}$:

$$H_1 = H_1^{++} + H_1^{--} \tag{55}$$

where

$$H_1^{++} = \sum_{\sigma\sigma'} \int \mathrm{d}^3r\, \Phi_\sigma^+(\mathbf{r})\, c \langle \sigma | \boldsymbol{\sigma} | \sigma' \rangle \cdot \left[\frac{\hbar}{i} \nabla - q\mathbf{A}_\perp(\mathbf{r}) \right] \Omega_{\sigma'}^+(\mathbf{r}) \tag{56.a}$$

$$H_1^{--} = \sum_{\sigma\sigma'} \int \mathrm{d}^3r\, \Omega_\sigma(\mathbf{r})\, c \langle \sigma | \boldsymbol{\sigma} | \sigma' \rangle \cdot \left[\frac{\hbar}{i} \nabla - q\mathbf{A}_\perp(\mathbf{r}) \right] \Phi_{\sigma'}(\mathbf{r}) . \tag{56.b}$$

In these expressions, $\langle\sigma|\boldsymbol{\sigma}|\sigma'\rangle$ represent the matrix elements of Pauli matrices between the eigenstates $|\sigma\rangle$ and $|\sigma'\rangle$ of σ_z. By using the expansions (50) for Φ and Ω^+, the terms H_1^{++} and H_1^{--} can be put in the form of discrete sums of matrix elements of the operator

$$\mathscr{U} = c\boldsymbol{\sigma} \cdot [\mathbf{p} - qA_\perp(\mathbf{r})] \tag{57}$$

between the states $|\mathbf{p}, \sigma\rangle$ and $|\mathbf{p}', \sigma'\rangle$ of Pauli particles:

$$H_1^{++} = \sum_{\mathbf{p}\sigma} \sum_{\mathbf{p}'\sigma'} C_{\mathbf{p}\sigma}^+ \langle \mathbf{p}\sigma | \mathscr{U} | \mathbf{p}'\sigma' \rangle B_{\overline{\mathbf{p}'\sigma'}}^+ =$$

$$= \sum_{\mathbf{p}\sigma} \sum_{\mathbf{p}'\sigma'} C_{\mathbf{p}\sigma}^+ \langle \mathbf{p}\sigma | \mathscr{U} | \overline{\mathbf{p}'\sigma'} \rangle B_{\mathbf{p}'\sigma'}^+ \tag{58.a}$$

$$H_1^{--} = \sum_{\mathbf{p}\sigma} \sum_{\mathbf{p}'\sigma'} B_{\overline{\mathbf{p}'\sigma'}} \langle \mathbf{p}'\sigma' | \mathscr{U} | \mathbf{p}\sigma \rangle C_{\mathbf{p}\sigma} =$$

$$= \sum_{\mathbf{p}\sigma} \sum_{\mathbf{p}'\sigma'} B_{\mathbf{p}'\sigma'} \langle \overline{\mathbf{p}'\sigma'} | \mathscr{U} | \mathbf{p}\sigma \rangle C_{\mathbf{p}\sigma} . \tag{58.b}$$

In this form, it is clear that H_1^{++} augments and H_1^{--} diminishes the number of electrons and positrons by 1. Thus, H_1^{++} and H_1^{--} have nonzero matrix elements between manifolds whose energy differs by $2mc^2$ and which correspond to the same value of Q.

In our nonrelativistic approximation of the energy states of the system, we will limit ourselves to terms of order 0 in $1/c$. To this end, it suffices to calculate the effect of H_R^T and V_{Coul} to first order in each manifold. Insofar as they are both diagonal with respect to the manifolds, their effect within the manifolds is described by the operators themselves.

Because of the presence of the factor c, H_1 is larger by one order of magnitude. It is necessary to examine its effect up to second order in perturbation theory in order to get terms of order 0 in $1/c$. Its effect in first order is zero, since H_1 is purely nondiagonal. To second order, the energy levels in a manifold are gotten by diagonalizing the effective Hamiltonian $\Delta^{(2)}$ derived from second-order perturbation theory. From the selection rules for H_1, and because of the simplicity of the spectrum of H_0, the energy denominators in the second-order terms are simply $\pm 2mc^2$. The expression for $\Delta^{(2)}$ is made up of two terms:

$$\begin{aligned}\Delta^{(2)} &= H_1^{++}\frac{1}{2\,mc^2}H_1^{--} + H_1^{--}\frac{1}{-\,2\,mc^2}H_1^{++}\\ &= \frac{1}{2\,mc^2}[H_1^{++}, H_1^{--}]\,.\end{aligned} \tag{59}$$

By using (58) for H_1^{++} and H_1^{--}, one finds for the commutator an expression of the form

$$\mathscr{C} = \left[\sum_{qr} C_q^+\,\mathscr{U}_{q\bar{r}}\,B_r^+\,,\ \sum_{st} B_s\,\mathscr{U}_{\bar{s}t}\,C_t\right]. \tag{60}$$

The indices q, r, s, t represent quantum numbers of the type $\mathbf{p}, \sigma$, and indices with bars above represent the opposite quantum numbers. The anticommutation relations of the operators B and C lead to

$$\begin{aligned}\mathscr{C} &= \sum_{qrst}\mathscr{U}_{q\bar{r}}\,\mathscr{U}_{\bar{s}t}(C_q^+\,B_r^+\,B_s\,C_t - B_s\,C_t\,C_q^+\,B_r^+)\\ &= \sum_{qrst}\mathscr{U}_{q\bar{r}}\,\mathscr{U}_{\bar{s}t}\,[(C_q^+\,C_t)(B_r^+\,B_s) - (-\,C_q^+\,C_t + \delta_{qt})(-\,B_r^+\,B_s + \delta_{rs})]\end{aligned} \tag{61}$$

Using the closure relation for the states $|\mathbf{p}, \sigma\rangle$, one can put $\mathscr{C}$ in the form

$$\mathscr{C} = \sum_{qt} C_q^+(\mathscr{U}^2)_{qt}\,C_t + \sum_{rs} B_r^+(\mathscr{U}^2)_{\bar{s}\bar{r}}\,B_s - \sum_q(\mathscr{U}^2)_{qq}\,. \tag{62}$$

To calculate the square of the operator $\mathscr{U} = \boldsymbol{\sigma}\cdot[\mathbf{p} - q\mathbf{A}(\mathbf{r})]$ we will use the properties of the Pauli matrices σ_i [Equation (8) of complement A_V and the anticommutation relation $\sigma_i\sigma_j + \sigma_j\sigma_i = 2\delta_{ij}$]. This gives

$$\begin{aligned}\frac{\mathscr{U}^2}{c^2} &= \sum_{ij}\left[\delta_{ij} + \mathrm{i}\sum_k \varepsilon_{ijk}\,\sigma_k\right][p_i - qA_i(\mathbf{r})]\,[p_j - qA_j(\mathbf{r})]\\ &= [\mathbf{p} - q\mathbf{A}_\perp(\mathbf{r})]^2 - \mathrm{i}q\sum_{ijk}\varepsilon_{ijk}\,\sigma_k[p_i\,A_j(\mathbf{r}) + A_i(\mathbf{r})\,p_j]\\ &= [\mathbf{p} - q\mathbf{A}_\perp(\mathbf{r})]^2 - \mathrm{i}q\sum_{ijk}\varepsilon_{ijk}\,\sigma_k[p_i, A_j(\mathbf{r})]\\ &= [\mathbf{p} - q\mathbf{A}_\perp(\mathbf{r})]^2 - q\hbar\,\boldsymbol{\sigma}\cdot\mathbf{B}(\mathbf{r})\end{aligned} \tag{63}$$

where $\mathbf{B}(\mathbf{r}) = \nabla \times \mathbf{A}(\mathbf{r})$ is the magnetic radiation field. The expression for $\Delta^{(2)}$ can then be written in the form of a sum of an electronic operator $\Delta_e^{(2)}$, a positron operator $\Delta_p^{(2)}$, and a constant which we will henceforth neglect:

$$\Delta^{(2)} = \Delta_e^{(2)} + \Delta_p^{(2)} \tag{64}$$

where

$$\Delta_e^{(2)} = \sum_{\substack{\mathbf{p}\sigma \\ \mathbf{p}'\sigma'}} C_{\mathbf{p}'\sigma'}^{+} \langle \mathbf{p}'\,\sigma' | \frac{[\mathbf{p} - q\mathbf{A}_\perp(\mathbf{r})]^2}{2\,m} - \frac{q\hbar}{2\,m}\boldsymbol{\sigma}\cdot\mathbf{B}(\mathbf{r}) | \mathbf{p}\sigma\rangle\, C_{\mathbf{p}\sigma} \tag{65}$$

and we have an expression of the same type for $\Delta_p^{(2)}$.

Finally, to order 0 in $1/c$, the new energy levels of the total Hamiltonian H are identical, to within constant terms and for "nonrelativistic" states of the system, to those of the effective Hamiltonian

$$H_{\text{eff}} = mc^2 \sum_{\mathbf{p}\sigma} C_{\mathbf{p}\sigma}^{+} C_{\mathbf{p}\sigma} + \Delta_e^{(2)} + mc^2 \sum_{\mathbf{p}\sigma} B_{\mathbf{p}\sigma}^{+} B_{\mathbf{p}\sigma} + \Delta_p^{(2)} + + V_{\text{Coul}} + H_R^T + \cdots \tag{66}$$

Remark

It is possible to get an expression for $\Delta_p^{(2)}$ lending itself to a simple physical interpretation. Denote by $\mathscr{U}^{(c)}$ the operator derived from $\mathscr{U}$ by changing q to $-q$, and by $|\mathbf{p}\sigma\rangle_c$ a spinor basis which differs from the basis $|\mathbf{p}\sigma\rangle$ by a phase factor $\pm \text{i}$:

$$\mathscr{U}^{(c)} = c\boldsymbol{\sigma}\cdot[\mathbf{p} + q\mathbf{A}_\perp(\mathbf{r})] \tag{67}$$

$$|\,\mathbf{p} \pm\,\rangle_c = \mp\, i\,|\,\mathbf{p} \pm\,\rangle\,. \tag{68}$$

A simple calculation of the matrix elements shows that

$${}_c\langle\, \mathbf{p}\sigma \,|\, \mathscr{U}^{(c)} \,|\, \mathbf{p}'\,\sigma'\,\rangle_c = \langle\, \overline{\mathbf{p}'\,\sigma'} \,|\, \mathscr{U} \,|\, \overline{\mathbf{p}\sigma}\,\rangle\,. \tag{69}$$

An analogous expression for $[\mathscr{U}^{(c)}]^2$ and $\mathscr{U}^2$ is immediate. The Hamiltonian for the positrons, $\Delta_p^{(2)}$, is then written

$$\begin{aligned}\Delta_p^{(2)} &= \sum_{rs} B_r^{+}\left(\frac{\mathscr{U}^2}{2\,mc^2}\right)_{\bar{s}\bar{r}} B_s \\ &= \sum_{rs} B_r^{+}\,{}_c\langle\, r \left| \frac{(\mathbf{p} + q\mathbf{A}_\perp)^2}{2\,m} + \frac{q\hbar}{2\,m}\boldsymbol{\sigma}\cdot\mathbf{B} \right| s\,\rangle_c\, B_s\,.\end{aligned} \tag{70}$$

It has the same form as $\Delta_e^{(2)}$ when q is replaced by $-q$, the operators C by B, and the changes (68) are made on the spinor basis. This transformation is actually a special case of the general charge conjugation transformation (*).

(*) See for example Berestetski, Lifschitz, and Pitayevski, §§23 and 26.

c) DISCUSSION

Consider the states containing only electrons. H_{eff} then reduces to

$$H_{\text{eff}}^{(e)} = \sum_{\mathbf{p}\sigma} mc^2\, C_{\mathbf{p}\sigma}^+ C_{\mathbf{p}\sigma} +$$

$$\sum_{\substack{\mathbf{p}\sigma \\ \mathbf{p}'\sigma'}} C_{\mathbf{p}'\sigma'}^+ \langle \mathbf{p}'\,\sigma' \,|\, \frac{[\mathbf{p} - q\mathbf{A}_\perp(\mathbf{r})]^2}{2\,m} - \frac{q\hbar}{2\,m}\boldsymbol{\sigma}\cdot\mathbf{B}(\mathbf{r}) \,|\, \mathbf{p}\sigma \rangle\, C_{\mathbf{p}\sigma} +$$

$$+ \frac{q^2}{8\,\pi\varepsilon_0}\iint \mathrm{d}^3 r\, \mathrm{d}^3 r' \,\frac{n_e(\mathbf{r})\, n_e(\mathbf{r}')}{|\,\mathbf{r} - \mathbf{r}'\,|} + H_R^T\,. \tag{71}$$

By construction, $H_{\text{eff}}^{(e)}$ is block diagonal, each block being relative to a manifold having a given number of electrons N_e. Equation (71) coincides in fact with the form taken in second quantization by the nonrelativistic Hamiltonian describing an ensemble of electrons coupled to the radiation field. More precisely, the matrix elements of (71) between two states containing the same number of electrons N and labeled by the occupation numbers (0 or 1) of the states $\mathbf{p}\sigma$ coincide with those of the following N-electron Hamiltonian:

$$\mathscr{H} = Nmc^2 + \sum_{\alpha=1}^{N} \frac{1}{2\,m}[\mathbf{p}_\alpha - q\mathbf{A}_\perp(\mathbf{r}_\alpha)]^2 - \sum_{\alpha=1}^{N}\frac{q\hbar}{2\,m}\boldsymbol{\sigma}\cdot\mathbf{B}(\mathbf{r}_\alpha) +$$

$$+ \sum_\alpha \sum_\beta \frac{q^2}{8\,\pi\varepsilon_0}\frac{1}{|\,\mathbf{r}_\alpha - \mathbf{r}_\beta\,|} + H_R^T \tag{72}$$

evaluated between two states formed by the antisymmetrized products of N two-component spinors relative to the occupied states $\mathbf{p}\sigma$.

The Hamiltonian (72) agrees with that introduced in Chapter III. The various terms of (72) describe respectively the rest mass energy of the N electrons, their nonrelativistic kinetic energy, the coupling of the spin magnetic moments of the electrons with the magnetic radiation field, the Coulomb energy of the N electrons, and the energy of the quantized transverse field. We have then justified, starting with relativistic quantum electrodynamics, the nonrelativistic theory in Coulomb gauge which has been the basis of most of this book. The advantage of such a treatment is that it introduces naturally the electron spin and the associated spin magnetic moment

$$\mathbf{M}_s = 2\,\frac{q}{2\,m}\,\frac{\hbar\boldsymbol{\sigma}}{2} \tag{73}$$

corresponding to a factor $g = 2$ [see (D.7) of Chapter III] as well as the Fermi–Dirac statistics which electrons obey. Actually, the anticommutation relations for the operators C^+ and C of (71) are a consequence of the anticommutation relations for the quantized Dirac field, which result from the positive-energy requirement (see §A_V. 2.*b*). We mention finally that it is possible to calculate the higher-order terms of $H_{\text{eff}}^{(e)}$ in $1/c$ and thereby get the expression for the first relativistic corrections (mass–velocity correction, spin–orbit interaction, Darwin term).

COMPLEMENT C_V

EXERCISES

Exercise 1. Other covariant Lagrangians of the electromagnetic field.
Exercise 2. Annihilation and creation operators for scalar photons: Can one interchange their meanings?
Exercise 3. Some properties of the indefinite metric.
Exercise 4. Translation operator for the creation and annihilation operators of a scalar photon.
Exercise 5. Lagrangian of the Dirac field. The connection between the phase of the Dirac field and the gauge of the electromagnetic field.
Exercise 6. The Lagrangian and Hamiltonian of the coupled Dirac and Maxwell fields.
Exercise 7. Dirac field operators and charge density. A study of some commutation relations.

1. Other covariant Lagrangians of the electromagnetic field

The purpose of this exercise is to study Lagrangians which, as with the Lagrangian used in this chapter, can serve as a departure point for a covariant quantization of the electromagnetic field. One looks to a set of Lagrangians $L_R(\lambda) = \int d^3r\, \mathscr{L}_R(\lambda)$ depending on a real parameter λ. The Lagrangian density in real space, $\mathscr{L}_R(\lambda)$, is written

$$\mathscr{L}_R(\lambda) = -\varepsilon_0 c^2\left[\sum_{\mu,\nu}\frac{1}{4}F_{\mu\nu}F^{\mu\nu} + \frac{\lambda}{2}\left(\sum_\mu \partial_\mu A^\mu\right)^2\right]. \tag{1}$$

a) Are two Lagrangians corresponding to different values of λ equivalent?

b) Give $L_R(\lambda)$ as a function of the transverse, longitudinal, and scalar components of the four-vector potential in reciprocal space $\mathscr{A}_\varepsilon$, $\mathscr{A}_{\varepsilon'}$, $\mathscr{A}_l$, and $\mathscr{A}_s = U/c$. Show that Lagrange's equations relative to $\mathscr{A}_\varepsilon$, $\mathscr{A}_l$, and $\mathscr{A}_s$ can be written

$$\ddot{\mathscr{A}}_\varepsilon + \omega^2\mathscr{A}_\varepsilon = 0 \tag{2.a}$$

$$\ddot{\mathscr{A}}_l + \omega^2\mathscr{A}_l - i\omega(\lambda - 1)(\dot{\mathscr{A}}_s + i\omega\mathscr{A}_l) = 0 \tag{2.b}$$

$$\ddot{\mathscr{A}}_s + \omega^2\mathscr{A}_s + (\lambda - 1)\frac{\partial}{\partial t}(\dot{\mathscr{A}}_s + i\omega\mathscr{A}_l) = 0\,. \tag{2.c}$$

Show that the equations relative to the longitudinal and scalar components of the potential are those of two independent oscillators when $\lambda = 1$.

c) Show that the equations (2) coincide with the Maxwell equations of the free field for all λ when the Lorentz condition is satisfied.

Solution

a) The difference between two Lagrangian densities corresponding to different values of λ is proportional to $(\Sigma_\mu \partial_\mu A^\mu)^2$, which is not a four-divergence. Thus these Lagrangians are not equivalent. In particular, the standard Lagrangian ($\lambda = 0$) and the Fermi Lagrangian ($\lambda = 1$) are not equivalent.

b) The Lagrangian $L_R(\lambda)$ is the integral in a reciprocal half space of the Lagrangian density $\bar{\mathscr{L}}_R(\lambda)$:

$$L_R(\lambda) = \int \mathrm{d}^3k\ \bar{\mathscr{L}}_R(\lambda) \tag{3.a}$$

where $\bar{\mathscr{L}}_R(\lambda)$ is gotten from (1) and (A.7) defining $F_{\mu\nu}$:

$$\begin{aligned}\bar{\mathscr{L}}_R(\lambda) = \varepsilon_0\Big[(\dot{\mathscr{A}} + \mathrm{i}\mathbf{k}\mathscr{U})\cdot(\dot{\mathscr{A}}^* - \mathrm{i}\mathbf{k}\mathscr{U}^*) - c^2\,|\,\mathbf{k}\times\mathscr{A}\,|^2 - \\ \lambda\left(\frac{\dot{\mathscr{U}}}{c} + \mathrm{i}c\mathbf{k}\cdot\mathscr{A}\right)\left(\frac{\dot{\mathscr{U}}^*}{c} - \mathrm{i}c\mathbf{k}\cdot\mathscr{A}^*\right)\Big]\\ = \varepsilon_0[\dot{\mathscr{A}}_\perp\cdot\dot{\mathscr{A}}_\perp^* - \omega^2\,\mathscr{A}_\perp\cdot\mathscr{A}_\perp^* + (\dot{\mathscr{A}}_l + \mathrm{i}\omega\mathscr{A}_s)(\dot{\mathscr{A}}_l^* - \mathrm{i}\omega\mathscr{A}_s^*) - \\ - \lambda(\dot{\mathscr{A}}_s + \mathrm{i}\omega\mathscr{A}_l)(\dot{\mathscr{A}}_s^* - \mathrm{i}\omega\mathscr{A}_l^*)]\,.\end{aligned} \tag{3.b}$$

To obtain Lagrange's equations, we calculate

$$\frac{\partial\bar{\mathscr{L}}_R(\lambda)}{\partial\mathscr{A}_\varepsilon^*} = -\,\varepsilon_0\,\omega^2\,\mathscr{A}_\varepsilon \tag{4.a}$$

$$\frac{\partial\bar{\mathscr{L}}_R}{\partial\dot{\mathscr{A}}_\varepsilon^*} = \varepsilon_0\,\dot{\mathscr{A}}_\varepsilon \tag{4.b}$$

$$\frac{\partial\bar{\mathscr{L}}_R(\lambda)}{\partial\mathscr{A}_l^*} = -\,\varepsilon_0\,\lambda(\omega^2\,\mathscr{A}_l - \mathrm{i}\omega\dot{\mathscr{A}}_s) \tag{5.a}$$

$$\frac{\partial\bar{\mathscr{L}}_R}{\partial\dot{\mathscr{A}}_l^*} = \varepsilon_0(\dot{\mathscr{A}}_l + \mathrm{i}\omega\mathscr{A}_s) \tag{5.b}$$

$$\frac{\partial\bar{\mathscr{L}}_R(\lambda)}{\partial\mathscr{A}_s^*} = \varepsilon_0(\omega^2\,\mathscr{A}_s - \mathrm{i}\omega\dot{\mathscr{A}}_l) \tag{6.a}$$

$$\frac{\partial\bar{\mathscr{L}}_R}{\partial\dot{\mathscr{A}}_s^*} = -\,\varepsilon_0\,\lambda(\dot{\mathscr{A}}_s + \mathrm{i}\omega\mathscr{A}_l)\,. \tag{6.b}$$

We get from (4.a) and (4.b) the Lagrange equation relative to $\mathscr{A}_\varepsilon$, which corresponds to (2.a). Equations (5.a) and (5.b) give

$$\ddot{\mathscr{A}}_l + \mathrm{i}\omega\dot{\mathscr{A}}_s + \lambda(\omega^2\,\mathscr{A}_l - \mathrm{i}\omega\dot{\mathscr{A}}_s) = 0\,. \tag{7}$$

By writing the term proportional to $\lambda\mathscr{A}_l$ in the form $(\lambda - 1)\,\mathscr{A}_l + \mathscr{A}_l$ and regrouping terms, we find (2.b). Likewise, (6.a) and (6.b) yield

$$\lambda(\ddot{\mathscr{A}}_s + \mathrm{i}\omega\dot{\mathscr{A}}_l) + \omega^2\,\mathscr{A}_s - \mathrm{i}\omega\dot{\mathscr{A}}_l = 0 \tag{8}$$

which can be transformed to give (2.c).

If we set $\lambda = 1$ in Equations (2), we find independent harmonic-oscillator equations for the four components $\mathscr{A}_\varepsilon$, $\mathscr{A}_{\varepsilon'}$, $\mathscr{A}_l$, and $\mathscr{A}_s$. This result is satisfactory, since the Fermi

Lagrangian ($\lambda = 1$) is equivalent to the Lagrangian $\bar{\mathscr{L}}_R$ (A.22), which describes, in the free-field case, four independent harmonic oscillators. Note also that by expanding (3.b) we get

$$\overline{\mathscr{L}}_R(1) = \overline{\mathscr{L}}_R + \frac{\partial}{\partial t} i\omega\varepsilon_0(\mathscr{A}_s\,\mathscr{A}_l^* - \mathscr{A}_l\,\mathscr{A}_s^*) \tag{9}$$

which demonstrates this equivalence in the case $\lambda = 1$.

c) The Lorentz condition (A.16.b) in reciprocal space is

$$ik\mathscr{A}_l + \frac{\dot{\mathscr{A}}_s}{c} = 0\,. \tag{10}$$

If this equation is satisfied, the last terms of (2.b) and (2.c) cancel and Equations (2) in real space become

$$\Box \mathbf{A} = 0 \quad (11.\text{a}) \qquad\qquad \Box U = 0 \quad (11.\text{b})$$

which are equivalent to the free-field Maxwell equations in the Lorentz gauge. All the Lagrangians $L_R(\lambda)$ then lead to the same equations if the subsidiary condition (10) is imposed on them. This is not really surprising, in that $L_R(\lambda)$ differs from the standard Lagrangian by a term proportional to the integral over all space of the square of a quantity whose value is taken as 0 when the subsidiary condition is imposed in classical theory.

Note finally that the conjugate momentum relative to $\mathscr{A}_s$, given in (6.b), is nonzero when $\lambda \neq 0$. Thus, it is possible to perform the free-field quantization and to treat the four components of the potential symmetrically, starting from $L_R(\lambda)$ with $\lambda \neq 0$. The physical results must evidently be independent of λ, but the intermediate calculations can depend on it.

2. Annihilation and Creation Operators for Scalar Photons: Can One Interchange Their Meanings?

We have seen in Part B of this chapter that, in the framework of covariant quantization, the state space is much larger than the space of physical states. These latter form a subspace defined by the following expression derived from the Lorentz condition:

$$[a_{il} - a_{is}]\,|\,\psi\,\rangle = 0\,. \tag{1}$$

To simplify the notation, we assume here box quantization so that the continuous index **k** is replaced by the discrete index i. The difficulty of the usual canonical quantization is that it leads to a commutator for the scalar field

$$[a_{is}, a_{js}^+] = -\,\delta_{ij} \tag{2}$$

whose sign differs from the usual sign obtained for the transverse and longitudinal field variables.

To remedy this difficulty, one might be tempted to reverse the roles of the annihilation and creation operators. One then takes

$$a_{is} = b_{is}^+ \quad (3.\text{a}) \qquad\qquad a_{is}^+ = b_{is} \quad (3.\text{b})$$

where b_{is} and b_{is}^+ satisfy the usual relations for the annihilation and creation operators. The purpose of this exercise is to examine the consequences of this change in the subspace of physical states defined by condition (1).

a) Consider a state $|\psi\rangle$ which is a tensor product of states relative to each mode of the field:

$$|\psi\rangle = |\psi_1\rangle \otimes \dots \otimes |\psi_i\rangle \otimes \dots \tag{4}$$

Write, using (3) and (4), the equation satisfied by $|\psi\rangle$. Is the vacuum of scalar and longitudinal photons a physical state from this new point of view?

b) Supposing $|\psi_i\rangle$ is expanded in the basis $\{|n_{il} = n,\ n_{is} = n'\}$, denoted simply $\{|n, n'\rangle\}$:

$$|\psi_i\rangle = \sum_{n,n'} b(n, n')\,|n, n'\rangle \tag{5}$$

find a recurrence relation between the coefficients $b(n, n')$. What is $b(n, 0)$ for $n > 0$?

c) Show that the component $|\psi_i\rangle$ of a physical state in this approach takes the form

$$|\psi_i\rangle = \sum_{m \geq 0} b(0, m) \sum_{n \geq 0} \frac{\sqrt{(n+m)!}}{\sqrt{n!}\sqrt{m!}}\,|n, (n+m)\rangle. \tag{6}$$

What is the value of $\langle\psi_i|\psi_i\rangle$? Discuss this result.

Solution

a) With the new operators introduced in (3), the physical states are solutions of

$$(a_{il} - b_{is}^+)|\psi\rangle = 0 \qquad \text{for all } i. \tag{7}$$

If $|\psi\rangle$ is of the form (4), this becomes

$$(a_{il} - b_{is}^+)\,|\psi_i\rangle = 0 \tag{8}$$

since a_{il} and b_{is}^+ act only on the mode i. It is clear that $|0,0\rangle$ is not a solution of (8), since $b_{is}^+|0\rangle = |1\rangle$. It follows that the "vacuum" state of scalar and longitudinal photons is not a physical state in this approach.

b) $a_{il}|\psi_i\rangle$ and $b_{is}^+|\psi_i\rangle$ are equal to

$$a_{il}\,|\psi_i\rangle = \sum_{n,n'} b(n, n')\sqrt{n}\,|(n-1),\ n'\rangle \tag{9.a}$$

$$b_{is}^+\,|\psi_i\rangle = \sum_{n,n'} b(n, n')\sqrt{n'+1}\,|n,\ (n'+1)\rangle. \tag{9.b}$$

The use of the condition (8) then gives

$$b(n+1, n'+1)\sqrt{n+1} = b(n, n')\sqrt{n'+1} \qquad (10.a)$$

$$b(n,0) = 0 \qquad \text{for } n > 0. \qquad (10.b)$$

c) First of all we find $b(n, n')$ in the case $n \leq n'$. Using (10.a), we get

$$b(n, n') = \frac{\sqrt{n'!}}{\sqrt{n!}\sqrt{(n'-n)!}}\, b(0, n'-n). \qquad (11)$$

In the case $n' < n$, an analogous calculation gives

$$b(n, n') = \frac{\sqrt{n'!}\sqrt{(n-n')!}}{\sqrt{n!}}\, b(n-n', 0) \qquad (12)$$

which is zero from (10.b). Using (11) and (12), one shows that the most general form of $|\psi_i\rangle$ is given by (6). Note that since $|\psi_i\rangle$ must be nonzero, one at least of the $b(0, m)$ is nonzero.

Using (6), we find

$$\langle \psi_i | \psi_i \rangle = \sum_m |b(0, m)|^2 \sum_n \frac{(n+m)!}{n!\,m!}. \qquad (13)$$

It is clear that the sum on n leads to an infinite result. In the new approach, the physical states then have infinite norm. Such a result clearly raises difficulties for the application of the basic postulates of quantum mechanics (see also the remark of §B.2.*b*).

3. Some properties of the indefinite metric

a) Let A be a Hermitian operator in the new metric $A = \bar{A}$, and let $|u\rangle$ be an eigenstate of A with eigenvalue λ. Show that if λ is not real, the new norm of $|u\rangle$, $\subset u|u\supset$, is zero.

b) Let $\{|u_n\rangle\}$ be an orthonormal basis in the usual sense, and let $c_n = \langle u_n|\psi\rangle$ and $\gamma_n = \subset u_n|\psi\supset$ be the components of the ket $|\psi\rangle$ on $|u_n\rangle$ defined respectively with the old and the new scalar product. Show that there exists between the two types of components a relationship involving the matrix elements of the operator M defining the new scalar product. Give the new norm $\subset\psi|\psi\supset$ as a function of c_n and γ_n^*.

Solution

a) Consider the eigenvalue equation

$$A\,|u\rangle = \lambda\,|u\rangle \qquad (1)$$

and the adjoint equation in the new sense, which is written

$$\subset u|\,A = \lambda^* \subset u| \qquad (2)$$

since $A = \bar{A}$. By taking the new scalar product of both sides of (1) with $|u\rangle$ one gets

$$\subset u \mid A \mid u \supset \; = \lambda \subset u \mid u \supset . \tag{3}$$

On the other hand, Equation (2) gives

$$\subset u \mid A \mid u \supset \; = \lambda^* \subset u \mid u \supset . \tag{4}$$

Subtraction of (3) and (4) gives

$$(\lambda - \lambda^*) \subset u \mid u \supset \; = 0 \tag{5}$$

which shows that if $\lambda \neq \lambda^*$ (that is, if λ is not real), $\subset u|u \supset$ is zero.

b) The relation

$$\subset u_n \mid \; = \langle u_n \mid M \tag{6}$$

[see (C.3.b)] implies

$$\begin{aligned} \gamma_n &= \; \subset u_n \mid \psi \supset \; = \langle u_n \mid M \mid \psi \rangle \\ &= \sum_p \langle u_n \mid M \mid u_p \rangle \langle u_p \mid \psi \rangle = \sum_p M_{np} \, c_p . \end{aligned} \tag{7}$$

We have used the usual closure relation

$$\sum_p \mid u_p \rangle \langle u_p \mid \; = \mathbb{1} \tag{8}$$

and put $M_{np} = \langle u_n | M | u_p \rangle$. Also, since $M^2 = 1$, one has

$$\begin{aligned} c_n &= \langle u_n \mid \psi \rangle = \langle u_n \mid MM \mid \psi \rangle \\ &= \sum_p \langle u_n \mid M \mid u_p \rangle \langle u_p \mid M \mid \psi \rangle \\ &= \sum_p M_{np} \subset u_p \mid \psi \supset \; = \sum_p M_{np} \, \gamma_p . \end{aligned} \tag{9}$$

Finally, because $M^2 = 1$, Equation (6) implies that

$$\langle u_n \mid \; = \; \subset u_n \mid M \tag{10}$$

so that (8) can then be written

$$\sum_p \mid u_p \supset \; \subset u_p \mid M = \mathbb{1} \tag{11}$$

[see (C.15.b)]. Now insert (11) between $\subset \psi |$ and $| \psi \supset$ in the new norm $\subset \psi | \psi \supset$. This gives, using (10),

$$\begin{aligned} \subset \psi \mid \psi \supset \; &= \sum_p \subset \psi \mid u_p \supset \; \subset u_p \mid M \mid \psi \supset \; = \sum_p \subset \psi \mid u_p \supset \langle u_p \mid \psi \rangle \\ &= \sum_p \subset u_p \mid \psi \supset^* \langle u_p \mid \psi \rangle = \sum_p \gamma_p^* \, c_p . \end{aligned} \tag{12}$$

4. Translation operator for the creation and annihilation operators of a scalar photon

Let $a_s(\mathbf{k})$ and $\bar{a}_s(\mathbf{k})$ be the annihilation and creation operators of a scalar photon with wave vector $\mathbf{k}$ satisfying the commutation relation

$$[a_s(\mathbf{k}), \bar{a}_s(\mathbf{k}')] = - \, \delta(\mathbf{k} - \mathbf{k}') . \tag{1}$$

Consider the operator

$$T = \exp\left\{ \int d^3k[\lambda(\mathbf{k})\cdot\bar{a}_s(\mathbf{k}) - \lambda^*(\mathbf{k})\, a_s(\mathbf{k})] \right\} \tag{2}$$

where $\lambda(\mathbf{k})$ is either a classical function of $\mathbf{k}$ or a particle operator which commutes with $\lambda(\mathbf{k}')$ and its adjoint $\lambda^*(\mathbf{k}')$.

a) Show that T is unitary with respect to the new norm

$$T\bar{T} = \bar{T}T = 1\,. \tag{3}$$

b) Show that T is a translation operator for a_s and $\bar{a}_s$, that is,

$$\begin{cases} Ta_s(\mathbf{k})\,\bar{T} = a_s(\mathbf{k}) + \lambda(\mathbf{k}) & \text{(4.a)} \\ T\bar{a}_s(\mathbf{k})\,\bar{T} = \bar{a}_s(\mathbf{k}) + \lambda^*(\mathbf{k})\,. & \text{(4.b)} \end{cases}$$

c) The wave vectors $\mathbf{k}$ are assumed discrete, and one considers a single scalar mode with annihilation and creation operators a_s and $\bar{a}_s$ with

$$[a_s, \bar{a}_s] = -1\,. \tag{5}$$

The states $|n_s\rangle$ with n_s photons are defined by

$$|\,n_s\,\rangle = \frac{(a_s^+)^{n_s}}{\sqrt{n_s\,!}}\,|\,0_s\,\rangle \tag{6.a}$$

where $|0_s\rangle$ is the vacuum of scalar photons and where

$$a_s^+ = -\,\bar{a}_s \tag{6.b}$$

is the usual adjoint of a_s in the old positive definite metric. One limits oneself here to the case where λ and λ^* are c-numbers in $T = \exp\{\lambda\bar{a}_s - \lambda^* a_s\}$.

Find the expansion in the basis $\{|n_s\rangle\}$ of the state

$$|\,\tilde{0}_s\,\rangle = \bar{T}\,|\,0_s\,\rangle\,. \tag{7}$$

Is this a coherent state? What are its old and new norms?

Solution

a) The new adjoint of $\lambda(\mathbf{k})\bar{a}_s(\mathbf{k}) - \lambda^*(\mathbf{k})a_s(\mathbf{k})$ is equal to $\lambda^*(\mathbf{k})a_s(\mathbf{k}) - \lambda(\mathbf{k})\bar{a}_s(\mathbf{k})$, that is, its negative. One then concludes that

$$\bar{T} = \exp\left\{ -\int d^3k[\lambda(\mathbf{k})\,\bar{a}_s(\mathbf{k}) - \lambda^*(\mathbf{k})\,a_s(\mathbf{k})] \right\} \tag{8}$$

which demonstrates (3).

b) Since a_s and $\bar{a}_s$ commute with their commutator (1), which is a *c*-number, and since λ and λ^* self-commute and commute with a_s and $\bar{a}_s$, it is possible to apply the identity (C.64) of Chapter III to (2) and then get

$$T = \exp\left[\int d^3k\, \lambda(\mathbf{k})\, \bar{a}_s(\mathbf{k})\right] \exp\left[-\int d^3k\, \lambda^*(\mathbf{k})\, a_s(\mathbf{k})\right] \exp\left[\int d^3k\, |\lambda(\mathbf{k})|^2/2\right]. \tag{9}$$

Now find the commutator $[a_s(\mathbf{k}), T]$. The first term of (9) is the only one not commuting with $a_s(\mathbf{k})$. Using (1), we get

$$\left[a_s(\mathbf{k}), \exp\left[\int d^3k'\, \lambda(\mathbf{k}')\, \bar{a}_s(\mathbf{k}')\right]\right] = -\lambda(\mathbf{k}) \exp\left[\int d^3k'\, \lambda(\mathbf{k}')\, \bar{a}_s(\mathbf{k}')\right] \tag{10}$$

so that

$$[a_s(\mathbf{k}), T] = -\lambda(\mathbf{k})\, T \tag{11}$$

so finally

$$T a_s(\mathbf{k}) = a_s(\mathbf{k})\, T + \lambda(\mathbf{k})\, T\,. \tag{12}$$

It is sufficient then to multiply both sides of (12) on the right by $\bar{T}$ to get (4.a), then to take the new adjoint of (4.a) to get (4.b).

c) In the case of a single mode, (9) becomes

$$T = e^{\lambda\bar{a}_s - \lambda^* a_s} = e^{\lambda\bar{a}_s}\, e^{-\lambda^* a_s}\, e^{|\lambda|^2/2} \tag{13.a}$$

from which one gets

$$\bar{T} = e^{-\lambda\bar{a}_s}\, e^{\lambda^* a_s}\, e^{|\lambda|^2/2}\,. \tag{13.b}$$

Note the difference with Equation (C.65) of Chapter III. One expands the second exponential of (13.b) in a series. Since $a_s|0_s\rangle = 0$, only the first term of this expansion (order 0) gives a nonzero result when $\bar{T}$ acts on the vacuum $|0_s\rangle$. One has then

$$\begin{aligned} |\tilde{0}_s\rangle = \bar{T}\,|0_s\rangle &= e^{|\lambda|^2/2}\, e^{-\lambda\bar{a}_s}\, |0_s\rangle \\ &= e^{|\lambda|^2/2}\, e^{\lambda a_s^+}\, |0_s\rangle\,. \end{aligned} \tag{14}$$

Also, using the series expansion of $\exp(\lambda a_s^+)$ and (6.a) gives

$$|\tilde{0}_s\rangle = e^{|\lambda|^2/2} \sum_{n_s=0}^{\infty} \frac{\lambda^{n_s}}{\sqrt{n_s!}}\, |n_s\rangle\,. \tag{15}$$

Since $[a_s, a_s^+] = 1$, the states $|n_s\rangle$ defined in (6.a) are the usual basis states of a harmonic oscillator. The expansion (15) is proportional to that of the coherent state $|\lambda\rangle$ defined by

$$a_s|\lambda\rangle = \lambda|\lambda\rangle\,. \tag{16}$$

[see (C.51) of Chapter III]. However, since the coefficient multiplying the sum on n_s in (15) is $\exp(|\lambda|^2/2)$ and not $\exp(-|\lambda|^2/2)$, the old norm of $|\tilde{0}_s\rangle$ is not equal to 1 but to

$$\langle\tilde{0}_s|\tilde{0}_s\rangle = e^{2|\lambda|^2} \tag{17}$$

In contrast, since T is unitary with respect to the new norm, one has

$$\subset \tilde{0}_s|\tilde{0}_s \supset \; = \; \subset 0_s|\,T\bar{T}\,|0_s \supset \; = \; \subset 0_s|0_s \supset \; = 1\,. \tag{18}$$

It is also possible to get (18) directly starting from the expansion (15) and the orthonormalization relations

$$\subset n_s \mid n'_s \supset \;=(-1)^{n_s}\,\delta_{n_s n'_s} \tag{19}$$

in the new metric [see (C.29)].

Note finally that the fact that $|\tilde{0}_s\rangle$ is a coherent state can be established directly from (4.a). If one applies both sides of this equation to $|0_s\rangle$ and if one uses $a_s|0_s\rangle = 0$, one gets

$$T a_s \bar{T} \mid 0_s \rangle = \lambda \mid 0_s \rangle . \tag{20}$$

It suffices then to apply $\bar{T}$ on the left to both sides of this equation and to use (3) and (7) to get

$$a_s \mid \tilde{0}_s \rangle = \lambda \mid \tilde{0}_s \rangle . \tag{21}$$

5. Lagrangian of the Dirac field. The connection between the phase of the Dirac field and the gauge of the electromagnetic field

The purpose of this exercise is to show that the Dirac equation can be derived from a Lagrangian. We will also see that every phase change in the Dirac field is equivalent to a gauge change in the electromagnetic field. Consider a complex classical field with four components $\psi_\lambda(\mathbf{r})$, $\lambda = 1, 2, 3, 4$, whose evolution is derived from a Lagrangian $L_D = \int \mathrm{d}^3 r\, \mathscr{L}_D$ with

$$\mathscr{L}_D = \frac{i\hbar}{2}\Bigg[\sum_\lambda (\psi_\lambda^* \dot{\psi}_\lambda - \dot{\psi}_\lambda^* \psi_\lambda) + c \sum_{\lambda,\lambda'} (\psi_\lambda^* \boldsymbol{\alpha}_{\lambda\lambda'} \cdot \nabla \psi_{\lambda'} - (\nabla \psi_\lambda^*) \cdot \boldsymbol{\alpha}_{\lambda\lambda'} \psi_{\lambda'})\Bigg] - mc^2 \sum_{\lambda,\lambda'} \beta_{\lambda\lambda'} \psi_\lambda^* \psi_{\lambda'} \tag{1}$$

where $\boldsymbol{\alpha}$ and β are the Dirac matrices (see §A_V.1).

a) Write the Lagrange equations associated with (1). Show that they coincide with the Dirac equation of a free particle.

Rewrite (1) using the covariant notation γ^μ for the Dirac matrices ($\gamma^0 = \beta$, $\boldsymbol{\gamma} = \beta\boldsymbol{\alpha}$) and the relativistic adjoint $\bar{\psi}(\mathbf{r}) = \bar{\psi}^*(\mathbf{r})\gamma^0$.

b) Consider the Lagrangian density

$$\mathscr{L} = -\frac{c}{2}\sum_\mu \left\{ \bar{\psi}\gamma^\mu\left(\frac{\hbar}{i}\partial_\mu + qA_\mu^e\right)\psi + \left[\left(\frac{\hbar}{i}\partial_\mu + qA_\mu^e\right)^* \bar{\psi}\right]\gamma^\mu \psi \right\} - mc^2\,\bar{\psi}\psi \tag{2}$$

where the A_μ^e are the components of an external four-vector potential. Show that the Lagrange equations associated with (2) correspond to the Dirac equation of a particle in an external field.

Show that $\mathscr{L}$ is the sum of $\mathscr{L}_D$ and an interaction term $\mathscr{L}_I$. Write the expression for $\mathscr{L}_I$, and derive the form of the current j^μ associated with the Dirac field.

c) Make the change

$$\psi_\lambda(\mathbf{r}) = \psi'_\lambda(\mathbf{r})\, e^{-iqF(\mathbf{r},t)/\hbar} \tag{3}$$

where $F(\mathbf{r}, t)$ is an arbitrary function of $\mathbf{r}$ and t (and independent of λ). Show that the phase change of ψ is mathematically equivalent to a gauge change on the external field potentials.

Solution

a) To establish the Lagrange equations, we calculate

$$\frac{\partial \mathscr{L}_D}{\partial \dot{\psi}_\lambda^*} = -\frac{i\hbar}{2}\psi_\lambda \tag{4.a}$$

$$\frac{\partial \mathscr{L}_D}{\partial \psi_\lambda^*} = \frac{i\hbar}{2}\left[\dot{\psi}_\lambda + c\sum_{\lambda'} \boldsymbol{\alpha}_{\lambda\lambda'} \cdot \nabla\psi_{\lambda'}\right] - mc^2 \sum_{\lambda'} \beta_{\lambda\lambda'}\, \psi_{\lambda'} \tag{4.b}$$

$$\frac{\partial \mathscr{L}_D}{\partial(\partial_j \psi_\lambda^*)} = -\frac{i\hbar}{2} c \sum_{\lambda'} (\alpha_j)_{\lambda\lambda'}\, \psi_{\lambda'}\,. \tag{4.c}$$

The Lagrange equation (A.52.b) of Chapter II then gives the following equation of motion:

$$i\hbar\dot{\psi}_\lambda = -i\hbar c \sum_{\lambda'} \boldsymbol{\alpha}_{\lambda\lambda'} \cdot \nabla\psi_{\lambda'} + mc^2 \sum_{\lambda'} \beta_{\lambda\lambda'}\, \psi_{\lambda'}\,. \tag{5}$$

One rewrites (5) using $\mathbf{p} = -i\hbar\nabla$ and the column vector $\psi(\mathbf{r})$ with components $\psi_\lambda(\mathbf{r})$, $\lambda = 1, 2, 3, 4$. One gets

$$i\hbar\dot{\psi} = c\boldsymbol{\alpha} \cdot \mathbf{p}\psi + \beta mc^2\, \psi \tag{6}$$

which is indeed the Dirac equation of a free particle.

The Lagrangian density (1) can be written in another form using the Dirac matrices γ^μ and the relativistic adjoint $\bar{\psi}(\mathbf{r}) = \tilde{\psi}^*(\mathbf{r})\beta$:

$$\mathscr{L}_D = \frac{i\hbar}{2} c \sum_\mu \left[\bar{\psi}\gamma^\mu(\partial_\mu\psi) - (\partial_\mu\bar{\psi})\,\gamma^\mu\,\psi\right] - mc^2\bar{\psi}\psi\,. \tag{7}$$

Besides being compact, this form is better adapted to the covariant formalism.

b) One expresses (2) in a form analogous to (1). For this, one introduces the components $(U/c, \mathbf{A})$ of the four-potential A^μ:

$$\mathscr{L} = \frac{i\hbar}{2}\sum_\lambda (\psi_\lambda^*\,\dot{\psi}_\lambda - \dot{\psi}_\lambda^*\,\psi_\lambda) - \frac{c}{2}\sum_{\lambda,\lambda'}\left\{\psi_\lambda^*\,\boldsymbol{\alpha}_{\lambda\lambda'} \cdot \left(\frac{\hbar}{i}\nabla - q\mathbf{A}_e\right)\psi_{\lambda'} + \right.$$

$$\left. + \left[\left(\frac{\hbar}{i}\nabla - q\mathbf{A}_e\right)^*\psi_\lambda^*\right] \cdot \boldsymbol{\alpha}_{\lambda\lambda'}\,\psi_{\lambda'}\right\} - qU_e\sum_\lambda \psi_\lambda^*\,\psi_\lambda - mc^2\sum_{\lambda,\lambda'}\psi_\lambda^*\,\beta_{\lambda\lambda'}\,\psi_{\lambda'}\,. \tag{8}$$

Equations (4.a), (4.b), and (4.c) are then replaced by

$$\frac{\partial \mathscr{L}}{\partial \dot{\psi}_\lambda^*} = -\frac{i\hbar}{2}\psi_\lambda \tag{9.a}$$

$$\frac{\partial \mathscr{L}}{\partial \psi_\lambda^*} = \frac{i\hbar}{2}\dot{\psi}_\lambda - \frac{c}{2}\sum_{\lambda'}\left[\boldsymbol{\alpha}_{\lambda\lambda'}\cdot\left(\frac{\hbar}{i}\nabla - q\mathbf{A}_e\right)\psi_{\lambda'} - q\mathbf{A}_e\cdot\boldsymbol{\alpha}_{\lambda\lambda'}\psi_{\lambda'}\right] - - qU_e\psi_\lambda - mc^2\sum_{\lambda'}\beta_{\lambda\lambda'}\psi_{\lambda'} \quad (9.\text{b})$$

$$\frac{\partial \mathscr{L}}{\partial(\partial_j \psi_\lambda^*)} = -\frac{i\hbar}{2}c\sum_{\lambda'}(\alpha_j)_{\lambda\lambda'}\psi_{\lambda'} \quad (9.\text{c})$$

which gives as the equation of evolution for the column vector ψ

$$i\hbar\dot{\psi} = c\boldsymbol{\alpha}\cdot(\mathbf{p} - q\mathbf{A}_e)\psi + qU_e\psi + \beta mc^2\psi \quad (10)$$

which is the Dirac equation of a particle in an external field described by the potentials $\mathbf{A}_e$ and U_e.

Starting from (2) and (7), we rewrite $\mathscr{L}$ in the form $\mathscr{L} = \mathscr{L}_D + \mathscr{L}_I$ with

$$\mathscr{L}_I = -\sum_\mu j^\mu(\mathbf{r})\, A_\mu^e(\mathbf{r}) \quad (11.\text{a})$$

$$j^\mu(\mathbf{r}) = cq\overline{\psi}(\mathbf{r})\,\gamma^\mu\,\psi(\mathbf{r})\,. \quad (11.\text{b})$$

c) The transformation (3) gives

$$\frac{\hbar}{i}\partial_\mu\psi_\lambda = \left[\frac{\hbar}{i}\partial_\mu\psi'_\lambda - q(\partial_\mu F)\,\psi'_\lambda\right]e^{-iqF/\hbar}\,. \quad (12)$$

By putting this result as well as (3) in (2) we get a new Lagrangian density. This density is identical with the one which would be obtained by making the gauge change $(A_\mu^e)' = A_\mu^e - \partial_\mu F$ on the Lagrangian density (2).

6. The Lagrangian and Hamiltonian of the coupled Dirac and Maxwell fields

Consider a field whose Lagrangian density is defined by (1) of Exercise 5.

a) By following the procedures of Exercises 6 and 7 of Chapter II, show that the Hamiltonian of the Dirac field is of the form

$$H_D = \int d^3r \sum_{\lambda,\lambda'} \psi_\lambda^*(\mathbf{r})\,(c\boldsymbol{\alpha}_{\lambda\lambda'}\cdot\mathbf{p} + \beta_{\lambda\lambda'}\,mc^2)\,\psi_{\lambda'}(\mathbf{r})\,. \quad (1)$$

What is the momentum $\mathbf{P}$ of such a field?

b) Consider the system made up of interacting Dirac and electromagnetic fields. Show that the Lagrangian of the coupled system can be taken equal to

$$\mathscr{L} = \left\{-\frac{c}{2}\sum_\mu\left[\overline{\psi}\gamma^\mu\frac{\hbar}{i}D_\mu\psi - \frac{\hbar}{i}(D_\mu^*\overline{\psi})\,\gamma^\mu\,\psi\right] - mc^2\,\overline{\psi}\psi\right\} - - \varepsilon_0\frac{c^2}{2}\sum_{\mu,\nu}(\partial_\mu A^\nu)(\partial^\mu A_\nu) \quad (2)$$

where $D_\mu = \partial_\mu + \mathrm{i}qA_\mu/\hbar$. What condition should be imposed on the components A_μ of the potential?

c) What is the Hamiltonian associated with (2)?

Solution

a) The Dirac Lagrangian has a form analogous to the Lagrangians studied in Exercises 6 and 7 of Chapter II. For these exercises the Lagrange equations are of first order in time, which proves that there are redundant dynamical variables. We have shown that for a Lagrangian density of the form

$$\frac{\mathrm{i}\hbar}{2}\sum_\lambda(\psi_\lambda^*\dot{\psi}_\lambda - \dot{\psi}_\lambda^*\psi_\lambda) - f(\psi_\lambda, \psi_\lambda^*) \tag{3}$$

the Hamiltonian density is $f(\psi_\lambda, \psi_\lambda^*)$. The same procedure applied to $\mathscr{L}_D$ leads to a Hamiltonian

$$H_D = \int \mathrm{d}^3 r \left\{\left(-\frac{\mathrm{i}\hbar}{2}\right) c \sum_{\lambda,\lambda'}[\psi_\lambda^*\boldsymbol{\alpha}_{\lambda\lambda'}\cdot\nabla\psi_{\lambda'} - (\nabla\psi_\lambda^*)\cdot\boldsymbol{\alpha}_{\lambda\lambda'}\psi_{\lambda'}] + mc^2\sum_{\lambda,\lambda'}\psi_\lambda^*\beta_{\lambda\lambda'}\psi_{\lambda'}\right\}\cdot \tag{4}$$

By integrating the second term in the brackets by parts, we find, since the fields are zero at infinity, a term equal to the first term in the brackets, so that

$$H_D = \int \mathrm{d}^3 r \sum_{\lambda,\lambda'}[(-\mathrm{i}\hbar)\, c\psi_\lambda^*\boldsymbol{\alpha}_{\lambda\lambda'}\cdot\nabla\psi_{\lambda'} + mc^2\psi_\lambda^*\beta_{\lambda\lambda'}\psi_{\lambda'}]\,. \tag{5}$$

Replacing $-\mathrm{i}\hbar\nabla$ by $\mathbf{p}$, we get the required expression (1).

The treatment of part *d*) of Exercise 7 in Chapter II can be immediately used here to find the momentum. Note that this treatment, which concerns the momentum of the Schrödinger field, does not involve the explicit form of the density $f(\psi_\lambda, \psi_\lambda^*)$ of (3). The demonstration given for the Schrödinger field can thus be applied immediately to the case of the Dirac field and leads to

$$\mathbf{P} = \int \mathrm{d}^3 r \sum_\lambda \psi_\lambda^* \frac{\hbar}{i}\nabla\psi_\lambda\,. \tag{6}$$

b) The Lagrangian density (2) depends on variables of both the Dirac and electromagnetic fields. Since the last term of (2) depends only on the radiation field, it does not play any role in the Lagrange equations for the Dirac field. Now the other term of $\mathscr{L}$ between brackets is identical to that given in (2) of Exercise 5. The same calculation as was made there shows that the Lagrange equation of ψ can be written

$$\mathrm{i}\hbar\dot{\psi} = c\boldsymbol{\alpha}\cdot(\mathbf{p} - q\mathbf{A})\,\psi + qU\psi + \beta mc^2\,\psi \tag{7}$$

and coincides with the Dirac equation in the presence of an electromagnetic field described by the potentials $\mathbf{A}$ and U.

Consider now the Lagrange equations for A_μ. We regroup the terms of (2) involving the electromagnetic field variables. This gives

$$-\sum_\mu j^\mu A_\mu - \varepsilon_0\frac{c^2}{2}\sum_{\mu,\nu}(\partial_\mu A^\nu)(\partial^\mu A_\nu) \tag{8}$$

with $j^\mu = cq\bar{\psi}\gamma^\mu\psi$. The density (8) is the same as that used in §A.1.*b* of Chapter V. The same procedure as the one followed in §A.1.*c* shows that the Lagrange equations for A_μ are

$$\Box A_\mu = \frac{1}{\varepsilon_0\, c^2}j_\mu\,. \tag{9}$$

They coincide with the Maxwell field equations in the presence of sources j_μ if the A_μ also satisfy the Lorentz condition $\sum_\mu \partial_\mu A^\mu = 0$.

Thus the Lagrangian density (2), supplemented by the Lorentz condition, leads to the Dirac equation for ψ in the presence of the field A_μ and to the Maxwell equations of the field A_μ in the presence of the sources $j^\mu = cq\bar{\psi}\gamma^\mu\psi$. It therefore correctly describes the dynamics of the coupled Maxwell and Dirac fields and can be taken as the starting point for the formulation of electrodynamics in the Lorentz gauge.

c) Since $\mathscr{L}_I$ does not depend on the velocities $\dot{\mathbf{A}}$ and $\dot{U}$ of the variables of the Maxwell field, the calculation for the conjugate momenta Π_j and Π_s of A_j and A_s made in §A.2.*a* remains valid and gives

$$\Pi_j = \varepsilon_0 \dot{A}_j \tag{10.a}$$

$$\Pi_s = -\varepsilon_0 \dot{A}_s . \tag{10.b}$$

It is then possible to reexpress $\dot{A}_j$ and $\dot{A}_s$ in the density (2) as functions of Π_j and Π_s. In addition, the density (2) has, as far as ψ_λ and ψ_λ^* are concerned, a structure analogous to that of (3):

$$\mathscr{L} = \frac{i\hbar}{2}\sum_\lambda (\psi_\lambda^* \dot{\psi}_\lambda - \dot{\psi}_\lambda^* \psi_\lambda) - g(\psi_\lambda, \psi_\lambda^*, \mathbf{A}, \mathbf{\Pi}, A_s, \Pi_s) . \tag{11}$$

By following the same procedure as in Exercises 6 and 7 of Chapter II, one then shows that the Hamiltonian density associated with (11) is

$$\begin{aligned}\mathscr{H} &= \mathbf{\Pi} \cdot \dot{\mathbf{A}} + \Pi_s \dot{A}_s + g(\psi_\lambda, \psi_\lambda^*, \mathbf{A}, \mathbf{\Pi}, A_s, \Pi_s) \\ &= \frac{1}{\varepsilon_0}(\mathbf{\Pi}^2 - \Pi_s^2) + g(\psi_\lambda, \psi_\lambda^*, \mathbf{A}, \mathbf{\Pi}, A_s, \Pi_s) .\end{aligned} \tag{12}$$

By expressing g using (2), (11), and (10) one then gets

$$H = H_D + H_R + H_I \tag{13}$$

where H_D is given in (1), where

$$H_R = \frac{\varepsilon_0}{2}\int \mathrm{d}^3r \left[\frac{1}{\varepsilon_0^2}\mathbf{\Pi}^2 + c^2(\nabla \times \mathbf{A})^2 - \frac{1}{\varepsilon_0^2}\Pi_s^2 - c^2(\nabla A_s)^2\right] \tag{14}$$

[which, after transformation into reciprocal space, is in agreement with the expression derived from (A.27) and leads to (A.64) when reexpressed in terms of normal variables], and where

$$H_I = \int \mathrm{d}^3r \sum_\mu j^\mu A_\mu = cq \int \mathrm{d}^3r \sum_\mu \bar{\psi}(\mathbf{r})\, \gamma^\mu\, \psi(\mathbf{r})\, A_\mu(\mathbf{r}) . \tag{15}$$

Note finally that, as in *e*) of Exercise 7 of Chapter II, canonical quantization is achieved by associating two operators with $\psi_\lambda(\mathbf{r})$ and $\psi_{\lambda'}^*(\mathbf{r}')$ whose commutator or anticommutator is $\delta_{\lambda\lambda'}\delta(\mathbf{r} - \mathbf{r}')$. The commutator leads to an energy unbounded below, so that it is necessary to take the anticommutator, which gives

$$[\Psi_\lambda(\mathbf{r}), \Psi_{\lambda'}^+(\mathbf{r}')]_+ = \delta_{\lambda\lambda'}\, \delta(\mathbf{r} - \mathbf{r}') . \tag{16}$$

We then justify, from the Lagrangian formalism, the Hamiltonian and the anticommutation relations used in §§A_V.2, A_V.3, serving as the starting point for relativistic quantum electrodynamics in the Lorentz gauge.

7. Dirac field operators and charge density. A study of some commutation relations

Let $\Psi_\lambda(\mathbf{r})$ be the components of the quantized Dirac field at point $\mathbf{r}$, and

$$\rho(\mathbf{r}) = q \sum_\lambda \Psi_\lambda^+(\mathbf{r})\, \Psi_\lambda(\mathbf{r}) \tag{1}$$

the charge density operator at point $\mathbf{r}$.

a) Using the anticommutation relations between the field operators $\Psi_\lambda(\mathbf{r})$ and $\Psi_{\lambda'}^+(\mathbf{r}')$, establish that

$$[\rho(\mathbf{r}), \rho(\mathbf{r}')] = 0\,. \tag{2}$$

b) Consider the operator

$$X = -\frac{c}{\hbar}\int \mathrm{d}^3 r\, \rho(\mathbf{r})\, S(\mathbf{r}) \tag{3}$$

where $\rho(\mathbf{r})$ is defined in (1) and where $S(\mathbf{r})$, which can be a radiation operator, commutes with the field operators $\Psi_\lambda(\mathbf{r})$ and $\Psi_{\lambda'}^+(\mathbf{r}')$. Show that

$$\begin{cases} [X, \Psi_\lambda(\mathbf{r})] = \dfrac{qc}{\hbar}\, S(\mathbf{r})\, \Psi_\lambda(\mathbf{r}) & (4.\text{a}) \\ [X, \Psi_\lambda^+(\mathbf{r})] = -\dfrac{qc}{\hbar}\, S(\mathbf{r})\, \Psi_\lambda^+(\mathbf{r})\,. & (4.\text{b}) \end{cases}$$

c) Assume that $S(\mathbf{r})$ commutes with $S(\mathbf{r}')$, and take

$$T = \mathrm{e}^{\mathrm{i}X}\,. \tag{5}$$

Show that

$$\begin{cases} T\Psi_\lambda(\mathbf{r})\, T^{-1} = \exp\left[\dfrac{\mathrm{i}qc}{\hbar}\, S(\mathbf{r})\right] \Psi_\lambda(\mathbf{r}) & (6.\text{a}) \\ T\Psi_\lambda^+(\mathbf{r})\, T^{-1} = \exp\left[-\dfrac{\mathrm{i}qc}{\hbar}\, S(\mathbf{r})\right] \Psi_\lambda^+(\mathbf{r})\,. & (6.\text{b}) \end{cases}$$

Solution

a) The anticommutation relations (32) of Complement A_V permit us to write

$$\rho(\mathbf{r})\, \rho(\mathbf{r}') = q^2 \sum_\lambda \sum_{\lambda'} \Psi_\lambda^+(\mathbf{r})\, \Psi_\lambda(\mathbf{r})\, \Psi_{\lambda'}^+(\mathbf{r}')\, \Psi_{\lambda'}(\mathbf{r}') =$$

$$= -q^2 \sum_\lambda \sum_{\lambda'} \Psi_\lambda^+(\mathbf{r})\, \Psi_{\lambda'}^+(\mathbf{r}')\, \Psi_\lambda(\mathbf{r})\, \Psi_{\lambda'}(\mathbf{r}') + q^2 \sum_\lambda \delta(\mathbf{r} - \mathbf{r}')\, \Psi_\lambda^+(\mathbf{r})\, \Psi_\lambda(r')\,. \tag{7}$$

We anticommute $\Psi_\lambda^+(\mathbf{r})$ and $\Psi_{\lambda'}^+(\mathbf{r}')$ as well as $\Psi_\lambda(\mathbf{r})$ and $\Psi_{\lambda'}(\mathbf{r}')$ in the first term of (7), which gives a double sign change and thereby no sign change, and then anticommute $\Psi_\lambda^+(\mathbf{r})$

and $\Psi_{\lambda'}(\mathbf{r}')$. This gives

$$\rho(\mathbf{r})\,\rho(\mathbf{r}') = \rho(\mathbf{r}')\,\rho(\mathbf{r}) - q^2 \sum_{\lambda} \delta(\mathbf{r} - \mathbf{r}')\left[\Psi_\lambda^+(\mathbf{r}')\,\Psi_\lambda(\mathbf{r}) - \Psi_\lambda^+(\mathbf{r})\,\Psi_\lambda(\mathbf{r}')\right]. \tag{8}$$

On account of the delta function $\delta(\mathbf{r} - \mathbf{r}')$, we can replace $\mathbf{r}'$ by $\mathbf{r}$ in the bracket of the last term of (8), which shows that it is zero and gives (2).

b) Since S commutes with the field operators,

$$[X, \Psi_\lambda(\mathbf{r})] = -\frac{c}{\hbar}\int d^3r'\, S(\mathbf{r}')\,[\rho(\mathbf{r}'), \Psi_\lambda(\mathbf{r})]. \tag{9}$$

Now, the anticommutation relations (32) of Complement A_V let us write

$$\rho(\mathbf{r}')\,\Psi_\lambda(\mathbf{r}) = q\sum_{\lambda'} \Psi_{\lambda'}^+(\mathbf{r}')\,\Psi_{\lambda'}(\mathbf{r}')\,\Psi_\lambda(\mathbf{r}) = -\,q\sum_{\lambda'} \Psi_{\lambda'}^+(\mathbf{r}')\,\Psi_\lambda(\mathbf{r})\,\Psi_{\lambda'}(\mathbf{r}')$$

$$= \Psi_\lambda(\mathbf{r})\,\rho(\mathbf{r}') - q\sum_{\lambda'} \delta_{\lambda\lambda'}\,\delta(\mathbf{r} - \mathbf{r}')\,\Psi_{\lambda'}(\mathbf{r}') \tag{10}$$

—that is, finally,

$$[\rho(\mathbf{r}'), \Psi_\lambda(\mathbf{r})] = -\,q\delta(\mathbf{r} - \mathbf{r}')\,\Psi_\lambda(\mathbf{r}). \tag{11}$$

It suffices then to substitute (11) in (9) and to integrate over $\mathbf{r}'$ to get (4.a). An analogous procedure lets us get (4.b).

c) Start with the identity

$$e^{iX}\,\Psi_\lambda(\mathbf{r})\,e^{-iX} = \Psi_\lambda(\mathbf{r}) + [iX, \Psi_\lambda(\mathbf{r})] + \frac{1}{2\,!}[iX, [iX, \Psi_\lambda(\mathbf{r})]] + \cdots \tag{12}$$

From (4.a), the first commutator of (12) is $iqcS(\mathbf{r})\Psi_\lambda(\mathbf{r})/\hbar$. Since $S(\mathbf{r})$ commutes with $S(\mathbf{r}')$, the double commutator is simply

$$\frac{1}{2\,!}\,\frac{iqc}{\hbar}\,S(\mathbf{r})\,[iX, \Psi_\lambda(\mathbf{r})] \tag{13}$$

as can also be found starting from (4.a). We get then

$$e^{iX}\,\Psi_\lambda(\mathbf{r})\,e^{-iX} = \psi_\lambda(\mathbf{r})\left\{1 + \frac{iqc}{\hbar}S(\mathbf{r}) + \frac{1}{2\,!}\left[\frac{iqc}{\hbar}S(\mathbf{r})\right]^2 + \cdots\right\}. \tag{14}$$

One recognizes in the braces the series expansion of $\exp[iqcS(\mathbf{r})/\hbar]$, which demonstrates (6.a). An analogous procedure gives (6.b).

References

AKHIEZER, A. I., and BERESTETSKI, V. B., *Quantum Electrodynamics*, Wiley-Interscience, New York, 1965.

BERESTETSKI, V., LIFSHITZ, E., and PITAYEVSKI, L., *Relativistic Quantum Theory*, Vol. IV, Pergamon, Oxford, 1971.

BJORKEN, J. D., and DRELL, S. D., *Relativistic Quantum Mechanics*, McGraw-Hill, New York, 1964.

BLATT, J. M., and WEISSKOPF, V. F., *Theoretical Nuclear Physics*, Wiley, New York, 1963.

BOGOLIUBOV, N. N., and SHIRKOV, D. V., *Introduction to the Theory of Quantized Fields*, Interscience, New York, 1959.

COHEN-TANNOUDJI, C., Introduction to Quantum Electrodynamics, in *Tendances Actuelles en Physique Atomique/New Trends in Atomic Physics*, Les Houches, Session XXXVIII, 1982, Grynberg, G., and Stora, R, eds., Elsevier Science Publishers B. V., 1984, p. 1.

COHEN-TANNOUDJI, C., DIU, B., and LALOË, F., *Quantum Mechanics*, Wiley and Hermann, Paris, 1977.

COHEN-TANNOUDJI, C., DUPONT-ROC, J., GRYNBERG, G., *Processes d'interaction entre Photons et Atomes*, InterEditions et Editions du C.N.R.S., Paris, 1988, *Interaction Processes between Photons and Atoms*, Wiley, New York, to be published.

DIRAC, P. A. M., *The Principles of Quantum Mechanics*, 4th ed., Oxford University Press, Oxford, 1958.

FEYNMAN, R. P., *Quantum Electrodynamics*, Benjamin, New York, 1961.

FEYNMAN, R. P., LEIGHTON, R. B., and SANDS, M., *The Feynman Lectures on Physics, Vol. II: Electromagnetism and Matter*, Addison-Wesley, Reading, MA, 1966.

GLAUBER, R. J., "Optical Coherence and Photon Statistics", in *Quantum Optics and Electronics*, Les Houches 1964, de Witt, C., Blandin, A., and Cohen-Tannoudji, C., eds., Gordon and Breach, New York, 1965, p. 63.

GOLDBERGER, M. L., and WATSON, K. M., *Collision Theory*, Wiley, New York, 1964.

GOLDSTEIN, H., *Classical Mechanics*, Addison-Wesley, Reading, MA, 1959.

HAKEN, H., *Light*, Vol. I, North Holland, Amsterdam, 1981.

HEALY, W. P., *Nonrelativistic Quantum Electrodynamics*, Academic Press, New York, 1982.

HEITLER, W., *The Quantum Theory of Radiation*, 3rd ed., Clarendon Press, Oxford, 1954.

ITZYKSON, C., and ZUBER, J.-B., *Quantum Field Theory*, McGraw-Hill, New York, 1980.

JACKSON, J. D., *Classical Electrodynamics*, 2nd ed., Wiley, New York, 1975.

JAUCH, J. M., and ROHRLICH, F., *The Theory of Photons and Electrons*, Addison-Wesley, Reading, MA, 1955.

KLAUDER, J. R., and SKAGERSTAM, B.-S., *Coherent States—Applications in Physics and Mathematical Physics*, World Scientific Publishing Co., Singapore, 1985.

KROLL, N. M., "Quantum Theory of Radiation", in *Quantum Optics and Electronics*, Les Houches, 1964, de Witt, C., Blandin, A. and Cohen-Tannoudji, C., eds., Gordon and Breach, New York, 1965, p. 1.

LAMB, E. and SCULLY, M. O., in *Polarisation, Matière et Rayonnement, Jubilee Volume in honour of Alfred Kastler, edited by La Société bFrançaise de Physique*, Presses Universitaires de France, 1969 p. 363.

LANDAU, L. D., and LIFSHITZ, E. M., *Mechanics*, Vol. I. Pergamon Press, Oxford, 1960.

LANDAU, L. D., and LIFSHITZ, E. M., *The Classical Theory of Fields*, Addison-Wesley, Reading, MA, 1951.

LOUDON, R., *The Quantum Theory of Light*, Clarendon Press, Oxford, 1973.

MESSIAH, A., *Quantum Mechanics*, North Holland, Amsterdam, 1962.

NUSSENZVEIG, H. M., *Introduction to Quantum Optics*, Gordon and Breach, New York, 1973.

POWER, E. A., *Introductory Quantum Electrodynamics*, Longmans, London, 1964.

ROMAN, P., *Advanced Quantum Theory*, Addison-Wesley, Reading, MA, 1965.

SARGENT, M., SCULLY, M. O., and LAMB, W. E., JR., *Laser Physics*, Addison-Wesley, Reading, MA, 1974.

SCHIFF, L. I., *Quantum Mechanics*, 2nd ed., McGraw-Hill, New York, 1955.

SCHWEBER, S. S., *An Introduction to Relativistic Quantum Field Theory*, Harper and Row, New York, 1961.

SOMMERFELD, A., *Electrodynamics*, Academic Press, New York, 1952.

Index

References to Exercises are distinguished by an "e" after the page number.

A

Absorption (of photons), 316, 325, 338e, 344e, 348e, 349e
Action:
 for a discrete system, 81
 for a field, 92
 functional derivative, 128
 principle of least action, 79, 81
 for a real motion, 134, 152e
Adiabatic (switching on), 299
Adjoint (relativistic), 411
Angular momentum, *see also* Multipole, expansion
 conservation, 8, 139, 200
 contribution of the longitudinal electric field, 20, 45
 eigenfunctions for a spin-1 particle, 53
 for the field + particle systems, 8, 20, 118, 174, 200
 for a general field, 152e
 for a spinless particle, 137
 for a spin-1 particle, 49
 of the transverse field, 20, 27, 47
Annihilation and creation operators, *see also* Expansion in a and a^+; Translation operator
 a_d and a_g operators, 394, 429
 a_μ and $\bar{a}_\mu$ operators, 391
 anticommutation relations, 163e, 414
 commutation relations, 121, 171, 391
 for electrons and positrons, 414, 433
 evolution equation, 179, 217, 249e, 420
 for photons, 33, 121, 294
 for scalar photons, 381, 391, 443e, 446e
Antibunching, 211
Anticommutation relations:
 for a complex field, 98
 for the Dirac field, 414, 415, 453e, 454e
 and positivity of energy, 99, 416, 440, 453e
 for the Schrödinger field, 99, 162e
Antihermiticity, *see* Scalar potential
Antiparticle, 187, 413, 433
Approximation:
 long wavelength, 202, 269, 275, 304, 342e
 nonrelativistic, 103, 122, 200
Autocorrelation, 229

B

Basis:
 in reciprocal space, 25, 36
 of vector functions, 51, 55
Bessel:
 Bessel functions, 345e
 spherical Bessel functions, 56, 71e
Born expansion, 300
Bose–Einstein distribution, 234e, 238e
Bosons, 99, 161e, 187
Boundary conditions, *see* Periodic boundary conditions

C

Canonical (commutation relations), *see also* Commutation relations; Quantization (general)
 for a discrete system, 89, 90, 147e, 155e, 258
 for a field, 94, 98, 148e, 158e, 380
Center of mass, 232e, 342e
Change, *see also* Gauge; Lagrangian (general); Transformation
 of coordinates, 84, 88
 of dynamical variables, 86, 260
 of quantum representation, 260, 262
Characteristic functions, 236e

Charge, *see also* Density
conservation, 7, 12, 108, 368, 411, 416, 421
total, 416
Charge conjugation, 438
Classical electrodynamics:
in the Coulomb gauge, 111, 121
in the Lorentz gauge, 364
in the Power–Zienau–Woolley picture, 286
in real space, 7
in reciprocal space, 11
standard Lagrangian, 100
Coherent state, *see* Quasi-classical states of the field
Commutation relations:
canonical commutation relations for an arbitrary field, 94, 98, 148e
canonical commutation relations for a discrete system, 89, 147e, 155e, 258
covariant commutation relations, 381, 382, 391
for electromagnetic fields in real space, 120, 173, 230e
for electromagnetic fields in reciprocal space, 119, 145, 380
of the fields with the energy and the momentum, 233e, 383, 417
for free fields in the Heinsenberg picture, 223, 355e, 382
for the operators a and a^+, 34, 171, 241e, 391, 394, 443e
for the operators a and $\bar{a}$, 391, 395
for the particles, 34, 118, 145, 171
Complex, *see* Dynamical variables; Fields (in general)
Compton:
scattering, 198
wavelength, 202
Conjugate momenta of the electromagnetic potentials:
in the Coulomb gauge, 115, 116, 143
in the Lorentz gauge, 369
in the Power–Zienau–Woolley representation, 289, 291, 294
Conjugate momenta of the particle coordinates:
in the Coulomb gauge, 20, 115, 143
in the Göppert–Mayer representation, 270
in the Henneberger representation, 276
for the matter field, 157e
in the Power–Zienau–Woolley representation, 289, 290, 293
transformation in a gauge change, 267
Conjugate momentum (general):
of a complex generalized coordinate, 88, 96, 154e
of a discrete generalized coordinate, 83, 147e, 256
of a field, 93, 96, 148e
in quantum mechanics, 258, 266
transformation in a change of generalized coordinates, 85
transformation in a change of Lagrangian, 257
Conservation:
of angular momentum, 8, 139, 200
of charge, 7, 12, 108, 368, 411, 416, 421
of energy, 8, 61e, 137, 200
of momentum, 8, 61e, 138, 200, 232e
Constant of the motion, 8, 61e, 134, 152e, 200, 370
Contact interaction, 42
Continuous limit (for a discrete system), 126, 147e
Convolution product, 11
Correlation function, 181, 191, 227. *See also* Intensity correlations
Correlation time, 191
Coulomb, *see also* Coulomb gauge; Energy; Scalar photons
field, 16, 122, 172, 295
interaction, 18, 122, 330, 401, 426, 435
interaction by exchange of photons, 403
potential, 16, 67e, 172, 407
self-energy, 18, 71e, 201
Coulomb gauge, *see also* Hamiltonian (total); Lagrangians for electrodynamics; Transformation
definition, 10, 113
electrodynamics in the Coulomb gauge, 10, 113, 121, 169, 439
relativistic Q.E.D. in the Coulomb gauge, 424, 431
Counting signals, *see* Photodetection signals
Covariant:
commutation relations, 391
formulation, 361
notation and equations, 10, 17, 364, 411, 449e
Covariant Lagrangians:
for classical particles, 106
for coupled electromagnetic and Dirac fields, 451e
for the Dirac field, 449e
for the electromagnetic field (standard Lagrangian), 106, 365

Fermi Lagrangian, 366
interaction Lagrangian, 106, 365
in the Lorentz gauge, 365, 369, 441e
Creation operator, *see* Annihilation and creation operators
Cross-section, *see* Scattering
Current:
density, 7, 101, 115, 410, 419
four-vector, 10, 365, 411
of magnetization, 284
of polarization, 284
Cutoff, 124, 190, 200, 287

D

d'Alambertian, 10, 367
Damping (radiative), 71e, 76e
Darwin term, 440
Delta function (transverse), 14, 36, 38, 42, 64e, 120, 173, 231e
Density, *see also* Quasi-probability density
of charge, 7, 101, 309, 410, 419, 434, 454e
of current, 7, 101, 115, 410, 419
Hamiltonian, 93, 106, 147e, 158e, 370
Lagrangian, 91, 101, 106, 113, 147e, 157e, 167e, 365, 369, 441e
of magnetization, 42, 284, 285, 292
of polarization, 281, 292, 308, 329
Diamagnetic energy, 290, 293
Dipole–dipole interaction:
electric, 313
magnetic, 43
Dipole moment, *see* Electric dipole; Magnetic dipole moment
Dirac, *see also* Matter field; Spinors
delta function, 94
equation, 408, 449e, 452e
Hamiltonian, 410
matrices, 409
Discretization, 31
Dispacement, 282, 291, 292, 308, 310
Dynamical variables:
canonically conjugate, 34, 86, 93, 257, 258, 369
change of dynamical variables in the Hamiltonian, 86, 260
change of dynamical variables in the Lagrangian, 84
complex dynamical variables, 87, 90
for a discrete system, 81
for a field, 90
redundancy, 109, 113, 154e, 157e, 362

E

Effective (Hamiltonian), 435, 438
Einstein, 204
Electric dipole:
approximation, 270
interaction, 270, 288, 304, 306, 312, 313, 342
moment, 270, 288, 306, 343
self-energy, 312
wave, 71e
Electric field, *see also* Electromagnetic field; Expansion
in the Coulomb gauge, 117, 122, 172
longitudinal, 15, 64e, 117, 172, 283
of an oscillating dipole moment, 71e, 353e
in the Power–Zienau–Woolley picture, 295
total, 66e, 117, 172, 291, 295, 310, 330, 355e
transverse, 21, 24, 27, 32, 64e, 117, 171, 287, 295, 310
Electromagnetic field, *see also* Expansion in normal variables; External field; Quantization of the electromagnetic field
associated with a particle, 68e
free, 28, 58, 181, 221, 230e, 241e
mean value in the indefinite metric, 396
in real space, 7
in reciprocal space, 12
tensor $F^{\mu\nu}$, 17, 106, 365, 378
Electromagnetic potentials, *see also* Free (fields, potential); Gauge
covariant commutation reactions, 382
definition and gauge transformation, 9
evolution equations, 9, 10, 366, 367
four-vector potential, 10, 364, 376
mean value in the indefinite metric, 396, 406
retarded, 66e
Electron, *see also* Matter field
classical radius, 75e
elastically bound, 74e
g-factor, 439
Electron–positron pairs, 123, 413, 417
Elimination:
of a dynamical variable, 85, 154e, 157e
of the scalar potential, 111
Emission (of photons), 344e, 348e, 349e
Energy, *see also* Hamiltonian; Self-energy
conservation of, 8, 61e, 137, 200
Coulomb energy, 18, 114, 173, 283, 401, 403, 426

Energy (*Continued*)
of the free field, 183, 378
negative energy states, 413
of the system field + particles, 8, 19, 116
of the transverse field, 26, 31
Equations, *see* Dirac; Hamilton's equations; Heisenberg; Lagrange's equations; Maxwell equations; Newton–Lorentz equations; Poisson; Schrödinger
Equivalence:
between the $\mathbf{A} \cdot \mathbf{p}$ and $\mathbf{E} \cdot \mathbf{r}$ pictures, 272, 296, 316, 321, 337e, 356e
between the $\mathbf{A} \cdot \mathbf{p}$ and $\mathbf{Z} \cdot \nabla V$ pictures, 349e
between relativistic Q.E.D. in the Lorentz and the Coulomb gauges, 424
between the various formulations of electrodynamics, 253, 300, 302
Expansion in a and a^+ (or in a and $\bar{a}$):
of the electric and magnetic fields, 171, 241e
of the four-vector potential, 391
of the Hamiltonian and momentum in the Lorentz gauge, 382, 391
of the Hamiltonian and momentum of the transverse field, 172
of the transverse vector potential, 171
Expansion in normal variables:
of the electric and magnetic fields, 27, 28, 32
of the four-vector potential, 372, 376
of the Hamiltonian and momentum in the Lorentz gauge, 378, 379
of the transverse field angular momentum, 27, 48
of the transverse field Hamiltonian, 27, 31
of the transverse field momentum, 27, 31
of the transverse vector potential, 29, 31
External field, 141, 172, 178, 180, 198. *See also* Hamiltonian for particles in an external field; Lagrangians for electrodynamics
External sources (for radiation), 24, 219, 314, 370, 372, 400, 418

F

Factored states, 207
Fermi:
golden rule, 323
Lagrangian, 366
Fermion, 99, 161e, 413, 414
Fields (in general), *see also* Angular momentum; Energy; Hamiltonian (general considerations); Lagrangian (general); Momentum; Quantization (general)
complex, 95
real, 90
transverse and longitudinal, 13, 37
Fierz, *see* Pauli–Fierz–Kramers transformation
Final, *see* Initial and final states of a process
Fock space, 31, 175
Fourier transform, 11, 12, 15, 56, 97
Four-vector:
current, 10, 365, 411
field energy-momentum, 379
potential, 10, 364, 376
Free (fields, potentials), 28, 58, 183, 205, 373, 376, 382, 414
Fresnel mirror, 208
Functional derivative, 92, 126

G

Gauge, *see also* Coulomb gauge; Lorentz gauge; Poincaré gauge
gauge transformation and phase of the matter field, 167e, 449e
invariance, 8, 17, 107, 269
transformation, 9, 13, 108, 255, 267, 270, 331, 368, 375, 397
Generalized coordinates:
change of, 86, 260
complex, 87, 88
real, 81, 84
Göppert–Mayer transformation, 269, 275, 304
Ground state:
of the quantized Dirac field, 417
of the radiation field, 186, 189, 252e, 385, 386, 394

H

Hamiltonian (general considerations), *see also* Effective, (Hamiltonian)
with complex dynamical variables, 88, 97, 154e, 157e
for a discrete system, 83, 147e
for a field, 93, 97, 148e
Hamiltonian and energy, 83, 136, 146e
in quantum theory, 89, 259
transformation of, 258, 261, 263
Hamiltonian of the particles:
Dirac Hamiltonian, 410
expression of, 144, 197
Pauli Hamiltonian, 432
physical meaning in various representations, 271, 297

of the quantized Dirac Field, 415
for two particles with opposite charges, 232e
for two separated systems of charges, 313, 328
Hamiltonian for particles in an external field:
for a Dirac particle, 410
electric dipole representation (**E** · **r**), 271, 304, 320
Henneberger picture, 277
for an ion, 342e
for the quantized Dirac field, 419
standard representation (**A** · **p**), 144, 198, 266, 317
Hamiltonian for radiation coupled to external sources:
in the Couilomb gauge, 218
in the electric dipole representation, 314, 353e
in the Lorentz gauge, 370, 400, 418
Hamiltonian (total):
in the Coulomb gauge, 20, 33, 116, 138, 173, 439
in the Coulomb gauge with external fields, 144, 174, 198
of coupled Dirac and Maxwell fields, 419, 431, 451e
in the Power–Zienau–Wooley picture, 289, 292, 295, 329
Hamilton's equations:
for a discrete system, 83
for a field, 94, 132, 371
Heaviside function, 226
Heisenberg:
equation, 89
equations for a and a^+, 179, 217, 249e, 420
equations for the matter fields, 99, 161e, 420
equations for the particle, 177
picture, 89, 176, 185, 218, 221, 382
relations, 241e, 248e
Henneberger transformation, 275, 344e, 349e
Hilbert space, 89, 387
Hole theory, 413
Hydrogen atom:
Lamb transition, 327
$1s$–$2s$ two-photon transition, 324, 338e

I

Indefinite metric, *see also* Scalar potential
definition and properties, 387, 391, 445e
and probabilistic interpretation, 390, 392
Independent variables, 95, 109, 121, 362. *See also* Redundancy of dynamical variables
Initial and final states of a process, 264, 271, 296, 300, 302, 317, 326, 337e
Instantaneous, *see also* Nonlocality
Coulomb field and transverse field, 16, 21, 64e, 67e, 122, 291, 292
interactions, 18, 122, 313, 330
Intensity correlations, 186
Intensity of light, 185
Interaction Hamiltonian between particles and radiation:
in the Coulomb gauge, 197, 232e
in the electric dipole representation, 271, 307, 312, 315
in the Power–Zienau–Woolley representation, 290, 292, 296, 329
in relativistic Q.E.D., 419
Interactions, *see* Contact interaction; Coulomb; Dipole–dipole interaction; Electric dipole; Instantaneous; Magnetic dipole moment; Quadrupole electric (momentum and interaction); Retarded; Hamiltonian
Interference phenomena:
with one photon, 208, 210
quantum theory of light interference, 204
with two laser beams, 208, 212
with two photons, 209, 211
Interferences for transition amplitude, 213
Invariance, *see also* Covariant
gauge invariance, 9, 107, 167e, 267
relativistic invariance, 10, 15, 106, 114
translational and rotational, 134, 153e, 200, 370
Ion (interaction Hamiltonian with the radiation field), 342e

K

Kramers, *see* Pauli–Fierz–Kramers transformation
Kronecker (delta symbol), 94, 148e

L

Lagrange's equations:
with complex dynamical variables, 87, 96, 154e
for a discrete system, 82, 129, 147e
for the electromagnetic potentials, 104, 142, 150e, 151e, 366
for a field, 92, 96, 131, 147e, 150e

Lagrange's equations (*Continued*)
for a matter field, 157e, 167e, 367, 449e
for the particles, 103, 142, 151e
Lagrangian (general), *see also* Density, Lagrangian; Functional derivative; Matter field
with complex dynamical variables, 87, 95, 154e, 157e
of a discrete system, 81, 147e
elimination of a redundant dynamical variable, 84, 154e, 157e
equivalent Lagrangians, 82, 92, 108, 256
of a field, 91, 95, 147e
formalism, 79, 81
linear in velocities, 154e, 157e
Lagrangians for electrodynamics, *see also* Covariant Lagrangians; Standard Lagrangian
in the Coulomb gauge, 113, 137
with external fields, 142, 143, 266, 271, 449e
in the Power–Zienau–Woolley picture, 287
Lamb:
shift, 191
transition, 327
Least-action principle, 79, 81
Light intensity, 185
Linear response, 221, 352e
Linear susceptibility, 221, 352e
Locality, 12, 14, 15, 21, 103, 291. *See also* Instantaneous; Nonlocality
Localized systems of charges, 281, 304, 307
Longitudinal:
basis of longitudinal vector functions, 53
contribution of the longitudinal electric field to the energy, momentum and angular momentum, 17, 19, 20
electric field, 15, 64e, 172, 283
normal variables, 374
photons, 384, 430
vector fields, 13
vector potential, 112, 255
Longitudinal vector potential:
in the Coulomb gauge, 16, 113
in the Lorentz gauge, 22
in the Poincaré gauge, 332
Lorentz equation, 104, 178. *See also* Lorentz gauge; Subsidiary condition
Lorentz gauge, *see also* Subsidiary condition
classical electrodynamics in the Lorentz gauge, 364
definition, 9
relativistic Q.E.D. in the Lorentz gauge, 361, 419, 424, 453e

M

Magnetic dipole moment:
interaction, 43, 288
orbital, 288
spin, 44, 197, 439
Magnetic field, 21, 24, 27, 32, 42, 118, 171. *See also* Expansion
Magnetization:
current, 284
density, 42, 284, 292
Mass:
correction, 69e
rest mass energy, 432
Matter field:
Dirac matter field, 107, 366, 408, 414, 433, 451e, 454e
quantization, 98, 161e, 361, 414
Schrödinger matter field, 157e, 161e, 167e
Maxwell equations, *see also* Heinsenberg; Normal variables of the radiation
covariant form, 17, 366
for the potentials, 9, 10, 366
quantum Maxwell equations, 179
in real space, 7
in reciprocal space, 12, 21
Mean value in the indefinite metric, 389, 396, 398, 406
Mechanical momentum, 20, 177, 271, 290
Mode, 24, 27, 374. *See also* Normal mode, Normal variables of the radiation; Expansion
Momentum, *see also* Commutation; Expansion in normal variables; Expansion in a and a^+ (or in a and $\bar{a}$)
conservation, 8, 61e, 138, 200
contribution of the longitudinal field, 19, 20
of the Dirac field, 451e
of the electromagnetic field in the Lorentz gauge, 370, 379
of a general field, 152e
momentum and velocity, 20, 177, 271, 290
for a particle, 20, 177
of the particle + field system, 8, 20, 118, 139, 174, 199
of the Schrödinger field, 158e
of the transverse field, 19, 27, 31, 172, 193, 188

Multiphoton amplitudes (calculations in various representations), 316, 325, 338e, 344e, 348e, 349e
Multipole:
expansion, 287
waves, 45, 55, 58, 60

N

Negative energy states, 413
Negative frequency components, 29, 184, 193, 422
Newton–Lorentz equations, 7, 104, 178
Nonrelativistic:
approximation, 103, 122, 200
limit, 424, 432, 439
Nonresonant processes, 325, 356e
Nonlocality, 14, 15, 21, 151e. *See also* Instantaneous; Locality
Norm:
in the indefinite metric, 388, 445e, 447e
negative, 385
Normal mode, 24, 27, 374. *See also* Normal variables of the radiation; Expanion
Normal order, 185, 195, 237e
Normal variables of the radiation, *see also* Expansion in normal variables
α_d and α_g normal variables, 375, 376, 378
analogy with a wavefunction, 30
definition and expression, 23, 25, 29, 371
discretization, 31
evolution equation, 24, 26, 32, 66e, 219, 371, 372
Lorentz subsidiary condition, 374
quantization, 33, 171
scalar and longitudinal normal variables, 372, 374, 379
transverse normal variables, 25, 29, 374

O

Observables, *see* Physical variables
Operators in the indefinite metric:
adjoint, 388
eigenvalues and eigenfunctions, 389, 445e
hermitian, 388, 445e
Order:
antinormal, 237e
normal, 185, 195, 238e

P

Parseval–Plancherel identity, 11
Particles *see* Conjugate momenta of the particle coordinates; Matter field; Hamiltonian for particles in an external field
Particle velocities:
in the Coulomb gauge, 117, 177
in the Göppert–Mayer approach, 271, 306
in the Henneberger approach, 277
in the Power–Zienau–Woolley approach, 290, 295
Pauli:
exclusion principle, 163e, 413, 416
Hamiltonian, 432
matrices, 410, 437
Pauli–Fierz–Kramers transformation, 278, 429
Periodic boundary conditions, 31
Phase:
of an electromagnetic field mode, 208, 212, 243e
of a matter field and gauge invariance, 167e, 449e
Photodetection signals, *see also* Interference phenomena
double counting signals, 185, 209, 214
single counting signals, 184, 188, 206, 213
Photon, *see also* Annihilation and creation operators; Bose-Einstein distribution; Interference phenomena; *S*-matrix; States of the radiation field; Wave–particle duality
as an elementary excitation of the quantized radiation field, 30, 187
longitudinal and scalar photons, 384, 392, 403, 425, 430, 443e, 446e
nonexistence of a position operator, 30, 50, 188
photon number operator, 187
single-photon states, 187, 205, 208, 210, 385
transverse photons, 186, 385
wavefunction in reciprocal space, 30
Physical meaning of operators:
general, 259, 269
in the Göppert–Mayer approach, 271, 306, 310
in the Henneberger approach, 277, 345e
in the Power–Zienau–Woolley approach, 290, 292
Physical states, 384, 394, 396, 405, 423, 430, 443e. *See also* Physical meaning of

Physical states (*Continued*)
operators; Physical variables; Subsidiary condition
Physical variables, *see also* Angular momentum; Electric field; Energy; Magnetic field; Momentum; Particle velocities; Photodetection signals; Physical meaning of operators; Position operator
in classical theory, 257
corresponding operators in various representations, 116, 117, 271, 277, 294, 306, 310
mean value in the indefinite metric, 396
in quantum theory, 259, 296
transformation of the corresponding operators, 260, 263
Planck, 1
Poincaré gauge, 331, 333
Poisson:
brackets, 86
equation, 10, 345e
Polarization:
current, 284
density, 281, 292, 308, 329
Polarization of the radiation:
polarization vector, 25, 376
sum over transverse polarizations, 36
Position operator, *see also* Photon; Translation operator
in the Henneberger approach, 276, 345e
for the particles, 33, 118, 258
Positive:
positive energy states, 412
positive frequency components, 29, 184, 193, 422
Positron, 408, 413
Potential, *see* Longitudinal vector potential; Scalar potential; Transverse vector potential
Power–Zienau–Woolley transformation, 280, 286, 328, 331
P-representation, 195, 206, 211, 236e, 251e
Processes, *see* Absorption (of photons); Emission (of photons); Multiphoton (amplitudes (calculations in various representations); Nonresonant processes; Resonant, processes; Scattering; *S*-matrix

Q

Quadrupole electric (momentum and interaction), 288
Quantization (general), *see also* Matter field
with anticommutators, 98, 162e, 453e
canonical quantization, 34, 89, 258, 380
for a complex field, 98, 99, 161e
for a real field, 94, 148e
second quantization, 414, 439
Quantization of the electromagnetic field:
canonical quantization in the Coulomb gauge, 119, 144
canonical quantization in the Power–Zienau–Woolley representation, 294
covariant quantization in the Lorentz gauge, 380, 383, 387, 391
elementary approach, 33
methods, 33, 34
Quantum electrodynamics (Q.E.D.):
in the Coulomb gauge, 169
in the Power–Zienau–Woolley picture, 293
relativistic Q.E.D. in the Coulomb gauge, 424, 431
relativistic Q.E.D. in the Lorentz gauge, 361, 419, 424, 453e
Quasi-classical states of the field, *see also* Photodetection signals; Quasi-probability density
definition, 192
graphical representation, 242e
interferences with, 207, 209
production by external sources, 217, 404
properties, 194, 447e
Quasi-probability density:
suited to antinormal order, 236e, 250e
suited to normal order, 195, 206, 211, 236e, 250e

R

Radiation emitted by an oscillating dipole, 71e, 352e
Radiation Hamiltonian:
eigenstates of, 186
as a function of a and a^+, 172, 197, 241e, 296, 382
as a function of a and $\bar{a}$, 391
as a function of the conjugate variables, 116, 144, 290, 296, 370
as a function of the fields, 18, 312
as a function of the normal variables, 27, 31, 378
in the Lorentz gauge, 370, 378, 382, 391, 398
physical meaning, 292, 312

Radiation reaction, 68e, 74e
Radiative damping, 71e, 76e
Raman scattering, 326
Rayleigh scattering, 75e, 198, 326
Reciprocal:
  half-space, 102
  space, 11, 36
Redundancy of dynamical variables, 109, 113, 154e, 157e, 362. *See also* independent variables
Relativistic, *see also* Covariant; Covariant Lagrangian; Quantum electrodynamics (Q.E.D.)
  description of classical particles, 107
  Dirac field, 366, 408, 414, 433, 451e, 454e
  modes, 123
Resonant:
  processes, 316, 326, 349e
  scattering, 75e
Retarded, *see also* Instantaneous
  field, 21, 310, 330
  potential, 66e

S

Scalar photons, 384, 392, 403, 425, 430, 443e, 446e
Scalar potential, *see also* Expansion in a and a^+ (or in a and $\bar{a}$); Expansion in normal variables
  absence of a conjugate momentum with the standard Lagrangian, 109, 362
  antihermiticity in the Lorentz gauge, 392
  conjugate momentum in the Lorentz gauge, 369
  in the Coulomb gauge, 16, 22, 67e
  elimination from the standard Lagrangian, 111
  in the Poincaré gauge, 333
Scalar product:
  in a Hilbert space, 387
  with the indefinite metric, 387, 395, 445e
Scattering, *see also* Compton; Raman scattering; Rayleigh scattering; Thomson scattering; Transition amplitudes
  cross section, 74e, 346e
  nonresonant scattering, 356e
  in presence of radiation, 344e
  process, 326
  resonant scattering, 75e
Schrödinger:
  equation, 89, 157e, 167e, 176, 261, 263
  representation, 89, 176, 219
Schrödinger field:
  Lagrangian and Hamiltonian, 157e, 167e
  quantization, 161e
Schwarzchild, 79
Second quantization, 414
Selection rules, 199, 233e
Self-energy
  Coulomb, 18, 71e, 201
  dipole, 312
  of the transverse polarization, 290, 329
S-matrix:
  definition, 299, 317
  equivalence in different representations, 298, 302, 321, 349e, 356e
  for one- and two-photon processes, 317, 349e
Sources (classical or external), 24, 217, 314, 370, 372, 400, 418
Spectral density, 191
Spin:
  magnetic moment, 44, 197, 439
  spin–statistics theorem, 99
Spin-1 particle, 49
Spin–orbit interaction, 440
Spinors:
  Dirac spinors, 409, 412, 433
  two-component Pauli spinors, 434
Squeezed states, 245e, 246e, 248, 250
Standard Lagrangian:
  difficulties for the quantization, 109
  expression, 100
  symmetries, 105
State space, *see also* Subsidiary condition
  in the Coulomb gauge, 175
  in the covariant formulation, 385
  for scalar photons, 392, 443e
States of the radiation field, *see also* Physical states; Quasi-classical states of the field; Vacuum
  factored states, 205, 207
  graphical representation, 241e
  single-photon states, 187, 205, 208, 210, 385
  squeezed states, 243e, 246e, 248e, 250e
  two-photon states, 211
Subsidiary condition:
  in classical electrodynamics, 9, 10, 22, 368, 370, 374, 442e, 443e
  in presence of interaction, 406, 421, 430
  for the quantum free field, 384, 386, 394
Sudden switching-on of the potential, 264, 336e

Symmetries
and conservation laws, 134
of the standard Lagrangian, 105

T

Thomson scattering, 75e, 198
Transformation, *see also* Physical variables; Unitary transformation; entries under Gauge; Hamiltonian; Lagrangian
of coordinates and velocities, 85
from the Coulomb gauge to the Lorentz gauge (or vice versa), 63e, 425
Göppert–Mayer transformation, 269, 304
Henneberger transformation, 275, 344e, 349e
Pauli–Fierz–Kramers transformation, 278, 429
Power–Zienau–Woolley transformation, 280, 287, 328, 331
of the state vector, 261, 263, 268
Transition amplitudes
definition and calculation, 176, 271, 316, 337e, 338e, 346e
identity in different pictures, 264, 269, 273, 297, 316, 321, 349e, 356e
interference between, 213
Transition matrix, 300, 356e
Transition rate, 323
Translation operator:
for the a and a^+ operators, 195, 308
for the a and $\bar{a}$ operators, 404, 425, 446e
infinitesimal generators, 163e, 199, 383, 417
for the momentum of a particle, 305
for the position of a particle, 276
Transverse, *see also* Expansion; Instantaneous; Nonlocality; Photon
basis of transverse vector functions, 25, 37, 53
commutation relation for the transverse field, 119, 223, 230e
delta function, 14, 36, 38, 42, 64e, 120, 173, 231e
displacement, 283, 291, 295, 310
energy, momentum and angular momentum of the transverse field, 18, 19, 20, 27, 47, 48, 174, 312
equations of motion of the transverse field, 21
electric field, 21, 24, 27, 32, 64e, 117, 171, 287, 295, 310
magnetic field, 21, 24, 27, 32, 42, 118, 171
projector onto the subspace of transverse fields, 37
summation over transverse polarizations, 36
vector field, 13, 50
vector potential, 17, 29, 31, 119, 171, 223, 294, 377, 396
Transverse vector potential, *see also* Expansion; Instantaneous; Nonlocality
commutation relations, 119, 223, 230e
conjugate momentum, 115, 289
gauge invariance, 17

U

Unitary transformation, *see also* Translation operator
associated with a change of Lagrangian, 260, 262, 296
associated with a gauge transformation, 268, 271
on the Hamiltonian, 262, 276, 304, 343e

V

Vacuum, 186, 189, 252e, 385, 386, 394
Vacuum fluctuations, 191, 199, 279
Vector potential, *see* Longitudinal vector potential; Transverse vector potential
Velocity, *see* Particle velocities

W

Wavefunction of the photon, 30, 50. *See also* Photon
Wavelength scale, 202. *See also* Approximation; Compton
Wave–particle duality, 204, 215
Waves:
multipole waves, 45, 55
traveling plane waves, 28
Woolley, *see* Power–Zienau–Woolley transformation

Z

Zienau, *see* Power–Zienau–Woolley transformation